Information System Development

María José Escalona • Gustavo Aragón
Henry Linger • Michael Lang
Chris Barry • Christoph Schneider
Editors

# Information System Development

## Improving Enterprise Communication

Springer

*Editors*
María José Escalona
Department of Computer Languages
and Systems
University of Seville
Seville, Spain

Henry Linger
Monash University
Caaulfield East, VIC, Australia

Chris Barry
National University of Ireland
Galway, Ireland

Gustavo Aragón
IWT2 Research Group FIDETIA
University of Seville
Seville, Spain

Michael Lang
National University of Ireland
Galway, Ireland

Christoph Schneider
City University of Hong Kong
Kowloon, Hong Kong SAR

ISBN 978-3-319-34630-4 ISBN 978-3-319-07215-9 (eBook)
DOI 10.1007/978-3-319-07215-9
Springer Cham Heidelberg New York Dordrecht London


Softcover reprint of the hardcover 1st edition 2014


Printed on acid-free paper

Springer is part of Springer Science+Business Media (www.springer.com)

# Preface

Translating developments of information systems from the academic setting to industry remains elusive. Closing the gap between academic results and their application in enterprises is critical. Tools that help address real improvements in information system development in industry by grounding them in research results is the goal of the International Conference on Information Systems Development. This volume contains the proceedings from the 22nd annual meeting (ISD2013) held in Seville, Spain, from September 2 to 4, 2013. The evolution of information system development was clearly important in recent years. Research groups and universities have worked in proposing new methods, techniques, and tools to improve this activity. In the enterprise environment, the necessity of suitable solutions to their information systems requirements is a pressing problem. Although big improvements were incorporated in the enterprise environment, the application of efficient methods is not usual practice. Information system development continues being too artisanal and mostly adapted for academia. Suitable solutions from the academic community are not applied in industry. This book is oriented to present ideas, works, and tools that help address real improvements in information system development in industry by grounding them in research results. The book is oriented to improve the relevance of academic research and to close the gap between academic results and their application in enterprises.

Seville, Spain — María José Escalona
Seville, Spain — Gustavo Aragón
Caaulfield East, VIC, Australia — Henry Linger
Galway, Ireland — Michael Lang
Galway, Ireland — Chris Barry
Kowloon, Hong Kong SAR — Christoph Schneider

# Preface

# Contents

# Chapter 1
# Measuring Business-IT Alignment, Framework Development and Case Study Results

**Jos J.M. Trienekens, Rob J. Kusters, and Llanos Cuenca**

**Abstract** Information technology has come to play a crucial role in the support, sustainability and growth of organisations. In order to accomplish good value for the organisation, IT projects need to be in connection with the business side of the organisation. The process of managing and improving this connection is Business-IT alignment. Organisations that have successfully aligned their business and IT outperform organisations that do not, for example by maximising the return on IT investments and an improved competitive position. However, selecting and prioritizing IT projects present a challenge to IT managers. For a number of years already, Business-IT alignment is regarded by IT executives as one of their key issues. Much research has been carried out on Business-IT alignment. This has resulted in different models of alignment factors, different approaches to measure and communicate alignment in enterprises. The objective of this paper is to develop a conceptual basis for alignment measurement on the basis of the existing models. This paper will present the development of a conceptual Business-IT model of five alignment factor categories. These categories are: Intention and Support, Working Relationship, Shared Domain Knowledge, IT Projects and Planning, and IT Performance. The conceptual model has been made operational by identifying measures and measurement scales. To validate the alignment factor measurement model a case study has been carried out at five organisations operating in the financial sector in The Netherlands. The final objective is to improve the communication in enterprises between business and IT professionals on alignment and subsequently the selection and prioritization of IT projects.

**Keywords** IT alignment • Business—IT expert communication • Case study

---

J.J.M. Trienekens (✉) • R.J. Kusters
University of Technology Eindhoven, Eindhoven, The Netherlands
e-mail: J.J.M.Trienekens@tue.nl; r.j.kusters@tue.nl

L. Cuenca
Universitat Politècnica de València, València, Spain
e-mail: llcuenca@cigip.upv.es

M. José Escalona et al. (eds.), *Information System Development: Improving Enterprise Communication*, DOI 10.1007/978-3-319-07215-9_1,

## 1.1 Introduction

In [1, 2] the following benefits of good Business-IT alignment are mentioned: "Maximising the return value from IT investments", "A better competitive position through information systems", "Providing direction and flexibility to react to change". The opposite however also holds: a lack of alignment between business and IT can seriously harm a business' performance. Many other studies on the effects of Business-IT alignment have been performed. In short, they support the hypothesis that organisations that have successfully aligned their business and IT outperform organisations that do not [3–6]. Even though alignment of business and IT has proved to be very important, many organizations still struggle with it. It appears that selecting and prioritizing is experienced as a burden by managers [7]. For a number of years already, business and IT alignment is regarded by IT executives as one of their 'key issues' [8, 9]. According to Luftman and Ben-Zvi [10] organizations need to recognize that it is not how IT is aligned with the business but it is how IT and business are aligned with each other and they need to go beyond just focusing on IT infrastructure. In order to better deal with this, organisations need a sound conceptual basis to deal with the different aspects of Business-IT alignment. Though many models have been developed for alignment, one of the main points of criticism on the existing research is that business-IT alignment models are too theoretical and fails to capture real life [4, 5].This gap between theory and practice is the focus of this research project, in particular on the measurement of Business-IT alignment. Section 1.2 addresses the research approach that has been followed, respectively the development of a conceptual alignment measurement model and the design of a multiple case study to validate this model. Section 1.3 resumes the related work in Business-IT alignment. Section 1.4 presents the development of the conceptual alignment measurement model. In Sect. 1.5 the conceptual measurement model is made operational for application in a case study. Section 1.6 presents the validation of the alignment measurement model in a realistic multiple business case study. Section 1.7 finalises the paper with conclusions.

## 1.2 Research Approach

### *1.2.1 Literature Study and Development of Conceptual Alignment Measurement Model*

The literature study is based on guidelines provided by [11], respectively with respect to: data gathering, data selection, data extraction and reporting, and validation. These guidelines ensure that a wide search is performed and provide criteria, e.g. for completeness and quality of the literature study. Several search strategies were used to gather relevant papers, such as a search for keywords, including a number of synonyms, a search for well-known authors of papers on Business-IT alignment, and a search to find relevant papers by making use of the so-called *snowball effect*. These different strategies resulted in a set of alignment definitions, models and factors.

The conceptual alignment measurement model has been developed in a stepwise approach of respectively:

1. Investigating and screening existing alignment factor models in literature
2. Analysing and labelling alignment factors
3. Merging umbrella (label) constructions
4. Classifying alignment factors and defining the conceptual alignment measurement model

### *1.2.2 Case Study Design*

The conceptual alignment measurement model will be put in practice to assess its validity. The goal of this case study is to answer the research question “Can the measurement of the alignment factors be used in practice to measure Business-IT alignment in organisations?” The case study will consist of multiple cases. For each of these cases the Business-IT alignment is examined for the organisation as a whole, making this a holistic multiple case design [12]. For each case, two interviews are held to rate all the defined measures of the alignment factors, together resulting in a rating for the different alignment factors. One interview is held with a business expert of the organisation, and one with an IT expert. In both cases these experts operate in the upper levels of the organisation's hierarchy to be knowledgeable about strategic planning, both for business and IT, meaning that they work at one hierarchical level below the CEO/CIO of their organisations. The positions of the experts across all of the cases should be comparable. Chan and Reich [4, 5] argue that alignment research should focus on specific firm sizes, types and specific industries. Taking this into account, the selected cases for this study will meet the following criteria. Regarding size: the organisation has a medium to large size and regarding specific industry: the organisation operates in the financial sector. Selecting and contacting potential cases was done via the network of KPMG professionals (which is a network of member firms, offering audit, tax and advisory services for their clients, in particular in the financial sector). Using their personal networks appointments could be made with 10 informants, i.e. 2 per organisation. Because this is a non-random sampling method, the possibility of a sampling bias deserves special attention. In order to reduce the risks of self-selection, the potential informants were only given very basic information about the case study in advance to an agreement to participate. Furthermore, there was no pre-screening of informants or organisations. The only information considered prior to the selection of cases was information concerning the selection criteria for this study. Finally, there was no exclusion of particular organisations and a variation of organisations sizes and structure were included in the study to improve the generalisability of the case study results. All of the interviews were recorded with a voice recorder. These recordings were used to write a report for each interview, answering each of the interview questions and rating the accompanying measures. All of the ratings and calculations are captured in a case study database.

## 1.3 Related Work

The Strategic Alignment Model (SAM) [3], the generic framework [1] and the unified framework [13] all describe different business domains that need to be connected in some way to achieve better Business-IT alignment. These models remain fairly 'high level', i.e. they describe business and IT domains that should be aligned but not in what way this can be achieved in practice. This is also one of the main points of criticism on existing business/IT alignment research, e.g. 'too theoretical and failing to capture real life' [5]. As a consequence there is a lack of support regarding the communication between business and IT professionals on business/IT alignment plans and activities, e.g. IT project selection and prioritization. However, one proposal on how to use the SAM and its derivatives in practice is given in a study by Avison et al. [1]. The goal of their study was to validate the SAM, and demonstrate its practical value. They propose a practical framework for managers to change future alignment as required. The Strategic Alignment Maturity Model (SAMM) [14] defines six criteria determining a number of different alignment maturity levels. In contrary to the other models for Business-IT alignment, de SAMM is based solely on practices and processes in organisations. It defines a number of criteria by which the alignment of an organisation can be rated. However, research results on this model provide until now a very limited empirical validation; see Sledgianowski et al. [15]. The lack of practical aspects causes the mentioned models to provide only a very limited basis for the construction of measurement. Therefore, the question remains how alignment can be theoretically defined and practically measured, in order to support the communication on the progress and/or the failures of alignment between business and IT professionals.

## 1.4 Development of the Conceptual Alignment Measurement Model

In conformance with [4, 13] Business-IT alignment is defined: "as the continuous process, involving management and design sub-processes, of consciously and coherently interrelating all components of the business-IT relationship in order to contribute to the organisation's performance over time". Using this definition, and the guidelines of Kitchenham and Charters [11], the literature research resulted in six sets or models of alignment factors. The available studies offer a wide spectrum of research methods, both qualitative and quantitative, to identify and apply alignment factors. E.g. some studies used predefined alignment factors, others started from scratch. In some studies the factors were analysed by having stakeholders rate them, others by analysing 'real-life' documents. The alignment factors from each of the six studies are presented in Fig. 1.1, grouped by their source. The first step towards a conceptual alignment measurement model is to condense the collected alignment factors into a core set of alignment factors, from hereon called alignment factor categories. The following steps were undertaken, (adapted from [16]): 1) **Screening** the alignment factors. A small number of alignment factors describe a characteristic of an organisation, instead of a process that interrelates business and IT. As these do not fall under the definition of Business-IT alignment used in this research project these alignment factors have been omitted. 2) For each individual alignment factor, a **label** is **assigned** to it that captures the topic of the alignment factor. 3) Some of the labels **umbrella** others. Factors that fall under such an umbrella are assigned the label of this umbrella. 4) For the resulting labels, the labels that touch on a similar aspect are **grouped** together (i.e. classified) and assigned a label that covers all content.

### *1.4.1 Step 1 and Step 2: Selection and Labelling of Alignment Factors*

Two of the alignment factors are omitted because they concern characteristics of the organisation, opposite to processes interrelating business and IT: Organisational size [17], and Information intensity of the value chain [18]. Subsequently each of the alignment factors has been assigned a label, capturing its topic (Fig. 1.1).

### *1.4.2 Step 3: Merging Umbrella Constructions*

In Fig. 1.2 labels that fall under another label are grouped. All factors under the umbrella of another factor are assigned the label of this umbrella. The labels that do not fall under another label remain and they are not included.

### *1.4.3 Step 4: Grouping Similar Labels*

When labels concern a similar aspect, they are grouped in this step of the construction process (Fig. 1.3).

The latter condensation of alignment factors resulted in five different alignment factor categories. These are shown in the conceptual alignment factor model in Fig. 1.4.

| ***Luftman et al. (1999)*** | ***Label*** |
|---|---|
| Senior executive support for IT | Management support |
| IT involved in strategy development | Participation |
| IT understands the business | IT's business knowledge |
| Business-IT partnership | Partnership |
| Well-prioritised IT projects | Project prioritization |
| IT demonstrates leadership | Leadership |
| Inhibitor: IT/business lack close relationships | Working relationship |
| Inhibitor: IT does not prioritise well | Project prioritisation |
| Inhibitor: IT fails to meet commitments | IT Commitments |
| Inhibitor: IT does not understand business | IT's business knowledge |
| Inhibitor: Senior executives do not support IT | Management support |
| Inhibitor: IT management lacks leadership | Leadership |
| ***Teo & Ang (1999)*** | *Label* |
| Top management is committed to the strategic use of IT | Commitment |
| IS management is knowledgeable about business | IT's business knowledge |
| Top management has confidence in the IS department | Confidence in IT |
| The IS department provides efficient and reliable services to user departments | Perception |
| There is frequent communication between user and IS department | Communication |
| The IS staff are able to keep up with advances in IT | IT advancement |
| Business and IS management work together in partnership in prioritizing applications development | Partnership |
| Business goals and objectives are made known to IS management | Information sharing |
| The IS department is responsive to user needs | Responsiveness |
| Top management is knowledgeable about IT | Business' IT knowledge |
| The IS department often comes up wint creative ideas on how to use IT strategically | Creativity |
| The corporate business plan is made available to IS management | Information sharing |
| Shared domain knowledge | Shared domain knowledge |
| Communication between business and IT executives | Communication |
| Connections between business and IT planning | Linkage |
| Successful IT history | IT history |
| ***Chan (2002)*** | *Label* |
| CEO and CIO have a strong working relationship | Working relationship |
| Business and IS plans are closely linked | Linkage |
| IS personnel participate in business planning | Participation |
| IS personnel make lateral short- or long-term transfers into business partner areas | Cross-over opportunities |
| IS projects have business sponsors | Sponsors |
| Incentive/compensation bonus schemes exist | Rewarding |
| ***Kearns & Lederer (2003)*** | *Label* |
| Information intensity of the value chain | Information Intensity |
| The CIO participates in business planning | Participation |
| The CEO participates in IT planning | Participation |
| The IT plan reflects the business plan | Linkage |
| ***Chan et al. (2006)*** | *Label* |
| Shared domain knowledge | Shared domain knowledge |
| Priror IS sucess | IT history |
| Organisational size | Organisational size |
| ***Huang & Hu (2007)*** | *Label* |
| Integrating IT planning with business planning | Linkage |
| Maintaining effective communication channels | Communication |
| Developing strong relationships between IT and business | Working relationship |
| Institutionalising the culture of alignment | Culture |

**Fig. 1.1** Labelled alignment factors, grouped by their source

| Label | Intermediate labelling steps | New label | Rationale |
|---|---|---|---|
| Business' IT knowledge | | Shared domain knowledge | One other topic in this list is "shared domain knowledge". The means business and IT are knowledgeable (to some extent) about each other's domain. Business' IT knowledge is a part of this. |
| Communication | | Working relationship | See: rationale working relationship |
| Confidence in IT | | Perception | Having confidence in the IT department is one example of how the IT department can be perceived by top management. Therefore, 'confidence in IT' is placed under 'perception'. |
| Culture | | Commitment | The alignment factor on culture concerns the fact that management has to promote alignment |
| Information sharing | IT's business knowledge | Shared domain knwoledge | If business plans and goals are not known to the IT management, they cannot devise IT plans that support these plans and goals. Therefore, 'information sharing' is placed under 'IT's business knowledge', and thereby also under 'shared domain knowledge'. |
| IT advancement | | Leadership | Keeping up with the advances in IT is explained by Teo and Ang (1999) as IT having to take on a leadership role in planning and implementing new technologies and systems. Therefore, 'IT advancement' is placed under 'Leadership" |
| IT commitments | | IT history | Chan et al. (2006): past implementation failures negatively influenee top management's on-going perceptions of IT. IT not meeting their commitments is closely related to this. Therefore, 'IT commitments' is placed under 'IT history'. |
| IT's business knowledge | | Shared domain knwoledge | Analogous to 'business' IT knowledge. |
| Participation | Partnership | Working relationship | The labels 'participation' and 'partnership' concern the same thing: business and IT work together. 'Participation' was placed under 'partnership' and thereby also under 'working relationship'. |
| Partnership | | Working relationship | See: rationale woring relationship |
| Responsiveness | | Perception | This aligment factor concerns the fact that IT deliverables correspond to needs existing in the business environment. Business should perceive these useful. Consequently, this label was merged with 'perception'. |

**Fig. 1.2** Umbrella labelling: grouping of labels that fall under another label

| Label | Intermediate labelling steps | New label | Rationale |
|---|---|---|---|
| **Commitment**<br>**Management support** | | Intention and support | Both these labels concern the way that management looks at IT department. There has to be the intention to use IT strategically and management should provide the ncessary support to achieve this. Accordingly, these labels are merged under the name 'intention and support'. |
| **Cross-over opportunities**<br>**Rewarding** | Opportunities | IT performance | These labels are concerned with the opportunities in place for IT personnel; is it possible for them to cross-over to a business-related function, and are bonus schemes in place. Having such opportunities stimulates the personnel, which on its turn contributes to the performance of the IT department. Therefore, these labels are merged under 'IT performance'. |
| **Creativity**<br>**Leadership** | Innovation | IT performance | These labels concern the technological innovation by the IT department. Innovation and creativity contribute to the performance of the IT department. Therefore, these labels are merged under 'IT performance'. |
| **IT history**<br>**Perception** | Reliability | IT performance | These labels are concerned with the reliability of IT department. IT department should be perceived as a reliable partner, providing useful services, and have a successful history to support this. This reliability is considered as a part of the performance of the IT department. Therefore, these labels are merged under 'IT performance'. |
| **Linkage**<br>**Project prioritisation**<br>**Sponsors** | | IT projects and planning | These labels both concern the actual projects and planning that exist within an organisation. Accordingly, theese items are merged under the label "IT projects and planning". |

**Fig. 1.3** Grouped labels that concern similar aspects

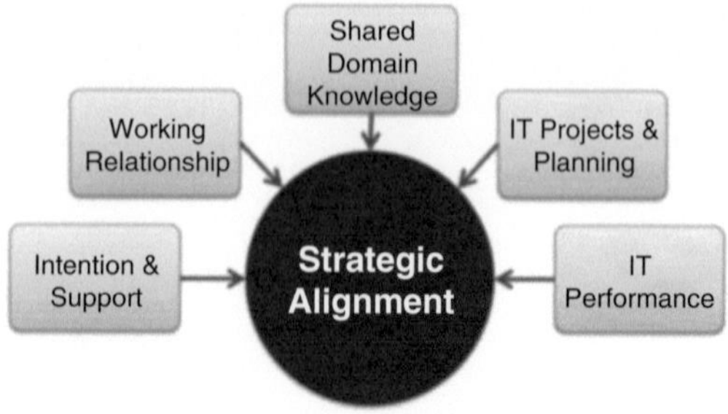


**Fig. 1.4** Conceptual alignment factor model

In order to apply the conceptual alignment factor model in a case study this model will be made operational in the following section by adding measurements and rating possibilities.

## 1.5 Making the Conceptual Alignment Model Operational

The five alignment factor categories will form the basis of the operational measurement model for Business-IT in the defined categories. In order to assign these ratings we need to know what measures determine the level of alignment for each category.

### *1.5.1 The Construction Process*

In order to construct these ratings, information from different sources is combined: the labelling structure resulting from the category construction, the studies used in finding the individual alignment factors and (where possible) information from the models on Business-IT alignment. This combination was done through the following process:

1. The labelling process followed to construct the alignment factor categories resulted in a labelling structure for each category. This structure can be visualised as a tree for each alignment factor category, with the category as the root of the tree and the alignment factors as leaves. The nodes directly subsidiary to the root (first level of the tree) are used as measures.
2. Each measure contains a number of alignment factors. Based on the information given in the detailing of the alignment factors of each measure, relevant information is extracted from three additional sources: alignment factors specifications in investigated literature, explicit criteria of SAMM [14], questionnaire items from the studies in the investigated literature, additional items related to identified factors (eventually).

3. The previous steps result in a list of measures for each alignment factor category, together with a list of items for each measure that can be rated directly from the input given in the interviews. Rating of the items:
   - where available, the ratings as defined in the SAMM are used;
   - if such a rating is not available one is constructed where possible, based on the available information;
   - the ratings for umbrella constructions are calculated by averaging the ratings of their subsidiaries.

This process results in a list of measures for each alignment factor category, along with the necessary information to rate each of the measures.

## *1.5.2 Carrying Out the Construction Process: An Example*

In this section the execution of the construction process will be presented for a particular part of the conceptual alignment measurement model, i.e. the alignment factor category IT projects & planning. The three steps of the construction process will be addressed briefly.

### 1.5.2.1 Step 1: Extracting Measures from the Labelling Structure

In Fig. 1.5 the labelling structure for alignment factors on IT projects & planning is presented. Based on this structure, sponsors, linkage and project prioritisation are used as measures for this alignment factor category.

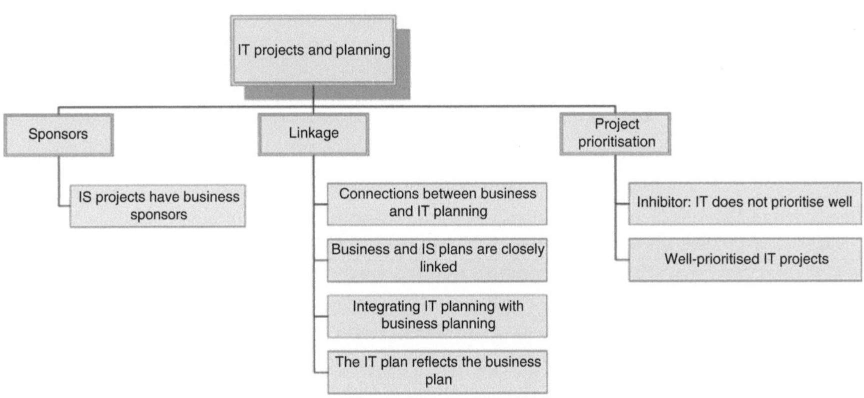


**Fig. 1.5** Labelling tree: IT projects and planning

| | Source | Reference | Information |
|---|---|---|---|
| Sponsors | Alignment factors from general literature | Chan (2002) | *Description: IS projects should have business sponsors* |
| | SAMM | Luftman (2000) | Question: Who sponsors IT project? (varying from none, to IT management to business management) |
| | Additional related items | | N/A |
| Linkage | Alignment factors | Reich and Benbasat (2000) | *Description: Connections between business and IT Planning* |
| | | Chan (2002) | *Description: Business and IS plans are closely linked* |
| | | Kearns and Lederer (2003) | *Description: The IT plan reflects the business plan*<br>*Description: IT Plan reflects business plan: mission, goals, strategies, external business environment forces, resource constraints.* |
| | | Huang and Hu (2007) | *Description: Integrating IT planning with business planning* |
| | SAMM | | N/A |
| | Additional related items | | N/A |
| Project prioritisation | Alignment factors | Luftman (1999) | *Description: Well-prioritised IT projects / Inhibitor: IT does not prioritise well* |
| | SAMM | | N/A |
| | Additional related items | SAM | Additional information: Avison et al. 2004 (SAM) proposed that projects are prioritised according to how they match the ideal alignment perspective of the organisation. The initiation of a project should follow the process set out by the organisation. |

**Fig. 1.6** Collected information from different sources for IT projects & planning

#### 1.5.2.2 Step 2: Gathering Available Information for Each Measure and Construct Grouping of Information

For each of the measures an overview of the available information is given in the figure below (Fig. 1.6).

The information on project prioritisation presents two different aspects: whether projects are well-prioritised and project initiation. Accordingly, project prioritisation has been divided in: adequacy of prioritisation, and project initiation process

#### 1.5.2.3 Step 3: Define Ratings

Finally the process leads to the following ratings for the measures on IT projects & planning (Fig. 1.7).

In order to rate all of the constructed alignment factor categories and measures, a set of interview questions was constructed as a basis for each interview. Examples of these questions are: In the case of sponsoring measure. "Who sponsors IT projects?". To linkage measure, "On a scale from 1 to 5, to what extent do the IT plans reflect the business plan?" The questions associated to project priorisation measurement are: On a scale from 1 to 5, how well-prioritised are the current IT projects?, On a scale

| Measure | Rating |
|---|---|
| **Sponsors** | Rated according to "Business sponsors/champions" of the SAMM |
| **Linkage** | Rating on extent of linkage provided by informants |
| **Project prioritisation** | Adequacy of prioritisation: rating on both technology incorporation and perceived adequacy given by informants<br>Project initiation process: rating on adherence to initiaion process provided by informants. |

**Fig. 1.7** Ratings for measures on IT projects and planning

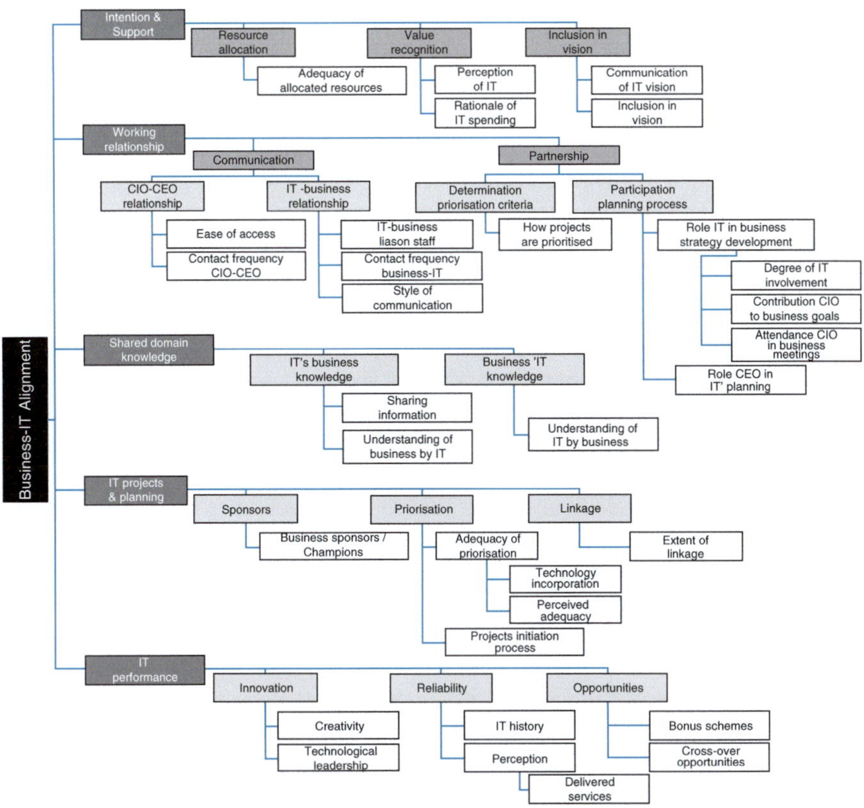


**Fig. 1.8** Overview of the operational alignment measurement model

from 1 to 5, to what extent do the priorisation criteria for this organization meet this requirement?, How are projects initiated? And do all projects follow this path?

The similar steps were carried out in the other four alignment factor categories. In Fig. 1.8 an overview is given of the operational measurement model.

## 1.6 Business-IT Alignment Measurement: Case Study Results

The operational alignment measurement model has been applied in a multiple case study in five organisations.

The main characteristics of these five organisations are summarised in Fig. 1.9.

Figure 1.10 shows the case study resulted in ratings for all the defined categories and measures for each of the participating organisations. When a value is followed by "+", it is understood that the value is slightly greater than the integer value. If the value is followed by "–", it is understood that the value is slightly smaller than the integer value.

Examples of answers from company A in IT projects and planning are: All projects have a business sponsor, according to prince2 (corporate level). On a strategic level, there is high alignment; business benefits for IT projects are defined in the European plans. In general the projects are looked upon as well-prioritised. However, the time needed for the projects is often underestimated. The manpower is still required by the delayed project and cannot be assigned to others. Because of this some (smaller) projects do not receive the attention they require, even though these projects can be very important. Critical developments are followed, but innovation is not a goal of the organisation. That which is necessary for the organisation is carried out. IT is there to support the business, sometimes this can be done by incorporating new technologies. Business should be leading for IT. When discussing the project portfolio it was pointed out that change-projects are only initiated when they have a clear business benefit and are approved by the executive board. For maintenance projects, business benefits are not considered.

The results for the alignment factor categories are displayed in Fig. 1.11.

Figure 1.11 i.e. the overall picture on the alignment factor category level, shows some interesting results, although all five organisations seem to be reasonably to well-aligned. The results were communicated individually with the participating organisations while making use of the underlying collected information. These discussions lead to fruitful insights on causes of particular weak and strong scores.

| Org. | Employees (in FTE) | | Budget (in EUR Million) | Structure |
|---|---|---|---|---|
| | Total | IT | Total [build-run] | |
| A | 1400 | 85-90 | 55 [33-22] | Subsidiary of larger organisation, with regional divisions. Each region has a decentralised IT group. |
| B | 323 | 60 | 25 [13-12] | Subsidiary of larger organisation, with regional divisions. Each region has a decentralised IT group. |
| C | 602 | 150 | 30 | Federal structure: back-end IT is centralised, front-end decentralised. The decentralised IT units are under direct control of the business divisions. |
| D | 6077 | 600 | 135 | Organisation operates under a number of brands. IT is centralised for the entire organisation. |
| E | 1066 | 180 | 30 [20-10] | Federal structure: software development is decentralised for the major business division (falls under control of IT). |

**Fig. 1.9** Main characteristics of the five case study organisations

| | Companies | | | | |
|---|---|---|---|---|---|
| | A | B | C | D | E |
| **Strategic alignment** | **4-** | **3.5** | **4+** | **3+** | **3.5** |
| ***Intention & Support*** | **4-** | **4-** | **4+** | **3** | **3+** |
| Value recognition | 2.5 | 3 | *5* | *2.5* | 3 |
| Inclusion in vision | 5 | 3 | 4 | 3.5 | 3 |
| Resource allocation | 4 | 5 | 4 | 3 | 4 |
| ***Working Relationship*** | **4.5** | **4-** | **4.5** | **3.5** | **3+** |
| Communication | 4.5 | 4 | *5-* | 4+ | 3.5 |
| Partnership | 5- | 3.5 | *4.5* | 3- | 3 |
| ***Shared Domain Knowledge*** | **3** | **4-** | **4-** | **3+** | **3+** |
| IT's business knowledge | 4 | 4.5 | 4.5 | 4 | 4 |
| Business' IT knowledge | 2 | 2 | 3 | 2.5 | 2.5 |
| ***IT Projects & Planning*** | **4+** | ***4*** | **5-** | **3.5** | **4** |
| Sponsors | 4 | 4 | 5 | 3.5 | 3 |
| Linkage | 5 | 4 | 5 | 3.5 | 5 |
| Prioritisation | 3.5 | 3.5 | 4.5 | 4- | 4+ |
| ***IT Performance*** | **3** | **2** | **4-** | **2.5** | **4** |
| Innovation | 3 | 2.5 | 3.5 | 2.5 | 4.5 |
| Reliability | 3.5 | 2 | 4 | 4- | 4 |
| Opportunities | 3 | 2 | 3.5 | 1.5 | 4 |

**Fig. 1.10** Alignment factor category scores in the five organisations

Take the case of IT projects and planning, the results obtained in the different companies are summarized as follow:

**Company A** IT projects and plans are very closely linked to the business. Projects are only initiated when they have business benefits, IT plans are linked to business plans, and all projects have business sponsors. The only problem that arose on this topic lies in the prioritisation of projects. Though in general, projects are well-prioritized, some projects do not get the attention they need due to the overrunning of other projects.

**Company B** Projects have business sponsors and there is a high level of linkage between business plans and IT plans. Also, when projects are initiated, they need to have a business driver. The prioritisation of projects was rated as reasonable, needing improvement.

**Company C** The head of a business division is its business sponsor, and business cases are defined. When there are no clear business benefits, projects are not initiated. Also, there is a close link between the business plans for the divisions and the IT plans, as these IT plans are set up by the business.

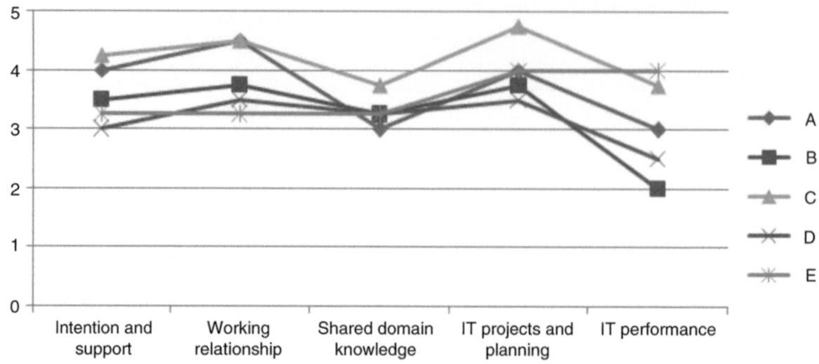


**Fig. 1.11** Alignment factor category scores of the five organisations

**Company D** Projects are sponsored by the director of a business division or a manager form the division. The IT plans are considered to be reasonably linked to the business plans. How well-prioritised projects are varies per business division, in general this was qualified as reasonable.

**Company E** Projects are considered to be well-prioritised and tightly linked to business plans. Projects do have business sponsors, but not at a corporate level; they are sponsored at a unit level.

A broad analysis has been carried out in every company per alignment factor category. In general the best-rated measures were Linkage, IT's business knowledge and Communication. And the worst-rated measures were Business' IT knowledge, Opportunities and Value recognition.

## 1.7 Conclusions

The goal of this research was to construct a measurement method for Business-IT alignment. This measurement method contributes to reduce the gap between theory and practice of Business-IT alignment, by improving the communication on alignment between business and IT experts. A further refinement of the measurement method has been developed, and the alignment factor categories form a good basis for the measurement of business-IT alignment. The categories show a relation to alignment in practice. First the factors have been investigated that determine how well-aligned an organisation is. A large set of factors could be condensed into a conceptual alignment factor model with five alignment factor categories, respectively: intention and support, working relationship, shared domain knowledge, IT projects and planning, IT performance. Measurement of the five alignment factor categories could be based on the underlying structure of the alignment factor category model, and on studies and research reports on (individual) alignment factors. In order to quantify the level of alignment each alignment factor has been rated on a scale of 1–5. The results are derived from interviews, both with business experts and with IT experts. The final results have been communicated with the participating experts and have resulted in interesting insights, both with respect to weak and strong aspects of particular alignment factors, and the selection and prioritization of their IT projects. A small case study was used to perform an initial inspection of the measurement method. This case study does not yet attempt to generalise the underlying model to a wider domain. Regarding future work we intend to apply the alignment framework and measurement method in other IT sectors (than the financial sector), and also to apply them to a larger number of organisations. Regarding future work we intend to apply the alignment framework and measurement method in other IT sectors (than the financial sector), and also to apply them to a larger number of organisations. The final goal is to clarify and to make operational, in a well-founded way for both business experts and IT experts, the factors that play a role in Business-IT alignment, and subsequently to improve IT project selection and prioritization.

**Acknowledgements** The authors wish to KPMG Advisory N.V., Management Consulting, The Netherlands for their contribution and support in the research project.

## References

1. Avison D, Jones J, Powell P, Wilson D (2004) Using and validating the strategic alignment model. J Strat Alignm Syst 13:223–246
2. de Haes S, van Grembergen W (2008) Best practices in IT governance and alignment. In: van Bon J (ed) IT service management global best practices. Van Haren Publishing, Zaltbommel, The Netherlands, pp 47–56
3. Henderson JC, Venkatraman N (1999) Strategic alignment: leveraging information technology for transforming organizations. IBM Syst J 38(2&3):472–484
4. Chan Y, Reich B (2007) IT alignment: an annotated bibliography. J Inform Technol 2007(22):316–396
5. Chan Y, Reich BH (2007) IT alignment: what have we learned? J Inform Technol 2007(22): 297–315
6. Cuenca L, Boza A, Ortiz A (2011) An enterprise engineering approach for the alignment of business and information technology strategy. Int J Comput Integrated Manuf 24(11): 974–992
7. Moonen H, de Jong-Elsinga D, Schut D, van den Hoek E (2008) Organizations no longer accept a low degree of project success. In: Donkers H, Beugelaar B (eds) IT governance, performance and compliance. Uitgeverij Kleine Uil, Groningen, The Netherlands, pp 144–154
8. Luftman J, Kempaiah R (2008) Key issues for IT executives 2007. MIS Quart Exec 7(2):99–112
9. Luftman J, Kempaiah R, Rigoni E (2009) Key issues for IT executives 2008. MIS Quart Exec 8(3):151–159
10. Luftman J, Ben-Zvi T (2010) Key issues for IT executives 2009: difficult ecenomy's impact on IT. MIS Quart Exec 9(1):203–213

11. Kitchenham B, Charters S (2007) Guidelines for performing systematic literature reviews in software engineering. EBSE, Durham
12. Yin RK (2003) Case study research, 3rd edn. Sage, Thousand Oaks, CA
13. Maes R, Rijsenbrij D, Truijens O, Goedvolk H (2000) Redefining business – IT alignment through a unified framework. Prima Vera Working Paper
14. Luftman J (2000) Assessing business-IT alignment maturity. Commun Assoc Inform Syst 4(14):1–49
15. Sledgianowski D, Luftman JN, Reilly RR (2006) Development and validation of an instrument to measure maturity of IT business strategic alignment mechanisms. Inform Resour Manage J 19(3):18–33
16. Baarda D, de Goede M, Teunissen J (2005) Basisboek Kwalitatief Onderzoek, 2nd edn. Noordhoff Uitgevers, Houten, The Netherlands
17. Chan YE, Sabhelwahl R, Thatcher JB (2006) Antecedents and outcomes of strategic IS alignment: an empirical investigation. IEEE Trans Eng Manage 53(1):27–47
18. Kearns GS, Lederer AL (2003) A resource-based view of strategic IT alignment: how knowledge sharing creates competitive advantage. Decis Sci 34(1):1–29

# Chapter 2
# Collaborative Health Informatics: A Multidisciplinary Approach

**Ovidiu Noran**

**Abstract** Modern healthcare is confronted with serious issues that are threatening its sustainability. Increasing costs and complexity, progressive population ageing and rapidly spreading pandemics triggered by new disease strains and by increased population displacements fuelled by conflicts and climate change are all major contributors to the healthcare quandary. In this context, effective cooperation and interoperability of the participants in the healthcare effort becomes paramount. Collaboration is an essential factor but also a major challenge, as typically healthcare institutions are hierarchical and heterogeneous due to various administrative, geographical and historical reasons. As the pressure on healthcare resources and management cost is constantly increasing, governments can no longer rely on information and organisational silo paradigms for managing population wellbeing. Innovative holistic and integrated models and procedures taking into account all essential aspects, elements, participants and their life cycle are necessary if these challenges are to be successfully met. Based on previous research and applications, this paper argues that such necessary artefacts can be built using a life cycle-based holistic paradigm enabled by advances in Information Systems, Interoperability, Collaborative Networks and Enterprise Architecture. This approach aims to provide a sound platform for sustainable solutions to both long and short-term challenges to population health and well-being.

**Keywords** Information Systems • Interoperability • Enterprise Architecture • Collaborative Networks • Health Informatics

O. Noran (✉)
School of ICT, Griffith University, Australia
e-mail: O.Noran@griffith.edu.au

M. José Escalona et al. (eds.), *Information System Development: Improving Enterprise Communication*, DOI 10.1007/978-3-319-07215-9_2,

## 2.1 Introduction

The healthcare environment is under escalating pressure from population ageing, risk of drug-resistant pandemics, increasing complexity and rising costs. In this context, legacy silo-type governance models have lost much of their relevance as collaboration is nowadays a mandatory requirement for survival and progress.

Unfortunately, due to complex regional, historical, organisational and political reasons, there are significant challenges in managing the internal and external collaboration and interoperation of the typically heterogeneous set of participants involved in the healthcare endeavour. This constitutes a particularly critical issue in handling acute health incidents (e.g. pandemics) that require prompt response and claim resources and capabilities beyond those of any particular individual healthcare organisation. New innovative and *integrated* models, methods and tools are required in order to enable proper inter-professional and inter-organisational cooperation, so as to meet these serious long and short term healthcare challenges.

Previous research [1, 2] has investigated the use of Collaborative Networks (CN) [3] and Enterprise Architecture (EA) [4] concepts and methodologies in supporting generic disaster management efforts. This paper aims to build on the previous results by extending this multidisciplinary approach and focusing it on the healthcare-specific Information Systems (IS) area—hereafter, considered synonymous to Health Informatics [5] (HI). It is hypothesised that this approach will allow addressing the above-mentioned issues in a multifaceted life cycle-based, holistic and integrated manner. Owing to this new approach, the resulting models are expected to enable a prompt and efficient response by agile and synergic teams to both acute and long-term challenges to population health and well-being.

## 2.2 Challenges in Healthcare Management Collaboration

Healthcare has made significant advances in the last century, such as the development and wide use of vaccines, eradication of serious diseases and large reductions in communicable disease epidemics and chronic diseases [5, 6].

While solving some very important problems, some of these advances have unfortunately also contributed to a new set of challenges faced by the public and private healthcare infrastructure and organisations. For example, population growth and ageing triggered by increased longevity [6], while reflecting mankind progress and providing benefits [7], also brings significant social security and healthcare challenges [8]. Another major concern are the increasingly complex health incidents such as pandemics, owing to new strains of diseases [9], population displacements fuelled by regional conflicts and climate change [10].

Whereas healthcare as a system has become somewhat more organised, it has also become more expensive, complex and difficult to manage. New technologies are making considerable progress towards supporting collaborative healthcare;

however, the intricate nature of the host organisations involved presents significant impediments to their successful transfer and diffusion [11] that includes interactional user resistance to the new systems [12].

Research in the field also confirms that the main barriers to healthcare cooperation are of organisational and cultural nature [13–16]. Thus, collaboration between participants in the healthcare effort does not automatically occur. It must be "constructed, learned, and once established, protected" [13]. Like most human-related processes, collaboration can neither be successfully forced on the participants nor achieved in a short time.

The divergent perceptions and expectations of the parties involved [15], owing to a traditionally strong hierarchy and marked difference in status between partners [16], can be best dealt with by the higher ranking participants. They can promote collaboration and trust by employing a participatory and inclusive approach [17] which will also build a beneficial sense of security [18].

Inter-professional and inter-organisational collaborative healthcare is encouraged in various medical and emergency response reports, conferences and journals (e.g. [19–24]) as well as in international projects. For example, the BRAID [25] project deliverables advocate the necessity for collaborative healthcare ecosystems [26] supported by integrated assistive services and infrastructure, as part of a 'healthy living and ageing' paradigm [24]. Unfortunately however, the extent of actual cooperation in healthcare is still limited.

In disaster management, often there is a tendency of the higher ranking and more powerful organisation(s) to override or exclude some participants, adopting a 'central command' approach in preference to a cooperative one [27]. This is not desirable as successful disaster management relies on a wide range of community, economic, social-psychological, and political resources. This cooperation brings communities together, gives them a sense of usefulness (ibid.) and thus also alleviates the negative psychological effects such as uncertainty, anguish, confusion, panic etc that are significantly augmented in pandemic-type situations.

## 2.3 A Combined Approach for Collaborative Healthcare

Efficient healthcare collaboration requires that organisational cultures, processes and resources of the participants acquire suitable preparedness [19, 28, 29], with ethics playing a prominent role [30, 31]. This endeavour requires access to a plethora of interdisciplinary information and knowledge not always easily accessible to planners and disaster managers. Therefore, multidisciplinary and participatory analysis and design [32] represent important collaborative healthcare enablers that helps integrate all necessary scientific, administrative, social and political aspects into a whole-system approach [20, 28, 33].

The following sub-sections briefly explain the potential contributions of the IS, Interoperability, CN and EA disciplines to the proposed combined approach. In addition, Fig. 2.1 synthesizes the main barriers to collaborative healthcare and solutions offered by these disciplines.

| Healthcare and Health Informatics Issue | Applic ability | Input from Collaborative Networks | Input from Information Systems | Input from Enterprise Architecture | Input from Interoperability |
|---|---|---|---|---|---|
| Divergent perceptions of the participants' roles | Long & Short Term | Clear, agreed roles for network and task force participants | Identify / address the root problems in divergent perceptions | Integrated modelling of all necessary aspects of collaboration | |
| Lack of trust between partiticipants | Long & Short Term | Trust building in time, within the network | | Promote trust by common understanding of models | Methods to tackle cultural and organisational interoperability |
| Poor life cycle management of task forces / collaborative healthcare IS | Long & Short Term | | Identified / addressed problems in healthcare management | Intrinsic life cycle context to the creation and operation of network and task forces | Interoperability reqs. and capabilities in respect to current life cycle phase/s |
| Difficulties setting up and operating Collaborative Healthcare (e.g. unclear rules, disagreement on the present and future situations) | Long / Short Term | Participatory design, inclusive approach by lead network partner. Agreed upon models of Networks as Collaborative Healthcare Ecosystems. | Participatory design methods and models | Integrated modelling of the creation and operation of complex projects | |
| Focus on a limited set of interoperability aspects | Long / Short Term | | Cooperative IS requirements | A whole-system approach integrating all relevant aspects | Identify all relevant aspects based on interop. frameworks |
| Information sharing and cooperation impeded by traditional hierarchy | Long / Short Term | Information and process interoperability achieved at network level and carried on in task forces created | Methods to improve HI cooperation in hierarchical organisations | | Methods to tackle cultural and organisational interoperability |
| Tendency to overrule rather than cooperate in task forces | Short Term | Cooperation previously agreed upon and built in the task forces created by the network | | | |
| Lack of preparedness to participate in a task force on short notice | Short Term | Participant preparedness built in advance within the network, ready for prompt taskforce / VO creation | | | Identify and address all required Interoperability aspects of network partners |
| Difficult discovery and assessment of suitable participants for an effective and agile task force | Short Term | Task forces created promptly using pre-qualified network partners implementing agreed upon processes. | | Previous research results in 'methods to build methods' for creation and operation of complex projects | Interoperability and agility of task force inherited from the network |

**Fig. 2.1** Sample barriers in establishing collaborative HI and some potential solutions offered by combining the CN, IS, EA and Interoperability disciplines selected in the proposed approach

### 2.3.1 *Healthcare Informatics as a Healthcare Information System*

Due to their close relationship, the area of Information Systems (IS) research provides a sound platform for the study of the more specific HI collaboration; therefore, throughout this research we have drawn on the rich and diverse field of IS research. Major IS issues such as politics, organisational culture, user resistance, difficulties of research results and information technology diffusion in the organisations, privacy, quality of information, ethics, etc all apply to collaborative HI to various degrees as shown throughout this paper (and somewhat more detailed in Sect. 2.2).

### 2.3.2 Interoperability as a Measure of Cooperation

The concept of interoperability is often used as a measure of IS cooperation capability (see e.g. the Levels of Information System Interoperability taxonomy in the Department of Defence Architecture Framework v1 [34]) and it is therefore also useful in the analysis of HI collaboration.

The analysis of interoperability in the HI domain must include some important aspects, such as extent, approach and aspects covered. As shown in previous research [2], too high an interoperability degree (close to total integration) would be detrimental as it would mean a significant loss of autonomy, which is not desirable (e.g. in crisis situations). On the other extreme, minimal IS interoperability (compatibility) of the healthcare or health crisis management effort participants would be only good as a starting point (which is often not met unfortunately). Thus, 'optimal interoperability' lies somewhere between total integration and minimal, depending on the specific healthcare or health crisis management endeavour [ibid.].

In relation to the interoperability approach, the full integration and federalisation options specified in ISO14258 [35] did not seem to achieve the desired results due to organisational heterogeneity and the impossibility to properly negotiate in the limited time available in the case of a disaster event. The apparently more suitable unified approach [ibid.] assumes that ontology is negotiated in advance. For this to happen however, the organisations need to 'spend time together' in order to agree on the meanings associated with the concepts used to exchange knowledge.

Interoperability aspects are provided by various standards [ibid.] and frameworks (e.g. European Interoperability Framework (EIF) [36], IDEAS project [37], ATHENA Interoperability Framework (AIF) [38], Chen's Interoperability Framework [39]). As all these frameworks have overlapping and complementary areas, a combined model has been constructed and applied in [2] for identifying the relevant aspects for generic disaster management. The results largely apply to HI interoperability as well; thus, the data and process areas are the most urgent in a disaster situation as the ability to extract and exchange data from heterogeneous sources providing high volume and often unreliable data is paramount to being aware of the conditions on the ground and avoiding unknown and potentially life-threatening situations for emergency crews. Prior agreements on data format and especially on meaning are essential. Note that 'process interoperability' here concerns the capability to perform joint operations but also to 'take over' and perform a process instead of a disaster management task force participant that may have been temporarily or permanently disabled.

The pragmatic interoperability aspect [40] relates to the capacity but also *willingness* of the participants to interoperate, suggesting that the human component of the HI needs attention prior to task force formation as to allow gaining trust and knowledge between the organisations.

Organisational interoperability is an important aspect in disaster management, as task force participants may often exhibit significant organisational structure diversity that is reflected in their IS. Issues identified by Chen [39] based on the EIF [36], such as responsibility and authority, seem to imply that the roles and hierarchy within a disaster management task force must be clearly understood and reflected in their IS so that the focus is kept on managing the disaster event.

Cultural interoperability [40] appears to be one of the hardest problems. Similar to obtaining pragmatic and semantic interoperability, the only current solution appears to be the regular immersion of the participant organisations in each other's cultures, which facilitates the transfer and conversion of tacit and explicit knowledge between the participants. This recurring 'co-habitation' concept could be facilitated by the Collaborative Network concept explained in the next section.

### *2.3.3 Collaborative Networks for Healthcare*

The concept of networks in disaster management and recovery as an alternative to a centralised command and control approach has been advocated, studied and applied to some extent for a number of years with mixed results (e.g. [27, 41–43]). While providing valuable data, such attempts appear to have two main shortcomings. Firstly, they appear to use untested models focusing on a specific aspect at a time, rather than employing a proven set of integrated models in a whole-system approach. Secondly, the life cycle aspect of the participant organisations, networks and other relevant entities (including the disaster event/s) appears to be less addressed. As all participants and their systems are evolving, it is essential that the interactions required for collaboration and interoperation be considered in an integrated life cycle context.

In attempting to tackle these issues, it has been observed that the healthcare challenges identified in the critical literature review describe a situation similar to that of commercial enterprises who, owing to a global business environment, find themselves compelled to tackle projects requiring resources beyond their own staff, knowledge and time capabilities. Their usual reaction to this problem is to set up or join so-called Collaborative Networks (CNs) that act as breeding environments for Virtual Organisations (VOs) who are promptly created in order to bid for and (if successful) complete projects requiring combined resources and know-how. The view of CNs as social systems composed of commitments, who absorb uncertainty and reduce complexity [44] also supports their use is healthcare and health disaster management projects that typically display such features.

The CNs and VOs set up for the healthcare domain would have specific features. For example, the competitive motivations of commercial CN participants that guide their decisions to create / join / remain / leave the network would transform into the need to cope with increasingly complex health challenges and healthcare systems. The use of reference models, customary in commercial CNs, could be limited by the diversity in scale and type of healthcare incidents [45]. The Health Management CN (HMCN) would create health management VOs (HMVO) for long term projects (e.g. as described in [46]), or task forces (HMTF) for shorter term and more intense events (e.g. pandemics).

Importantly, for a HMCN to function, the lead partner/s (here, government emergency management / healthcare agencies) need to take a participatory and inclusive approach. Thus, scientific, faith and community representatives and all relevant non-governmental and volunteer organisations must also be included in the setup and operation of the HMCN, in addition to the typical participants such as hospitals, allied healthcare [47], fire and rescue services, etc.

Adopting a CN approach for health disaster management provides benefits going beyond mere technical and syntactic-type interoperability. Thus, the participants in a HMCN have the time and suitable environment to overcome the previously described hierarchical, organisational and cultural interoperability [40] barriers and achieve the required preparedness. This is essential in the prompt and successful setup of HMTFs for disasters and in the creation and operation of continuing HMVOs for long term healthcare challenges such as population ageing.

### *2.3.4 The Enterprise Architecture Perspective*

IS and HI collaboration requirements are inherently linked to the current life cycle phase(s) of the host organisations; it is therefore essential that the analysis of possible cooperation improvements is performed in a life cycle context. It is hereby argued that an optimal way to integrate the life cycle aspect in a collaborative HI scenario is by using EA approach.

EA is seen in this context as a holistic change management paradigm that bridges management and engineering best-practice, providing the "[...] key requirements, principles and models that describe the enterprise's future state. [...] EA comprises people, processes, information and technology of the enterprise, and their relationships to one another and to the external environment" [4]. This EA definition reinforces the view of CNs as social systems composed of commitments [44] and IS / HI as socio-technical systems [48] with voluntaristic people [49] in a complex organisational, political and behavioural context [12, 50, 51]. As such, EA is capable of providing a framework integrating all necessary aspects in a life cycle-based set of models ensuring the consistency and sustainability of complex projects. The fundamental role played by EA in this approach and use of EA artefacts is exemplified within in a typical DMCN scenario in the next section.

## 2.4 Life Cycle Integration Modelling for Collaborative Healthcare

Integration modelling of collaborative HI will only be successful if accomplished collaboratively, with all the network participants [33]. The proposed approach supports this audience variety by graphical models and complexity management. While several EA frameworks would have been suitable for this example, we have selected the modelling framework (MF) provided by GERAM (Generalised Enterprise Reference Architecture and Methodology), described in ISO 15704:2005 [35]. This MF provides a large set of aspects, importantly including life cycle, management, organisation and human. For example, Fig. 2.2 right shows the sample use of the GERA MF life cycle viewpoint to define and map the life cycle phases of a health incident on typical health disaster management activities [52].

Figure 2.3 left shows a modelling construct based on a subset of the GERA MF containing orthogonal life cycle, management and information viewpoints. Further on, a projection of this construct is used in Fig. 2.3 to depict an information-based dynamic business model of HMCN and HMTF / HMVO creation and operation.

The arrows in Fig. 2.3 show influences and contributions among the entities involved in the long and short term healthcare endeavour. Thus, healthcare organisations HO (e.g. hospitals), allied health professionals (AHP) and scientific, faith and other communities representatives (CSFR) all contribute to the design and operation of a HMCN in its various life cycle phases. These contributions may also extend directly to the design and operation of the HMTFs/HMVOs created by the HMCN, and to the health management projects (HMPs) created by the HTMF/HMVOs. Influences and contributions also come from 'non-physical' artefacts such as emergency management laws (EML), pandemic preparedness (PPF), or e-health strategies/frameworks (EHF) [53]. Access to properly aggregated, understandable information [54] is provided by HTMFs / HMVOs. Population, organisations and community representatives' feedback flows to Government agencies (GDMAs) and the HMTFs/ HMVOs and may result in changes at various levels.

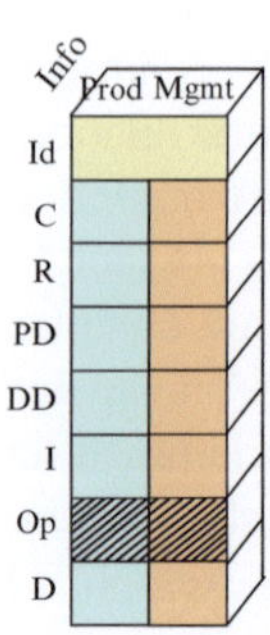


| Health Incident Life Cycle Phase (GERA MF) | Health Disaster Management Phase | Comment |
|---|---|---|
| Identification | Prevention | Identification of the Health Hazard |
| Concept | Prevention | Response Required? Why / why not? |
| Requirements | Preparation | Response Requirements |
| Preliminary Design | Preparation | Response Solution (Principles, Policies) |
| Detailed Design | Preparation | Detailed Response Solution; Prepare Partners for fast Task Force Implementation |
| Implementation | Response | Create Health Disaster Management Task Force |
| Operation | Response | Deploy, Respond |
| Decommisssioning | Recovery | Decommissison the Disaster Management Task Force or reconfigure it for Recovery |

**Fig. 2.2** Mapping a health incident on disaster management using GERA MF life cycle phases

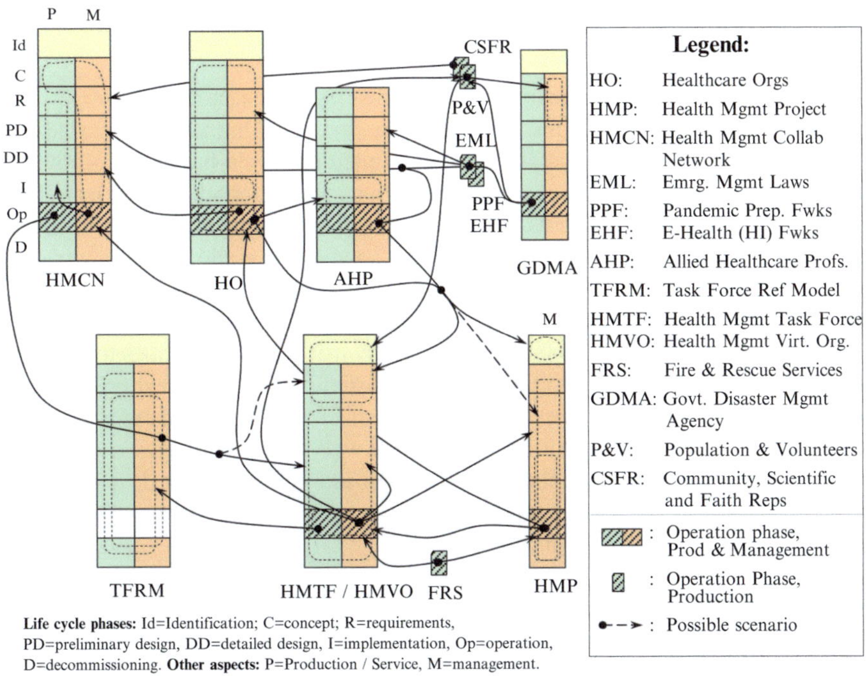


**Fig. 2.3** Sample HI dynamic business model integrating life cycle, management and information viewpoints in a possible health management collaborative network and task force scenario

The arrow from HMTF/HMVO's Management side of the Operation life cycle phase to some of its upper phases represents a very important (if limited) 'self redesign' capability, showing a need for the HMTF to be *agile* and adapt in real time in the face of rapidly changing conditions on the ground that are typical of some disaster events. However, any major HMTF/HMVO reconfiguration (e.g. involving Requirements or Architectural Design life cycles) must involve the HMCN participants and the influence of the other entities on HMCN, as shown in Fig. 2.3.

Note that a high-level model such as shown in Fig. 2.3 does not aim to provide all the details necessary for actual HI implementation. Rather, its main purpose is to facilitate stakeholder common understanding and consensus on the present state and support the selection of the optimal future state. Such models can provide checklists of the 'things' that need to be considered in the collaborative healthcare endeavour and spell out the interactions between them in the context of their life cycles. They can be used to build scenarios representing various degrees of autonomy and agility of the participants and their systems. Once consensus on present and future has been achieved these models can be evolved into design and implementation blueprints. Note that a complete analysis (not possible here due to space limitations) should include an integrated set of models depicting all required aspects, such as process, resource, organisation, decision, etc.

## 2.5 Conclusions and Further Work

Healthcare and HI need to adopt a collaborative approach in order to cope with major contemporary challenges. Politics, hierarchy, diverging perceptions, lack of trust, dissimilar organisational cultures and limited life cycle-based perspective of the healthcare participants' roles and interactions are collaboration barriers. This paper has argued and attempted to demonstrate that an optimal way to address these issues is to adopt a combined interdisciplinary approach that allows drawing upon a rich repository of Information Systems, Collaborative Networks, Enterprise Architecture and Interoperability research state-of-the-art results.

The paper makes a theoretical contribution by using four disciplines to advance collaborative healthcare research and a practical contribution by providing an example of how CN concepts can be employed from an EA perspective in order to model a collaborative healthcare solution to health and well-being challenges.

This is just the beginning; the proposed approach will be further developed and tested in a variety of healthcare management and health disaster case studies in order to verify, validate and refine it.

## References

1. Noran O (2011) Towards a collaborative network paradigm for emergency services. In: Camarinha-Matos LM, Pereira-Klen A, Afsarmanesh H (eds) Adaptation and value creating collaborative networks, vol 362, IFIP advances in information and communication technology. Springer, Berlin, pp 477–485
2. Noran O, Bernus P (2011) Effective disaster management: an interoperability perspective. Lecture Notes in Computer Science 7046:112–121
3. Camarinha-Matos L, Afsarmanesh H (2005) Collaborative networks: a new scientific discipline. J Intell Manuf 16:439–452
4. Gartner Research (2012) IT glossary. http://www.gartner.com/technology/it-glossary/enterprise-architecture.jsp [cited 2012]
5. Fielding JE (1999) Public health in the twentieth century: advances and challenges. Annu Rev Publ Health 20:xiii–xxx
6. World Health Organization (1998) The world health report 1998—life in the 21st century: a vision for all. http://www.who.int/whr/1998/en/whr98_en.pdf [cited 2013 April]
7. Healy J (2004) The benefits of an ageing population. Discussion Paper 63. The Australia Institute. http://www.tai.org.au/documents/dp_fulltext/DP63.pdf [cited 2013 April]
8. International Labour Organisation (2009) Ageing societies: the benefits, and the costs, of living longer. http://www.ilo.org/global/publications/magazines-andjournals/world-of-work-magazine/articles/WCM_041965/lan--n/index.htm [cited 2013 March]
9. Kilbourne ED (2006) Influenza pandemics of the 20th century. Emerg Infect Dis 12(1):9–14
10. Donohoe M (2003) Causes and health consequences of environmental degradation and social injustice. Soc Sci Med 56(3):573–587
11. Southon FC, Sauer C, Grant CN (1997) Information technology in complex health services: organizational impediments to successful technology transfer and diffusion. J Am Med Inform Assoc 4(2):112–124
12. Markus ML (1983) Power, politics, and MIS implementation. Comm ACM 26:430–444

13. Wilson K et al (2005) Nurse practitioners' experience of working collaboratively with general practitioners and allied health professionals in NSW, Australia. Aust J Adv Nurs 23(2):22–27
14. Braude RM (1997) People and organizational issues in health informatics. J Am Med Inform Assoc 4(2):150–151
15. Krogstad U, Hofoss D, Hjortdal P (2004) Doctor and nurse perception of interprofessional cooperation in hospitals. Int J Qual Health Care 16(6):491–497
16. Ramanujam R, Rousseau DM (2006) The challenges are organizational, not just clinical. J Organ Behav 27(7):811–827
17. Baker D, Day R, Salas E (2006) Teamwork as an essential component of high reliability organizations. Health Serv Res 41(4):1577–1598
18. Nembhard IM, Edmondson AC (2006) Making it safe: the effects of leader inclusiveness and professional status on psychological safety and improvement efforts in health care teams. J Organ Behav 27(7):941–966
19. Kapucu N, Arslan T, Demiroz F (2010) Collaborative emergency management and national emergency management network. Disaster Prev Manage 19(4):452–468
20. Utah Department of Health (2007) Governor's task force for pandemic influenza preparedness–finalreporttoGovernor.http://pandemicflu.utah.gov/docs/PandInfluTaskforceFinalReport.pdf [cited 2013 April]
21. Waugh WL, Streib G (2006) Collaboration and leadership for effective emergency management. Public Adm Rev 66(Suppl 1):131–140
22. Hughes RG (2008) Professional communication and team collaboration, Chapter 33. In: Agency for Healthcare Research and Quality (ed) Patient safety and quality: an evidence-based handbook for nurses. Agency for Healthcare Research and Quality, Rockville, MD
23. Institute of Medicine (2000) To err is human: building a safer health system. National Academy Press, Washington, DC
24. Sansoni J et al (2012) An assessment framework for aged care. Centre for Health Service Development, University of Woolongong
25. BRAID (2011) Bridging research in ageing and ICT development – consolidated vision of ICT and ageing. http://auseaccess.cis.utas.edu.au/sites/default/files/D4.2%20Final.pdf [cited 2013 April]
26. Holzman TG (1999) Computer-human interface solutions for emergency medical care. Interactions 6(3):13–24
27. Waugh WL (1993) Coordination or control: organizational design and the emergency management function. Int J Disaster Prev Manage 2(4):17–31
28. World Health Organisation (2011) Pandemic influenza preparedness framework. http://whqlibdoc.who.int/publications/2011/9789241503082_eng.pdf [cited 2013 May]
29. U.S. Dept of Health and Human Services (2005) HHS pandemic influenza plan. http://www.flu.gov/planning-preparedness/federal/hhspandemicinfluenzaplan.pdf [cited 2013 May]
30. Thompson AK et al (2006) Pandemic influenza preparedness: an ethical framework to guide decision-making. BMC Med Ethics 7:12
31. NZ National Ethics Advisory Committee (2006) Ethical values for planning for and responding to a pandemic in New Zealand. http://neac.health.govt.nz/sites/neac.health.govt.nz/files/documents/publications/pandemicplanning-and-response.pdf [cited 2013 May]
32. Kristensen M, Kyng M, Palen L (2006) Participatory design in emergency medical service: designing for future practice. In: Conference on human factors in computing systems – CHI 2006, Montréal, Québec, Canada
33. Moghadas SM et al (2008) Managing public health crises: the role of models in pandemic preparedness. Influenza Other Respir Viruses 3(2):75–79
34. DoD Architecture Framework Working Group (2004) DoD architecture framework ver 1.0. http://www.dod.mil/cio-nii/docs/DoDAF_v1_Volume_I.pdf, http://www.dod.mil/cio-nii/docs/DoDAF_v1_Volume_II.pdf [cited 2007 Feb]
35. ISO/IEC (2005) Annex A: GERAM. In: ISO/IS 15704:2000/Amd1:2005: industrial automation systems – requirements for enterprise-reference architectures and methodologies

36. EIF (2004) European interoperability framework for pan-European eGovernment services. Interoperable Delivery of European eGovernment Services to public Administrations, Businesses and Citizens (IDABC), Luxembourg
37. IDEAS (2003) IDEAS project deliverables (WP1–WP7), Public reports. http://www.ideas-roadmap.net [cited 2011 July]
38. ATHENA (2004) State of the art of enterprise modelling techniques and technologies to support enterprise interoperability. Deliv D.A1.1.1. http://www.athena-ip.org [cited 2011 Mar 30]
39. Chen D (2006) Framework for entrerprise interoperability. http://www.fines-cluster.eu/fines/jm/Download-document/53-Framework-for-EnterpriseInteroperability-Chen.html [cited 2011 July]
40. Whitman L, Panetto H (2006) The missing link: culture and language barriers to interoperability. Annu Rev Contr 30(2):233–241
41. Australian Psychological Society (2013) Disaster response network (DRN). http://www.psychology.org.au/medicare/drn/ [cited 2013 April]
42. Bard Y et al (2011) Multi-level social networking to enable and foster collaborative organisations. In: Camarinha-Matos L, Pereira-Klein A, Afsarmanesh H (eds) Adaptation and value creating collaborative networks. Springer, Berlin, pp 477–485
43. Cooper S et al (2007) Collaborative practices in unscheduled emergency care: role and impact of the emergency care practitioner – qualitative, quantitative and summative findings. Emerg Med J 24:625–633
44. Neumann D, de Santa-Eulalia LA, Zahn E (2011) Towards a theory of collaborative systems. In: Adaptation and value creating collaborative networks. Proceedings of the 12th IFIP working conference on virtual enterprises – PROVE 11, São Paulo, Brazil. p 362
45. Tierney K, Quarantelli EL (1989) Needed innovation in the delivery of emergency medical services in disasters: present and future. Disaster Manage 2(2):70–76
46. Noran O (2013) Collaborative networks in the tertiary education industry sector: a case study. Int J Comput Integrated Manuf 1–2:29–40
47. Queensland Health (2012) Allied health career structure. http://www.health.qld.gov.au/allied/career-structure.asp [cited 2013 April]
48. Pava C (1983) Managing new office technology, an organisational strategy. Free Press, New York, NY
49. McGregor D (1960) The human side of enterprise. McGraw-Hill, New York, NY
50. Keen PGW, Scott Morton M (1978) Decision support systems: an organisational perspective. Addison-Wesley, Reading, MA
51. Iivari J (1991) A paradigmatic analysis of contemporary schools of IS development. Eur J Inform Syst 1(4):249–272
52. Australian Government (2011) Attorney's General's Office – emergency management in Australia. http://www.ema.gov.au/ [cited 2011 Mar 30]
53. Council of Australian Governments (2008) National e-health strategy. http://www.ahmac.gov.au/cms_documents/national%20e-health%20strategy.pdf [cited 2013 April]
54. Alpay L et al (2009) Current challenge in consumer health informatics: bridging the gap between access to information and information understanding. Biomed Inform Insights 2(1):1–10

# Chapter 3
# Understanding Feasibility Study Approach for Packaged Software Implementation by SMEs

**Issam Jebreen, Robert Wellington, and Stephen G. MacDonell**

**Abstract** Software engineering often no longer involves building systems from scratch, but rather integrating functionality from existing software and components or implementing packaged software. Conventional software engineering comprises a set of influential approaches that are often considered good practice, including structured programming, and collecting a complete set of test cases. However, these approaches do not apply well for packaged software (PS) implementation; hence this phenomenon requires independent consideration. To explore PS implementation, we conducted ethnographic studies in packaged software development companies, in particular, to understand aspects of the feasibility study approach for PS implementation. From an analysis of these cases, we conclude that firstly; the analyst has more of a hybrid analyst-sales-marketing role than the analyst in traditional RE feasibility study. Secondly; the use of a live scenario software demonstration in order to convince the client to buy into the PS may lead to increased perceived feasibility and reduced resistance to PS implementation. Thirdly; the assessment criteria that are used to estimate the effort and time needed for PS implementation are new features, level of customization, software 'output', and technical needs. Fourthly; the feasibility study for PS implementation differs strongly from traditional RE as the analyst mainly considers how to deal with requests for modifications to existing functions.

**Keywords** Requirements engineering • Packaged software implementation • Feasibility study

I. Jebreen (✉) • R. Wellington • S.G. MacDonell
SERL, School of Computing and Mathematical Sciences, AUT University, Auckland 1142, New Zealand
e-mail: issam.jebreen@aut.ac.nz; rwelling@aut.ac.nz; stephen.macdonell@aut.ac.nz

M. José Escalona et al. (eds.), *Information System Development: Improving Enterprise Communication*, DOI 10.1007/978-3-319-07215-9_3,

## 3.1 Introduction

Often, Small to Medium Sized Software Development Companies (SMSDCs) or small medium enterprises (SMEs) are unable to apply Requirements Engineering (RE) methods and techniques without modification [1]. In addition, shortcomings in applying RE methods due to time constraints or limited resources may arise [2]. Biirsner and Merten [3] noted that researchers need to intensify the investigation of RE practices in SMEs, otherwise SMEs will waste effort in searching for methodical orientation and dedicated tool support. Normally, the people responsible for requirements in SMEs are ambitious, but suffer from a scarcity of resources, and their time for undertaking research and trying different methods is very limited. They need quick methodical improvement of requirements elicitation, documentation, communication and traceability as well as more continuity of requirements management through the whole software lifecycle.

Karlsson [4] provides us with a summary of RE issues from studies into software development companies. However, none of these focus primarily on Packaged Software (PS) development and implementation. Furthermore, in most of these studies, the projects and organizations under consideration are large, in terms of the number of persons, the requirements involved, and the duration of the projects. Quispe [5] highlighted that there is a lack of knowledge about RE practices in SMEs. This lack of knowledge is particularly apparent when it comes to PS companies. It is difficult for researchers to gain much knowledge about how SMEs carry out RE given that SMEs seldom request external support. However, RE research should eventually enable those companies to become aware of state of the art or innovative RE techniques and to be able to improve their RE practice without external help [2]. Several questions remain unanswered. An important one being: How does requirements engineering in packaged software implementation contexts differ from traditional RE? In particular, there is a need for greater understanding of the feasibility study approach for PS implementation at SMEs.

The paper is organized as follows. In Sect. 3.2 we review literature related to our work. In Sect. 3.3 we briefly describe the research method. In Sect. 3.4 we present our findings and results, which are then discussed in Sect. 3.5. Finally, Sect. 3.6 sets out our conclusion and considers future work.

## 3.2 Literature Review

The concept of PS is defined as a ready-made software product that can be obtained from software companies, and which generally requires modification or customization for specific markets. They are often exemplified by enterprise resource planning (ERP) systems [6]. Previous researchers have highlighted that there is a lack of knowledge about the RE practices that assist PS implementation in these types of companies, and due to the particular characteristics of SMEs, several software engineering researchers have argued that most current RE practices are unsuitable for SMEs [5].

The poor use of RE practices (or the use of unsuitable practices) has often been identified as one of the major factors that can jeopardize the success of a software project [1, 7]. Whereas, it has also been recognized that following appropriate RE practices contributes to the success of software projects [8]. For example, Aranda [1] stated that gathering and managing requirements properly are key factors when it comes to the success of a software project. There is a general consensus that RE practices plays a very important role in the success or failure of software projects [2]. However, it is not possible to improve RE practices until areas that need improvement in an organization's current RE practice have been identified [3, 5].

The conclusion of the first Workshop on Requirements Engineering in Small Companies (2010) [3] was that existing RE techniques are not sufficient for small companies. However, size is not the only measure to categorize smaller companies and be the focus of research, that tacit knowledge and social structures in place in SMEs may play an important role in RE research, that introducing RE methods designed for larger companies may actually be harmful to the specific features of an SMSDC, and that RE methodologies need to be made more lightweight.

Much previous research on RE practices at SMEs has investigated the development of bespoke software. The majority of research in this area has related to software development studies [2, 7]. On the other hand, there are some studies that relate to the software development of packaged solutions, such as those by Daneva [9], Barney [10], Daneva and Wieringa [11]. However, these studies address PS development rather than the RE practices involved with PS implementation. Little attention has been paid to the phase of software package implementation from the perspective of the RE practices that are involved. Further research addressing the topic is needed. Knowledge about this topic could be broadened and enhanced by carrying out intensive and in-depth work place studies.

This paper presents a study about SMEs, carried out in PS implementation and, more specifically, the RE practices perspective, highlighting some of the dynamics and complexity that these SMEs face, as well as their reactions to the challenges. Putting the organization and organizational practices at the centre of attention, this research advances our understanding of PS implementation from a work organization point of view, and in terms of RE practices.

## 3.3 Research Approach

An ethnographic study was conducted for 7 months across two software development companies. The business of the companies considered in this study is dominated by the provision of PS solutions. Data was collected throughout the research during field work. The three data collection methods, namely, interviews, participant

Table 3.1 Coding process

| Data extract | Codes for |
|---|---|
| Question:<br>Tell about software demonstration?<br>The software we present is based on the notes from the sales team about the user's interest in potential software. Then we present the functions of the software that supports the user's business.... I think that helps the user know what their expectations could be for the software functions | Present software<br>Sales team report<br>User's Interest<br>Explain software functions Users' business Support Help users<br>Users' expectations |
| It is good for us to make a software demonstration, in which we start to present a possible solution for users' issues. The flexibility that we want to have during software demonstration was constrained by a time limit since we only have one hour and a half to present our software....so we have to do our best to explain our software functions to the users. | Benefits of software demonstration<br>Present a possible solution Users' Issues Constraints<br>Time limitation<br>Present software<br>Explain software functions |

observation, and focus groups, were used due to their suitability for qualitative research. Due to the hermeneutic characteristics of the analysis, the cases are not presented separately here, rather, every piece of data was analyzed in context, and the data relates not just to the context of the company, but to the specific individuals, projects, roles, process, and so on.

### 3.3.1 Data Analysis

Inductive analysis, used in this study, refers to an approach that primarily uses detailed reading of raw data to derive concepts, themes, and models through the researcher's interpretations of the raw data [12, 13]. During ethnographic research the ethnographer goes through a learning process at the same time as conducting their research. This process allows any prior presumptions that are found to be false to be redefined, reformed, or discarded [12]. We were then more open to experiencing what was going on around us, to paying attention to the details of the process, and to observe what was actually happening in the companies, rather than trying to search for relevant data. The first author conducted the field research and began to reflect on what actually occupied the analysts rather than on his own ideas or presumptions as a researcher. In field notes, he has noted information about everyday activities as well as conversations with the employees in the companies.

In this study, an initial round of field observations was conducted to find interesting topics involved with company's practices for PS implementation. We wished to discover the situations in which PS implementation occurs and understand the process that participants apply. After the initial field work, an initial round of coding was conducted in order to single out the descriptive and interpretive codes [13] (Table 3.1).

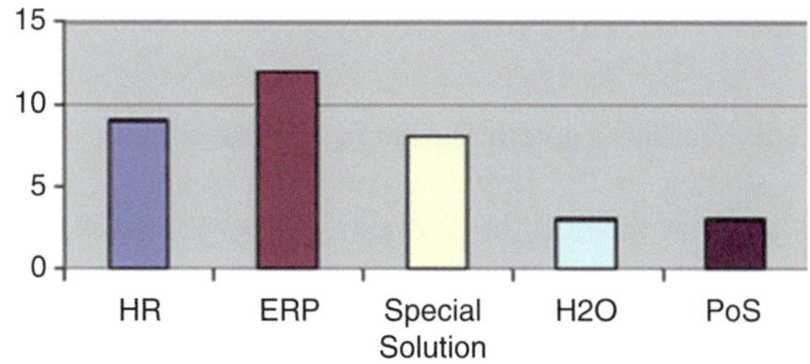


**Fig. 3.1** Number of project cases observe

Excerpts from the text of interview/focus group transcripts or specific phrases from the transcripts were assigned interpretive codes in the initial coding. Further analysis of the initial codes allowed the grouping of similar descriptive and interpretive codes in order to form categories with common themes. The theme names were derived from the concepts in the organizations' projects. This concluded the second round of coding. For example, 'Present software', 'Explain software functions', 'Help users', and 'Users' business Support' were categorized under software demonstration. A third round of analysis was conducted to derive higher-level concepts that would comprise the theoretical constructs in a model of PS implementation. Engaging in a third round of analysis aids the researcher to reach a higher level of abstraction [14]. For example, 'live scenario software demonstration' was conceptualized as a strategy that was intended to help analysts convince clients about a software solution, as they demonstrate, and discuss, a possible solution.

### 3.3.2 *Participants*

The two software development companies that participated in this research were established in 1997 and 1998. There were a total of 40 employees across the two companies, including people working in marketing and sales, analysts, developers, and management teams. The services they offer include software development, systems integration, and software localization.

There were a total of 16 research participants that included analysts and developers. The team leaders made up seven of the participants. Most of the participants were system analysts and developers at same time. The majority of the participants had a total experience of 3–10 years in the field, while a few had over 10 years' experience. Most of the participants had experience working as analysts, designers, and developers at the same time. Some participants had experience as analysts only, and some as developers only. Most participants had experience with business application software and database system software.

A total of 35 project cases were observed across the two companies. The PS types were Human Resource (HR) software, Enterprise Resource Planning (ERP) software, Special Solution software (such as a school management system), Restaurant Management software, and Point of Sale (PoS) software. Figure 3.1 below visually represents the number of cases observed.

## 3.4 Findings and Results

This section is an ethnographic account of the events taking place when feasibility is studied for PS implementation. In doing so, we address a pre-implementation stage that includes (1) analysts' roles; (2) software demonstration utilizing a live scenario; and (3) mechanisms of scoping and creating packaged software offer. We also make a comparison between traditional RE and pre-implementation PS RE, given that the pre-implementation stage in this study resembles such feasibility studies as those used in traditional RE at a high abstract level. This is because feasibility studies in traditional RE and the pre-implementation stage of PS discussed here are similar in terms of their purpose, such as dealing with software objectives, time and budget.

### *3.4.1 Analysts' Roles*

The analysts always carried out a software demonstration, wherein the analysts' roles have changed from just collecting requirements and undertaking software analysis to also engaging in the pre-sale of the software. It seems that this was likely due to the belief that analysts were the only ones who knew how the software was built, how it works, and how to explain it. It also appears that it was thought more likely that the client would buy the software if he/she feels the software solves their problems and helps them achieve their objectives — and the analysts may be better at explaining such issues than the sales team.

> "Before, the sales team made a software demonstration. But we found out that they just talked without understanding the software so sometimes they mentioned the wrong things to our client. The sales team is useless when making a software demonstration because they did not develop the software". [General Manager]

Analysts approached the software demonstration from the perspective of convincing the client to buy the software, and from the perspective of identifying misalignments. This requires skill, knowledge, and experience in packaged software functionality and how components of the software relate to each other. It appears that, in one organization at least, the sales team did not possess such capabilities.

### *3.4.2 Software Demonstration Utilizing a Live Scenario*

It can be observed that analysts consider the importance of a software demonstration from two dimensions: the business dimension which consists of presenting a possible solution to the client's issues and convincing the client to buy the software; the other dimension relates to software analysis, as the client might recognize new requirements and features in addition to those they initially perceived as requirements. The team leader of the analysts explained that the software demonstrations were usually capable of delivering the possible solution that the client needs, and the analysts believed that the system capabilities could be most effectively demonstrated with a live case.

> "We represent our software to a client by developing a real case scenario that can simulate and cover various aspects of a real situation within the client's work environment. It really helps us to explain our software functions and connect these functions to a real case". [Team Leader]

One observed case of a live scenario was when a team of analysts demonstrated the H2O software 'Restaurant management system' for clients. The reason behind the live scenario was that H2O software included hardware functionality, such as using a personal digital assistant (PDA) to take customer orders, and software functionality, such as representing menu items.

During the demonstration process analysts explained what the software could do in order to solve the client's issues, such as their inability to follow an order, missing orders in the kitchen, and difficulties with the inventory of items. The analysts listed a set of functions that the software provided, such as making it easy to take an order, making sure the kitchen receives the order, helping the cashier to receive the payment required, and creating a record of the payment. The team of analysts then sought to represent this functionality by using a live scenario that allowed them to link software functionality to the business case to help convince the client about the software product. Meanwhile, initial requirements were also collected. In this case, the client had two different types of menu: one for local customers and another for tourists. The client also had three types of service: takeaway, delivery, and internet service. After the live scenario demonstration the client asked for each different kind of service to have its own printer in the kitchen, so that when the PDA was used it would not send all of the orders to the same printer.

It was clearly beneficial to plan the software demonstration by focusing on the client's specific business issues and to develop a live scenario, as this best showed that the analysts had a possible solution to the client's business process needs, and in this case, the client accepted the software offer after the demonstration.

### *3.4.3 Mechanisms of Scoping and Creating a Packaged Software Offer*

The scoping process involves software analysis through discussion of high level modification requirements and new features. Without scoping, the software implementation time frame might expand to the degree where this would impact negatively on the cost and time involved in PS implementation. A participant explained:

> "It was a big issue, and it had required a lot of effort and time to work with that undefined scope to understand business practices and business requirements. Therefore, our strategy now is to define the scope of the software early on so that we will be prepared for the next step in the case that clients accept the software offer". [Team Leader]

For example, a team of analysts demonstrated a Human Resource Management System (HRMS) to their clients. HRMS deals with demographic data, current employment information, employment history, qualification tracking, and salary information. The analysts' team and first author went to meet the client to discuss her needs. The client already had HR software and was experienced in its use. In this case, the analysts started by asking her about the issues that she and her employees faced with the current software. We discussed the client's issues and the possible solutions.

After the visit, the analysts' team provided an assessment report to the general manager. This report included a discussion of the client's issues, the modifications required to existing functions, and new features to be added. The client's issues were categorized into various types: transaction issues, such as employees' bonuses mechanism and costs related to the provision of uniforms, and output format issues, such as the software reports format. The general management and the analysts' team met to discuss possible ways of improving the already existing software in order to fit with the initial requirements, and to discuss the analysts' expectations of the development time-frame involved. After this, they also resolved client issues related to the price of the software (this is to make the software offer). Hence it is clear that one part of creating a software offer is scoping through software analysis. During this process of creating a PS offer, core requirements are emphasized but detailed requirements are neglected.

When creating a packaged software offer the analysts consider the scope of the offer according to the initial requirements, the number of modifications requested by the client, the nature and extent of the modifications, and the technical requirements involved. By focusing on such elements of a PS offer, the analysts were able to work towards their main goal of accurately estimating the cost and time requirements of PS implementation.

> "After we understand clients' need if it transaction function modifications that require 'customization' or new features, we will assess the cost and time require to develop such required". [Team Leader]

When creating a PS offer, the software company also considers issues related to various kinds of 'assessment criteria' that were used to measure the level of impact on effort needed to develop, customize, and modify the PS. When we outline the relationship between the type of elements and the assessment level, this type of relation also highlights the software company's duties and the client's duties. The general manager and the analysts have defined the following assessment criteria for PS offers:

- New Features required, that consists of developing new functions that change the existing package
- Customization, which consists of modifying the existing functions to fill the gaps between the functions offered by a software package and the client's needs
- Software output, that consists of creating new reports or modifying existing reports or screens
- Technical needs, which consists of assessing the client's infrastructure requirements such as hardware and software.

**Table 3.2** Traditional RE feasibility study vs. pre-implementation RE feasibility study

| Elements | Traditional RE: feasibility study | Pre-implementation PS: feasibility study |
|---|---|---|
| Goals | Are the overall objectives of the organization satisfied by the proposed system?<br>Can the system be developed with the proposed budget and timeline? | What are the client issues?<br>What is the possible solution?<br>Is the possible solution within the scope of the software company's domain?<br>What are the cost and time required for a possible solution? |
| Business dimension | Worthiness of proposed system. | Instilling confidence in the client, securing business, and creating a software offer. |
| Software analysis dimension | Information gathering to assist the assessment of proposed system. | Information gathering to identify client's issues, new requirements and new features needed (if any) to assess cost and time for proposed solution implementation. |
| Domain of knowledge | The development organization and the customer can cooperate to ensure that the domain is understood. | The development organization has to be an expert in the domain. |
| Assessment criteria | Objectives of the organization are satisfied by proposed system.<br>System is developed with the proposed budget and timeline. | A new features level, customization level and output level. |
| Critical decision | Considers the worthiness of the proposed system, or regards changes, development decisions, seclude and budget. | The possible solution is within the software company's domain. |
| Scoping factors | Budget, timeline, technical and development issues. | Packaged software assessment level, elements, and limitation of work domain, client organization size, and client's issues. |

### *3.4.4 Traditional RE vs. Pre-implementation*

Table 3.2 characterizes the feasibility study in traditional RE [15] vs. pre-implementation PS. There are a number of differences in practice, and differences of purpose, between the elements of feasibility studies carried out in traditional RE and for pre-implementation PS RE. Traditional RE and pre-implementation PS RE share similarities as both can be seen as comprised of the same kinds of elements, and as, to some degree they sharing similar objectives and being influenced by similar business concerns and technical concerns. For example, both processes can be considered in terms of the same dimensions, which involve goals, the business dimension, the software analysis dimension, domain knowledge, assessment criteria, critical decision, and scoping factors. However, within these dimensions, important differences appear.

The analyst concerned with carrying out RE for PS implementation will be concerned with accessing different information and meeting different objectives than the analyst concerned with building custom-made software. For example, when building a bespoke system, traditional RE will focus on identifying whether the timeline and budget that have been proposed are feasible, and then with making sure that the organization's objectives can actually be met by the system that has been proposed. With pre-implementation of PS, however, the analyst must instead think about what the client's specific issues are and identify whether any existing packages offered by the analysts' company can offer a solution. The analysts engaging in pre-implementation PS RE must also consider the possibility of refusing a request for a particular solution if that solution falls outside the scope of the company's capabilities or outside the scope of the company's current products. Part of the process of identifying whether the solution is within the company's scope may involve thinking about the time and cost involved with implementing a particular package or with making requested changes to that package.

With traditional RE, the main goal of the 'business dimension' of RE is concerned with establishing whether the proposed system is 'worthy': whether it can be created and whether it will actually satisfy the demands of the business and be the best possible system for the business. The analyst carrying out pre-implementation PS RE, however, will be engaged with different concerns, such as actually selling the proposed packaged system to the client by showing them how the package operates and how it could fulfill their requirements. The analyst carrying out pre-implementation PS RE must actively instill confidence in the client, secure his/her company's business, and create a software product offer.

The software analysis dimension in traditional RE and pre-implementation PS RE is quite similar. The analysts in both forms of RE carry out a range of activities that find out the client's issues that need solving and that help them to find initial requirements. They will later need to follow up on such requirements by checking in case new requirements are needed or new features need to be added to the proposed solution. If new features are required, they will again need to assess the cost and time involved with such requirements. However, there are some differences between the two forms of RE. In pre-implementation PS RE, analysts need to consider the modifications to existing functions that have been requested by clients. However, such considerations do not concern analysts practicing traditional RE.

The level of domain knowledge required for the analyst engaging in these different forms of RE also differs. With traditional RE, the analyst can gain sufficient knowledge of the client's domain by interacting with and listening to the client. The client is more active in advising the analyst what is needed in the system. With pre-implementation PS RE, however, the client will expect the development organization to already be an expert in the domain and to offer them the best possible solution or a range of viable solutions.

The assessment criteria used to develop and implement the software also differ between traditional RE and pre-implementation RE. With traditional RE, the feasibility of the system is seen to depend on whether the objectives of the organization will be satisfied by the proposed system. If it is considered that they will be, the system will then be developed in accordance with the proposed budget and timeline. Pre-implementation PS RE involves its own set of assessment criteria, as outlined above.

With traditional RE, the main critical decision that needs to be made usually relates to confirming the worthiness of the proposed system. Other critical decisions, or factors, may relate to changes to the proposed system, or to budgetary factors or company developments. The analyst engaging in pre-implementation PS RE will make a critical decision when deciding whether the solution needed by the potential client is within the domain of the analyst's company.

The last element of comparison between feasibility studies in traditional RE and pre-implementation PS RE are scoping factors. Again, the scoping factors involved in the two different forms of RE are not the same. In traditional RE, scoping is guided mainly by the budget that has been set for the project, and by its timeline, and also by technical and development issues. Pre-implementation PS RE practice differs from this, as scoping for packaged software is influenced by a number of elements, including assessment criteria, the packaged software offer, and the limitation of the work domain, the client's organization size, and the client's issues.

## 3.5 Discussion

This study describes activities that should help analysts conduct or manage a feasibility study for PS implementation in terms of RE practices by highlighting, in particular: (1) analysts' roles during pre-implementation; (2) software demonstration utilizing a live scenario; (3) mechanisms of scoping and creating a packaged software offer; and (4) traditional RE vs. pre-implementation PS RE.

### *3.5.1 Analysts' Roles During Pre-implementation*

The main new finding of this study in terms of the role played by analysts in the feasibility study/pre-implementation stage, was that they rather than the sales team usually carried out the task of conducting a software demonstration for the client. In fact, in one of the cases observed, the sales team had previously carried out a software demonstration and had provided wrong information about the software to the client. Therefore, adopting the policy of analysts carrying out software demonstrations has been a company strategy to reduce the risk related to sales team members trying to sell features or accept to add new features to customers that had not actually been developed yet — an action that basically forced the company to include those features [16, 18].

One major difference, therefore, between the role of the analyst in pre-implementation PS RE and in traditional RE is that in pre-implementation RE the analyst has greater involvement in demonstration of the solution, being expected to conduct such software demonstrations. The analyst involved in pre-implementation

PS RE is also expected to have some understanding of business concerns and how to engage in marketing. While the analyst in traditional RE is generally limited to developing the software and then stepping back and allowing sales and marketing teams to pitch the product to clients, the analyst doing pre-implementation PS RE has more of a hybrid analyst-sales-marketing role and is required to have the soft skills needed for a software demonstration presentation, such as presentation skills, communication skills, and sales skills. The analyst is no longer only concerned with software analysis but also with the business dimension of creating software [7, 19].

### *3.5.2 Software Demonstration Utilizing a Live Scenario*

One of the factors differentiating pre-implementation PS RE practices from traditional RE practices is that analysts have a choice of how to conduct a feasibility study. Analysts may be able to offer more than one solution to the client. In such a case, analysts then need to choose which solution is the preferred one to offer to the client [7]. In order to make such a decision, analysts hold meetings that involve themselves and the sales and marketing teams. The limitations of the work domain of the analysts' company are considered. Other factors that are considered relevant to making a decision about the solution to offer include the size of the client's organization, the number of users at the organization, the kinds of departments the organization has, and the kinds of transactions the client organization will need to carry out.

The live scenario was used in cases where it was decided that this provided the best option for showing the capabilities of a PS. The live scenario aims to simulate a situation that could occur in the client company's real work environment. It was found, therefore, that contrary to simply describing or demonstrating a software package during a meeting, the live scenario involved analysts creating an environment that simulates the client company's site in order to show the client a real-time and live example of how the system could function within their operational context.

It appeared that using a live scenario helped the analysts to better understand and respond to the needs of the client company. For example, by conducting the live scenario the analysts were better able to see the challenges actually faced by the company, it was found that this kind of software demonstration could have a strong influence on whether the client would purchase the solution [7, 18].

It was also found that the planning of such a live scenario software demonstration relied on the analysts developing soft skills such as being able to present in a way that is personable and convincing, not merely to display knowledge about software.

The use of the live scenario for pre-implementation PS RE is different from any procedure used by analysts in traditional RE because the analyst conducting traditional RE is not typically concerned with showing the client company how a software solution may work for them through the use of imagined scenarios or by analogies showing how the solution worked for another company. In traditional RE the analyst is only concerned with building a solution to meet the demands of the client and to test them, during the software's development, that it does meet their requirements or fix their problems, and adjust the software until it is correct. The use of the live scenario during software demonstration, therefore, is a method unique to pre-implementation PS RE.

### 3.5.3 Mechanisms of Scoping and Creating a Packaged Software Offer

The analysts attempted to define the scope of the software during discussions with potential clients about their needs. Analysts believed it was important to carry out this scoping process early on since this would help them to construct a software offer, since such scoping would help everyone involved to maintain control of the time taken for implementation. Collecting such information not only provided analysts with details about what the new software needed to do, but also helped them to see what its limitations would be and what new features or modifications would be necessary. This step therefore helped them significantly with designing a PS that would suit the client.

It was also observed that the steps the software company took related to software scoping were generally limited to finding out information about only the core requirements of the system or solution to be implemented. This was found to involve transaction issues and output format issues [17]; during the scoping process the analysts were not concerned with discovering detailed requirements.

### 3.5.4 Traditional RE vs. Pre-implementation PS RE

The scoping factors involved when creating bespoke software and therefore conducting traditional RE are; budget constraints, timeline issues and constraints, technical issues, and development issues. Analysts conducting traditional RE will consider whether the timeline and budget that have been set are feasible, and must also ensure that the client organization's objectives can be met by the proposed software. Their main concern is whether the system that is developed will be 'worthy' for use.

Analysts engaging in pre-implementation RE must think about the client's specific issues and decide whether any existing packages offered by their company can offer a solution. They will need to consider the time and cost involved with implementing a particular package and with making requested changes to that package, and they may well decide to refuse a request for a particular solution if that solution falls outside the scope of the company or outside the scope of their current products. In this regard, pre-implementation PS RE differs strongly from traditional RE as analysts practicing traditional RE do not need to consider how to deal with requests for modifications to existing functions. Neither does the analyst practicing traditional RE need to engage in scoping with the aim of creating a software offer or actively take part in selling the proposed packaged system to the client.

## 3.6 Conclusion

This ethnographic account of pre-implementation PS RE, for the first time, shows how a software development company of this size (small-medium) approaches the challenge of managing the pre-implementation process at this point of the packaged software implementation life cycle. We have highlighted elements and assessment criteria involved with creating a software offer that is based on modification and customization of existing packaged software.

In this paper, we have outlined the importance of the software demonstration, and as such that the role of the analyst engaged in pre-implementation PS RE differs from their respective role in traditional RE. In pre-implementation PS RE, the analyst is likely to be the staff member delivering the software demonstration. Because analysts know how the software is built, how it works, and how to explain it, clients may be more likely to buy the software when the analyst explains it to them. Clients will buy the software if they feel that the software will help them to achieve their objectives and to solve their problems. Therefore, the analysts' presentation of the software is approached from the perspective of convincing the client of the package's suitability, and from the perspective of identifying misalignments, within the client's context. This presentation requires knowledge related to packaged functions and how functions are related to each other, as well as communication skills and the ability to persuade the client.

Future work should be undertaken to discover the differences in philosophy behind release plans for packaged software between large packaged software development companies and companies that are SMEs. From observations and from previous literature, it appears that large packaged software development companies tend to have very detailed release plans and schedules for future packaged software products mapped out months or years in advance, while SMEs may take a more ad hoc approach to release planning that instead involves continuous improvement of their product in response to clients' requirements and clients' responses to their product.

## References

1. Aranda J, Easterbrook S, Wilson G (2007) Requirements in the wild: How small companies do it. In Requirements Engineering Conference, 2007. RE'07. 15th IEEE International (pp. 39–48). IEEE.
2. Merten T, Lauenroth K, Biirsner S (2011) Towards a new understanding of small and medium sized enterprises in requirements engineering research. In: Requirements engineering: foundation for software quality. Springer, Berlin, pp 60–65

3. Bürsner S, Merten T (2010) RESC 2010: 1st Workshop on Requirements Engineering in Small Companies. econstor, 128.
4. Karlsson L et al (2007) Requirements engineering challenges in market-driven software development—an interview study with practitioners. Inform Software Technol 49(6):588–604
5. Quispe A, Marques M, Silvestre L, Ochoa SF, Robbes R (2010) Requirements engineering practices in very small software enterprises: A diagnostic study. In Chilean Computer Science Society (SCCC) (pp. 81–87). IEEE.
6. Xu L, Brinkkemper S (2007) Concepts of product software. Eur J Inform Syst 16:531–541
7. El Emam K, Madhavji NH (1995) A field study of requirements engineering practices in information systems development. In: Proceedings of the second IEEE international symposium on Requirements Engineering, 1995. IEEE
8. Solemon B, Sahibuddin S, Ghani A (2009) Requirements engineering problems and practices in software companies: An industrial survey. In Advances in Software Engineering (pp. 70–77). Springer Berlin Heidelberg.
9. Daneva M (2004) ERP requirements engineering practice: lessons learned. IEEE SOFTWARE
10. Barney S, Aurum A, Wohlin C (2008) A product management challenge: Creating software product value through requirements selection. J Syst Architect, 54(6), 576–593.
11. Daneva M, Wieringa RJ (2006) A requirements engineering framework for cross-organizational ERP systems. Reqs Eng 11:194–204
12. Hammersley M, Atkinson P (2007) Ethnography: principles in practice. Routledge, London
13. Miles MB, Huberman AM (1994) Qualitative data analysis: an expanded sourcebook. Sage, Thousand Oaks, CA
14. Khoo HM, Robey D (2007) Deciding to upgrade packaged software: a comparative case study of motives, contingencies and dependencies. Eur J Inform Syst 16:555–567
15. Sommerville I (2004) Software engineering, 7th edn. Addison-Wesley, Boston, MA
16. Jantunen S (2010) The benefit of being small: exploring market-driven requirements engineering practices in five organizations. In: Proceedings of the 1st workshop on RE in Small Companies (RESC), Essen, Germany
17. Dittrich Y, Vaucouleur S, Giff S (2009) ERP customization as software engineering: knowledge sharing and cooperation. IEEE Software 26(6):41–47
18. Jebreen I, Wellington R (2013) Understanding requirements engineering practices for packaged software implementation. In: 4th IEEE international conference on software engineering and service sciences, Beijing, China
19. Soh C, Kien SS, Tay-Yap J (2000) Enterprise resource planning: cultural fits and misfits: is ERP a universal solution? Comm ACM 43(4):47–51

# Chapter 4
# Towards P Systems Based Approach for Evolutionary Enterprise Application

**Gunnar Piho, Jaak Tepandi, and Viljam Puusep**

**Abstract** Development of enterprise applications is expensive, takes time and requires knowledge, tools and techniques. Contemporary enterprise applications must be dependable as well as customizable in the evolutionary way according to changes in the enterprise business processes. The wider goal of our research is to develop techniques for development of enterprise applications that software end users, in collaboration with software developers, are able to change safely and easily according to changing requirements. In accordance to the software engineering triptych: to write software, the requirements must be prescribed; to prescribe the requirements, the domain must be understood; to understand the domain, we must study one. We present and exemplify P systems based enterprise domain model. We treat an enterprise as a membrane-computing structure and utilize P system notions, notations and formalisms in modelling of enterprises and enterprise business processes. In our understanding this P systems based enterprise model can provide a practically usable framework for development of evolutionary enterprise applications.

**Keywords** Evolutionary Enterprise Applications • P Systems • Model Driven Engineering • Laboratory Information Management Systems

---

G. Piho (✉)
Clinical and Biomedical Proteomics Group, University of Leeds,
Beckett St, Leeds LS9 7TF, UK

Department of Informatics, Tallinn University of Technology,
Akadeemia St. 15A, Tallinn 12618, Estonia
e-mail: g.i.piho@leeds.ac.uk; gunnar.piho@ttu.ee

J. Tepandi • V. Puusep
Department of Informatics, Tallinn University of Technology,
Akadeemia St. 15A, Tallinn 12618, Estonia
e-mail: jaak.tepandi@ttu.ee; viljam.puusep@ttu.ee

M. José Escalona et al. (eds.), *Information System Development: Improving Enterprise Communication*, DOI 10.1007/978-3-319-07215-9_4,

## 4.1 Introduction

The goal of our research is to develop practically usable techniques for development of enterprise applications that software end users, in collaboration with software developers, are able to change safely and easily according to changing requirements. We use a case-study-based research methodology. The case is Laboratory Information Management System (LIMS) software development in Clinical and Biomedical Proteomics Group (Cancer Research UK Clinical Centre, Leeds Institute of Molecular Medicine, St. James University Hospital at University of Leeds). LIMS represents a class of computer systems designed to manage laboratory information [1].

According to *software engineering triptych*, in order to develop software we have to: (a) informally and/or formally describe a domain ($D$); (b) derive requirements (R) from these domain descriptions; and (c) finally from these requirements we have to determine software design specifications and implement the software ($S$), so that $D, S \models \mathcal{R}$ (meaning the software is correct) holds [2]. The term *domain* or *application domain* can be anything to which computing can be applied [3]. In our studies the application domain is business domain in general (producing, buying and selling either products or services) and clinical research laboratory domain in particular.

In research laboratories, like CBPG, business processes are changing constantly and different research groups within the same research laboratory, sometimes even different investigators in one and the same research group, require different business processes and different or differently organized data. While standardized in some ways, such system for scientists has to be flexible and adaptable so, that there are customizable possibilities to describe data, knowledge and also research methods.

In addition to the three principal layers (presentation, domain and data source), in contemporary distributed enterprise application architecture there is also a communication (containing and connecting logic) layer [4, 5], illustrated on Fig. 4.1.

In our understanding the communication layer and the presentation layer are similar in their nature. The presentation layer gives humans an interface (forms, documents, etc.) to the defined logic (domain model). Similarly, the communication layer gives artificial agents (services, software systems, etc.) an interface (communication protocols, etc.) to the defined logic.

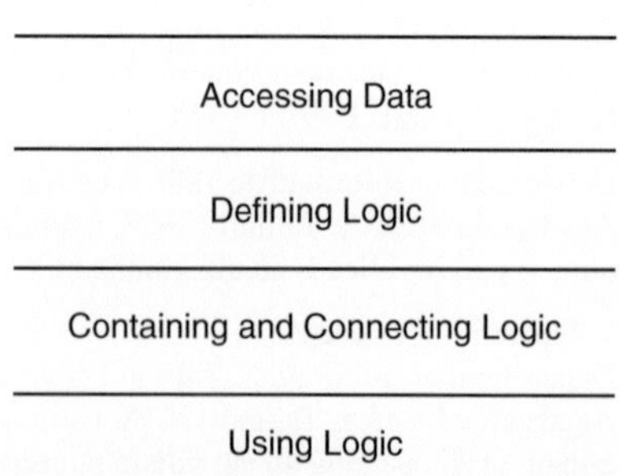


**Fig. 4.1** 4-Tier architecture of distributed enterprise application architecture

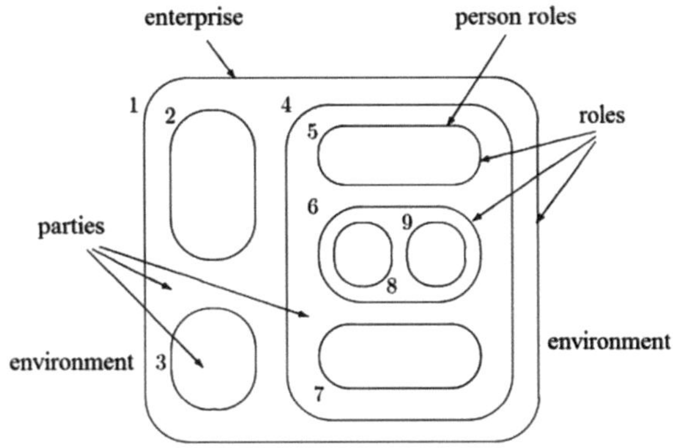


**Fig. 4.2** A cell-like enterprise structure

We are looking for ways to minimize (better to completely avoid) changes in the domain logic and in the data source (and therefore also in data access) layers as these changes are risky and time consuming. We are trying to find possibilities to fulfil user requirements only by making changes in the presentation or in the communication layers. It would be nice if these changes can be made by end users even at run-time.

P system (membrane computing) [6] is the model for the distributed computing where multisets of symbol objects encapsulated into membrane structure are processed in the maximum parallel manner. P system aims to mimic the structure and the functioning of living cells. In [7] we presented and exemplified P systems based enterprise domain model. We treated an enterprise as a membrane-computing structure (Fig. 4.2) and utilized P system notions, notations and formalisms in modelling of enterprises and enterprise business processes.

In the current paper we describe, how in our understanding this P systems based enterprise domain model can provide a practically usable framework for development of evolutionary enterprise applications.

In Sect. 4.2 we describe the related works. In Sect. 4.3 we shortly describe the P systems based enterprise domain model we presented in [7]. We propose and illustrate the evolutionary criteria and explain the possible UI based evolution of the enterprise applications in Sect. 4.4. These criteria are derived from the P systems based enterprise domain model. We conclude in Sect. 4.5.

## 4.2 Related Works Towards Evolutionary Information Systems

Searching in the document titles of the IEEE Xplore digital library the words "evolution", "information" and "system", we found 30 papers. Eight of them we found were related to the evolutionary information systems research topic.

Layzell and Luocopoulus [8], in 1988, described the collaborative European project RUBRIC, whose aim was to improve the practice of constructing large, sophisticated information systems by avoiding the practice of embedding business policy within program logic. The target was to separate not only business data, but also business policy from the operational procedure. They proposed a rule-based approach to the construction and evolution of business information systems. The structural components were based on entity-relationship model consisting of entities, which are any concrete or abstract things from the universe of discourse.

Clark, Lobsitz and Shields [9], in 1989, described the documenting process of the TASC-EDGETM (The Analytic Science Corporation—Effective Development through Growth and Evolution) methodology for the evolutionary software development. Based on the classical waterfall model and on the manual change of the code, the main models (process model, information model, and development model) of the TAS C-EDGE methodology are eligible also today. By combining development, operation (includes user feedback) and requirements analysis this TASC-EDGE evolutionary software development was based on the principle of listening to the system users and to responding to their needs.

Oei, Proper and Falkenberg [10], in 1992 *discussed the need for information systems capable of evolving to the same extent as organization systems do* and proposed a formal model for the evolution of information systems. Informally this model transformed the requests from users of the information system into update and retrieval actions of the application.

In 1992, Lui and Chang [11] were working for evolutionary information systems. They proposed a Visual Specification Model for specification design, change and redesign. In this method they focussed on the database schema changes in a single relation and on their effects to other components in the specification. The wider goal of their research was to develop a unified methodology to construct conversion functions, to maintain change history, and to detect inconsistency when multiple relations involved in a change.

Shifrin, Kalinina and Kalinin [12], in 2002, described the MEDSET technology for development, deployment and support of information systems in clinical medicine and other poorly formalized subject domains. The basic statements they pointed out were: (a) realization data structures in compliance with the structure of business process; (b) stability of the database model; and (c) user interface consistency to support some definite part of one business process.

Wang, Liu and Ye [13], in 2008, pointed out the potential inconsistencies among the ontology and the dependant applications in case of the ontology evolution and analysed the approaches of maintaining the consistency and keeping the continuousness of the dependant applications during the evolution. They proposed two scenarios for maintaining the consistency: property split and property range changes. By splitting they mean the situation, when for instance the text property of employee's home address is split into more specific properties of city, district, street, etc. In splitting the business process itself does not change: only the data used in this business process becomes more accurate. By range changes they mean the situation when the business process itself will change. For example the funding: what was only for full time students is now available also for part time students.

Aboulsamh and Davies [14], in 2010, proposed a metamodel-based approach to information systems evolution and data migration. They expand model-driven engineering to facilitate the evolution of information systems. They claim that in the domain of information systems, after the initial development and delivery, the automatic change of the systems is useful only if the data held in the system can be migrated easily to the next version and therefore their proposed method is focussed on migration of the information held in the system data.

Ralyté, Arni-Bloch and Léonard [15], in 2010, proposed a process model for integrating new services with the information system. They claim, that *before publishing a service to be reused in some composition, the validation of the data consistency, rules soundness, process compatibility and organizational roles compliance have to be guaranteed.*

## 4.3 P Systems Based Enterprise Domain Model

Cells and enterprises have both very intricate structure and functioning. Similarities can be seen in their internal activities as well as in their interactions with the neighbouring cells/enterprises, and with the environment in general.

Both cells and enterprises have a way to organize and control their processes. Cells have biological processes, enterprises have business processes and both have informational processes. Any cell means membranes. Any enterprise means enterprise structure. Both cells and enterprises are clearly separated from their environment and both have compartments (membranes/enterprise structure units) where specific processes take place.

Similarities can be seen also in sending messages from compartment to compartment. Cells can transport packages of molecules from one part to other part through vesicles enclosed by membranes in a way that transported molecules are not "aggressed" during their journey by neighbouring chemicals. Enterprises have internal or use external post and transportation services to transport messages, documents and products from location to location in a secured way.

Similarities can be seen also in creating and destroying compartments. Cell is able to create and destroy membranes in order to make biochemical processes more efficient. Enterprise is able to create and liquidate their structural units in order to make business processes more efficient.

In summary, cells and enterprises contain many fascinating similarities at various levels starting with similarities in structure, in communication/cooperation among neighbouring cells/enterprises, and ending with the interaction with the environment.

In Fig. 4.2 the membrane like enterprise structure is illustrated. Membranes are the roles [16] the parties (persons or enterprise units) can play in order to fulfil some enterprise related tasks. The external membrane (skin) we call enterprise. The elementary membrane is the role only persons can play. Patients, managers, doctors, students, etc. are the examples of such person only roles. The regions in the context of enterprise are parties (persons and/or organizations—groups of persons) playing the particular role. In P systems based enterprise model (differently from the P systems) the distinction must be made between the notion role (membrane) and the notion party (region), because for instance, the manager's role and Mr John Smith being the manager from the date to the date do not have the same meaning. However, sometimes we can use these notions also interchangeably because of one-to-one correspondence; especially in case of roles only the organizations (e.g. hospital, laboratory, etc.) can play.

Every structural unit in an enterprise receives documents (d) (information received in speech can also be modelled as a document). In every structural unit there are also descriptions of these documents i.e. formats (*f*) of these documents. For instance, a human resources (HR) department knows exactly what should be the format ($f_i$) of a document ($d_i$) to compose an employment contract document ($d_j$) according to the contract document format ($f_j$). HR department also knows, what kind of information ($a_i$, e.g. name, date of birth, address, etc.) should be recorded into the company records ($d_k$). Therefore the business rules for the HR department can be notated as follows.

$$d_i f_i f_j \rightarrow d_j f_i f_j a_i$$

$$a_i f_k \rightarrow d_k f_k$$

This means that according to a document $d_i$ (e.g. order), the recruitment of an employee is started. As a result the employment contract ($d_j$) is concluded and some employee data ($a_i$) are created. The employee data is then recorder into a specific document $d_k$ (e.g. record in a list of workers) of the company. Both of illustrated rules are catalytic rules where the document formats $f_i$, $f_j$ and $f_k$ are catalysts describing precisely the documents ($d_i$, $d_j$, $d_k$) the company uses. This means, that documents and therefore document formats are company specific. The employee data *ai* (e.g. archetypal domain model describing persons) we consider to be written in universally understandable language of business archetypes and archetype patterns. We describe them in Sect. 4.4.1.

We can also use indicators h*ere*, i*n* and *out*. For instance the same rules with these indicators can be as $d_i f_i f_j \rightarrow (d_j, out) f_i f_j$ $(a_i, here)$ and $a_i f_k \rightarrow (d_k, out) f_k$. This means for example that documents $d_j$ and $d_k$ created by HR department are going to the "surrounding" enterprise unit.

Membrane dissolving ($\delta$) in P system based enterprise model means either concentrating of the post or eliminating of the structural unit of organization. Differently from the basic P system, in the P system based enterprise domain model, in case of dissolving, not only the objects and child members, but also the rules are left free in the surrounding enterprise unit. We also expect that in P system based enterprise domain model the enterprise itself (skin membrane) can be dissolved.

Similarly to base P system, also in the P system based enterprise model the rules are processed non-deterministically in the maximally parallel manner. Non-determinism can be seen for instance as HR department's possibility to choose the order of rules (creating employment contract or recording employee data) and objects (whatever document first) deliberately.

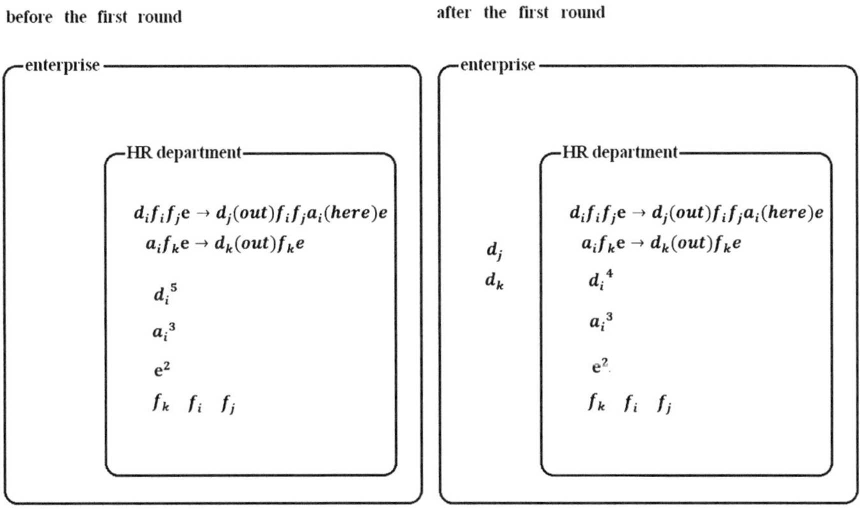


**Fig. 4.3** Processing of rules in P systems based enterprise domain model

For the maximum parallelism we probably have to determine the employees (catalyst e) in the HR department rules, e.g. $d_i f_i f_j e \rightarrow d_j(out) f_i f_j a_i\ (here) e$ and $a_i f_k e \rightarrow d_k(out) f_k e$.

For instance (illustrated in Fig. 4.3), in case there are five orders ($d_i^5$), three units of employee data ($a_i^3$) and two HR department employees ($e^2$) in HR department, then after the first round of processing with the maximum parallel manner the HR department objects can be $d_i^3 a_i^5 e^2$ (both employees were choosing the first rule), $d_i^5 a_i^1 e^2$ (both were choosing the second rule) or $di^4 ai^3 e^2$ (shown in picture, one employee choose the first rule and other the second rule).

Concluding the paper [7], in our understanding, we can formally define the enterprise using the formal definition ($\Pi = (O, C, \mu, \omega_1, \omega_2, \ldots, \omega_m, R_1, R_2, \ldots, R_m, i_0)$ of P systems of degree $m$. Looking for the Fig. 4.1, the enterprise application architecture (also some model of the enterprise) we have *data* ($\omega_1, \omega_2, \ldots, \omega_m$ in formal definition where $\omega_i$ means particular set of data in particular compartment, e.g. in HR department, Fig. 4.3); finite set of *classes* ($O$ in formal definition) describing business domain logic objects (describing any data) and workflows (using the data); different business domain *rules* ($R_1, R_2, \ldots, R_m$ in formal definition where $Ri$ means particular set of rules in particular compartment, Fig. 4.3); different *document formats* (catalysts, $C$) for converting the data defined by business domain logic to and from man or machine readable documents like printable reports, web pages, standalone client forms, communication protocols for remote logic (another computer or software) etc. We also have the enterprise structure ($\mu$) and the compartment ($i_0$) which contains (can contain) the result of the calculation. Informally it means that any enterprise is a "computer" calculating its budget. Result of this "calculation" is kept in the accounting department (compartment, $i_0$) data.

## 4.4 Towards Evolutionary Enterprise Applications

We see the concept of P-systems $\Pi = (O, C, \mu, \omega_1, \omega_2, \ldots, \omega_m, R_1, R_2, \ldots, R_m, i_o)$, described by classes ($O$), catalysts ($C$), structure ($\mu$), data ($\omega_1, \omega_2, \ldots, \omega_m$) and rules ($R_1, R_2, \ldots, R_m$), as a roadmap towards evolutionary information systems.

In our understanding, the enterprise application is evolutionary, if software end users, in collaboration with software developers, are able to change safely and easily (according to changing requirements) the following: (1) classes ($O$) describing domain concepts; (2) document formats (catalysts, $C$) that the system uses; (3) the structure ($\mu$) of the system: (4) data ($\omega_1, \omega_2, \ldots, \omega_m$) can be changed anyway (we can add as many lines as needed to documents or send as many documents as needed); (5) calculation rules ($R_1, R_2, \ldots, R_m$); and (6) the location ($i_o$) where the calculation results are held.

We describe here (because of limited number of pages) only how we change classes and document formats in current LIMS for CBPG project and let the other two for our later papers.

### 4.4.1 *Classes Are Changed Using Archetypes and Archetype Patterns*

$O$ is the finite and non-empty set of all possible *classes* (names of concepts and groups of concepts as well as logical and computational models of these concepts and groups of concepts) used in modelling of enterprises (then names and logical models of concepts) and in enterprise application implementations (then computational models of these concepts). Let as suppose, that we have at least two subsets of classes in $O$: (a) classes describing archetypes and archetype patterns ($O_A \subset O$), (b) classes describing business domain logic ($O_B \subset O$; $O_A \subset O_B$).

Business archetypes and archetype patterns are originally designed and introduced by Jim Arlow and Ila Neustadt [16]. Business archetype patterns (product, party, order, inventory, quantity and rule), composed by business archetypes (person's name, address, phone number, etc.) are information models and describe the universe of discourse of businesses as it is, neither referring to the software requirements nor to the software design. For instance, in Fig. 4.4 the *party archetype pattern* is illustrated.

The party archetype pattern represents a (identifiable, addressable) unit that may have a legal status and has some autonomous control over its actions. Persons and organizations are types of parties. Party has different properties like *party contacts* (phone number, e-mail, web address, and postal address), *registered identifiers* (passport, VAT number, domain name, stock exchange symbol, etc.) etc. Each party can play different *roles* (patient, clinician, employee, customer, etc.) and can be in different *relationships* (e.g. being married, having a child, studying in school). Both roles and relationships are time limited (property *valid* in role and

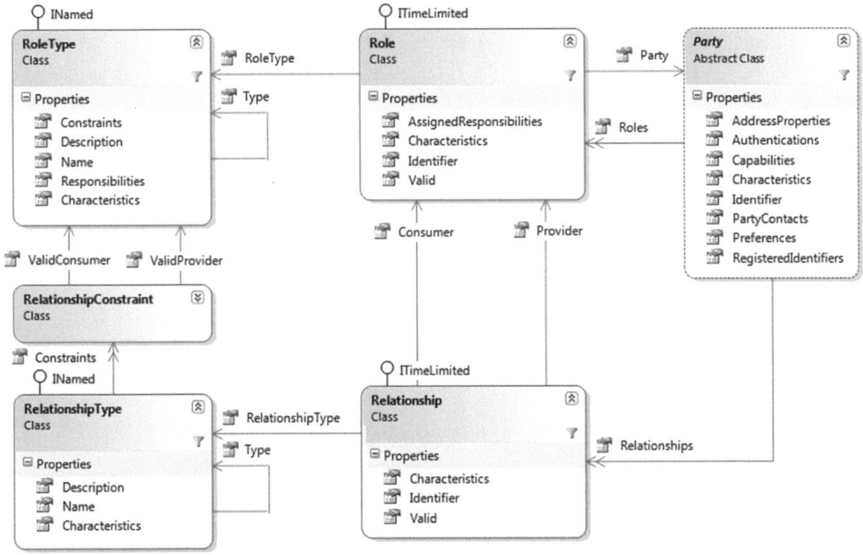


**Fig. 4.4** Party archetype pattern

relationship archetypes). *Role type* is used to store common information (e.g. constraints describing parties who can play the role) for a set of similar role instances and *relationship type* is used to store common information (e.g. constraints describing which roles can form the relationship) for a set of a similar relationship instances. In the current model only binary (more flexible and cleaner than n-ary) relationships are used, which means that one *relationship* binds exactly two related roles conditionally called as "provider" and "consumer".

Let us suppose, that classes illustrated in Fig. 4.4 are realized in code, and that we also have a database schema support for such classes (this is what we have in CBPG LIMS project). This means, that we have database tables for recording *party* instances (persons and organizations), *role* instances, *relationship* instances, *relationship type* instances and *role type* instances.

In a clinical laboratory, for instance, the common roles for persons are patient (whose blood will be tested), clinician (who ordered the blood testing), and medical technical assistant (MTA, who performed the blood testing). Normally in the domain logic layer of enterprise applications in such case we have to realize *Patient*, *Clinician* and MTA classes in code. In archetypes and archetype patterns based approach, we use, we do not. We have only one *RoleType* class for all classes and one *Role* class for all possible class instances (e.g. objects) of all possible classes. Such an approach gives for us the possibility to add new "classes" even at runtime. For instance, to add a new "class" named *laboratory manager*, the only thing to do is to add a new record to the *RoleType* database table.

To "add new properties" in such an archetypes and archetype patterns based approach, we have the property *Characteristics* (Fig. 4.4) in all our programmed

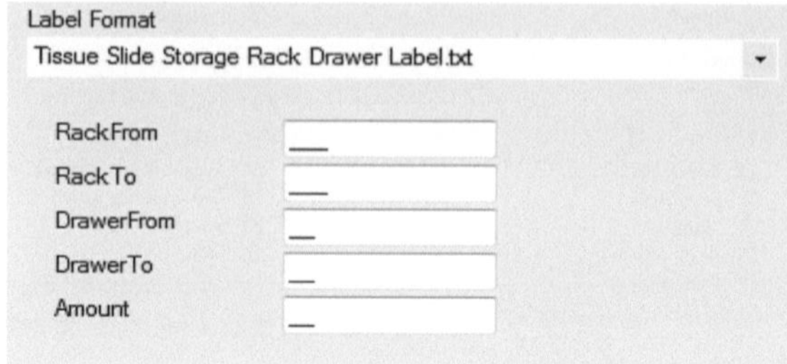


**Fig. 4.5** Fragment of generated barcode printing dialog

**Fig. 4.6** Example of generated barcode

classes. These are constructs similar to the *RDF* (Resource Description Framework) *triplets*. This means, that the *Characteristics* property holds a collection of {*category*, *name*, *value*, *authorized by*, *valid from*, *valid to*} records. For example the notation "person is 176 cm tall, measured by Dr Smith at 3rd of May 2000" is a characteristic with "body metrics" denoting *category*, "is tall" denoting *name*, "176 cm" denoting *value*, "Dr Smith" denoting *authorized by*, and "3rd of May 2000" denoting *valid from*.

## 4.4.2 Document Formats Are User Editable

We use document formats (e.g. end user editable files, DB records or values of properties) in number of places in current LIMS for CBPG project. For example, the following end user editable script (content of a file)

```
;{a} SLR
;{b} I RackFrom = 990
;{c} I RackTo = 990
;{d} C Rack = Foreach(b,c)
;{e} I DrawerFrom = 90
;{f} I DrawerTo = 90
;{g} C Drawer = Foreach(e,f)
;{0} C Barcode = a:3|d:3|g:2
;{1} I Amount = 90
```

first describes the automatically generated dialog, shown in Fig. 4.5, and then prints the barcode (Fig. 4.6) according to entered, using this dialog, values.

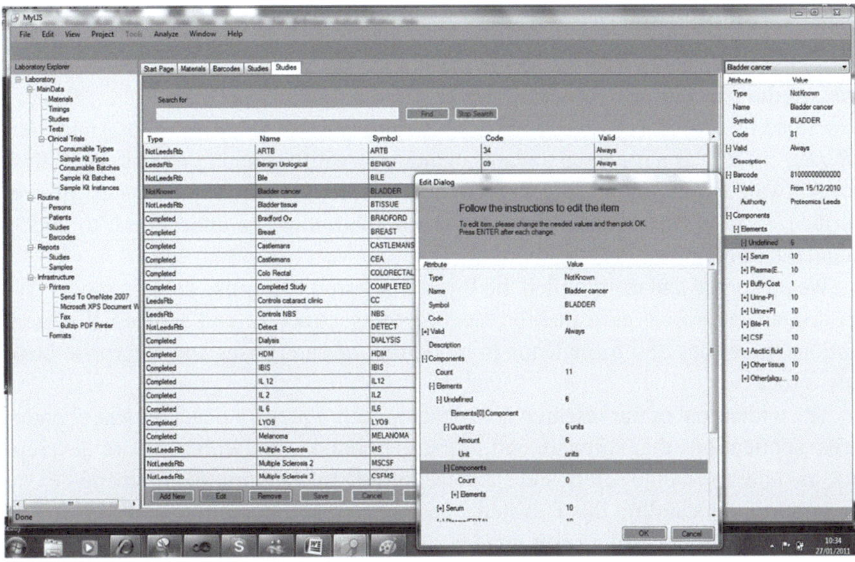


**Fig. 4.7** Screenshot of MyLIS (LIMS for CBPG) user interface

The other example is how we generate UI at runtime. For example, the UI, illustrated in Fig. 4.7, is generated according to the following scripts.

```
static string[] GridRange = new[] { "Type","Name","Symbol","Code","Valid" };
static string[  ]PropertyRange = new[] {"Type","Name","Symbol","Code",
      "Valid","Description","Barcode","Components" };
static string[] EditRange = new[] {"Type","Name","Symbol","Code","Valid",
      "Description","Components" };
```

First, the *GridRange* lists the properties, by their names, the master grid shows. Next, the *PropertyRange* lists the properties the detail panel (left side panel of main form) shows. Finally, the *EditRange* lists the properties the edit dialog shows.

This means that we have document formats (GridRange, PropertyRange, EditRange) which describe documents (user interfaces). When we change document formats, the user interfaces, and therefore the information system, will change. As document formats are properties, it is possible to change the values of these properties at runtime using, for example, reflection technology. Generally speaking we use the document formats as catalysts (*C*) for mapping data described by business domain logic classes to user or machine readable documents. Different mappings require varying levels of knowledge and authority from the end users. Therefore an organisation deploying such a system should determine the ability of users to perform a specific mapping, for example by specifying their roles and responsibilities with the aid of a RACI (Responsible, Accountable, Consulted, Informed) matrix.

## 4.5 Conclusion

We see the concept of P-systems ($\Pi = (O, C, \mu, \omega_1, \omega_2, \ldots, \omega_m, R_1, R_2, \ldots, R_m, i_o)$), described by objects ($O$), catalysts ($C$), structure ($\mu$), data ($\omega_1, \omega_2, \ldots, \omega_m$) and rules ($R_1, R_2, \ldots, R_m$), as a roadmap towards evolutionary information systems. We follow the software engineering triptych: to write software, the requirements must be prescribed; to prescribe the requirements, the domain must be understood; to understand the domain, we must study one.

We presented and exemplified the P systems based enterprise domain model. We treated an enterprise as a membrane-computing structure and utilized P system notions, notations and formalisms in modelling of enterprises and enterprise business processes.

The wider goal of our research is to develop techniques for development of enterprise applications that software end users, in collaboration with software developers, are able to change safely and easily according to changing requirements.

In our understanding this P systems based enterprise domain model can lead us towards evolutionary enterprise applications. In our understanding, the enterprise application is evolutionary, if software end users, in collaboration with software developers, are able to change safely and easily (according to changing requirements) the following:

1. Classes ($O$) describing domain concepts. (For example being able to define a new domain concept *patient*, using some archetypal language (Sect. 4.4.1), so that *patient* is a *role* that only *party* who is a *person* (Fig. 4.4) can play).
2. Document formats (catalysts, $C$) that the system uses.
3. The structure ($\mu$) of a system.
4. Data ($\omega_1, \omega_2, \ldots, \omega_m$) can be changed anyway. We can add as many lines as needed to documents or send as many documents as needed.
5. Calculation rules ($R_1, R_2, \ldots, R_m$).
6. Location ($i_o$) where calculation results are held.

**Acknowledgments** This work is supported by Estonian Ministry of Education and research (SF0140013s10); by Tallinn University of Technology (Estonia); by University of Leeds (United Kingdom); by Cancer Research UK.

## References

1. ASTM (2006) E1578-06 Standard guide for laboratory information management systems (LIMS). ASTM International. http://www.astm.org/Standards/E1578.htm. Accessed 4 June 2014
2. Bjørner D (2007) Domain theory: practice and theories (a discussion of possible research topics). Macau SAR, China. http://www.imm.dtu.dk/~dibj/ictac-paper.pdf. Accessed 4 June 2014
3. Bjørner D (2006) Software engineering. In: Abstraction and modelling. Texts in theoretical computer science. The EATCS series, vol 1. Springer, Heidelberg
4. Fowler M (2003) Patterns of enterprise application architecture. Addison-Wesley, Boston, MA

5. Chappell D (2006) Comparing .NET and Java: the view from 2006. Microsoft TechEd Developers, Barcelona
6. Paun G (2004) Introduction to membrane computing. http://psystems.disco.unimib.it/download/MembIntro2004.pdf. Accessed 30 Aug 2011
7. Piho G, Tepandi J, Puusep V (2013) P systems based enterprise domain model. Research Report. Tallinn University of Technology
8. Layzell P, Loucopoulos P (1988) A rule-based approach to the construction and evolution of business information systems. In: Software maintenance
9. Clark P, Lobsitz R, Shields J (1989) Documenting the evolution of an information system. In: Aerospace and Electronics Conference, NAECON 1989. Proceedings of the IEEE 1989 National
10. Oei J, Proper H, Falkenberg E (1992) Modelling the evolution of information systems. Department of Information Systems, University of Nijmegen, Nijmegen, The Netherlands
11. Liu C, Chang S (1992) A visual specification model for evolutionary information systems. In: Computing and information, 1992. Proceedings. ICCI '92
12. Shifrin M, Kalinina E, Kalinin E (2002) MEDSET – an integrated technology for modelling, engineering, deployment, support and evolution of information systems. In: Computer-based medical systems 2002 (CBMS 2002)
13. Wang Y, Liu X, Ye R (2008) Ontology evolution issues in adaptable information management systems. In: e-Business engineering, 2008. ICEBE '08
14. Aboulsamh M, Davies J (2010) A metamodel-based approach to information systems evolution and data migration. In: Software engineering advances (ICSEA), 2010 fifth international conference
15. Ralyté J, Arni-Bloch N, Léonard M (2010) Information systems evolution: a process model for integrating new services. In: AMCIS 2010 proceedings
16. Arlow J, Neustadt I (2003) Enterprise patterns and MDA: building better software with archetype patterns and UML. Addison-Wesley, Boston, MA

# Chapter 5
# Data-Oriented Declarative Language for Optimizing Business Processes

**Luisa Parody, María Teresa Gómez-López, and Rafael M. Gasca**

**Abstract** There is a significant number of declarative languages to describe business processes. They tend to be used when business processes need to be flexible and adaptable, being not possible to use an imperative description. Declarative languages in business process have been traditionally used to describe the order of activities, specifically the order allowed or prohibited. Unfortunately, none of them is worried about a declarative description of exchanged data between the activities and how they can influence the model. In this paper, we analyse the data description capacity of a variety of declarative languages in business processes. Using this analysis, we have detected the necessity to include data exchanged aspects in the declarative descriptions. In order to solve the gap, we propose a Data-Oriented Optimization Declarative LanguagE, called DOODLE, which includes the process requirements referred to data description, and the possibility to include an optimization function about the process output data.

**Keywords** Business Processes • Declarative Language • Data-Oriented Optimization

## 5.1 Introduction

A business process, henceforth referred to as BP, consists of a set of activities that are performed in coordination within an organizational and technical environment. These activities jointly perform a business goal [1]. In management theory during the last years, a process-oriented perspective has been considered the shell in organizational (re)structuring. Nowadays, organizations still experience difficulties in

L. Parody (✉) • M.T. Gómez-López • R.M. Gasca
Computer and Language Systems Department, ETSII, University of Seville, Avda. de la Reina Mercedes s/n, 41012 Seville, Spain
e-mail: lparody@us.es; maytegomez@us.es; gasca@us.es

M. José Escalona et al. (eds.), *Information System Development: Improving Enterprise Communication*, DOI 10.1007/978-3-319-07215-9_5,

applying this process-oriented perspective to the design and maintenance of their information systems. Currently, the deployment of more complex systems has put forth new requirements for flexible and dynamic specification. Several languages propose an imperative representation of business processes, whose specification allows business experts to describe relationships between activities, and to transform the process into an executable model. Therefore, an imperative description defines exactly how activities have to be performed, and how to handle the data-flow. However, one of the disadvantages of imperative languages comes from the use of unsuitable information for computer systems, since they do not provide flexible and adaptable business processes.

A declarative representation takes into account the business concerns that govern the BP. A BP may be exposed to different environments and subjected to many conditions in which the order of activities, or the data exchanged, cannot always be described at design time or in an easy way. This is the reason why several authors have proposed languages to define BP as declarative models, since sometimes the process cannot be completely defined at design time. One of the reasons why it is not possible to create an imperative model at design time, is related to the data that flow through the process. Depending on the data instantiated, the creation of one model or another could be better, since, for example, the model will influence the selection of the best data input of the activities to satisfy the process requirements. The role of data in declarative languages has not been very relevant, mostly limited to describe the execution or not of an activity, depending on the value of a variable of the data-flow. In this paper, we present a new point of view of declarative language focused on data, where we highlight the significance of the information that flows through the process and between the activities to reach an optimal model according to the user requirements.

Sometimes, the user of an application supported by a business process has to decide about the values to introduce at runtime, for example, the dates in the organization of a trip (booking flights, hotel room and renting a car). Frequently, if the user can choose between different dates in order to minimize the price, (s)he needs to search by hand the combinations of dates with the activities this mean searching for flights, hotel rooms and car rental. For this example, the model is known, being possible to execute the activities in a parallel manner. However, the process goal is to book a trip, and the objective function is to minimize the price of the trip. Then, in order to achieve the user requirements (to minimize the sum of the prices for the services, for a given set of dates), the best combination of data input for the activities needs to be found. To the best of our knowledge, none of the declarative languages permit the inclusion of data output optimization in this sense. For this reason, we propose a data-oriented decision language, called DOODLE. Although there is already a declarative language called DOODLE, presented by Cruz in [2], it is a visual and declarative language for object-oriented databases and our proposal is focused on a definition of a business process in a declarative way. Since DOODLE is oriented to data perspective, when in this paper we use the term ‘optimize the process’, it means that there is a function to be optimized and where the data-flow involved in the process are related with the aim to optimize the data output.

To meet this challenge, we have analysed the existing declarative languages and how they have addressed data management, and the features that have not yet been analysed.

The rest of the paper is organized as follows: Sect. 5.2 motivates and explains, through an illustrative example, the necessity in some cases to find the data input values to optimize the process execution. Section 5.2 introduces a motivating example where a declarative description oriented to data is necessary. Section 5.3 details the proposed language based on the description of a declarative subprocess that can be combined with an imperative description, such as BPMN. Section 5.4 studies some of the most relevant proposals of declarative languages and their contributions to data management. The motivating example described by means of this language has been included and a comparison with the studied languages is presented. And finally, conclusions are drawn and future work is proposed in Sect. 5.5.

## 5.2 On-Line Book Store: A Motivating Example

In this section, we introduce an example to motivate the necessity to include in declarative languages some aspects related to data that have influence in the model. The example is based on a sale and delivery process that has been used in many papers before [3–5] and [6]. The example is the on-line Book Store (BS), that represents a company that collaborates with two services in order to sell and deliver books (see Fig. 5.1). Both services inform the customer about the final price of buying and delivering a number of units of a book. However, it could be cheaper to buy this quantity of books in different packages by obtaining a discount. For example, if there is a discount depending on the number of units for a maximum, or the price of shipping depends on both the weight and volume of the boxes to send. In this case, it is cheaper to send two small boxes (e.g. 8€ each one) than a bigger one (e.g. 25€). Another example is if the customer wants to buy 5 units of a book, the cheapest option could be to buy them in two packages (3 + 2), since there is an offer "buy 3 pay 2", and although two deliveries are paid, the delivery cost is increased considerably due to the weight and volume. The sale terminates when the books are delivered to the customer and the payment is made.

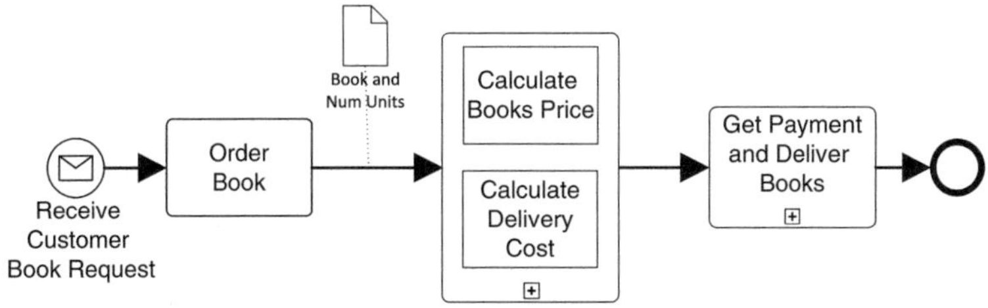


**Fig. 5.1** On-line book store example

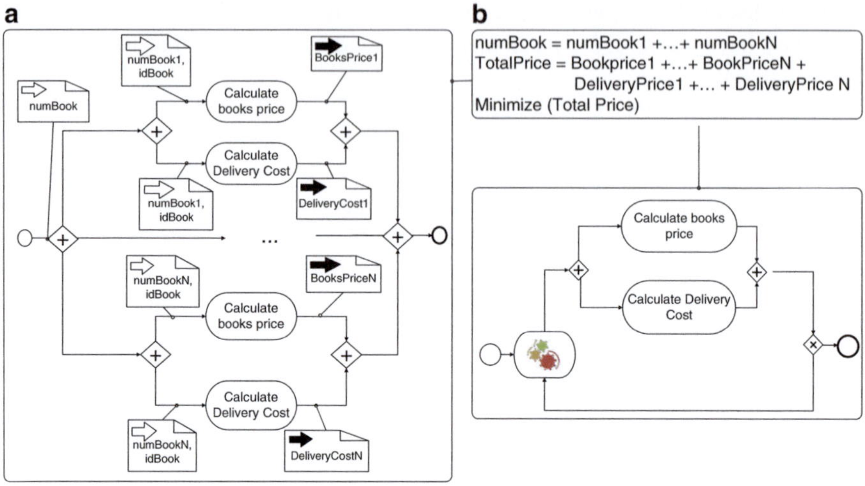


**Fig. 5.2** Imperative models for the on-line book store example. (**a**) Deployed imperative model. (**b**) Imperative model with implementation of tasks

In order to model the business process which support the BS example including the requirement to minimize the final price, neither imperative nor declarative languages can be used. For imperative languages, Fig. 5.2 depicts two possibilities of implementation to satisfy the constraints, which are also included in the figure. On the one hand, Fig. 5.2a presents a possible imperative model in BPMN to describe the BS example. The model has several activities to execute the purchase and the delivery for each package, but the problem is that the *numBooki* variables are unknown, even at runtime, since they are determined to minimize the total cost in each instance. Moreover, the number of activities to buy and deliver the books are unknown at design time. The difficulty in this case is that the specific values of the variables and the number of activities executed to minimize the objective function are unknown at design time. On the other hand, another solution could be that shown in Fig. 5.2b: various executions of the activities can be made by means of a loop, and the values of the *numBooki* variables will be determined programmatically in a new activity included in the model. The problem with this solution is the significant difficulty in implementing this activity, being necessary to delegate their programming to a computer science expert, not to a business process expert.

In relation to the existing declarative languages, as is analysed in Sect. 5.4, to the best of our knowledge, the current declarative languages are centred on the declarative description of the order of activities. However, none of them includes data input and output of the activities to optimize the object obtained from the business process.

## 5.3 DOODLE: Data-Oriented Optimization Declarative LanguagE

In order to include the optimization function and the constraints into the business process model, we have defined a declarative language called DOODLE (Data-Oriented Optimization Declarative LanguagE). This language combines the imperative description of BPMN with a declarative description of the parts of the process that need more flexibility (declarative subprocesses). The declarative description includes the data subprocess, data activities, objective function and the numerical constraints that let the designer describe the possible data values in a declarative way. These numerical constrains are defined by the following grammar where `Variable` and `Value` can be defined as Integer, Natural or Float domain:

```
Constraint := Atomic_Constraint BOOL_OP Constraint
| Atomic_Constraint
| '¬'Constraint
BOOL_OP:= '∨' |'∧'
Atomic_Constraint:= function PREDICATE function
function:= Variable FUNCTION_SYMBOL function
| Variable
    | Σ Variable
    | Value
  PREDICATE:= '=' |'<' |'≤' |'>' |'≥'
  FUNCTION_SYMBOL:= '+' |'−' |'*'
```

These constraints make it easier and more precise when handling numeric data (that can be represented as variables) that represent relations between variables.

In order to introduce the language, we have divided the description into two parts: (i) the internal description of the components associated with the activities of the declarative subprocess (Table 5.1), and (ii) the external description of the declarative subprocess, that implies the relation of the activities with subprocess data input and output also with constraints (Table 5.2). The language proposes to describe the order of the activities using imperative or declarative languages, depending on the necessity of the process. For the BS example, a parallel execution is possible, then it can be described imperatively.

In order to understand the example better, the language has been used to model the BS problem, as shown in Fig. 5.3. In order to transform this declarative model into an imperative model that supports any value of input variables of the process (idBook and number Of Books for example), we propose the use of Constraint Programming paradigm (explained in Sect. 5.4.1) [7] and domain local analysis of the variables [8]. The created model will depend on the knowledge of the relationship between input and output data of the activities, specifically, if this relationship

**Table 5.1** Internal components associated to the activities of the declarative subprocess

| Symbol | Name | Description | Parameters |
|---|---|---|---|
| 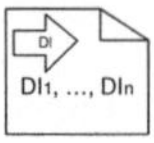 | Data input (DI) of the activity | Set of data input of each activity | List of variables |
| 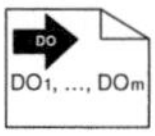 | Data output (DO) of the activity | Set of data output of each activity | List of variables |
| 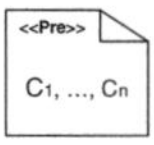 | Precondition | Set of constraints that represents the values of the DI that satisfy the execution of the activity | Numerical constraints |
|  | Postcondition | Set of constraints that represents the values of the DO that are satisfied after the execution of the activity | Numerical constraints |
| 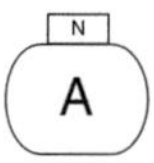 | Repetition of an activity A | Representation of the number of times that an activity can be executed. The value can be numeric (e.g. 5), or symbolic (e.g. N) | Integer or string |

**Table 5.2** External components of the declarative subprocess

| Symbol | Name | Description | Parameters |
|---|---|---|---|
| 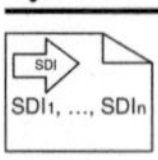 | Subprocess data input (SDI) | Data input of the subprocess that describe the user requirement in each process execution | List of variables |
| 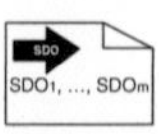 | Subprocess data input (SDI) | Data input of the subprocess that describe the user requirement in each process execution | List of variables |
| 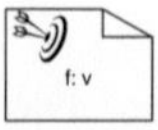 | Objective function | An optimization function in terms of a data | Minimize or maximize and the objective variable |
| 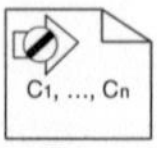 | ID input constraints (IC) | Set of constraints that relates the SDI with the DI of each activity of the subprocess | Numerical constraints |
| 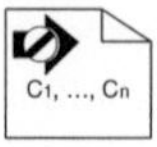 | ID output constraints (OC) | Set of constraints that relates the SDO with the DO of each activity of the subprocess | Numerical constraints |
| 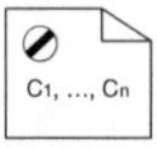 | Internal constraints | Set of constraints that relates the DI and DO of the activities among them | Numerical constraints |

can be known with or without executing the activities. For the BS example, the *totalPrice* can only be known if the activities are executed, although if the pre and post-conditions of the activities could relate input and outputs, the execution of the activities would be not necessary.

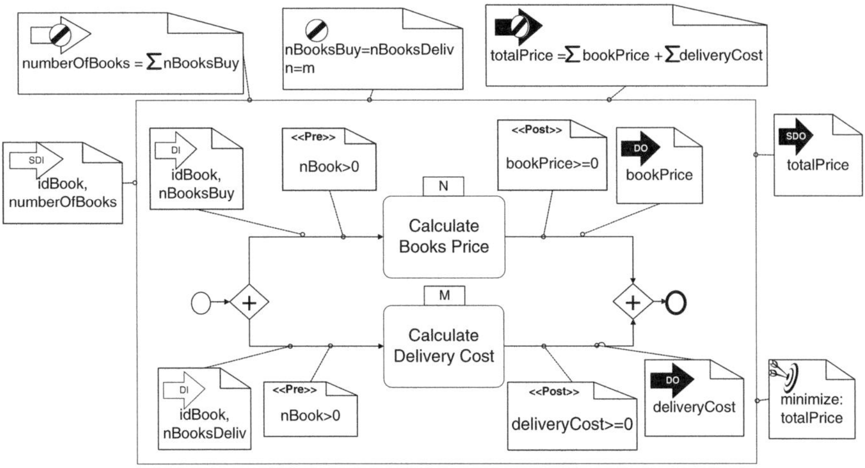


**Fig. 5.3** Example of buy books described using DOODLE

## 5.4 Related Work

There are many languages that enable the description of business processes in a declarative way. Generally, the common idea of declarative business process modelling is to model a process as a trajectory in a state space, and declarative constraints are used to define the valid movements in that state space [9]. Therefore, the differences between declarative process languages can, in part, be understood as a different perception of the meaning of 'state'. Analysing the different proposals, we have found some characteristics that we consider interesting to be compared. The characteristics are studied in Sect. 5.4.1 and compared in Sect. 5.4.2.

### *5.4.1 Declarative Language Characteristics*

Analysing the different proposals related to declarative languages, the main characteristics that we think should be analysed and compared are:

- **Formalism for reasoning:** The proposals use different formalism for reasoning. Sometimes, although we show the most relevant in each case, they can combine more than one and be improved with made-to-measure algorithms. Only the most relevant have been included in the paper due to space limitations.
  - **Linear Temporal Logic (LTL).** As demonstrated by Chomicki [10], Bacchus and Kabanza [11] and Pesic and van der Aalst [12], LTL expressions can be used to represent desirable or undesirable patterns within a history of events. LTL is a modal temporal logic that allows the expression of temporal constraints

on infinite paths within a state space. LTL formula can be evaluated by obtaining the Büchi automaton that is equivalent to the formula and checks whether a path corresponds to the automaton. Unfortunately most LTL checking algorithms assume infinite paths and construct non-deterministic automata [12]. Another disadvantage is that LTL does not allow the expression of the effect that results from a particular transition in a state space. For these reasons, it is not evident to express a goal state in LTL nor to construct automata for planning an execution scenario to obtain a goal state [11] as is needed in an optimization function.

- **The Event Calculus.** In first-order logic there is a formalism that elegantly captures the time-varying nature of facts, the events that have taken place at given time points and the effect of these events reflecting on the state of the system. This formalism is called the Event Calculus. The Event Calculus, introduced by Kowalski and Sergot [13], is a logic programming formalism to represent and reason about the effect of events on the state of a system. The Event Calculus is appealing for several reasons, as it builds on a first-order predicate logic framework, for which efficient reasoning algorithms exist. In addition the Event Calculus not only has the ability to deductively reason about the effects of the occurrence of events (leading to the coming into existence of fluency or the ceasing to hold), most importantly, it also has the ability to reason abductively. Abductive reasoning over the event calculus has been shown to be equivalent to planning. In particular, abductive reasoning produces a sequence of transitions (denoted by events) that must happen for a particular instance to hold in the future [14–16].
- **Constraint Programming (CP).** Constraint Programming [17] is a paradigm that permits the description of the model by means of the variables and the constraints that relate the variables. The model is called a Constraint Satisfaction Problem (CSP), that represents a reasoning framework consisting of variables, domains and constraints. Formally, it is defined as a tuple $(X, D, C)$, where $X = \{x_1, x_2, \ldots, x_n\}$ is a finite set of variables, $D = \{d(x_1), d(x_2), \ldots, d(x_n)\}$ is a set of domains of the values of the variables, and $C = \{C_1, C_2, \ldots, C_m\}$ is a set of constraints. Each constraint $C_i$ is defined as a relation $R$ on a subset of variables $V = \{x_i, x_j, \ldots, x_l\}$, called the *constraint scope*. The relation $R$ may be represented as a subset of the Cartesian Product $d(x_i)$ x $d(x_j)$ x … x $d(xl)$. A constraint $Ci = (V_i, R_i)$ simultaneously specifies the possible values of the variables in $V$ in order to satisfy $R$. Let $V_k = \{x_{k1}, x_{k2}, \ldots, x_{kl\}}$ be a subset of $X$, and an l-tuple $(x_{k1}, x_{k2,} \ldots, x_{kl)}$ from $d(x_{k1}), d(x_{k2}), \ldots, d(x_{kl})$ can therefore be called an *instantiation* of the variables in $V_k$. An instantiation is a solution if and only if it satisfies the constraints $C$.

- **Imperative and Declarative:** This is the capacity of a language to describe imperative and declarative aspects in the same model, since sometimes a part of the process is completely unknown, and other parts are totally known.

- **Use of the language:** The existing proposals that we have analysed are focused on different objectives: *Validation* of the model for a trace of events, *Construction* of automatons to generate a possible sequence of activities, or *Assistance* to the user to decide which is the best activity to execute at runtime.
- **Data perspective:** The possibility to include the values of the data-flow variables in the rules that describe the declarative model.
- **Pre and Post-condition:** The inclusion of a description of the system before and after it is instantiated by means of pre and post-conditions. This is a relevant aspect since it allows the modeller to describe the data before and after the process execution, without the inclusion of details about the internal description of the process.
- **Optimization Function:** The possibility to include an optimization function in the declarative description that is taken into account in the model.

### 5.4.2 Analysis of Declarative Languages

Some of the most important declarative languages have been included and compared in this section.

- **Pocket of flexibility.** This solution is based on constraint specification of the business process workflow. The constraint specification framework [18] represents the workflow as a directed graph where there are two types of nodes: activity nodes and coordinator nodes. In the framework, it is possible to combine activities whose relation is known with activities whose relation is unknown (called pocket of flexibility) that include a set of constraints for concretizing the pocket with a valid composition of work-flow fragments. It includes different types of constraints (Serial, Order, Fork, Inclusion, Exclusion, Number of executions for activity or in parallel). The constraints relate the number of times that each activity can be executed and the order and/or mandatory execution of each activity. The proposal defines a set of algorithms to find possible discrepancies between the constrains that describe the process and an instance. The implementation is based on a made-to-measure algorithm that uses the graph to represent the constraints. The implementation has been included in the prototype called *Chameleon*. The data aspect has not been included in this proposal.
- **DeCo.** Irina Rychkova et al. in [19], [4] and [20] presented a declarative BP specification language that enables designers to describe the actions that a business process needs to contain, but not where their specific sequence can be postponed to the instance time. They improve the alignment of the BP with the business strategy of an organization by giving a synthesis of a set of business processes (abstracting the control flow), while maintaining a rigorous relationship with the detailed process. These specifications complement the traditional

(imperative) business process model by specifying the process independently from a particular environment, (e.g. a process can be executed). This technique includes checking the conformance of the imperative and the declarative specifications, using the case handling paradigm [21]. For every action of the working object they define a precondition and a postcondition. A precondition specifies a set of states where the action can be executed, and postcondition specifies the possible set of states after the action was executed. The pre and postcondition represent how the different actions can modify the state of the objects transformed during the process execution, they do not define the order of the actions, as different imperative description for the same declarative descriptions are possible. Thereby this proposal focuses on the problem from the working object point of view, and data values is one of the analysis.

- **Compliance Rule Graphs.** The Compliance Rule Graphs (CRGs) [5, 22, 23] focus their challenge on finding an appropriate balance between expressiveness, formal foundation, and efficient analysis. For these reasons, the authors propose a language based on a graph representation where the order of the activities and the occurrence or absence of activities can be included as well. The proposal verifies the correctness of the process analysing the compliance rules and the events monitored. The description of the order of activities can be enriched including conditions to the rules that will be satisfied or not depending on the data value for each instance. The analysis is done using pattern matching mechanisms, and is included in a prototype called SeaFlow.
- **Em-Bra²Ce.** The Enterprise Modeling using Business Rules, Agents, Activities, Concepts and Events (Em-Bra²Ce) Framework [24, 25] presents a declarative language based on SBVR (Semantics Of Business Vocabulary And Business Rules) to describe the vocabulary of the process, and an execution model to represent the control flow perspective based on Colored Petri Nets. The use of SVBR allows the description of data aspects in the business process that can be included in the ECA (Event Condition Action) rules, used as a pattern to write the rules.
- **Penelope.** The language Penelope (Process ENtailment from the ELicitation of Obligations and PErmissions) [6] expresses temporal rules about the obligations and permissions in a business interaction using Deontic logic. This language is supported by an algorithm to generate compliant sequence-flow-based process models that can be used in business process design. This language uses the Event Calculus to model the effects of performing activities with respect to the coming into existence of temporal deontic assignments. The only type of data that is included in the definition is related to the execution time of the activities, but the data managed during each instance is not an object of the proposal.
- **ConDec.** The ConDec [12] language was designed for modelling and enacting dynamic business processes. The language defines the involved activities in the process and the order relations between them. This order relation is expressed using LTL to represent desirable or undesirable patterns within a history of

events. However, LTL formulas are difficult to read due to the complexity of expressions. Therefore, the authors have defined a graphical syntax for some typical constraints that can be encountered in workflows. ConDec initially defined three groups of templates to make the definition of activity relations easier: (1) existence, (2) relation and (3) negation templates. An automaton can be built in accordance with the ConDec model, where the automaton can be used to validate a sequence of events. Declare tool [26] is a prototype of a workflow management, that supports the ConDec language. This tool has been used for frameworks such as *Mobucon* [27, 28] for runtime validation. This framework allows for the continuous verification of compliance with respect to a predefined constraint model. ConDec has been enlarged to include the resource perspective (ConDec-R) and the data-aware constraints in Declare, both analysed in the following items.

- **ConDec-R.** This is an extension of the ConDec language to include a description of the resources necessary during process execution. The implementation extension, called ConDec-R [29], assists the user by means of recommendations to achieve an optimized plan for one or multiple objectives [30]. In order to obtain the plan, a CP paradigm is used, combined with a set of algorithms to minimize evaluation time. Although this proposal incorporates the resource perspective which is a type of data, this type of information is not oriented to activity input and output data.
- **Data-aware Constraints in Declare.** This is an extension of the Declare framework [31] that permits the representation of the input, internal and output data of the activities in a declarative representation of the order of the activity. Event calculus has been used to formalize the language and to validate if the traces of events are in accordance with the declarative model. Although the data aspect is included, only input and output data relations between activities can be described.

Although all these declarative languages include some information about data, none of them include the data input and output of the activities with the aim to optimize the object obtained from the business process.

### *5.4.3 Declarative Languages Comparative*

Compared with declarative languages, our proposal DOODLE includes data-oriented aspects in a declarative manner. It is done by means of a set of constraints that describe the data exchanged among the activities, when their relations cannot be defined explicitly at design time. The model and the reasoning framework use Constraint Programming, in order to infer the possible values of the data and achieve the optimization function at run-time (Table 5.3).

**Table 5.3** Declarative languages comparative

| | Formalism | Imper. and Decl. | Use of model | Data perspective | Pre and post | Opt. function |
|---|---|---|---|---|---|---|
| Sadiq | Graph theory | ✓ | Validation | | | |
| DeCo | First Order Logic | ✓ | Validation | ✓ | ✓ | |
| CRGs | Pattern matching | | Validation | ✓ | | |
| Em-Bra$^2$Ce | Color Petri Net | ✓ | Validation | ✓ | | |
| Penelope | Event calculus | ✓ | Construction | | | |
| ConDec | LTL | | Validation | | | |
| ConDec-R | Const. programming | | Assistance | ✓ | | ✓ |
| Data-aware | Event calculus | | Validation | ✓ | | |
| Doodle | Const. programming | ✓ | Constr. and Assist. | ✓ | ✓ | ✓ |

## 5.5 Conclusions

In this paper, we have analysed some of the most relevant declarative languages in business processes. From this analysis we have detected that none of them permit the data declarative description in the business processes, and how it can influence in the obtained model. In this paper, a declarative language called DOODLE is described, which permits the description of the data exchanged among the activities of the process in a declarative way by means of constraints, and obtains an optimization of the business process objects.

For future work, we plan to enlarge the language to support more complex semantics and data relations. We also consider interesting the development of a framework for the transformation from the declarative model to an imperative model implemented in a Business Process Management Systems, and supporting different types of technologies in a transparent way for the business modeller.

## References

1. Weske M (2007) Business process management: concepts, languages, architectures. Springer, Secaucus, NJ
2. Cruz IF (1992) Doodle: a visual language for object-oriented databases. SIGMOD Rec 21(2):71–80
3. Sadiq S, Governatori G, Namiri K (2007) Modeling control objectives for business process compliance. In: Proceedings of the 5th international conference on business process management, BPM'07. Springer, Berlin, pp 149–164
4. Rychkova I, Regev G, Wegmann A (2008) High-level design and analysis of business processes the advantages of declarative specifications. In: IEEE international conference on research challenges in information science, RCIS'08, Marrakech, Morocco. pp 99–110. ISBN: 978-1-4244-1677-6

5. Knuplesch D, Ly LT, Rinderle-Ma S, Pfeifer H, Dadam P (2010) On enabling data-aware compliance checking of business process models. In: 29th International conference on conceptual modeling, ER 2010, Vancouver, BC, Canada, vol 6412. Springer, Berlin, pp 332–346. ISBN: 978-3-642-16372-2
6. Goedertier S, Vanthienen J (2006) Designing compliant business processes with obligations and permissions. In: Business process management workshops, vol 4103. Springer, Berlin, pp 5–14. ISBN: 978-3-540-38444-1
7. Parody L, Gomez-Lopez MT, Gasca M (2013) Decision-making sub-process to obtain the optimal combination of input data in business processes. In: Proceedings of the IX Jormadas de Ciencia e Ingeniería de Servicios, JCIS '13, pp 17–31. ISBN: 978-84-695-8351-7
8. Parody L, Gomez-Lopez MT, Martinez Gasca R, Varela-Vaca AJ (2012) Improvement of optimization agreements in business processes involving web services. Communications of the IBIMA 2012, vol 2012, Article ID 959796, 15 pp. doi:10.5171/2012.959796
9. Bider I, Khomyakov M, Pushchinsky E (2000) Logic of change: semantics of object systems with active relations. Autom Softw Eng 7:9–37
10. Chomicki J (1995) Depth-bounded bottom-up evaluation of logic program. J Log Program 25(1):1–31
11. Bacchus F, Kabanza F (2000) Using temporal logics to express search control knowledge for planning. Artif Intell 116(1–2):123–191
12. Pesic M, van der Aalst WMP (2006) A declarative approach for flexible business processes management. In: Proceedings of the 2006 international conference on business process management workshops, BPM'06. Springer, Berlin, pp 169–180
13. Kowalski RA, Sergot MJ (1986) A logic-based calculus of events. New Generat Comput 4(1):67–95
14. Eshghi K (1988) Abductive planning with event calculus. In: Proceedings of the fifth international conference and symposium on logic programming, Seattle, Washington. MIT Press. pp 562–579. ISBN: 0-262-61056-6
15. Shanahan M (1997) Event calculus planning revisited. In: Steel S, Alami R (eds) ECP. Lecture notes in computer science, vol 1348. Springer, Heidelberg, pp 390–402
16. Nuffelen BV, Kakas AC (2001) A-system: declarative programming with abduction. In: Eiter T, Faber W, Truszczynski M (eds) LPNMR. Lecture notes in computer science, vol 2173. Springer, Heidelberg, pp 393–396
17. Rossi F, Beek PV, Walsh T (2006) Handbook of constraint programming (foundations of artificial intelligence). Elsevier, New York, NY
18. Sadiq SW, Orlowska ME, Sadiq W (2005) Specification and validation of process constraints for flexible workflows. Inform Syst 30(5):349–378
19. Rychkova I, Regev G, Wegmann A (2008) Using declarative specifications in business process design. IJCSA 5(3b):45–68
20. Rychkova I, Nurcan S (2011) Towards adaptability and control for knowledge-intensive business processes: declarative configurable process specifications. In: 44th Hawaii international conference on system sciences (HICSS). pp 1–10. ISSN: 1530-1605
21. van der Aalst WM, Weske M, Grnbauer D (2005) Case handling: a new paradigm for business process support. Data Knowl Eng 53:129–162
22. Ly LT, Rinderle S, Dadam P (2008) Integration and verification of semantic constraints in adaptive process management systems. Data Knowl Eng 64(1):3–23
23. Ly LT, Rinderle-Ma S, Knuplesch D, Dadam P (2011) Monitoring business process compliance using compliance rule graphs. In: 19th International conference on cooperative information systems (CoopIS 2011). LNCS, vol 7044. Springer, Heidelberg, pp 82–99
24. Goedertier S, Haesen R, Vanthienen J (2007) Em-bra$^2$ce v0. 1: a vocabulary and execution model for declarative business process modeling. FETEW Research Report KBI 0728, K.U. Leuven
25. Roover WD, Caron F, Vanthienen J (2011) A prototype tool for the event-driven enforcement of sbvr business rules. In: Daniel F, Barkaoui K, Dustdar S (eds) Business process management workshops (1). Lecture notes in business information processing, vol 99. Springer, Heidelberg, pp 446–457

26. Maggi FM, Westergaard M, van der Aalst W, Staff F, Pesic M, Schonenberg H. Declare tool. http://www.win.tue.nl/declare/
27. Maggi FM, Montali M, Westergaard M, van der Aalst W (2011) Monitoring business constraints with linear temporal logic: an approach based on colored automata. In: Proceedings of BPM. LNCS. Springer, Heidelberg
28. Maggi F, Westergaard M, Montali M, van der Aalst W (2011) Runtime verification of LTL based declarative process models. In: Proceedings of RV. LNCS. Springer, Heidelberg
29. Barba I, Weber B, Valle CD, Ramrez AJ (2013) User recommendations for the optimized execution of business processes. Data Knowl Eng 86:61–84
30. Ramrez AJ, Barba I, Valle CD, Weber B (2013) Generating multi-objective optimized business process enactment plans. In: 25th International conference on advanced information systems engineering. CAISE'13. Lecture notes in computer science, vol 7908. pp 99–115. ISBN: 978-3-642-38708-1
31. Montali M, Chesani F, Mello P, Maggi FM (2013) Towards data-aware constraints in declare. In: Proceedings of the 28th annual ACM symposium on applied computing, SAC '13. ACM, Coimbra, Portugal, pp 1391–1396

# Chapter 6
# A Set of Practices for the Development of Data-Centric Information Systems

**Erki Eessaar**

**Abstract** We present a set of practices for the development of data-centric information systems (IS's). The practices are about the architectural envisioning of systems and model-driven analysis of their certain subsystems. The goal of the architectural envisioning is to quickly achieve a suitable (in terms of well-established patterns) decomposition of the system into relatively autonomous subsystems to facilitate better understanding of it as well as support its design, implementation, maintenance, and future evolution. The goal of the model-driven analysis is to treat models as first-class citizens to facilitate better understanding of the system as well as the automation of its development. The practices can be used in the context of different development methodologies.

**Keywords** Business architecture • Subsystems • Model-driven development • Data-centric systems

## 6.1 Introduction

Databases are the core of many information systems (IS's). Recently data-centric IS development approaches, in case of which informational entities have a driving role in the context of business process design, have gained interest [1, 2]. Independently of this line of research, we have over the years developed and mastered a set of practices for the development of data-centric IS's. The practices are based on a methodological framework for the Enterprise IS strategic analysis and development [3, 4]. The practices use data to drive the modeling (including process modeling) of systems. There is a wide body of literature about data modeling patterns. One can

---

E. Eessaar (✉)
Department of Informatics, Tallinn University of Technology, Tallinn, Estonia
e-mail: Erki.Eessaar@ttu.ee

M. José Escalona et al. (eds.), *Information System Development: Improving Enterprise Communication*, DOI 10.1007/978-3-319-07215-9_6,

use the enterprise integration pattern (a pattern of patterns) [5] as the basis of data-centric system decomposition and assembly patterns [5] as the basis of development of the identified data-centric subsystems.

We have developed these practices based on our years of experience of work in the area of teaching development of IS's to university students with the focus on database development. Students have to develop a part of the software of a data-intensive transactional enterprise IS in their course projects. In these projects databases are implemented as SQL databases. However, the presented practices do not prescribe the use of a particular database platform. Like in real life projects, the goal is to achieve the best possible result in the context of limited time and limited number of participants, who may have different interests as well as different level of knowledge and practical experience. During the last 10 years more than 1,000 student projects have been created based on the practices.

The rest of the paper is organized as follows. Firstly, we explain the context of using the practices and refer to some related work. Next, we explain and discuss our practices of developing data-centric IS's. We use the IS of a university as an example. However, the practices are general and can be used in case of IS's from other domains as well. We cannot present a thorough and complete example of using the practices due to space restrictions. Therefore, we use smaller examples based on the university system to facilitate understanding of the practices.

## 6.2 The Context of Using the Practices

The practices describe two sets of activities. One is *architectural envisioning* and another is *model-driven development of subsystems*. In case of the model-driven development, we only consider the *analysis* part of the development in this paper due to space restrictions. One could perform both sets of activities or use only the architectural envisioning practices to find and prioritize subsystems, which constitute the business architecture of the IS. The use of the architectural envisioning practices facilitates the quick creation of a business architecture specification of an IS, which is a critical stabilizing force [6] during the development and allows us to ask and answer important questions about the scope, schedule, and cost of the system development. One must be able to use (data) modeling patterns for this purpose in order to achieve the speed of development and the quality of the result.

We propose a set of practices rather than a complete methodology or a methodology framework, meaning that the practices can be used completely or partially in the context of different methodologies, which follow different development models (waterfall, iterative etc.) and prescribe different sets of roles, software languages, supporting tools etc. For instance, the Disciplined Agile Delivery (DAD) [7], which is a non-proprietary, agile process *framework* that covers the entire delivery life cycle by adopting and tailoring strategies from a variety of sources, specifies the *inception phase* as the first phase of system development. During this phase, among other things, architectural envisioning and initial release planning take place, which could employ the practices that are proposed in the paper.

The proposed practices are domain-specific because they are designed to be used for the creation of *data-centric transactional systems*, the main task of which is to manage data. These systems employ window-on-data applications [8], the user interface of which consists of windows (or web-pages) that allow users to read and modify data through the entire life cycles of the interconnected main entity types. Therefore, we see a similarity between our approach and data-centric approaches to business process design [1, 2] because in both cases data has a foundational role in the context of business process design. For instance, the practices suggest the use of the data-centric decomposition of a system as the basis for finding the functional decomposition of the system. In addition, life cycles of the main entity types guide the creation of process models, which is similar to existing object lifecycle-based data-centric approaches [2]. The acronym BALSA—"Business Artifacts with Lifecycle, Services, and Associations" is used to refer to the main building blocks of data-centric workflow models [1]. Similar concepts are used in the context of our proposed practices as well (see Sect. 3.5).

The practices do not prescribe a particular modeling language. We have used the UML in the projects that apply the practices. The reason is that UML is nowadays seen as standard language for the IS modeling and there are many CASE tools that support the use of UML and to some extent the generation of documentation and code based on models. However, one could use, for instance, some ERD notation or Object-Role Modeling for the creation of conceptual data models during the model-based development of subsystems. The proposed practices are independent of logical platforms (like the type of the underlying data model of a database) and physical platforms (like the database management systems that will be used to implement databases) that will be used to create the systems. The practices are not used to describe the technical architecture of an IS but the business architecture, which can be implemented by using different technical means. Hence, for instance, the practices do not prescribe a particular type of communication (synchronous vs. asynchronous) between the presentation layer (created based on the specifications of areas of competence and/or functional subsystems), application layer (created based on the specifications of functional subsystems), and the database layer (created based on the specifications of registers).

The practices try to increase *agility* by suggesting how to *quickly* produce *high quality solutions*. The practices combine, in our view novel manner, ideas from different techniques, methods, and strategies. For example, like the DAD they suggest using initial architectural envisioning in the early stages of development [7] and like Giles [5] and Piho [9] they suggest the data-centric use of *patterns* to speed up this process as well create specifications that would be better than *barely good enough*. In addition, it should ensure that evolution of the system causes as little as possible changes in the structure of subsystems as well as within the subsystems. It should also ensure that the models and implementations of subsystems have to be refactored as little as possible. The change rate of the structure of subsystems is generally slower than the change rate within the subsystems.

## 6.3 The Practices

A well-known strategy for solving a complex problem is to divide the problem into smaller and more manageable pieces (subproblems) and to solve the big problem step-by-step by solving the subproblems [10]. The task of developing an IS is a complex problem that can be solved by using this problem-solving strategy. One can divide a system into subsystems to manage complexity and to specify different types of subsystems to separate concerns. Of course, this kind of approach is not new in the context of IS development. For instance, the BALSA methodologies require that business processes should be designed in terms of artifacts that encapsulate business-relevant data and have an associated macro-level life cycle [1]. Similarly, Engels et al. [11] present an approach, according to which one should find business services of an organization and their corresponding domains in the IS by using a *functional decomposition technique*. The approach structures the system based on its functionality but does not structure it based on the organizational structure or data. Therefore, it does not give the full picture of the system.

### *6.3.1 Structure Elements of the Business Architecture of an IS*

Based on our approach, one must describe the layered business architecture of each IS in terms of three types of interconnected subsystems: areas of competence (organizational subsystems), functional subsystems, and registers (data-centric subsystems) [3]. The architecture should take into account the expected evolutionary path of the organization (with the perspective of about 5 years) to make the architecture more stable and able to withstand short-term changes.

Each area of competence corresponds to a role in the organization that will use the IS in order to achieve its goals. The role can be internal to the organization or external. External roles correspond to different types of customers who use the services provided by the organization. Each functional subsystem corresponds to one or more business processes. Each register corresponds to a main business entity type. Among other things, registers are used to store the current state of known main business entities and possibly also the history of state changes. Each register provides a set of services, through which one can access the data in the register. We specify the services by using the notion of *design by contract* (see Sect. 3.5).

We differentiate between administrative and business functional subsystems and registers. Each business functional subsystem (top-level core business service according to the terminology in [11]) directly helps an organization to achieve its business goals. These goals (and tasks that correspond to the goals) are the reason why the organization currently exists. For instance, the main goals of a university are to teach students and facilitate research and development. Hence, some of the business subsystems of its IS are *student management*, *curriculum management*, and *scientific projects management*. Each administrative subsystem (top-level management & support service according to the terminology of [11]) helps an organization to perform

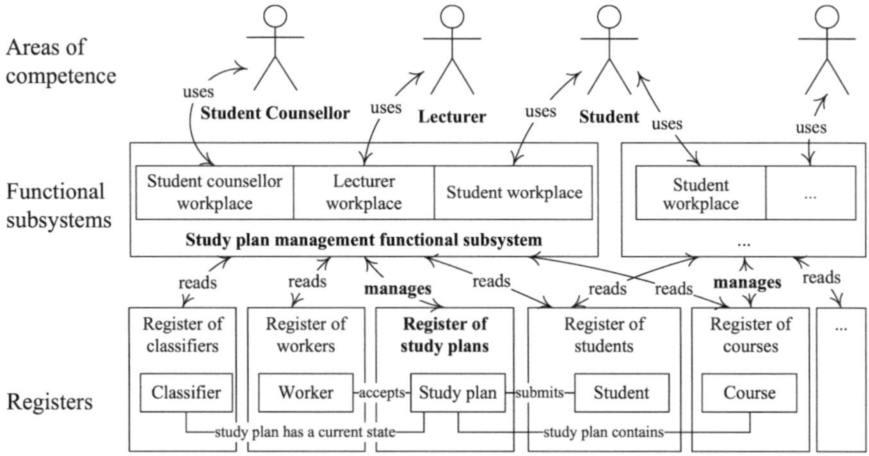


**Fig. 6.1** An example of different types of subsystems and their interconnections in an university IS

its administrative tasks. These tasks are needed to *support* achieving the business goals of the organization but the tasks (and their corresponding goals) are not the reason why this organization currently operates. For instance, a university needs to have buildings for teaching and research as well as contracts with its partners and service providers. Hence, examples of administrative subsystems of its IS are *building management* and *contract management*. A subsystem, which is a business subsystem in an IS, can be an administrative subsystem in another and vice versa, depending on the current business goals of the organizations. It is also possible that if a business evolves and its focus changes, then some of its business subsystems become administrative and vice versa.

Each area of competence uses the services of one or more functional subsystems to achieve its goals in the context of the organization (see Fig. 6.1). Each functional subsystem uses the services of one or more registers to provide required services to the areas of competence (see Fig. 6.1).

## 6.3.2 *Working Out the Business Architecture of an IS*

Next, we describe practices for *quickly* finding subsystems of data-centric IS's. Firstly, one has to find the areas of competence, each of which corresponds to a role. A possibility to find internal roles of an organization is to treat organizational units and positions that are currently used in the organization as roles. A downside of the approach is that it forces us to closely follow the current practices of an organization and carry these over to the new IS. It can very well happen that the business will be changed together with the development of the IS and it leads to a new organizational structure and set of positions with changed responsibilities.

Optionally, representatives of areas of competence can participate in the development work [3]. They could describe their goals, problems with an existing system (if such system exists), and requirements to the new system from their perspective. This approach is based on the meta-design theory [12], which means active end-user participation in the system development. Representatives of areas of competence can even themselves create models that describe their requirements to the new system. The models of areas of competence are a source to the next step, the goal of which is to find the main business entity types of the organization. These are the basis for identifying registers and subsequently the functional subsystems that read and manage data in the registers. One can skip the modeling from the perspective of areas of competence, for instance, if there are strong time restrictions, system is simple, or the developers who start to specify functional subsystems and registers know the organization and its domain well enough.

The main business entity types become the basis of a stakeholder vocabulary. One can use the enterprise integration pattern [5], which refers to the main concepts of organizations (party & role, location, agreement etc.), to *systematically* identify the main business entity types. Each such concept has a corresponding assembly pattern [5], which can be used to quickly start the creation of the conceptual data model of the corresponding register. For each such concept, one has to think whether the IS of the organization has to manage data that corresponds to the concept or some domain-specific subtype of the concept. For instance, a university IS has to manage data about *parties and roles* (students, workers, partners etc.) as well as about the *locations* where the university is situated. These concepts are presented in the integration pattern [5]. On the other hand, the integration pattern [5] does not mention the concept *study plan*. However, a study plan is a kind of *agreement* between a student and the university and the concept *agreement* is present in the integration pattern. Each student who submits a study plan agrees to study a set of courses and the university agrees to teach the courses.

A concept that requires a corresponding administrative register and functional subsystem in the architecture of most of the IS's is "classifier". Classifier values (reference data) are used to categorize other data in a database or relate data in a database with the data that is outside the control of the organization [13]. Our proposed practices require the use of state classifiers in order to be able to record the current state (and possibly the history of state changes) of main business entities. In general, there would be a separate state classifier for each main business entity type. Therefore, for instance, if *study plan* is a main business entity type, then it is associated with the business entity type *study plan state type*, which is a subtype of the main business entity type *classifier*. It would also mean that data in the *register of study plans* is associated with data in the *register of classifiers*.

One can use other frameworks like the business model canvas [14] and columns of the Zachman framework [15] as tools that facilitate *systematic* discovery of the main business entity types. Each column of the Zachman framework focuses on a single, independent phenomenon [9]. Table 6.1 presents an example of using the framework in case of a university IS. The technique is similar to the use of conceptual class category list during domain modeling [16]. Piho [9] demonstrates that each Zachman framework column has corresponding archetype patterns.

**Table 6.1** An example of the use of the Zachman framework columns for discovering and classifying the main business entity types in case of a university IS

| What (Things) | How (Processes) | Where (Locations) | Who (Persons) | When (Events) | Why (Strategies) |
|---|---|---|---|---|---|
| Course | Assessment | Structural unit | Student | State changes of courses, assessments, buildings, students, study plans etc. | Curriculum |
| Gallup | Graduation | Building | Worker | | Study plan |
| Document | Gallup answering | Room | Partner | | Timetable |
| Contract | | | Competitor | | |
| Classifier | | | Supplier | | |

For instance, the pattern *Party* [17] corresponds to the column "Who". These assembly patterns help us to quickly start the modeling of the corresponding registers.

We propose to define the business architecture so that there is a separate register for each main business entity type. Conceptually it means that data about an entity that belongs to a main entity type is clustered together. In general, we suggest to have for each register a separate functional subsystem that *manages* (creates, reads, and possibly updates and deletes) data in the register. Therefore, for instance, if there is a *register of study plans*, then there should also be a separate *study plan management functional subsystem* (see Fig. 6.1). We find the list of functional subsystems based on the list of registers. We derive the names of functional subsystems and registers from the names of main business entity types.

However, in case of our approach, the management relationship between the functional subsystems and registers is not bijective. There are situations, in which case a functional subsystem is permitted to manage data in more than one register and more than one functional subsystem could manage data in a register.

Let us assume that there is a main business entity type $e$ that has a corresponding register $r$. Let us also assume that the data in $r$ is managed by the functional subsystem $f$. One does not have to create a separate functional subsystem to manage data in the register $r'$ that is created in order to keep track of the history of the state changes of $e$. Instead, the business architecture can be defined in a way that $f$ manages data in both $r$ and $r'$. We note that if it is not necessary to keep track of the history of state changes of $e$, then the register $r'$ will not be needed. Tracking of history of state changes is needed in case we need to know who, when, and why changed the state of a main entity (question of responsibility).

### 6.3.3 Desired Properties of the Business Architecture of an IS

Market requirements, laws and regulations that influence the business as well as technical platforms evolve. Therefore, IS's (structure and business processes of organizations as well as their supporting computer systems) must also evolve to meet the needs of their owners and users. IS's should be evolutionary in the sense that they must be able to withstand and deal with evolutionary changes. If an organization evolves, then new subsystems could be added to the architecture of its IS, existing subsystems could be modified, or possibly some subsystems have to be phased out. The latter is much more probable in case of business subsystems (because the organization can change its business focus over time) rather than administrative subsystems. In order to facilitate the development of a system in a piecemeal fashion and simplify making changes in the system, its decomposition into subsystems must follow the advice of patterns of assigning responsibilities (General Responsibility Assignment Software Patterns, GRASP) like *Low Coupling* [16] and *High Cohesion* [16]. Low Coupling means in this context that a subsystem should not depend too much on other subsystems. It helps us to achieve low dependency of subsystems, low change impact of subsystems, and increased reuse of the artifacts that are created during the development of subsystems. If one changes the internal structure or behavior of a subsystem *s* so that it can still provide the services that have been previously agreed, then one does not have to change the subsystems that use the services of *s*. It is similar to the concept "data independence" [18] in the database world and simplifies the evolution of subsystems. High Cohesion means in this context that the responsibilities of a subsystem must be strongly related in order to reduce the complexity of subsystems.

The use of the architecture guidelines ensures that the resulting subsystems have *high cohesion*, meaning that all the responsibilities of a subsystem are related to a main business entity type. For instance, the *study plan management functional subsystem* has responsibilities that are associated with managing study plans during their entire life cycle (register a new study plan, submit the study plan, search study plans etc.). On the other hand, it does not have responsibilities to manage timetables or assessments because these are the responsibilities of the *timetable management* and *assessment management* functional subsystems, respectively. It is consistent with the GRASP pattern *Information Expert* [16], according to which a responsibility must be assigned to the class (or in this case a subsystem) that has the information necessary to fulfill the responsibility. However, one has to bear in mind that a functional subsystem may have to *read* data from more than one register in order to provide required services to the areas of competence. For instance, the *study plan management subsystem* has to read data from the *register of workers* in order to allow students to select a teacher in case of a course.

The approach that each main business entity type has generally a corresponding functional subsystem and register also ensures that the resulting subsystems have a relatively similar size, which can be measured as the number of responsibilities. If a subsystem has unrelated responsibilities (for instance, manage study plans as well as curricula), then it may be too big and hence more difficult to develop. The use of the guidelines also helps us to achieve *low coupling* between functional subsystems and registers in the sense that each functional subsystem manages (creates, reads, and possibly updates and deletes) data in as few registers as possible. However, the interconnection of the life cycles of the main business entity types contributes to the increase of coupling between subsystems. For instance, if one deactivates certain number of courses, then one may also have to deactivate a curriculum that includes the courses.

One can solve the problem by letting the *course management functional subsystem* invoke an operation (service) that is provided by the *register of curricula*. Another solution is to extend the responsibilities of the operations that are provided by the registers (in this case the operation *deactivate course* of the *register of courses*) so that they could directly modify data in other registers or invoke operations (services) of other registers. The first solution increases coupling between functional subsystems and registers. The second solution increases coupling between registers.

### 6.3.4 The Development Plan of an IS

Nowadays there is practically no green-field software development [6] meaning that usually new systems have to replace or extend existing systems. But even in this case the architectural envisioning practices are useful because the resulting architecture is a framework, based on which to continue the development of the system. Users are usually satisfied with some implemented subsystems, are not satisfied with others, and some subsystems are not implemented at all. In case of the latter, one has to decide whether to implement these by adapting Commercial Off-The-Shelf software packages or create them from scratch. Regardless of the decision, one has to integrate the existing software with the implementations of new subsystems by using appropriate enterprise integration patterns [19]. Data in a register usually has to refer to data in other registers to make the business context fully understandable. Hence, the creation of a *shared database* [19] that implements all the registers is a suitable selection in case of the use of the practices.

The business architecture allows planners to create the development plan by prioritizing the subsystems and determining their development order. Development of the system consists of a series of projects, each of which includes development of one or more subsystems. The proposed practices are flexible enough to allow the development of the entire system by using only one project and iteration—essentially development based on the waterfall model. However, in this case the advantage of solving a big (system development) problem in a piecemeal fashion will be lost. The planners should use the requirements of stakeholders, information about the currently implemented parts of a system as well as information about the classification of subsystems and interconnections between the subsystems to determine the most important subsystems that should be firstly designed and implemented. Importance in this context could mean importance to the customer in the sense that the subsystems in question support the most critical parts of its business process or are technically the most challenging. Importance could also mean that even the first versions of some subsystems cannot be implemented without implementing the first version of some other subsystem. By starting the development with such subsystems, we tackle the risks and try to prove the architecture as early as possible (similarly to the goals of DAD [7]). One should develop registers, the data of which is read by many different functional subsystems in the early stages of development in order to deal with possible risks in the early stages of development as well as prepare for the development of functional subsystems that read data from such registers. Many functional subsystems need to read data from the administrative registers. Therefore, development of administrative subsystems (for instance, the *register of classifiers*) must start in the early stages of the overall system development. Because many IS's have similar administrative subsystems there is a greater chance than in case of domain-specific business subsystems that for these subsystems there are available patterns that "can give the modeler proven, robust, extensible, and implementable designs" [5, p. 38].

There is freedom to decide whether to use the waterfall model or the iterative model in the resulting projects. In the projects, one could use the proposed model-driven development (MDD) approach (see Sect. 3.5) and possibly do it in an agile manner as described by [20]. If the decision was made to develop the system or parts of it from scratch, then the use of MDD would allow developers to achieve semiautomatic creation of the system or its parts by using transformations.

### 6.3.5 Model-Driven Analysis of Subsystems

For the development of subsystems from scratch, we suggest developers to use a model-driven development approach, according to which one has to model a system from different aspects and for different audiences to finally implement the system with the help of model transformations. Based on the models of functional subsystems one will finally reach to the application software that will automate business processes. Based on the models of registers one will finally reach to the databases. In this paper, we do not consider optional modeling of areas of competence. A usual practice would be to develop a functional subsystem together with its managed registers. This could produce new requirements to the registers that are read by the functional subsystem. The first versions of the latter registers must be ready or developed in parallel with the functional subsystem. Developers should start with the modeling of registers. Later the modeling of the functional subsystem and registers can continue in parallel. Information captured in the models of registers is an important input for the modeling of functional subsystems.

Developers should at least define objectives as well as create a conceptual data model, state machine diagram of the main business entity type, and contracts of database operations in order to model registers from the analyst's perspective (it corresponds to the business management's perspective in [15]). Objectives should be defined in a way that it would be later possible to measure their fulfillment. Conceptual data model consists of entity-relationship diagrams and textual specifications of entity types and attributes. It is a non-technical model that is used to capture requirements to the database. It must be free of implementation considerations (like for instance, which data model will be used to implement the database). Similarly to [18], we use a subset of the UML class diagram notation in order to create entity-relationship diagrams. Conceptual data model also contains declarative specifications of constraints to data. These are derived from the business rules of the organization. One could specify constraints by using some formal language like OCL [21] or a natural language that is inherently more imprecise but more understandable to the human users. These constraints are invariants that cannot be violated by any operation (service) that is provided by the registers.

We model the entire life cycle of the main business entity type by using a state machine diagram. In addition, in the conceptual data model one must foresee that it must always be possible to find the current state of a particular main business entity based on the data in the database. One can use state classifiers for this purpose. One has to specify the operations (services) that are provided by a register by using contracts of database operations based on the principles of design by contract. Pre- and postconditions of the contracts have to be specified in terms of the attributes/relationship types/entity types that are defined in the conceptual data model. We usually write only contracts of operations that modify data. Each state transition that is specified in the state machine diagram must have a corresponding operation. For instance, a study plan can move from the state "created" to the state "submitted". It is modeled by using a state machine diagram that is a part of the specification of the *register of study plans*. The register must provide an operation (service) for changing the state of a study plan from "created" to "submitted". Hence, there must be a contract that specifies the operation.

Each functional subsystem corresponds to one or more business processes. We suggest to model functional subsystems from the analyst's perspective by defining at least their objectives and creating fully dressed use cases [16]. It must be possible to measure the fulfillment of objectives. Each use case must help users to achieve one or more of these objectives. The textual specification of use cases must contain references to the contracts of database operations. If we firstly identify the life cycle of the main entity type and based on that identify the operations (services) that are provided by the registers, then these can be used as the basis to discover use cases that need the services. If we firstly describe use cases independently of the definition of the registers, then we still have to check whether each operation is invoked by at least one use case. If not, then it means that we have to create a new use case or modify an existing use case. In this sense data is "a first-class modeling primitive" [1] that helps us to guide the process modeling.

## 6.4 Conclusions

In this paper, we presented practices for the architectural envisioning of the data-centric transactional IS's and model-driven analysis of the identified functional subsystems and registers. We explained the context of using the practices and acknowledged that the practices can be used together with different development methodologies as well as technical architectures and software languages.

Future work must include specification of a detailed metamodel of the practices. It can be used to compare the practices with the similar approaches like the different data-centric approaches of business process design. Although we have used the practices successfully with the existing CASE tools it is worth to consider the creation of a specialized CASE tool that supports these practices. The metamodel is needed to create such a tool.

## References


1. Bhattacharaya K, Hull R, Su J (2009) A data-centric design methodology for business processes. In: Cardoso J, Aalst W (eds) Handbook of research on business process modeling. Information Science Reference, IGI Publishing, Hershey, PA, pp 503–531
2. Garcia HDD (2011) Evaluation of data-centric process modeling approaches. Master thesis, Eindhoven University of Technology
3. Roost M, Kuusik R, Veskioja T (2001) A role-based framework for information systems self-development. In: IFIP TC8/WG8.2 working conference on realigning research and practice in information systems development: the social and organizational perspective, pp 95–105
4. Roost M, Kuusik R, Rava K, Veskioja T (2004) Enterprise information system strategic analysis and development: forming information system development space in an enterprise. In: International conference on computational intelligence, pp 215–219
5. Giles J (2012) The nimble elephant. Agile delivery of data models using a pattern-based approach. Technics Publications, Denville, NJ
6. Jarke M, Loucopoulos P, Lyytinen K, Mylopoulos J, Robinson W (2010) The brave new world of design requirements: four key principles. In: Pernici B (ed) Proceedings of CAiSE 2010, LNCS, vol 6051. Springer, Berlin, pp 470–482
7. Ambler SW, Lines M (2012) Disciplined agile delivery: a practitioner's guide to agile software delivery in the enterprise. IBM, Indianapolis, IN
8. de Haan L, Koppelaars T (2007) Applied mathematics for database professionals. Apress, New York
9. Piho G (2011) Archetypes based techniques for development of domains, requirements and software. Towards LIMS software factory. Dissertation, Tallinn University of Technology
10. Polya G, Sloan S (2009) How to solve it: a new aspect of mathematical method. Ishi, Kagawa
11. Engels G, Hess A, Humm B, Juwig O, Lohmann M, Richter J-P, Voß M, Willkomm J (2008) A method for engineering a true service-oriented architecture. In: 10th international conference on enterprise information systems, pp 272–281
12. Fischer G, Giaccardi E, Ye Y, Sutcliffe AG, Mehandjiev N (2004) Meta-design: a manifesto for end-user development. Commun ACM 47(9):33–37
13. Chisholm M (2000) Managing reference data in enterprise databases: binding corporate data to the wider world. Morgan Kaufmann, Burlington, MA
14. The Business Model Canvas. http://www.businessmodelgeneration.com/canvas. Accessed 5 May 2013
15. Zachman JA (2013) The Zachman Framework for Enterprise Ontology™. The Enterprise Ontology™, version 3.0. http://www.zachman.com/about-the-zachman-framework. Accessed 30 Mar 2013
16. Larman C (2002) Applying UML and patterns: an introduction to object-oriented analysis and design and the Unified Process, 2nd edn. Prentice-Hall, Englewood Cliffs, NJ
17. Fowler M (1997) Analysis patterns: reusable object models. Addison-Wesley, Menlo Park, CA
18. Connolly TM, Begg CE (2002) Database systems: a practical approach to design, implementation and management, 3rd edn. Addison-Wesley, Menlo Park, CA
19. Trowbridge D, Roxburgh U, Hohpe G, Manolescu D, Nadhan EG (2004) Integration patterns. Microsoft, USA
20. Ambler SW (2012) Approaches to agile model driven development (AMDD). http://www.agilemodeling.com/essays/amddApproaches.htm. Accessed 3 Mar 2013
21. Miliauskaitė E, Nemuraitė L (2005) Taxonomy of integrity constraints in conceptual models. In: IADIS virtual multi conference on computer science and information systems, pp 247–254

# Chapter 7
# Towards an Ontology of SoS Interoperability: Proposition of a SoS Interoperability Framework and a SoS Conceptual Interoperability Model

**Houda Benali, Narjès Bellamine Ben Saoud, and Mohamed Ben Ahmed**

**Abstract** Systems engineering is currently evolving to extend beyond the scope of a single system. Recently, the increasing number of interacting systems in society and industry generated a growing interest in a class of complex systems, called Systems of Systems (SoS), whose components are complex and independent. There is a growing interest for the creation of synergies between these component systems to achieve the desired performance of the overall system (SoS). The SoS approach is intended to build and analyze complex, independent and heterogeneous systems working (or designed to work) in cooperation. The SoS concept presents a high level perspective that explains the interactions between the independent systems. In the context of SoS engineering, interoperability is one of the most challenging problems. The aim of this paper is to present an interoperability framework and a conceptual model to formally structure knowledge about SoS Engineering and build an Ontology of SoS Interoperability (OoSoSI).

**Keywords** Interoperability • System of systems • Ontology • Framework

H. Benali (✉) • M. Ben Ahmed
Laboratoire RIADI, Ecole Nationale des Sciences de l'Informatique,
Université de la Manouba, Manouba, Tunisia
e-mail: houda_benali@yahoo.fr; houda.benali@ensi.rnu.tn; mohamed.benahmed@riadi.rnu.tn

N. Bellamine Ben Saoud
Laboratoire RIADI, Ecole Nationale des Sciences de l'Informatique,
Université de la Manouba, Manouba, Tunisia

Institut Suprieur d'Informatique, Université Tunis El Manar, Tunis, Tunisia
e-mail: narjes.bellamine@ensi.rnu.tn

M. José Escalona et al. (eds.), *Information System Development: Improving Enterprise Communication*, DOI 10.1007/978-3-319-07215-9_7,

## 7.1 Introduction

The issue of interoperability is crucial especially when integrating multiple heterogeneous and independently governed systems. In recent years the concept System of Systems (SoS) has emerged as a new approach to solving complex problems [1]. In [2], the author has derived, from a variety of SoS definitions, this definition: a "System of Systems is any system that is a relatively large and complex, dynamically evolving, and physically distributed system of pre-existing, heterogeneous, autonomous, and independently governed systems, whereby the system of systems exhibits significant amounts of unexpected emergent behavior and characteristics" [2]. SoS are complex socio-technical systems that interconnect multiple components, not only hardware and software but also organizations, processes and human resources systems. To control this complexity and this multidisciplinary aspect, several researches have been devoted to the construction of SoS architectures in terms of components, relationships between these components and the evolution of the resulting structure over time. Mastering and understanding the nature of the relationships between SoS' components make the treatment of the problem of interoperability much easier. This problem is widely studied in many fields and several solutions have already been developed: [3–6], and many others. Although these solutions do not cover all aspects of SoS engineering, they represent a solid basis to propose and build new solutions covering all levels of interoperability according to many of the existing interoperability models of reference. However, the proposal of such new solutions requires an enormous effort to master knowledge (usually non-formal) on SoS and interoperability. This knowledge is related to concepts describing the field of interoperability for SoS and thus include concepts related to the description of a SoS (this is equivalent to describe the elements of its architecture) and the concepts related to the field of interoperability (characterization of the problems and solutions). Having a knowledge sharing in the community of SoS engineering, while providing a mechanism for the implementation of this knowledge, greatly facilitates the rapid and efficient development of SoS. This is the main concern of the discipline of Knowledge Management (KM). KM is an area of research suggesting methods, techniques and tools assisting the creation, acquisition, sharing and use of knowledge in an organization [7]. New formalisms are introduced to represent the underlying concepts in a domain of knowledge, the relationships between them, and the semantics of these relations, regardless of the use of this knowledge. The Ontology Engineering is a methodology that can be used to generate knowledge representation and knowledge reasoning models for Knowledge Based Systems (KBS) [8]. In the scope of this paper, we tackle the problem of SoS interoperability knowledge description (i.e. the description of problems and solutions related to Interoperability in the context of SoS Engineering). To structure this description, we propose a stratified and multi-views framework of interoperability. Based on this framework, we formally structure SoS interoperability knowledge using an Ontology Engineering approach.

Our paper is structured as follows. First, we present some definitions of interoperability and some existing Interoperability Frameworks. Afterwards, we describe the proposed SoS Interoperability Framework (SoSIF) and the Interoperability Conceptual Model to formally define an Ontology of SoS Interoperability (OoSoSI). In the last section, we conclude and present an outlook on future work.

## 7.2 Preliminaries

Interoperability is generally seen as "the ability of two or more systems or components to exchange information and to use the information that has been exchanged" [9]. Another definition of interoperability is "the capability to communicate, execute programs, or transfer data among various functional units in a manner that requires the user to have little or no knowledge of the unique characteristics of those units" [10].

From these two definitions, the two communicating systems necessarily use the same communication protocol, i.e. ***technical interoperability***. Indeed, they should use the same format for the representation of information, i.e. ***syntactic interoperability***. Also, the properly use of the exchanged information requires accurately interpretation of the meaning, i.e. ***semantic interoperability***. Some other levels have been proposed by [11] to define properly use of Data (i.e. ***Pragmatic Interoperability***), the effect of exchanged information (i.e. ***Dynamic Interoperability***) and a formal description of information, processes, contexts, and modeling assumptions (i.e. ***Conceptual Interoperability***).

A more general definition of interoperability is given by The Commission of the European Communities [12]: "Interoperability is like a chain that allows information and computer systems to be joined up both within organizations and then across organizational boundaries with other organizations, administrations, enterprises or citizens". In this definition, another dimension of interoperability is performed, i.e. ***Organizational Interoperability***. Organizational interoperability "is concerned with modeling business processes, aligning information architectures with organizational goals and helping business processes to co-operate" [12]. Many other definitions were associated with organizational interoperability [13–15]. These proposed definitions of organizational "mix methods and standards for the technical linkage of business processes (process organization) with questions about the organization of support functions" [16].

In the context of SoS Engineering [17], Interoperability is "the ability of a set of ***communicating entities*** to (1) exchange specified ***state data*** and (2) ***operate on that state data*** according to ***specified, agreed-upon***, ***operational semantics***". The abstract definition "communicating entities" could cover three levels: programmatic (i.e. entities engaged in the acquisition management of systems), constructive (i.e. entities engaged in the development and maintenance of systems) and operational (i.e. entities engaged in the operation of systems). The use of "state data" could also cover the same three levels: programmatic, constructive and operational. ***Programmatic interoperability*** is "the ability of a set of communicating entities engaged in acquisition management activities to (1) exchange specified acquisition management information and (2) operate on that acquisition management information according to a specified, agreed-upon, operational semantics" [18]. In the same technical report, another definition is given to programmatic interoperability that emphases the importance of this level to achieve an effective collective behavior. In this context, authors define the programmatic interoperability as "the ability of a set of communicating entities to achieve a shared understanding with respect to acquisition management functions to allow them to engage in effective collective behavior". The acquisition includes all activities that are needed to develop, deploy, operate, maintain, replace, and dispose of a SoS. The two other aspects of interoperability (constructive and operational) are presented in [17]: (1) ***Constructive interoperability*** "reflects the degree to which the different system design, engineering, and production processes and tools are able to exchange information in an understandable manner"; (2) ***Operational interoperability*** "is closely aligned with the traditional definition of interoperability (the ability of systems to exchange relevant information), but it adds the notion of compatible (or complementary) operational concepts".

An interoperability framework is "the set of policies, standards and guidelines describing the way in which organizations have agreed, or should agree, to do business with each other" [19]. This definition accentuates the need of a standardized solution that guides organizations to achieve effective collaboration. Common elements are needed to be specified when defining an interoperability framework such as "vocabulary, concepts, principles, policies, guidelines, recommendations, standards, specifications and practices" [20].

Concerned by the interoperability in the context of SoS, we propose to adopt almost the same definition of interoperability frameworks. Based on the given definition of SoS and SoS I, we propose this definition of an interoperability framework: "The set of common elements—such as vocabulary, concepts, principles, policies, guidelines, recommendations, standards, specifications and practices—describing the way in which programmatic entities achieve a shared understanding with respect to acquisition management functions to allow them to engage in effective collective behavior."

Many e-government frameworks were proposed. We present some examples of frameworks proposed to support interoperability in e-Government (Readers are recommended to see [17, 21–23] for a larger survey on e-Government Inter-operability Frameworks) in the field of e-Government.

In the field of Enterprise Interoperability, The Framework for Enterprise Interoperability (FEI) was initially developed under INTEROP NoE and now is considered as a coming ISO standard (CEN/ISO 11354). Three main dimensions of the framework (FEI) are presented in [24, 25]: (1) Interoperability barriers concerns incompatibility problems between enterprise systems; (2) Interoperability concerns describes areas in which interoperability takes place; (3) Interoperability approaches defines the ways in which solutions can be developed to remove interoperability barriers.

## 7.3 SoSIF: A Framework for SoS Interoperability

SoSIF is intended to be used by people who are concerned by managing interoperability in the context of SoS Engineering and by detecting what might need to be improved to meet high level of interoperability between interacting entities. For this purpose, the proposed framework helps researchers properly collect knowledge covering multiple aspects and views of SoS interoperability. The SoSIF forms a basis for experts to organize and structure interoperability knowledge in the context of SoSE. The overall goal is to ease the task of bringing together problems and solutions when dealing with SoS interoperability.

We firstly present an overview of the proposed framework, SoSIF. Then we present a conceptualization of interoperability knowledge in accordance with SoSIF. This conceptualization is the first step towards the development of a SoS Interoperability Ontology (SoSIO).

### *7.3.1 An Overview of the Proposed Framework*

The interoperability framework, presented in Fig. 7.1, is proposed to structure knowledge on SoS interoperability that can help solve interoperability problems. The purpose of the proposed framework is not the assessment of interoperability when engineering SoS. However, the objective of SoSIF is to structure knowledge about Interoperability in the context of SoS and knowledge about existing solutions

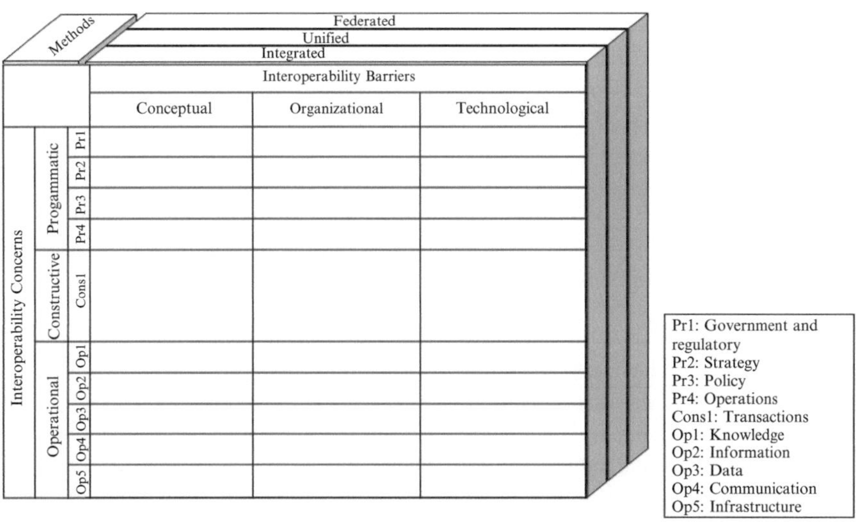


**Fig. 7.1** SoSIF: a framework to structure interoperability knowledge in the context of SoSE

to solve interoperability related problems. Our proposition is elaborated on the basis of INTEROP FEI [27]. We propose to redefine interoperability concerns to support those in the SoS Context. The proposed framework shows three (3) views: Interoperability concerns, Interoperability approaches and Interoperability barriers.

#### 7.3.1.1 Interoperability Concerns

These concerns are classified by stratum (programmatic, constructive and operational concerns). As defined in [26] there is ten concerns (levels) in the context of SoSE. We propose to structure these concerns by interoperability level:

- Programmatic Level: concerns in this layer are defined based on [26]

  1. Government and regulatory—At this level, the three main considerations are the legal framework that ensures fair competition, the governance of SoS, and the implications for society in general.
  2. Strategy—This level is concerned with the strategy of the SoS itself. Its strategy will depend on its environment, the volatility of that environment, its geographic spread across different legislations in different regions, and the nature of its business domain and its perceived risks.
  3. Policy—This general level is more concerned with how the SoS regulates and constrains what and how it executes its business. It is at this level where considerations such as inter-organizational trust, partnering, service level agreements, and contracts are important. Various fundamental issues such as the potential clash between lean operations and the need for resilience must be addressed, and plans for risk mitigations are prepared.
  4. Operations—It is at this level where organizational and ITC considerations overlap. The ideal for the SoS is smooth operations within the established strategic and policy constraints of each of the individual organizations and of the SoS in which they are embedded. The quality of each interface between the organizations is critical in achieving this ideal.

- Constructive Level: concerns in this layer are defined based on [26]

  1. Transactions—At this level, IT&C plays a critical part, and the lower NIF (NCOIC (Network Centric Operations Industry Consortium) Interoperability Framework) layers are very significant.

- Operational Level

  1. Knowledge/awareness of actions [27]—In this layer, knowledge represents a form of information transmitted from a provider to a consumer with sufficient semantic content for understanding. Sufficient semantic content allows the consumer to understand and take ownership of the knowledge in the sense that the consumer is then a "knower" of the knowledge. The knowledge is actionable by the knower.

2. Semantic/information [27]—In this layer, information represents data combined with semantics. The semantics may include a relation to a larger environment in which the data can be used. The context in which the data is applied is enlarged to include an application context. Information is a set of data in context that has meaning or relevance to an "actor." That actor may be a human or machine that supports business processes and business objectives.
3. Data/object model [27]—This layer is responsible for the representation of facts, concepts, or instructions in a formal manner suitable for communication, interpretation, or processing by humans or automatic means. This representation is for characters or analog quantities to which meaning is or may be assigned.
4. Connectivity & network [28]—This layer is responsible for forwarding packets from source to destination node based on end-to-end routing information. The network layer is responsible for translating logical addresses, or names, into physical addresses.
5. Physical Interfaces [28]—This layer gives the Data-Link layer its ability to transport a stream of serial data bits between two communicating systems it conveys the bits that move along the cable or over the air.

#### 7.3.1.2 Interoperability Approaches

As defined by the ISO 14258, there are three basic ways to relate entities together to establish interoperability [25]:

- The ***integrated approach*** characterized by the existence of a common format for all the component systems. This format is not necessarily a standard but must be agreed by all parties to elaborate models and build systems.
- The ***unified approach***, also characterized by the existence of a common format but at a meta-level. This meta-model provides a mean for semantic equivalence to allow mapping between diverse models and systems.
- The ***federated approach***, in which no common format is defined. This approach maintains the identity of interoperating systems; nothing is imposed by one party or another and interoperability is managed in an ad-hoc manner.

#### 7.3.1.3 Interoperability Barriers

We adopt the same barriers identified in [25]:

- The conceptual barriers are related to the syntactic and semantic differences of information to be exchanged as well as the expressivity of the information.
- Technological barriers are related to the incompatibility of information technologies (architecture & platforms, infrastructure). A technological barrier is the lack of a set of compatible technologies which prevent collaboration between two or more systems. Examples of technological barriers are: Communication barriers (incompatibility in the protocols being used to exchange information), Content

barriers (different standards used to represent information, or incompatibility in the tools used to encode/decode the information being exchanged) and Infrastructure barriers (use of different incompatible middleware platforms).

- The organizational barriers are related to the definition of responsibility and authority so that interoperability can take place under good conditions. These can be seen as 'human technologies' or 'human factors' and are concerned with human and organization behaviors which can be incompatible to interoperability.

### *7.3.2 A Conceptual Model to Structure SoS Interoperability Knowledge in Accordance with SoSIF*

Some research efforts attempting to define the interoperability domain were presented in [29]. These efforts lead, first, to a UML formalization of the Interoperability problem based on three meta-models: decisional meta-model, resource composition meta-model and systemic meta-model. The Ontology of Interoperability (OoI) [30] was developed based on these research efforts. The operationalisation of this ontology was done using the Ontology Web Language (OWL).

Based on the proposed framework SoSIF, we have proposed an integration of SoS related concepts in the OoI allowing to add a specific specialization of the OoI in the SoSE domain. A widely adopted definition of Ontology is given by Gruber: "an ontology is aformal, explicit specification of a shared conceptualization" [31]. As depicted in this definition, the first step towards building an ontology is the conceptualization that should be shared. In this paper, we propose a conceptualization of SoS Interoperability knowledge in the frame of SoSIF and by specialization of the OoI in the SoSE domain. Interoperability knowledge concerns the domain definition of interoperability (aspects of interoperability) and the proposed solutions related to the problem of interoperability. The first elaboration of this conceptualization is presented in Fig. 7.2. The conceptual model in this figure represents the concepts as declared in the SoSIF. A central concept is "SoS Interoperability" that is defined by levels, barriers, concerns and implementation approaches.

The "SoS Interoperability" concept is then defined as a specialization of the concept "Interoperability" of the OoI. The "Interoperability Barriers" are presented as an interoperability "Incompatibility" that is an interoperability existence condition defined in OoI. The proposed model is then improved to include relation between "Interoperability approaches" and the concept "Solution" from the OoI. This same relation was proposed by [32]. This relation represents the way the interoperability barriers could be removed. A proposed solution should cover one or more interoperability levels (and related concerns).

The proposed Conceptual model is then enriched by concepts describing a SoS characteristics, Fig. 7.3, adding a systemic dimension to the conceptualization of SoS Interoperability knowledge. The detailed description of the conceptual model that we have proposed to describe SoS is out the scope of this paper. We only present concepts that are directly concerned by or that directly affect the problem of inter-operability. "Heterogeneity" concept from the OoEI (Ontology of Enterprise

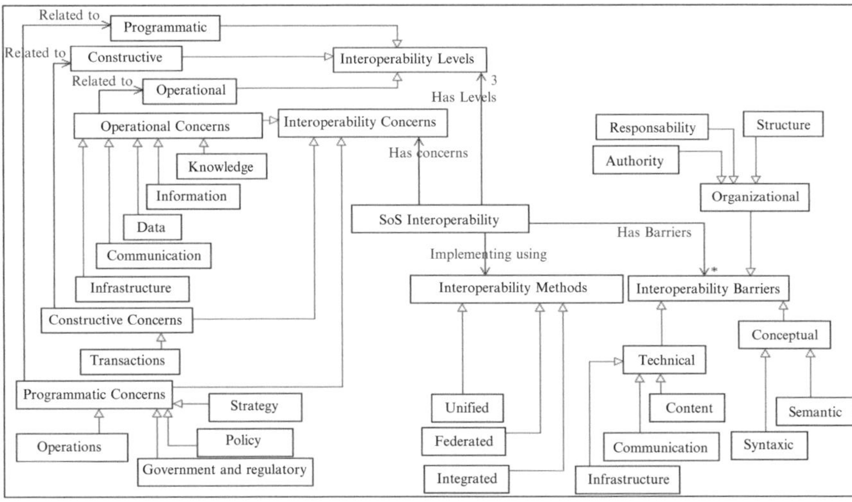


**Fig. 7.2** A conceptual model describing different views of the SoSIF

Inter-operability), proposed by [32], is equivalent to the "Heterogeneity" concept from the SoS Conceptual Model. Another interoperability existence condition proposed by [32] is "Problematic Relation". In our proposal, this problematic relation concerns the "Relations" between SoS components. Any proposed solution should be appropriate to the type of SoS. So, we add a relation "DependsOn" that depicts that a proposed solution depends on the "Type" of the "System of Systems".

As mentioned earlier, our contribution consists on developing an ontology that formally describes the SoS Interoperability knowledge. Interoperability domain concerns both interoperability domain definition (levels, concerns and barriers) and solutions solving interoperability problem. So, we propose to instantiate the concept "Solution" from proposed SoSICM using existing solutions supporting interoperability in different levels. We start by structuring these solutions in the frame of the proposed SoS Interoperability Framework of Fig. 7.1. To illustrate this process of solutions structuring, we present in Fig. 7.4 some solutions proposed by the Software Engineering Institute (SEI) SoS Development Group [3, 4] that we structure in the frame of our proposed SoSIF.

## 7.4 Conclusion and Future Work

The SoS interoperability is a complex problem that can be studied from multiple facets. In this paper, we proposed a stratified and multi-views framework that helps solving this complexity. The proposed framework, called SoSIF, presents three views: interoperability concerns, interoperability barriers and interoperability approaches. Interoperability concerns are grouped by stratum. Every stratum presents a level of interoperability: operational, constructive or programmatic.

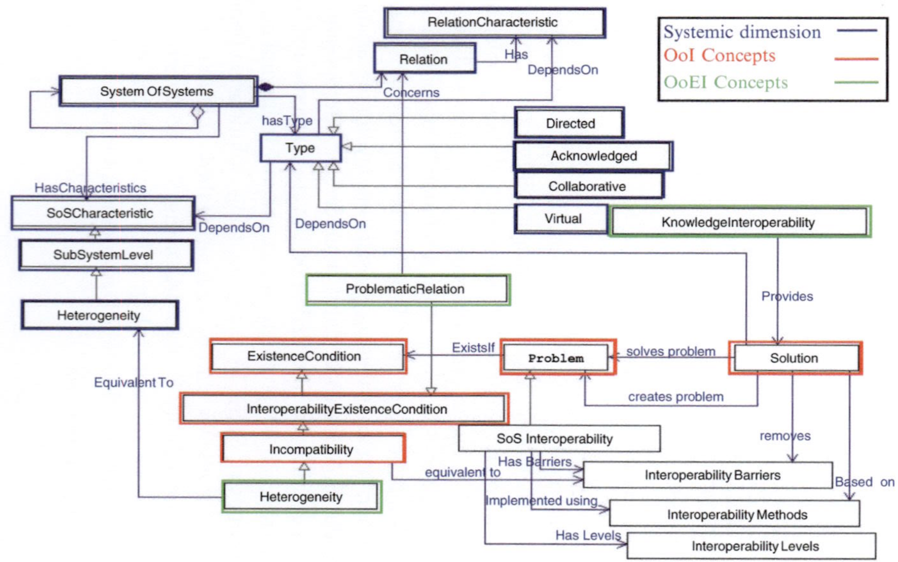


**Fig. 7.3** Enrichment of SoS interoperability conceptual model (SoSICM) of Fig. 7.2 by specialization of OoI and OoEI concepts and by adding a systemic dimension

| Approaches to Constructive Interoperability [3] | | | | | | | | | | | | | | |
|---|---|---|---|---|---|---|---|---|---|---|---|---|---|---|
| | Levels | | | Barriers | | | | | | | | Methods | | |
| | | | | Con. | | Org. | | | Tech. | | | | | |
| **Solution** | Pro. | Con. | Op. | Svn. | Sem. | Resp. | Auth. | Struc. | Inter. | Com. | Cont. | Integrated | Unified | Federated |
| Model-Driven Architecture | | | | | | | | | | | | | | |
| Service Oriented Architecture | | | | | | | | | | | | | | |
| Web services | | | | | | | | | | | | | | |
| Semantic Web Services | | | | | | | | | | | | | | |
| Open Grid Services Architecture | | | | | | | | | | | | | | |
| Component Frameworks | | | | | | | | | | | | | | |
| SoS Navigator 2.0 [4] | | | | | | | | | | | | | | |
| | Levels | | | Barriers | | | | | | | | Methods | | |
| | | | | Con. | | Org. | | | Tech. | | | | | |
| **Solution** | Pro. | Con. | Op. | Svn. | Sem. | Resp. | Auth. | Struc. | Inter. | Com. | Cont. | Intégrée | Unifiée | Fédérée |
| SoS Navigator | | | | | | | | | | | | | | |

**Fig. 7.4** structuring solutions proposed by the SEI in the frame of SoSIF

Interoperability barriers concern any kind of incompatibility that presents an obstacle to attend interoperability between different systems. To solve interoperability problem and remove related barriers, one approach from ISO approaches (integrated, unified or federated) could be applied. Based on the SoSIF we propose a conceptual model, SoSICM, that conceptualizes knowledge related to the SoS interoperability domain in the frame of SoSIF.

We are working on other existing solutions that should be mapped to the proposed framework. The proposed SoSICM and the associated OoSoSI should then be validated by experts. Another future work concerns the operationalization of the proposed ontology in a domain of application. We started by instantiating the SoSO concepts (the SoS ontology is out of the scope of this paper) in the field of Emergency Management and we intend to use the proposed ontology, OoSoSI, in the same field. Our ultimate goal is the use of the proposed ontology in a case based reasoning (CBR) system. Using CBR approach, new solutions to interoperability barriers covering all levels of interoperability could be defined and tested.

## References

1. Wells GD, Sage AP (2009) Engineering of a system of systems. In: Jamshidi M (ed) System of systems engineering: innovations for the 21st century. Wiley, New York, pp 44–76
2. Firesmith D (2010) Profiling systems using the defining characteristics of systems of systems (SoS). Technical note: CMU/SEI-2010-TN-001, Carnegie Mellon University

3. Lewis GA, Wrage L (2004) Approaches to constructive interoperability. Technical report: CMU/SEI-2004-TR-020, Carnegie Mellon University, Pittsburgh
4. SEI-SoS (2008) SoS navigator 2.0: a context-based approach to system-of-systems challenges. Technical note: CMU/SEI-2008-TN-001, Carnegie Mellon University
5. Barot V (2012) T-Area-SoS. Available from tareasos. https://www.tareasos.eu/index.php. Accessed 10 Apr 2013
6. Payne R (2011) COMPASS. Available from compass-research. http://www.compass-research.eu/. Accessed 10 Apr 2013
7. Grundstein M (2002) Management des connaissances de l'entreprise: problmatique, axe de progrs, orientations. MG Conseil, France
8. Mizoguchi R (2003) Tutorial on ontological engineering. Part 1: introduction to ontological engineering. New Generation Comput 21(4):365–384
9. Institute of Electrical and Electronics Engineers (1990) IEEE standard computer dictionary: a compilation of IEEE standard computer glossaries, New York, NY
10. ISO/IEC JTC 1 N7409 (2004) Information technology: fundamental terms. http://std.dkuug.dk/jtc1/sc2/wg2/docs/n2733. Accessed 5 Mar 2013
11. Tolk A, Turnitsa C, Diallo S (2008) Implied ontological representation within the levels of conceptual interoperability model. Intell Decis Technol 2(1):3–19
12. CEC (2003) Commission of the European communities: communication from the commission to the council, the European Parliament, the European Economic and Social Committee and the Committee of the Regions, COM (2003) 567 final—"The Role of eGovernment for Europe's Future", Brussels
13. EPAN (2004) European public administration network, "e-government working group: keyprinciples of an interoperability architecture", Brussels
14. Pardo TA, Nam T, Burke GB (2011) E-government interoperability: interaction of policy, management and technology dimensions". Soc Sci Comput Rev 30(1):7–23. doi:10.1177/0894439310392184
15. Lynn LE Jr, Heinrich CJ, Hill CJ (2000) Studying governance and public management: challenges and prospects, J-PART 10:2, pp 233–261
16. Kubicek H, Cimander R, Scholl HJ (2011) Organizational interoperability in e-government: lessons from 77 European good-practice cases, ISBN 978-3-642-22501-7 e-ISBN 978-3-642-22502-4. Springer, Heidelberg. doi:10.1007/978-3-642-22502-4
17. Morris E, Levine L, Meyers C, Place P, Plakosh D (2004) System of systems interoperability (SOSI): final report. CMU/SEI-2004-TR-004 ESC-TR-2004-004
18. Meyers BC, Smith DJ II (2007) Programmatic interoperability. Technical note: CMU/SEI-2008-TN-012
19. European Communities (2004) European interoperability framework for pan-European e-government services (EIF 1.0)
20. Commission of the European communities: communication from the commission to the European Parliament, the council, the European Economic and Social Committee and the Committee of the Regions (2010) towards interoperability for European Public Services, COM(2010) 744 final (including EIF 2.0), Brussels
21. Office of the e-Envoy (2004) e-Government interoperability framework. Technical specification, version 6.0
22. Pankowska M (2008) National frameworks' survey on standardization of e-Government documents and processes for interoperability. J Theor Appl Electron Commerce Res 3(3):64–82. Universidad de Talca—Chile. ISSN 0718-1876 electronic version
23. The New Zealand Government, State Services Commission, E-Government Unit (2002) A New Zealand e-government interoperability framework (e-GIF), version 0.9, release 8
24. Chen D, Dassisti M, Tsalgatidou A (2005) Interoperability knowledge corpus: intermediate report DI.1. University Bordeaux 1, Bordeaux
25. Chen D (2006) Enterprise interoperability framework. In: Proceedings of enterprise modelling and ontologies for interoperability

26. Siemieniuch C, Sinclair M (2012) Socio-technical considerations for enterprise system interfaces in systems of systems. In: 7th international conference on system of systems engineering, IEEE SoSE, pp 59–64
27. Jones AS, Sonnenberg J, Newman KB, Anderson L, Schmidt M, Brooks R (2011) Net-centric information framework. Technical report, NCOIC-network centric operations industry consortium
28. Miller FW, Bound J, Yanosy J, Polzer H, Creps D, Ross J et al (2006) Interoperability framework (NIF) communications. Technical Report, Network-centric operations industry consortium
29. Rosener V, Latour T, Dubois E (2004) A model-based ontology of the software interoperability problem: preliminary results. In: CAiSE04 workshops EMOI 04, pp 241–252
30. Naudet Y, Latour T, Guedria W, Chen D (2010) Towards a systemic formalisation of interoperability. Comput Ind 61(2):176–185
31. Gruber T (1995) Towards principles for the design of ontologies used for knowledge sharing. Int J Hum Comput Stud 43(5–6):907–928
32. Guedria W (2012) A contribution to enterprise interoperability maturity assessment. Ecole Doctorale Des Sciences Physiques Et De L'ingénieur, Luxembourg

# Chapter 8
# Involving End-Users in the Design of a Domain-Specific Language for the Genetic Domain

**Maria Jose Villanueva, Francisco Valverde, and Oscar Pastor**

**Abstract** The participation of end-users in software development ensures a better representation of their requirements in the resulting software product. With this aim, Domain-specific Languages (DSL) were proposed for abstracting the programming complexity of a domain and to provide an understandable tool for end-users. However, how end-users must be involved in the DSL development itself is also important, but it is not usually taken into account. As a solution, we propose a DSL development process for involving end-users; concretely, in this paper, we focus on their involvement during the design of the DSL syntax. For achieving this goal, we identify the decisions that developers must address to design the syntax, and we discuss how end-users can participate in those decisions. As a proof of concept, we apply the proposal to design the syntax of a DSL for genetic disease diagnosis software with the collaboration of geneticists and bioinformaticians from two SMEs.

**Keywords** Domain-specific language development • End-user involvement • Agile development • Genetics

## 8.1 Introduction

The importance of involving end-users in software development has been previously noticed by several authors [1, 2]: it reduces potential miscommunications of requirements, and, therefore, avoids the frustration that end-users feel when software engineers develop products that do not fulfil their needs.

M.J. Villanueva (✉) • F. Valverde • O. Pastor
Centro de Investigación en Métodos de Producción de Software, Universitat Politècnica de València, Camí de Vera sn, 46022 València, Spain
e-mail: mvillanueva@pros.upv.es; fvalverde@pros.upv.es; opastor@pros.upv.es

M. José Escalona et al. (eds.), *Information System Development: Improving Enterprise Communication*, DOI 10.1007/978-3-319-07215-9_8,

Domain-specific Languages [3] have been proposed as a solution to involve end-users in software development by abstracting development issues as domain concepts. Using them, end-users are able to understand the language and contribute in the development of tools that really satisfy them. However, this situation is only possible if the DSL fulfils their requirements. For this reason, end-user involvement in the DSL development itself is also essential; especially in domains not related with IT, usually less familiar to software engineers.

The problem of current DSL development methodologies or processes [4, 5] is that the participation of end-users in DSL development is not taken into account. One example is the design stage, usually carried out by developers only.

As a solution, our goal is to provide the suitable mechanisms, agile and easy to understand for end-users, so they can contribute in DSL development decisions. However, we also focus on avoiding their direct participation in low-level tasks that are outside their domain knowledge, such as metamodelling, grammar specification or technological discussion.

In previous work, we presented a first proposal of an agile DSL development process to involve end-users [6]: we provided an overview of the complete process and we addressed specifically the decision and the analysis stages.

In this paper, we extend that proposal discussing how end-users are involved in the design stage: we identify the decisions to be made to design the DSL syntax and we explain how end-users' ideas are included in those decisions.

We apply the proposal to design the syntax of a DSL for genetic disease diagnosis software. The development of a DSL in this domain is a very good example to appreciate the proposal benefits: because the genetics domain is complex and heterogeneous, involving domain experts in the development of bioinformatic software and genetic information systems is very important if not essential.

The main contributions of this work are: (1) a real-world analysis regarding DSL design decisions to identify the potential involvement of end-users; (2) a first version of a DSL for specifying genetic disease diagnosis software, created in close collaboration with end-users from two SMEs [7, 8], whose expertise is the genetic domain; and (3) a discussion about the main findings while applying the proposal in a real environment and about the opinions of the end-users who participated.

The rest of the paper is structured as follows. Section 2 overviews the full DSL development process to establish the context of this work. Section 3 explains all the decisions to make in the design stage and the involvement of end-users in those decisions. Section 4 details the application of the proposal to design the syntax of a DSL together with experts from the genetics domain. Section 5 discusses the main findings while applying the proposal in a real scenario. Section 6 presents and compares other proposals that involve end-users in DSL design, and finally Sect. 7 provides the conclusions and the future work.

## 8.2 An Agile DSL Development Process: Overview

In previous work, we proposed an agile DSL development process [6] to encourage end-user involvement. We combined the methodological guidelines for DSL development proposed by [4], with good practices from the agile methods Extreme Programming [9], Scrum [10] and Agile Modeling [11]. Briefly, the process is configured as an iterative cycle (Fig. 8.1) to create the DSL incrementally by means of small releases or "sub-DSLs" that satisfy a set of requirements.

We addressed how to involve end-users in the decision stage and in the analysis stage: enumerating the technological solutions of the domain, and specifying the set of user stories, acceptance tests, and scenarios, which developers will use to create the models of the analysis (a feature model that describes the commonalities and variabilities of the DSL, and a conceptual model that describes the concepts of the DSL and the relationships among them).

Each *iteration* of the cycle (Fig. 8.2) *addresses* a set of user stories, *tested* by a set of acceptance tests, in order to *accomplish* a set of scenarios, which *group* a set of acceptance tests:

- A *user story* is an end-user's demand to the system: a user with a *role* executes an *action* to accomplish a specific *goal*. For example: "As user (role) I want to choose a VCF file from the computer (action), so that a set of variations are read (goal)".
- An *acceptance test* is the checking that a user story works with a real example: a user with a *role* provides an *input* to the system and expects a specific *system*

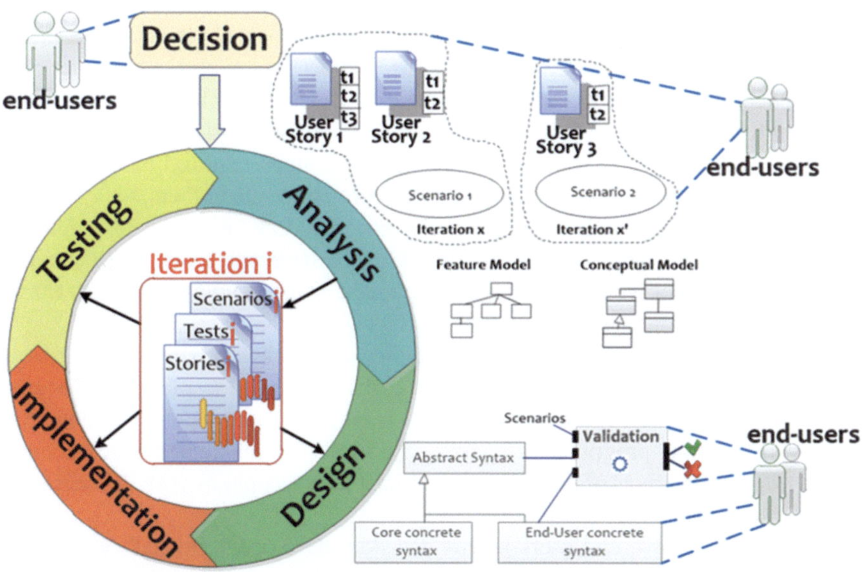


Fig. 8.1 Overview of the DSL development process proposed

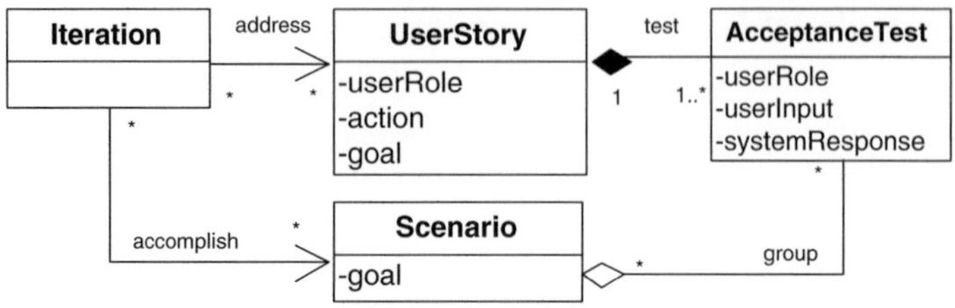


**Fig. 8.2** Conceptual model describing the proposal concepts

*response*. For example: "As user (role), when I choose the VCF file C:/File1 .vcf (input), the variation v1 = g.1234A > C is read (response)".

- A *scenario* is also the real checking that a set of user stories work together to accomplish a *goal.* For example: "To diagnose the Achrondoplasia disease (goal), when I choose the VCF file "C:/file.vcf", the variations {g.1234A > C, p.Gly380Arg} are read (first acceptance test); then, when I filter those variations by notation="p.", the variation {p.Gly380Arg} is obtained (second acceptance test).

To go on with the proposal, in this paper we address the involvement of end-users in the design stage: (1) designing the syntax of the DSL; and (2) validating the design taking into account the scenarios defined in the analysis stage.

## 8.3 How to Involve End-Users in DSL Design

Designing the syntax of a DSL implies four main decisions: (1) to design an external or an internal DSL [4]; (2) to express the abstract syntax with a grammar or a metamodel [12]; (3) to choose the representation (textual, graphical, wizard, table, etc.) and the style of the concrete syntax (imperative, declarative) [12]; and (4) to validate the design [13]. In Fig. 8.3, we show the steps required to make those decisions and the mechanisms proposed to involve end-users in them.

The first decision—only required in the first iteration—is about designing an internal or external DSL (1). An internal DSL is a language based in an existing language or, in other words, a particular way to use a base language, while an external DSL is a new language with its own syntax.

This decision is important for end-users because each approach has their own advantages and disadvantages that will affect them [14]: (1) availability of a programming context (f1): if it is possible to add new functionality not included in the DSL while creating programs; (2) syntax flexibility (f2): the degree of restrictions to express the constructs of the DSL; and (3) language cacophony (f3): the necessity to learn a new language.

In order to involve end-users in this decision, developers will talk with them about their preferences (taking into account the aforementioned three features). This way, developers can identify the weight of each feature (f1, f2 and f3) for each end-user. With this weighting, developers identify which approach—internal or external—best fits with their preferences.

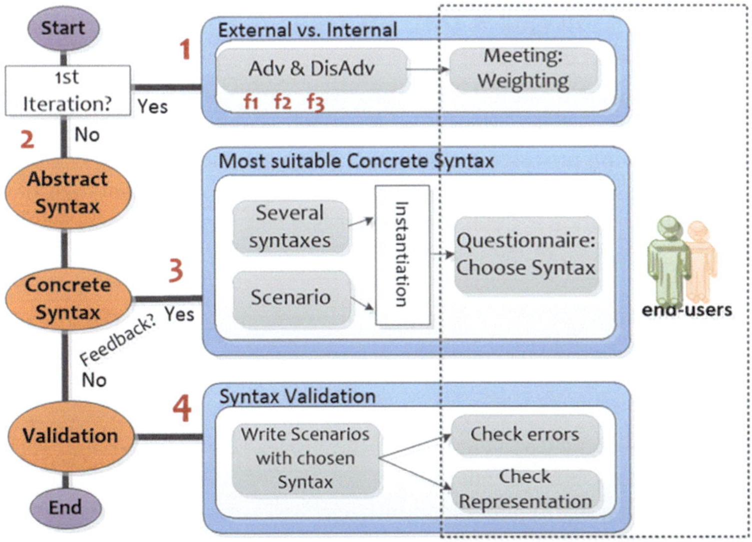


**Fig. 8.3** Proposal overview: syntax design steps

The second decision is about the abstract syntax. In order to design it, developers take into account the domain and feature models created in the analysis stage. Briefly, the domain model defines the entities that must be instantiated using the DSL, whereas the feature model establishes the optionality of those entities and their set of possible instantiations. Although the way to express the abstract syntax—grammar or metamodel—does not affect end-users' view, the decision to create the as the abstract syntax before the concrete syntax (2) allows generating several concrete syntaxes using the same abstract syntax.

The third decision is choosing the representation and the style of the concrete syntax most suitable for end-users. In the first iteration, end-users are provided with a set of different syntaxes so they can express their preferences (3). In following iterations, developers assess if the inclusion of new user stories from the iteration implies a meaningful change in the syntax structure. If that is the case, end-users will provide new feedback about the most suitable concrete syntax for them: choosing a new one or staying with the current one with minor updates.

In order to show the concrete syntax options to end-users in an understandable way, developers: (1) create a set of syntaxes with different representations and styles; (2) instantiate an scenario of the iteration with them; and (3) provide end-users with a questionnaire to obtain which syntax they like most.

The last decision is about the syntax validation. Developers specify all the scenarios of the iteration with the chosen syntax and show them to end-users. In order to assess the syntax, first they check if any scenario contains errors or misunderstandings in requirements—validating the abstract syntax—and then, they check if they agree with the DSL constructs—validating the concrete syntax.

## 8.4 Proof of Concept: A DSL for Genetic Disease Diagnosis

As a way to assess the feasibility of the proposal, we applied it to design the syntax of a DSL for genetic disease diagnoses. First, we explain the purpose of the DSL from the information provided by IMEGEN [7], a genetic company that works on genetic diagnosis, and GEMBiosoft [8], a bioinformatics company that develops software and information systems for geneticists. Then, we explain how we involved geneticists in every step of the syntax design.

### *8.4.1 Genetic Disease Diagnosis*

A genetic disease diagnosis (Fig. 8.4) is the process of: (1) searching variations located in patients' DNA; (2) analysing if any of them is the cause of the disease; and (3) generating a report with the results.

The term variation names the content of the patient's DNA in a specific position when it is different from the content of a "disease-free" reference DNA in the same position. Pointing a variation as the cause of a disease is possible when the variation is located in the genetic knowledge base: a researcher previously published a research work that confirmed the association.

Geneticists have established a common workflow to execute the three steps described to diagnose a genetic disease. However, each disease—and also each different procedure to diagnose the same disease—has its own properties, hence the set of possibilities is very wide. For example: the number or type of variations to search, the localization of the variations inside the DNA, or the technique used to identify the variations, etc.

In this scenario, the idea to create a DSL is justified because of: (1) the geneticists' need to continuously develop software to support their analysis; and (2) their difficulties to manage the technological details to specify diagnosis tools.

The first iteration of the DSL process addresses six User Stories, with their correspondent Acceptance Tests in order to accomplish two diagnosis Scenarios: the "Achrondoplasia diagnosis" and the "Multiple Cavernous Haemangioma diagnosis". Some examples are (the full set is available in [15]):

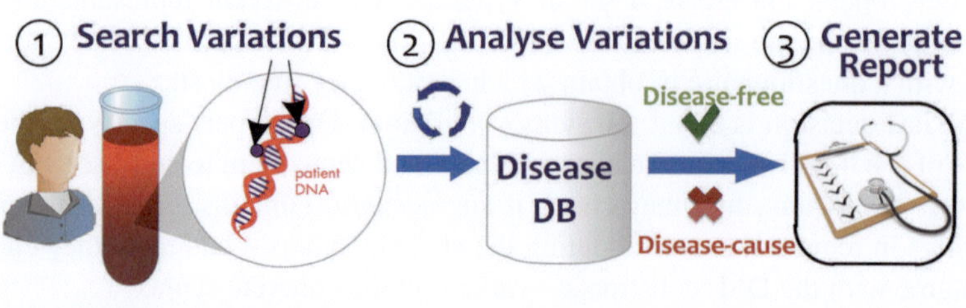


**Fig. 8.4** Genetic disease diagnosis process

User story 1: "I want to choose a file with VCF format from the computer, so that the reference and the patient's variations in chromosomal positions are read"
Acceptance Test 1.1: "When I choose the VCF file 3Variations.vcf, the reference "1000GenomesPilot-NCBI36" and the three variations {g.6438366A>G, g.6438499C>T, g.7387205A>G} are read"
User story 2: "I want to get the variations' rsIdentifier from the dbSNP database, so that patient's variations can be identified"
Acceptance Test 2.1: "When I search the rsIdentifier of the variation {g.1234A>C} in the dbSNP database, I see that the rsIdentifier is rs2000023.

### *8.4.2 Involving Geneticists in DSL Design*

In order to address the first decision (external vs. internal), we met with geneticists and asked about the three features described in the previous section:

- Availability of programming context (f1): Geneticists manifested that programming is something they want to avoid as much as possible. They know the existence of programming libraries, such as Biojava [16] and the possibility to execute program pipelines by writing Perl-based scripts. However, they feel their use is responsibility of informaticians or bioinformaticians with a strong programming background. Regarding this feature, they prefer an external DSL.
- Syntax flexibility (f2): Geneticists want to specify their own analyses and they don't mind so much the structure of the language as long as it only contains words related with their diagnoses and words they could understand. Regarding this feature, they don't mind if the DSL is external or internal.
- Language cacophony (f3): Geneticists know languages like Taverna [17] to create workflows, which can be used to specify genetic diagnoses, but they do not like its visual notation. They do not want to learn a workflow language, but they are willing to make the effort as long as it is easy and also if they are provided with a usable editor that guides them in the task of writing the language constructs. Regarding this feature they don't mind if the DSL is external or internal.

In [18], Cuadrado et al. compare both approaches of developing and internal or external DSL and conclude that in regards with the target audience: (1) end-users tend to perceive that an internal DSL is more complicated to learn because it implies learning a new general purpose language; (2) an external approach is recommended if end-users may feel intimidated; and (3) the freedom offered by an external DSL may allow to satisfy end-users when they request changes or new specific constructs. Together with the feedback provided by geneticists and the previous remarks, the external decision is the best solution in our context.

In order to address the second decision (grammar vs. metamodel) we were guided by the technology chosen for defining our DSL editor. We selected EMFText [19] to design the syntax and to generate the syntax editor because it allows first to define the abstract syntax, and then specify the concrete syntax (one or several)

**Table 8.1** Average punctuation of syntaxes by geneticists

| | Syntax 1 | Syntax 2 | Syntax 3 | Syntax 4 | Syntax 5 |
|---|---|---|---|---|---|
| Average (scale 1–5) | 4 | 4.7 | 2 | 1 | 3.1 |

```
diagnosis ::= "Diagnose " name[]
                  patientData (analyses)+ report;
patientData ::="Load file" source[]
          "Transform the " format[VCF:"VCF "]
          "to " type[Variations:"Variations"];
filter ::= "Choose " "{"((",")* punctual)*"}";
hgvs_dna ::= "g."description[];
search ::=  "Search in " dbSNP;
dbSNP ::= "dbSNP "rsIdentifier["the rsIdentifier" : ""]
       (" and ")* isSNP["if is SNP" : ""];
report ::= "Report " reportVariations;
reportVariations ::=  chosen["chosen " : ""] "variations"
             | known["known " : ""] "variations";
```

```
Diagnose Multiple Cavernous Hemangioma
Load File C:/File1.vcf
Transform the VCF to Variations
Choose {g.91851272C>T}
Search in dbSNP the rsIdentifier and if is SNP
Report known Variations
```

**Fig. 8.5** Grammar (*left-side*) of the chosen syntax the selected scenario (*right-side*)

based on the previously defined abstract syntax. Using EMFText, the abstract syntax (available in [15]) is transformed in an underlying metamodel using the Ecore metamodelling language.

In order to address the third decision (allow end-users choosing the most suitable syntax) we generated five textual syntaxes with EMFText (available in [15])—using the Concrete Syntax Specification Language based on EBNF—and we use them to write the scenario of the “Multiple Cavernous Haemangioma diagnosis”.

We have created a questionnaire (available in [15]) with five sections (one per syntax) to illustrate the scenario using a concrete syntax and to ask about the specific DSL constructs associated with the user stories of the iteration, which geneticists must rank in a scale from 1 (least favourite) to 5 (most favourite). Also, we added a section that gathers the scenario written with each of the syntaxes, where geneticists indicate their favourite, and an open answer to express the same example with their own syntax. In addition, we added another section to ask about the questionnaire itself: their opinion about the type of questions and the examples provided. Geneticists ranked each of the DSL constructs according to their preferences and provided their comments about the syntaxes and the questionnaire (Table 8.1 shows the average punctuation of each syntax). The results of the questionnaire showed that geneticists prefer the Syntax 2: the imperative syntax written in natural language. Figure 8.5 shows the specification of the syntax (left-side) and the instantiation of the scenario (right-side).

Finally, in order to address the fourth decision (validation), we also wrote the “the Achrondoplasia disease” scenario and show both of them to geneticists. Regarding the correctness of the syntax, they did not find any error in requirements, however, regarding the concrete syntax of the DSL constructs they proposed to change some words, like Load to Open (Open File C:/File.vcf), and Transform to Read (Read variations in VCF format), and manifested their preferences to remove or change the symbols “{” and “}”.

## 8.5 Discussion

After applying the proposal to design the first version of the syntax together with geneticists, we confirmed the essential role of end-users in the decision making of the DSL design. We observed the importance of involving end-users in the described decisions: without their involvement in the design, developers can make decisions, implement them, and delay mistakes until testing. For example, geneticists found the XML-like syntax and the one organized by modules complex and incomprehensible, while developers thought they were the most straightforward. Also, one geneticist manifested that she finds very confusing seeing parenthesis or brackets, contrarily with developers, who are very used to them because their experience with programming languages.

Another important finding is that we realized, thanks to end-users comments, how the different feedback mechanisms of the proposal—meetings, questionnaires, brainstorming sessions—improve their collaboration enthusiasm and feedback quality, without perceiving the underlying metamodel or the different grammars.

Regarding the external/internal decision, although we expected that the decision was going to be external, we discover that geneticists knew another language for specifying workflows that could be used as the base language for an internal approach. Although in our case geneticists were not interested in the extension of that language, we realized that we cannot assume that end-users always require an external, even a visual-oriented, approach.

Regarding the selection of the syntax, geneticists found the questionnaire easy to respond, and although they clearly chose one syntax as their favourite, they appreciate the opportunity to transmit developers their own way to express the scenario. In the validation of the syntax, they understood their role and provided the suitable feedback to change the syntax according to their preferences. However, they emphasized their worries about the necessity to remember very specific details of the syntax: (1) the use of capital/lower letters; (2) punctuation symbols; (3) the order of different words; and (4) to face a white page where they must write the diagnosis without mistakes (their strongest issue).

In summary, we realized they do not have strong preferences about the syntax but they really need assistance using it: by means of toolbars, examples, code completion, wizards, etc.

Finally, we must remark that there is a limitation in the proposal because we have only generated textual syntaxes. Textual representations provide very interesting advantages over graphical ones [20] such as: (1) more information content; (2) better speed of creation; (3) better language integration; (4) better speed and quality of formatting; (5) platform and tool independency; (6) better version control; and (7) easiness of editor creation. Based on these advantages, we left the definition of graphical syntaxes as future work, and we created only textual syntaxes in the first iterations in order to get quick feedback from geneticists. Usually, it is assumed that graphical representations provide a better usability for end-users. However, some geneticists manifested that they don't like graphical symbols because they tend to be ambiguous and interpretation free. They also explained their experience with Taverna, which graphical workflow notation was useless for them in practice.

In this iteration we avoided the effort of generating graphical syntaxes and focused to represent and validate the domain in our design, if domain experts require a graphical syntax afterwards, we will generate them in next iterations.

## 8.6 Related Work

Several authors have also notice the importance of end-user involvement in DSL development.

First, Canovas et al. [13] propose a collaborative infrastructure to track the assessment of the DSL syntax by end-users. DSL developers create a first version of the syntax and provide end-users a set of examples of its usage; then, end-users provide their comments so developers can provide solutions to them.

Second, Kuhrmann [21] proposes the description of domain elements together with domain experts using an informal panel. In the panel, it is defined the set of entities, the set of relationships among them and the visual representation of each entity. Their goal is to translate all these informal drawings to the formal descriptions of the DSL.

Third, Sanchez-Cuadrado et al. [22] propose an approach to develop DSL metamodels that involves end-users. They use a tool to draw sketches that represent examples of the DSL. A translator obtains textual representations of this drawings taking into account heuristics and labels, to identify class names of objects and references to other objects. Then, those entities can be annotated by developers with design and domain concerns. After that, the metamodel is obtained by induction. This approach allows end-users to interact with drawings they are able to understand, but the underlying meaning of this drawings represent concepts related with modeling, a task that end-users are not used to accomplish.

Fourth, Cho et al. [23] propose the inference of syntax and static semantics by providing examples to end-users' and getting their feedback.

These four approaches address the involvement of end-users in some design steps of the DSL: tracking syntax assessment, designing the syntax from scratch or inferring the syntax from drawings. Our work pursues the same goal but focuses on involving end-users as much as possible in design decisions: both in syntax design and syntax assessment. Also, we stress the necessity to define understandable mechanisms for end-users providing useful design feedback, so we avoid their interaction with low-level artifacts such as metamodels or grammars.

And finally, Hermans et al. [24] address an empirical study based on questionnaires to ask end-users about their opinions in the context of DSL development.

While the goal of this work is to find the factors that contribute to the success of a DSL, our goal is to introduce end-users feedback into the DSL development.

## 8.7 Conclusions and Future Work

In this paper we have analysed how to involve end-users in the decisions of designing a DSL. First, we have analysed the decisions required in the design of the syntax; then, we proposed a set of steps to involve end-users' in those decisions, and finally, we applied the proposal in a real scenario to see their feasibility, the potential drawbacks, and to get end-users feedback about the proposal.

We applied the proposal to design the syntax of a DSL to specify genetic disease diagnosis. Concretely, we have defined a first version of the syntax, that is, the syntax obtained in the first iteration of the DSL development process followed. This syntax is supported using EMFText as the DSL implementation technology.

As a result, we observed that it is interesting to involve end-users in DSL decisions apart from requirements gathering and testing. The approach we followed to involve them in design was to provide user friendly mechanisms, such as informal meetings and questionnaires, that abstracted the DSL design decisions to their domain knowledge.

Although we obtained very interesting benefits in this first iteration, these are not thoroughly validated. In future work, we plan to assess those findings performing additional iterations of the DSL design.

Also, as future work, we will research how to involve end-users in decisions of the implementation and testing phases of the DSL development process. And, after completing the DSL development proposal, we will run additional iterations to get more feedback from end-users and provide to them a full DSL implementation.

**Acknowledgments** The authors would like to thank IMEGEN for all these years of collaboration, providing both useful genetic knowledge and a real environment for research, and also GEMBiosoft, especially to Dr. Ana M. Levin, for its support in the development of this paper. This work has been developed with the support of MICINN under the FPU grant AP2009-1895, the project PROS-Req (TIN2010-19130-C02-02), and co-financed with ERDF.

## References

1. Ko A, Abraham R, Beckwith L, Blackwell A, Burnett M, Scaffidi C et al (2011) The state of the art in end-user software engineering. ACM Comput Surv (CSUR) 43(3):21:1–21:44
2. Fischer G, Giaccardi E, Ye Y, Sutcliffe AG, Mehandjiev N (2004) Meta-design: a manifesto for end-user development. Commun ACM 47(9):33–37
3. Fowler M (2010) Domain-specific languages. Addison-Wesley Professional, Boston
4. Mernik M, Heering J, Sloane A (2005) When and how to develop domain-specific languages. ACM Comput Surv (CSUR) 37(4):316–344
5. van Amstel M, van den Brand M, Engelen L (2010) An exercise in iterative domain-specific language design. In: Proceedings of the joint ERCIM workshop on software evolution (EVOL) and international workshop on principles of software evolution (IWPSE), pp 48–57
6. Villanueva MJ, Valverde F, Pastor O (2012) Involving end-users in the design of a domain-specific language for the genetic domain. In: ENASE 2013, Angers
7. Instituto de Medicina Genomica, IMEGEN (2013). www.imegen.es
8. GemBiosoft (2013) www.gembiosoft.com

9. Beck K, Andres C (2004) Extreme programming explained: embrace change. Addison-Wesley Professional, Boston
10. Schwaber K, Beedle M (2002) Agile software development with Scrum 18. Prentice Hall PTR, Upper Saddle River
11. Ambler S (2002) Agile modeling: effective practices. Wiley, London
12. Langlois B, Jitia C-E, Jouenne E (2007) DSL classification. In: OOPSLA 7th workshop on domain specific modeling
13. Canovas-Izquierdo JL, Cabot J (2012) Community-driven language development. In: 2012 ICSE workshop on modeling in software engineering (MISE), pp 29–35
14. Fowler M (2005) Language workbenches: the killer-app for domain specific languages? White Paper, ThoughtWorks
15. www.pros.upv.es/technicalreports/PROS-TR-2013-01.pdf
16. Holland RCG, Down TA, Pocock M, Prlic A, Huen D, James K et al (2008) BioJava: an open-source framework for bioinformatics. Bioinformatics 24(18):2096–2097
17. Hull D, Wolstencroft K, Stevens R, Goble C, Pocock MR, Li P, Oinn T (2006) Taverna: a tool for building and running workflows of services. Nucleic Acids Res 34(Suppl 2):W729–W732
18. Cuadrado J, Molina J (2013) Formal and practical aspects of domain-specific languages: recent developments. IGI Global, Hershey, PA, pp 109–131
19. Group, T (2008) EMFText. http://emftext.org
20. Gronniger H, Krahn H, Rumpe B, Schindler M, Völkel S (2007) Text-based modeling. In: 4th international workshop on software language engineering
21. Kuhrmann M (2011) User assistance during domain-specific language design. In: ICSE 2011 workshop on flexible modeling tools (flexi tools)
22. Sanchez-Cuadrado J, De Lara J, Guerra E (2012) Bottom-up meta-modelling: an interactive approach. In: Model driven engineering languages and systems. Springer, Berlin, pp 3–19
23. Cho H, Gray J, Syriani E (2012) Creating visual domain-specific modeling languages from end-user demonstration. In: 2012 ICSE workshop on modeling in software engineering (MISE), pp 22–28
24. Hermans F, Pinzger M, Deursen A (2009) Domain-specific languages in practice: a user study on the success factors. In: Schürr A, Selic B (eds) Model driven engineering languages and systems, vol 5795. Springer, Berlin, pp 423–437

# Chapter 9
# Improving Information System Interoperability in Social Sector Through Advanced Metadata

**Francisco Delgado, Salvador Otón, Raúl Ruggia, José Ramón Hilera, and Roberto Barchino**

**Abstract** Information exchange constitutes a key operation in social security systems. In this context, semantic interoperability is fundamental in the development of joint definitions and interpretations concerning the data which is to be processed by various organisms. One of the main methods to handling of semantic conflicts are controlled vocabularies and metadata. This paper proposes a new metadata standard based on Dublin Core elements, focus on social security information systems problems and supported by controlled vocabularies. Future work should define testing scenarios to refine this new standard.

**Keywords** Metadata • Interoperability • Social security • Semantics

## 9.1 Introduction

Interoperability techniques for the integration of computer and information systems play an increasingly important role in the implementation of social security systems [1]. Social Security data vary from country to country even from sector to sector in the same country. Data have different formats, types, processing methods and requirements. These variations are reflected on how data are used and understood by different institutions, which have to implement adequate interoperability mechanisms.

---

F. Delgado (✉)
Instituto Nacional de la Seguridad Social, Madrid, Spain
e-mail: francisco.delgado@seg-social.es

S. Otón • J.R. Hilera • R. Barchino
University of Alcalá. Polytechnic School, Alcalá de Henares, Madrid, Spain
e-mail: salvador.oton@uah.es; jose.hilera@uah.es; roberto.barchino@uah.es

R. Ruggia
International Social Security Association, Genève, Switzerland
e-mail: ruggia-frick@ilo.org

M. José Escalona et al. (eds.), *Information System Development: Improving Enterprise Communication*, DOI 10.1007/978-3-319-07215-9_9,

In the context of Social Security systems, semantic interoperability is fundamental in the development of joint definitions and interpretations concerning the data which is to be processed by various organisms. Social Security operations involve a wide range of concepts which in spite of having the same name, may be interpreted differently (e.g. family group, members of the same household, unemployed person, old age pension, health benefits, social security contributions, etc.). Thus, in order to improve understanding of the concepts it is useful to be able to represent the relations between them, principally those of "sub-groups" (e.g. rural workers are a sub-group of workers, etc.).

The best way to avoid semantic conflicts is the mandatory use of standardized data schema by all participants of a data exchange. In this sense, metadata schemas can be of fundamental importance in achieving semantic interoperability, together with vocabulary, taxonomies and anthologies. Metadata are sets of basic data for the development of e-government applications and systems.

There isn't a worldwide standard for metadata schemas in social security information systems. Consequently, institutions carrying out information exchange have to specify and implement such systems case-by-case, which increases the involved complexity and costs. The growing of international social security agreements worldwide motivates the specification of standards that contribute to systematize data exchange.

In this paper, current metadata standards and situation in different countries are revised. Finally the development, test and adoption of one specific standard for social security information systems are proposed. Due to space limitations implementation aspects are not addressed in this paper.

## 9.2 Context

### *9.2.1 Social Security Information Systems*

The corporate use of ICT in Social Security institutions involves the implementation of social security functions and required resources, notably: benefit administration, contribution collection, financial management, and compliance control on one hand; internal services like Human Resources and Internal Audit on the other hand, and also corporate information systems and ICT platforms as corporate resources to be used by the former ones. The application of technologies such as Interoperability, Data Security and Privacy, and Mobile, play a key role in the effective and efficient implementation of high-performance social security systems.

Institutions need to use interoperability techniques to integrate the information systems of individual social programmes if they are to put into practice current trends towards the integration of social policies and programmes now appearing in all areas of social security, as well as to construct information services and systems to be shared between the different institutions or bodies which operate social policies [2].

In the same way, applying these interoperability techniques internally within institutions can bring improvements in the quality of their internal management. A more detailed analysis can be found in [3].

In the context of information systems, interoperability literally denotes the ability of a system to work with or use parts of others systems [4].

Interoperability can be approached from different angles which determine the types, aspects, focus and dimensions of interoperability.

The application of interoperability, particularly in complex organizational contexts, means taking into account the different organizational levels or sectors involved in the interaction. The general concept of interoperability can thus be subdivided into various dimensions.

Interoperability, whether applied to e-government as in defense and strategy systems, or in digital libraries and information services, is characterized by the following dimensions:

- Political: Collaborators must have visions which are compatible, priorities which are in line with each other and must focus on the same objectives.
- Legal: Adequate synchronization of the legislation in collaborating countries must include the fact that the electronic data generated in one country must be properly recognized when used by the receiving country.
- Organizational: This refers to the definition of business objectives based on business procedures and tries to facilitate collaboration between administrations or institutions which wish to exchange information and which may have different structures and internal procedures.
- Semantic: Ensures that the precise significance of the information exchanged is comprehensible for any other application not initially developed for this purpose. Semantic interoperability enables systems to combine the information received with information from other sources and to process it in a coherent way.
- Technical: Addresses critical aspects of linking computer and service systems. Includes key aspects such as open interfaces, interconnected services, the integration of data and middleware, the presentation and exchange of data, security and the accessibility of services.

In particular, this is the approach adopted by the European ISA Programme (Interoperability Solutions for European Public Administrations) and thus the approach of electronic public administrations. The European Interoperability Framework v 2.0 (EIF) [5] constitutes the current reference document.

The fact that the information generated by a computer must be processed by another system which must interpret its meaning correctly, leads to a series of additional complications which affect both the source of the information and its recipient, and constitutes the central theme of semantic interoperability. Achieving it requires agreement, for example, on the way in which information is represented and its context. This is what will enable automatic tools to share and process the information, even when it has been registered independently. The objective of semantic interoperability is not only to enable the interconnection of information resources, but also to enable them to be understood automatically and as a result to be reused by computer applications not involved in their creation.

An essential requirement for interoperability, particularly to carry out information exchange, is a single language to describe the meaning and structure of the subjacent data, for example a mark-up language. In the current technological and market environment, this language is XML. However, XML alone cannot guarantee or provide semantic interoperability. It requires joint semantic development initiatives based on XML. The subsequent introduction of XML schemas and related artefacts (metadata, anthologies, etc.) enables the integration of services developed using different vocabularies and with different perspectives on data.

### *9.2.2 Information Exchanges in Social Security*

Information exchange constitutes a key operation in social security systems. This is nowadays "a must", not only to integrate social security programmes, but also to implement of international social security agreements, which enable institutions of different countries to put into practice common social security regulations. A more detailed description can be found in [6].

The implementation of international social security agreements poses specific challenges as requires that institutions in different countries share an interoperability framework. Unlike intra-institution and national integration projects, international information exchange cannot take advantage of national interoperability policies and platforms. Therefore, the interoperability approaches to implement international information exchange have to be much more flexible and capable of integrating parties with different technological development and interoperability policies.

This section goes over systems implementing social security international agreements and the techniques used to address semantic interoperability.

- European Union [7]
  In the European Union, the EESSI project (EESSI) aims at supporting the rights to social protection and free movement of persons. To this, it implements the electronic exchange between the Member States on social security data related to citizens that move around in Europe. The project establishes common technical architecture which provides public access to an institutional data base (Master Directory), the exchange of information and a secure communication protocol. EESSI enables the exchange of information between Social Security Competent Institutions in the form of Structured Electronic Documents (SEDs).

- MERCOSUR [8]
  In South America, the MERCOSUR countries (Argentina, Brazil, Paraguay and Uruguay) have introduced a system to electronically manage the pension requests of individuals who have worked in these countries. The system, called SIACI, follows a federation architecture, in which institutions' information systems interoperate to exchange the business data. The information is exchanged in

XML archives and communications are based on "Web Services". Beneficiary identification is managed by the institutions of each country and the SIACI establishes the correspondences between them. The information transmitted is digitally signed by users, using external measures with processing capacity (tokens or smart-cards).

- Gulf Countries [9]
  In the Gulf Countries, the Gulf Cooperation Council (GCC) aims at the creation of a common market and keeping regional security. The current members are: Saudi Arabia, Bahrain, United Arab Emirates, Kuwait, Qatar and Oman. The High Council of the GCC ruled on the implementation of regulations ensuring the equality among citizens of the GCC countries in the field of social insurance and retirement. Accordingly, every state of the GCC should ask employers to pay the contributions of outside-state workers in the social insurance or civil retirement scheme. In 2006, the Saudi Arabian social security institution (GOSI) adopted national regulations to put into practice these agreements through operational mechanisms to facilitate the movement of the labour force among the GCC states.

- Republic of Korea [10]
  The National Pension Service of the Republic of Korea has concluded a memorandum of understanding (MOU) with the social security institutions of seven countries which enables us to Exchange data on beneficiaries residing in partner countries. The main objective of data exchange on overseas beneficiaries with social security institutions of other countries was to promptly confirm any changes in the beneficiary's status (such as death, divorce, etc.) which would terminate pension eligibility or necessitate an adjustment in its amount. Data exchanges between NPS and Social Security Administration (USA) are performed through the Government Services Online (GSO) website, while exchanges between NPS and Centrelink (Australia's S SI), are conducted through the EXCpert website. NPS has minimized the scope of personal identification information provided for partner institutions when requesting data on its beneficiaries. This information currently includes name, date of birth and social security number.

- Australia
  Australia is one of the countries that has put into practice more international social security agreements. The involved data exchange is implemented through the EXCpert system and it based on the Australian Government Locator Services [11], which is an Australian metadata standard (AS5044-2010) aligned with international Dublin Core Metadata Initiative (DCMI). It's used to describe online Government resources. The usage of AGLS metadata is mandatory for Australian Government Agencies. Summarizing, the implementation of international social security data exchange still constitutes a challenge, even for countries with high developed eGovernment systems (e.g. Australia, Korea). Most of the current operational systems implement "technical interoperability" mechanisms (mainly based on XML). Up to our knowledge, only Australia is implementing "semantic interoperability" for international data exchange through the AGLS Metadata.

## 9.3 Solution Approach

### 9.3.1 Controlled Vocabularies and Metadata

One of the main methods to handling of semantic conflicts are controlled vocabularies [12]. A controlled vocabulary is an organized arrangement of words and phrases used to index content and/or to retrieve content through browsing or searching. It typically includes preferred and variant terms and has a defined scope or describes a specific domain [13].

The general purpose of glossaries and code list is to provide a common understanding of the terms (and codes) within the currently applicable domain. Obviously, no structure is predetermined for code list, but the ability to represent synonyms and homonyms clearly supports the detection and resolution of conflicts related to these constructs. Another purpose of glossaries is to provide support for detecting homonymy conflicts whenever a qualification construct for representing homonyms is provided. For instance, if two schemata use the same term at different locations, these can be represented by surrounding names (context). In order to detect such homonymy, one can consult a glossary with the context as a parameter so that different qualifiers can be found.

Glossaries in the age of computing are often related to artifacts stored on computers. In this case, these glossaries are also called data dictionaries. A data dictionary can be seen as a collection of identifying descriptions of items within a data model. Its purpose is to avoid semantic mismatches when using data model for application development.

As it was mentioned before, the solution generally proposed for systematic treatment of semantic problems is the use of controlled vocabulary, i.e. in order of increasing complexity: glossaries, thesauruses, taxonomies and ontologies [14]. Although the use of ontologies would be a very important contribution towards guaranteeing the quality of electronic exchanges, the definition and maintenance of an ontology involves a very high level of complexity which is often off-putting.

Stronger than the term vocabulary, metadata schema emphasizes that metadata elements have a technical grounding in information systems, which is a significant aspect when metadata interoperability should be achieved also on a technical level [4]. The European Interoperability Strategy [15] aims to reach an agreement on the definition of metadata as a way to solve semantic interoperability problems.

The electronic concept of metadata came from the early theorist in database management who needed to be able to reference the aspects of the data models they were using for administrative purposes. Today metadata is still one of the most effective ways to describe "things" (resources or artefacts) and their relationship to one another.

Ideally, metadata would be written and rendered in one way that all computers and software applications could share. The programming world is not ideal so as well as standards we also use programming methods to cross-match information we need to combine.

Currently, to achieve interoperability we need to ensure that the data and metadata we intend to use can be matched (use the same language to talk to one another), no matter where the information comes from. We achieve this partly through standards but also through special files that help us compare metadata types—crosswalks and alias files (i.e. Data maps between one dataset and another to match common data properties).

### *9.3.2 Standard Metadata Specification*

Standardization is a strong way of establishing an agreement by means of consensus building and intuitive, technically effective and economically well-recognized way to achieve interoperability [4]. It requires accredited institutions for building consensus, setting a standard, and eventually assuring its uniform implementation. Regarding the building blocks of metadata, standardization can cover the language level (standardized language), the schema level (standardized metadata schema), the instance level, or several (hybrid metadata systems).

The choice of metadata schema is largely influenced by practices at peer institutions and compatibility with a content management system [16]. In the case of social security information systems metadata should describe data involved in information exchanges between social security institutions. The primary factor is selecting metadata standard in their suitability for describing the most common type of resources handled by the exchanges between information systems participants.

The second and third most common criteria, "target users/audience" and "Subject matters of resources", also seem to reflect how domain-specific standard are applied. Nowadays, social security is extended all over the world. However, existing technological infrastructure and resource in different institutions and constrains also determine options. So the elected standard should be based in a worldwide specification.

It was found that studies on metadata development for reference resource systems mostly had proposed Dublin Core metadata [17], which also have been widely used in existing reference systems [18]. Moreover, according with the experiences analyzed the fifteen Dublin Core elements could be an excellent starting point for a social security standard definition. Extending Dublin Core also meant developing a philosophy of extension and methods to extend the vocabulary without breaking any uses of the original set of data elements. Since 2003, Dublin Core is also an ISO standard [19].

In the eGovernment area, AGLS [11] is an Australian metadata standard (AS5044-2010) aligned with international Dublin Core Metadata Initiative (DCMI). It's used to describe online Government resources.

Another well-known standard is IEEE/LOM, Learning Object Metadata [20]. The LOM standard is a "conceptual data schema" for the description of learning objects (by a broad definition of the term). LOM was developed and formalized through the IEEE and their Learning Technology Standards Committee.

**Table 9.1** Mapping between DC elements and social security elements

| DC element | ESSIM | Obligation | Value |
|---|---|---|---|
| Title | Dataname | Mandatory | GOT |
| Creator | Creator | Mandatory | |
| Subject | Dataset | Optional | Core vocabularies |
| Description | Description | Mandatory | GOT |
| Publisher | Competent institution | Mandatory | MD |
| Contributor | | Mandatory | |
| Date | Date | Mandatory | |
| Type | | | |
| Format | Format | Mandatory | GOT |
| Identifier | | | |
| Source | | | |
| Language | Language | Mandatory | ISO 63 9-1 |
| Relation | Relation | Optional | GOT |
| Coverage | Scope | Mandatory | GOT |
| Rights | | | |
| | Mandate | Extension | Euro-lex |
| | References | Extension | |
| | Modified | Extension | |

In turn, the ISO/IEC 19788 Metadata Learning Resource [21] proposition is under development. Its first part is a recognized standard [22] and the second part is still a rough draft [23]. The set elements of MLR match up with Dublin Core elements.

In the case of AGLS, only three elements are mandatory (Creator, Date and Title). The others could be Conditional (present under certain circumstances), recommended ore optional. The metadata elements set, including extensions, consists of 60 elements.

According to these established standards, we propose the following schema (Table 9.1) for Exchanges of Social Security Information Metadata (ESSIM):

It's important to highlight that the identification of the standard's elements is based on social security information exchange systems and the underlying agreements. This analysis has been done by the authors as part of their professional duties.

Values always make reference to the particular European Union case. GOT, MD and Core Vocabularies make reference to controlled vocabularies in the social security field and are proposed as possible examples.

The Glossary of Terms (GOT) was defined [24] in the context of project TESS (precedent of EESSI project). The purpose was to permit a more efficient communication between the European Social Security Institutions. The GOT defined each data from the information repository (as per 1995) with a generic definition, followed by a definition specific to each Member states with a view to enable an institution to best process the data received from abroad.

**Table 9.2** Institution name metadata in social security context

| ESSIM element: institution name | Value | Remarks |
|---|---|---|
| Dataname | INSS ORG. ENLACE | From MD |
| Creator | xxx@seg-social.es | Responsible for creation |
| Subject | Competent institution | |
| Description | Competent institution in Spain | |
| Publisher | European Commission | |
| Date | 19781001 | Date of creation |
| Format | String | |
| Language | SPANISH | |
| Relation | Working life | Part of working life dataset |
| Scope | Pension, health care | From GOT |
| Mandate | European regulation | From Euro Lex |
| Reference | SED's P-xxx and H-xxx | Specific documents for these sectors |
| Modified | | Not modified yet |

Master Directory (MD) is an institutional data base. It is a tool developed in the project EESSI that permit that any institution is able to contact any other institutions thanks facilities of Directory Services and mechanisms established at both international and local level by EESSI Routing [25]. The MD only contains information about European Union social security institutions, but the idea could be worldwide extended.

To implement this standard it is needed to decide on:

- Which elements should be mandatory.
- Which elements should be optional.
- Which elements should be conditional.
- Which elements can content sub-elements.
- Which elements should be added.
- Which control vocabularies should support values.

In Table 9.2 there is a real example for the field "Institution Name" in the case National Institute of Social Security of Spain. The correspondent date area is now published in the public part of Master Directory [26].

## 9.4 Conclusion and Future Work

This paper addressed the issues related to information exchanges between social security information systems and the involved semantic interoperability problems.

While most accepted solutions for semantic interoperability problems are based on controlled vocabulary and metadata, a wide implementation of such techniques in the social security community requires their practical applicability in institutions from different parts of the world.

According to the literature, the Dublin Core metadata is the most widely employed. Different standards (LOM, AGLS, MLR) have been defined based in Dublin Core taking into account the particularities of every case.

This article proposes a new metadata standard based on Dublin Core elements and focusing on social security information systems as part of a more comprehensive solution for social security information exchange. Ten of 15 Dublin Core metadata elements are included in the new standard. Eight of them are considered mandatory and two optional. Three new elements are proposed for this especial case of social security exchanges. In turn, data values corresponding to the metadata should be supported by controlled vocabularies. In the case of the European Union some tools has been found for this propose (glossary of terms, Master Directory and Core Vocabularies). But if the standard is going to be used all over the world, extension of these tools should be considered.

Finally, the standard should be tested in a suitable scenario. The exchange of electronic documents (e.g. XML) between two or more social security institutions would supply feedback for deciding on open issues.

**Acknowledgements** This research is funded by the University of Alcalá (grant UAH49/2012). Authors also want to acknowledge support from the Master in Software Engineering for the Web and the TIFyC research group.

## References

1. ISSA (2012) 13th international conference on information and communication technology in social security. http://www.issa.int/News-Events/Events/13th-International-Conferenceon-Information-and-Communication-Technology-in-Social-Security
2. Durán-Valverde F (2012) Coordinate social protection programmes in the context of the creation of social protection floors. In: 13th international conference on ICT in social security, Brasilia, 2012. http://issa.int/web/event-13th-international-conference-on-information-and-communicationtechnology-in-social-security/documents
3. Interoperability in social security (2012). ISSA Technical Commission on ICT. In: 13th international conference on ICT in social security, Brasilia
4. Haslhofer B, Klas W (2010) A survey of techniques for achieving metadata interoperability. ACM Comput Surv 42(2): article 7. http://eprints.cs.univie.ac.at/79/1/haslhofer08_acmSur_final.pdf
5. EIF (2010) European Commission. Interoperability solutions for European public administrations—ISA. European interoperability framework (EIF). 2010. http://ec.europa.eu/isa/documents/isa_annex_ii_eif_en.pdf
6. Kounowski G (2012) Integration of social security programmes and the role of ICTs: case studies. In: 13th international conference on ICT in social security, Brasilia
7. EESSI (Electronic Exchange of Social Security Information). url:http://ec.europa.eu/social/main.jsp?catId=869
8. MERCOSUR (2009) http://www.issa.int/News-Events/News2/Good-practices-in-socialsecurity-The-case-of-MERCOSUR/(language)/eng-GB
9. GCC (2005) http://www.issa.int/Observatory/Country-Profiles/Regions/Asia-and-the-Pacific/Saudi-Arabia/Reforms/Extension-of-social-protection-between-Gulf-Cooperation-Council-member-states

10. NPS (2012) http://www.issa.int/Observatory/Good-Practices/Data-exchange-on-overseas-beneficiaries-between-social-security-institutions-for-the-prevention-of-illegal-pension-payments
11. AGLS (2010) http://www.naa.gov.au/records-management/agency/create-capture-describe/describe/index.aspx
12. Fraunhofer (2009-2) Study of multilingualism. European commission—IDABC. 2009
13. Gill T, Gilliland AJ, Whalen M, Woodley MS (2008) In: Baca M (ed) Introduction to metadata: pathways to digital information. Online version 3.0. J. Paul Getty Trust, Getty Research Institute, Los Angeles. http://www.getty.edu/research/publications/electronic_publicatons/intrometadata/setting.html
14. Hilera JR, Pagés C, Martínez JJ, Gutiérrez JA, de Marcos L (2010) An evolutive process to convert glossaries into ontologies. Information Technology and Libraries. December 2010
15. EIS (2012) European interoperability strategy. http://ec.europa.eu/idabc/servlets/Doc64c0.pdf?id=32207
16. Park J-R, Tosaka Y (2010) Metadata creation practices in digital repositories and collections: schemata, selection criteria, and interoperability. Inf Technol Libr
17. Dublin Core (2009) http://dublincore.org/metadata-basics/
18. Noh Y (2010) A study on metadata elements for web-based reference resources system developed through usability testing. Libr Hi Tech 29(2), 2011
19. ISO (2003) ISO/IEC 15836:2003. Dublin core metadata element set. International Standards Organization, 2003
20. LOM (2002) LOM. http://ltsc.ieee.org/wg12/
21. MLR (2008) http://www.ariadne.ac.uk/issue55/currier
22. ISO (2011) ISO/IEC 19788. Information technology for learning, education and training. Metadata learning resource. International Standards Organization (working draft), 2011
23. Pons D, Hilera JR, Pagés C (2011) ISO/IEC 19788 MLR: a new metadata standard for learning resources (in Spanish). IEEE-Rita 6m(3). http://rita.det.uvigo.es/201108/uploads/IEEERITA.2011.V6.N3.A8.pdf
24. Delgado (1995) The experience of the SOSENET project, communication as a prerequisite for the integration of technological resources. Communication as a prerequisite for the integration of technological resources. ISSA, Geneva, 1995. ISBN 92-843-3087-4
25. Delgado F, Hilera JR, Ruggia R (2012) Solutions for electronic exchange of social security information in an international context (in Spanish). El profesional de la Información. 2012, Julio-agosto.
26. Master Directory (2013) http://ec.europa.eu/employment_social/social-security-directory/welcome.seam?langId=en

# Chapter 10
# Using Process-Level Use Case Diagrams to Infer the Business Motivation Model with a RUP-Based Approach

**Carlos E. Salgado, Ricardo J. Machado, and Rita S.P. Maciel**

**Abstract** Building a business model for a software-based information system in ill-defined contexts is close to an impossible mission. Our recent work in eliciting and managing requirements from a process-perspective allows deriving a logical architecture but fails to provide a business and strategic view of the system. Adaptation of standard techniques to infer goals and requirements from scenarios and process-like diagrams, mapping backwardly the traditional business to process workflow, could help in building a business model and defining a strategy for the information system. We extend our V-Model approach with a Business Motivation Model representation, guided by a RUP-based backward transformation from process to business, so allowing for better and continuous alignment between Business and IS/IT, with improved traceability.

**Keywords** Requirements elicitation • Business process • Strategy • Business model • Business and IS/IT alignment

## 10.1 Introduction

Many studies state the importance of the business model and strategy definition in the early specification and during the evolution of a software-based information system [1]. Despite the many methods and techniques available to build them, it is not easy to do it in a clear way, due to a number of factors, especially in ill-defined contexts.

However, process-oriented approaches as the V-Model [2], allow eliciting requirements and obtaining a process-view of an information system in such

C.E. Salgado (✉) • R.J. Machado
Universidade do Minho, Guimarães, Portugal
e-mail: carlos.salgado@algoritmi.uminho.pt

R.S.P. Maciel
Universidade Federal da Bahia, Salvador, Brazil

M. José Escalona et al. (eds.), *Information System Development: Improving Enterprise Communication*, DOI 10.1007/978-3-319-07215-9_10,

contexts. Results normally consist in scenarios and processes that describe the expected behavior of the organization, the V-Model uses UML sequence diagrams and use cases, later derived into a logical architecture and product-level system information. Albeit the good results achieved, much information is lost by not uncovering the business vision and strategy information artifacts, so handicapping the obtained solution and its future evolution.

To overcome this insufficiency, some studies have been researching on the importance of integrating scenarios and business information, whether by deriving requirements models from organizational models by means of goal modeling [3] or by using organizational models to improve Use Cases development [4]. As usual, these studies follow a traditional development sequence, from business and requirements elicitation to process definition and detail.

On the other hand, opposite approaches that focus on the discovery of goals from scenarios, to directly help in the requirements engineering activity [5] or in the process of inferring formal specifications of goals and requirements from scenario descriptions [6], also have been explored for some time. In a more recent example, [7] proposed the validation of mappings that describe the realization of goals by activities (supported by models), considering tasks as goal-oriented requirements.

Relying on model driven methodologies to better handle requirements, the transformation from business to process can be performed backwardly, but as process information is not as rich as business information, other methods and techniques are required to help in filling some holes. So, we propose to combine the advantages of visual modeling with a well-established process, following the business modeling guidelines from the Rational Unified Process (RUP) [8] and the business plans representation of the Business Motivation Model (BMM) [9], in order to complete the business model and strategy definition for the referred information system.

Despite the promising results already obtained regarding the V-Model process modeling and alignment issues, there is still much room to improve in its associated requirements elicitation, and traceability management. Relevant contributions are expected from the strategy and business model topics, so, we propose to extend it with a BMM representation, guided by a RUP-based backward transformation from process to business.

Section 2 goes through the BMM and RUP main details, and also the V-Model topics. Section 3 presents our proposed transformation technique from the V-Model Sequence Diagrams (Scenarios) and Use Cases to a BMM Business Model, guided by backwardly applied RUP techniques. In Sect. 4, we discuss the application of the technique, its current status and work ahead. Section 5 concludes this paper.

## 10.2 Background

This section presents related background information regarding traditional approaches for business modeling, in the context of the design of an information system. First, the RUP framework, which has an entire discipline devoted to the topic of business modeling, detailing the steps involved in its definition. Next, the

BMM business plans representation, which stands as a growing reference in the area, and profits from its openness and flexibility.

Both approaches rely heavily on modeling and are standard-oriented, as modeling is crucial for the continuous alignment between business and IS/IT, and the improved traceability between both. With relation to BMM, it is already full modeled and UML compliant, while regarding RUP, there is already a solution modeling its textual artifact, the document for business vision, in UML [10], but needs some updates.

Also, we present the V-Model, a process-oriented approach that was able, in an ill-defined context, to derive a logical architecture of a system, and that we aim to extend, in order to provide a related business and strategic view.

### *10.2.1 Business Modeling with the Rational Unified Process*

The Rational Unified Process (RUP) is a comprehensive process framework that is iterative, use-case driven and architecture-centric. It has a dynamic structure, comprised of the phases inception, elaboration, construction, and transition, and a static structure, with disciplines such as business modeling, requirements, analysis and design, etc. With strong relationships with the Object Management Group (OMG), it integrates the use of the Unified Modeling Language (UML) and is rapidly becoming a de facto standard approach for developing software.

RUP supports the definition of requirements at multiple levels, with the goals (early requirements) being captured very early in the development lifecycle, during the Inception phase, as part of the Business Modeling discipline. Goals are subsequently refined and captured in additional models, functional goals are refined and captured in Use Case Diagrams and non-functional goals are refined and captured in the Special Requirements section of the Use Cases.

The main activity in the Business Modeling discipline is the Assess Business Status, which covers the work around assessing the status of the organization and setting the business modeling objectives. Between its delivered artifacts are the Business Vision, Business Goal and Business Rule documents.

In the Business Vision document, using natural text, the Product Features section describes the functional capabilities of the system, while the non-functional features reside in other specific sections. As the software development progresses, functional and non-functional goals are subsequently refined, now in a model and graphical-view, into Use Cases and in the special requirements of the Use Cases, respectively. Business rules are declarations of policies or conditions, and might apply always (invariants) or under a specific condition. Their purpose is to define specific constraints or invariants that must be satisfied by the business.

The second part of the Assess Business Status activity focuses on defining a candidate business architecture and in establishing Business Use Case realizations, linking Business Goals to Use Cases, aligning with the business strategy and goals, and establishing traceability relationships. Business Goals are seen as the requirements that the business must satisfy, directing the operation of the business mainly through the continuous improvement of business processes. Their purpose is to translate the business strategy into measurable, concrete steps, with which the business operations are aligned and steered in the right direction, with regular, iterative improvements. On the other hand, Business Use Cases (BUC) provide support to the Business Goal(s) and maintain the traditional relationship between business actors and use cases. Any given BUC should support one or more business goals and therefore support the strategic vision of the business. Also, through the BUC realization, there are a set of workers, entities, events, and rules, discovered and specified in an iterative process.

For better representing and visualizing these concepts, RUP has a component, a UML profile, which presents a UML language for capturing Business Models. The UML diagram presented in [11], acts as a guide to the profile, and demonstrates its important concepts and the relationships between them.

### *10.2.2 Business Plans with the Business Motivation Model*

In 2010, the Object Management Group's (OMG) Revision Task Force (RTF) completed its work on Version 1.1 of the Business Motivation Model (BMM) and published its updated specification [10], after in 2005 the BMM had become an adopted standard of the OMG, with the Finalization Task Force (FTF) completing its work in Sept. 2007.

BMM can be seen as a conceptual tool for engineering the business or as a tool for organizing and clarifying business plans; it provides a scheme or structure for developing, communicating, and managing business plans in an organized manner. Its constituents are the Ends and Means, its Influencers, and the Assessments made about the impacts of such Influencers on Ends and Means.

Conjunction of Ends and Means (being and doing) provide the core concepts of the Model. The term End refers broadly to any of the 'aspiration' concepts (Vision, Goal, Objective) and the term Means to any of the 'action plan' concepts (Mission, Strategy, Tactic). Assessments (usually in result of a SWOT analysis) of the impact of Influencers on the stated Ends and Means, connect with Internal Influencers (SW) and with External Influencers (OT), which are analyzed as a part of the business plan development.

As an Assessment identifies relevant Influencers in terms of their impact on Ends and Means, Directives (Business Policies and Business Rules) are put in place to govern and guide the enterprise Courses of Action. Directives keep the enterprise on course and moving toward its Desired Results, nevertheless, in contrast to Courses of Action, Directives cannot stand on their own. It is also possible for the Courses of Action to be formulated based on Directives, with the Directive serving as its source, and occasionally, a Directive can be defined to support the achievement of a Desired Result directly.

Business Rules sharpen the Business Tactics making Courses of Action concrete at the operational level, and also provide specific remedies when a Course of Action fails, as well as specific resolutions to conflicts. In its ways, Courses of Action are not Business Processes, but are realized by them.

Business Policies provide the basis for Business Rules and also govern Business Processes, by selecting from a variety of alternatives in response to one or more Assessments. That is, a Business Rule is a Directive, intended to govern, guide or influence business behavior, in support of a Business Policy that has been formulated in response to a SWOT. Often, a Business Rule is derived from Business Policy, and also, an Influencer may lead to the creation of a Business Policy through an Assessment that identifies some potential impact.

Regarding Ends, the Desired Results are composed of Goals, supported by Strategies, and Objectives, achieved by Tactics. The metrics for an Objective are established by the measures of performance of the Goal that the Objective quantifies. The Vision represents the enterprise's aspirations, i.e., what it tries to accomplish, what the organization wants to be or become. A Vision is supported by a Mission and amplified by Goals. Like a Vision, the Mission also indicates a correspondingly long-term approach and is not very specific, it is something broadly stated in terms of the overall functioning of the enterprise.

Designing business plans is not an easy exercise, and less an exact science, for instances, determining whether a statement is a Vision or a Goal, or a Course of Action is a Strategy or a Tactic, may be impossible without in-depth knowledge of the context and intent of the business planners. Also, an End does not include any indication of how it will be achieved, nor Means indicate the steps (business processes and workflow) necessary to exploit it or the responsibility for such tasks, only the capabilities that can be exploited to achieve the desired Ends.

Recursion and detail are present in BMM, there can be a 'parts explosion' of Desired Results, Courses of Action and Business Policies, but this connection should only be used to associate like instances, i.e., Goals only to other Goals, Strategies only to other Strategies, etc. Moreover, a Course of Action can provide basic support to make a former Course of Action viable; again, this connection should only be used to associate like instances.

The fundamental issue in BMM is motivation, why each element of the business plan exists and what is needed to achieve what the enterprise wishes. The basic idea is to develop a business model for the elements of the business plans before system design or technical development begins, so connecting system solutions to their business intent. Other concepts (Organization Unit, Business Process and Business Rule) have roles in the structure of BMM but belong in other OMG standards (respectively OSM, BPMN and SBVR), where they are defined and associated with related concepts needed for detailed business modeling.

### 10.2.3 The V-Model Approach

Recently, [2] presented a V-Model based approach (Fig. 10.1) to derive logical architectural models from a process-level perspective, instead of the traditional product-level perspective. This V-Model approach encompasses the initial definition

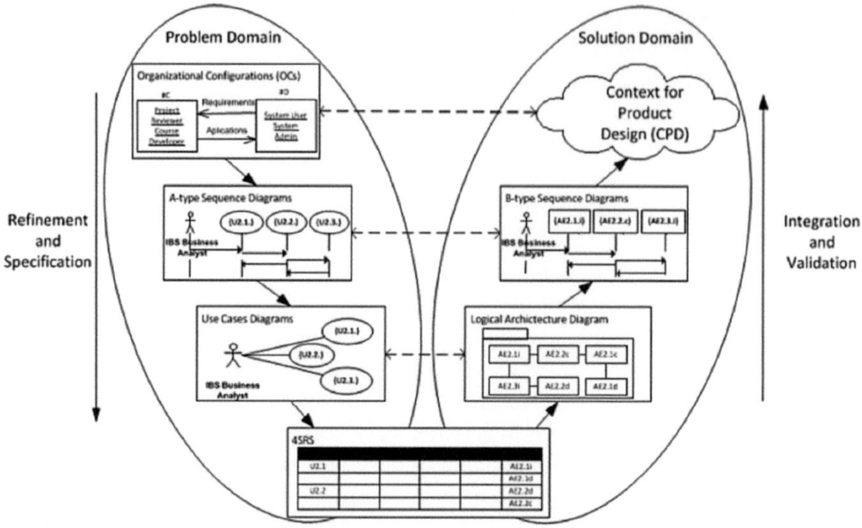


**Fig. 10.1** V-Model adaption for business and it-alignment [2]

of the project goals through organizational configurations, and the analysis and design of artifacts that result in a process-level perspective of the system's logical architecture, originating a validated process level structure and behavior architectural models that creates a context for eliciting requirements in a difficult, ill-defined context.

Instead of an architecture based on user requirements traditionally defined in a product-level perspective, the V-Model proposes the use of a process level perspective for the requirements definition and design of the logical model of the system architecture. It represents an IT solution by its logical architecture in order to achieve a structured view of the system functionalities, resulting from a process of transforming business-level and technological-level decisions, and requirements, into a representation (model) which is fundamental to analyze and validate a system but is not enough for achieving a full transformation of the requirements into a model able to implement business stakeholders' decisions. To achieve such representativeness, there are added artifacts that represent requirements at different levels of abstraction and promote an alignment between them and with the logical architecture.

The proposed V-Model approach is framed in the Analysis phase of the lifecycle model, according to traditional development processes. It uses Organizational Configurations (OC), A-type sequence diagrams and use case models, artifacts generated based on the rationale and in the information existing in previously defined artifacts, i.e., A-type sequence diagrams are based on OC and use case models are based on A-type sequence diagrams.

The OC model is a high-level representation of the activities (interactions) that exist between the business-level entities of a given domain, with each OC containing information on the performed activities, the participating professional profiles, and the exchange of information or artifacts. The set of interactions are based on business requirements and, in conjunction with the entities and the stakeholders, are represented with the intention of describing a feasible scenario that fulfills a business vision.

A-type sequence diagrams, a UML stereotyped sequence diagram representation, describe the interactions gathered in early analysis phase of system development, and are used afterwards as an elicitation technique for modeling the use cases. Next, the use cases are modeled and textually described, and used as input for the 4SRS. The execution of the 4SRS results in a logical architecture with a direct relation with the process-level use cases assured by the method's execution.

The process continues, with models placed on the left hand side of the path representation properly aligned with the models placed on the right side. The final resulting sets of transformations along the V-Model path provide artifacts properly aligned with the organization's business needs (formalized through OC).

## 10.3 Overview of Our Approach

Our approach for inferring business and strategy information from scenarios and process-like diagrams uses guidance from the RUP Business Modeling discipline, backwardly performed, and the business plans modeling representation capability of BMM.

Analyzing the RUP and BMM metamodels with regard to the three concepts of Business Use Case/Process, Business Goal and Business Rule, we first build a unified relation between them. Goals and Use Cases are directly connected in RUP, while Rules and Use Cases are connected through an Entity. In relation to BMM, Rules are directly connected to Processes, while Goals are connected to Processes through an Organization Unit. As at this phase we are not handling any Entity nor Organization Unit, our proposed relationship to unify those three concepts is depicted in Fig. 10.2.

Notwithstanding, this plain representation is insufficient for today's demanding needs, and for a sound and detailed business model. So, expanding the contribute from BMM and in a more business-like representation, we detail the previous schematics with Business Rule and Business Goal specific elements, maintaining the Business Use Case, Business Rule and Business Goal relation (Fig. 10.3).

With this representation in mind, the first step will be to perform the elicitation of Goals (Desired Results) and Rules (Courses of Action and Directives) for each root Use Case in the V-Model structure. This elicitation follows backwardly the RUP guidelines, starting from the V-Model's Sequence Diagrams and Use Cases scenarios information, to obtain the associated Goals and Rules.

According to the RUP guidelines there should be at least one Goal for each Use Case, with their classification being assisted by the use of the BSC technique [12]. Regarding Rules, the same process applies, but these being categorized by constraint and derivation types. BMM guidelines regarding the elicitation of Goals and Rules are also of great importance in this phase.

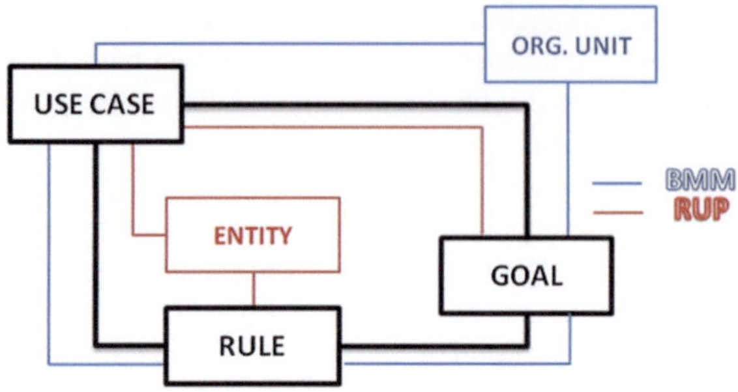


**Fig. 10.2** RUP and BMM unified relation for use cases, goals and rules

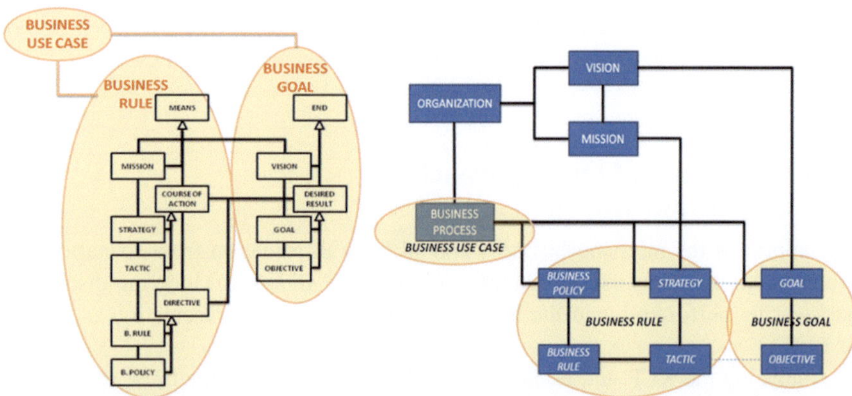


**Fig. 10.3** BMM business rule/business goal detail and the simplified business model

In a second step and going deeper into the leaf Use Cases of V-Model, the refinement of goals, including objectives at the lower leafs, and rules, or associated strategies, tactics and policies, will raise conflicts, new goals, reflexive changes upward, etc. [13]. This process initiates with a top-down approach, but soon intertwines with bottom-up techniques, using How and Why questions respectively to go up and down the Use Case/Goal tree (Fig. 10.4). The entire process should take several iterations until the business users are satisfied with the results.

Additional associated business strategy techniques and methods (SWOT and Porter's competitive analysis, among others), alongside the previous referred BMM guidelines, stakeholders knowledge contribute and specialists' heuristics [14], are welcome in this step to aid in building a well balanced, agile and realistic Goal/Rule hierarchy.

At this point, in this fourth step, we are ready to infer Vision and Mission statements for the information system, the higher elements of Ends and Means. This is done through abstraction of the Goals and Rules of the V-Model root Use Cases, following Why questions, and aided by the BSC classification technique and

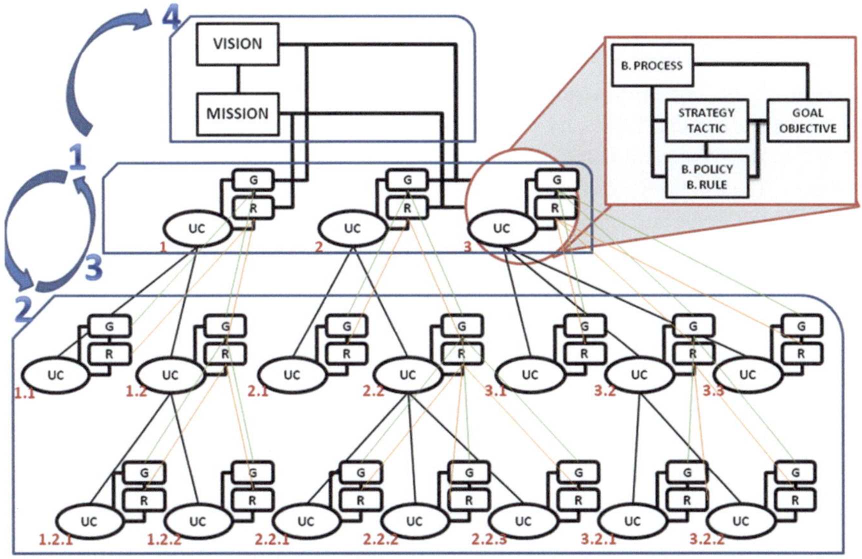


**Fig. 10.4** Use case tree with added goals and rules relations

information gathered from the RUP Business Vision guidelines. If pertinent, additional intermediate Rules and Goals can be defined to strengthen the connection between the high Vision and Mission statements and the intermediate root Use Cases Goals and Rules.

After this four-step process, a simplified business plan (according to Fig. 10.3) is ready to integrate the remainder of the V-Model process, with strong implications to the heart of it, the 4SRS. The entire dynamic of the 4SRS can be strengthened with the inclusion of the Goal and Rule elements, aiding and supporting the decisions taken within. Nevertheless, this is a thematic for future research and will not be discussed here.

In order to obtain a more complete business model, a fifth step is needed, besides uncovering the Ends and Means for the information system, there is a need to define the Assessments and Influencers. These BMM elements usually rely on a SWOT analysis to be elicited, but can additionally be guided by the backward execution of the RUP's business modeling related tasks, based on all the elements already available from previous steps.

This approach is not exhaustive nor are all steps mandatory for every project. Its aim is to stand as a flexible set of steps and techniques, for in conjunction with the V-Model process, to be tailored to each project, and according to each situation. Some might go deeper in the Goals and Rules derivation, others could emphasize the high-order statements, so taking different paths and making distinct choices.

Also, the number of iterations to perform, influenced by the time and resources available and influencing the depth and detail to obtain, and the techniques to use for elicitation, whether relying on simple heuristics or recurring to the user most preferred advanced method, are completely user-free to decide.

## 10.4 Discussion and Future Work

Currently, we are empirically evaluating this approach, applying the described guidelines to some projects, by starting from the already existing V-Model Sequence Diagrams (A-type) and Use Cases, reflecting those to RUP's BUC and then eliciting Business Goals and Rules for each one of them. Next, we iterate to infer higher and derive lower goals and rules, aggregating and hierarchyzing them, respectively, thus generating the strategy information, business vision and mission, to represent in the presented simplified BMM notation.

The RUP guidelines are not exhaustive, neither for the Business Goals and Rules elicitation nor for the Business Vision statements, being sometimes even dissonant, with some of the decisions having to rely on stakeholder contributions or specialists' heuristics. These are points to further explore in future enhancements and cross-disciplinary contributions.

The simplified BMM modeling representation supports all the information uncovered from the V-Model and generated from the RUP guidelines, contributing also with a few pertinent suggestions for elements elicitation and model completion. Until now no extension to the BMM metamodel was needed, as the connection between RUP and BMM is quite straightforward, with BMM showing a wider array of coverage on the modeled elements, while RUP still presents some natural language, document-spread information.

After the BMM representation is finished, next steps (4SRS) will recover the initial V-Model Sequence Diagrams and Use Cases scenarios and, together with the new business plan information structure, iterate through the 4SRS method, taking into account this new models in the decisions to be made inside it. Naturally, these new contributions from RUP and BMM will be integrated in the V-Model approach, strengthening it and broadening its scope.

As future work we intend to add the handling of Business Value related issues, already supported both by BMM and RUP, in order to close the cycle on the business modeling topic. Another point in consideration regards the error-prone situations of manually mapping from RUP to BMM and the benefits of a tool-based support for a larger community to be able to perform the steps described, with guidance and a similar language for different stakeholders. Added to this, a possible metamodel association, from RUP to BMM, recurring to the QVT/ATL model transformation would strengthen our proposed solution, all supported by the development of an Eclipse-based tool [15].

## 10.5 Conclusion

As business processes are increasingly becoming the focus of information systems, even as the primary choice for its requirement elicitation, they cannot be disconnected from the organizational environment.

Aiming to provide a business and strategic view of an information system, we adapt standard techniques to infer goals and requirements from scenarios and process-like diagrams, mapping backwardly the traditional business to process workflow.

The ongoing work on the extending of the V-Model approach with a BMM representation, guided by a RUP-based backward transformation, looks promising as an adequate approach for projects where business strategy is difficult to agree on.

Our pretension is to continue to empirically evaluate these solutions on real-world applications, in an industrial setting, to assess its scalability, practicality, and ease of use comparing to the previous results obtained. Association to other researchers and practitioners could improve and increase the maturity of current research, in order to enhance the V-Model alignment issues, all associated requirements elicitation and its traceability management.

**Acknowledgements** This project is financed with FEDER funds by the Programa Operacional Fatores de Competitividade—COMPETE and by national funds through FCT—Fundação para a Ciência e Tecnologia for the project: FCOMP-01-0124-FEDER-022674.

## References

1. Al-Debei MM, Avison D (2010) Developing a unified framework of the business model concept. Eur J Inf Syst 19:359–376
2. Ferreira N, Santos N, Machado RJ, Gasevic D (2013) Aligning Domain-Related models for creating context for software product design. Software quality. Increasing Value in Software and Systems Development. Springer, pp 168–190
3. González JD, la V, Díaz J (2007) Business process-driven requirements engineering: a goal-based approach. In: Proceedings of the 8th workshop on business process modeling, pp 1–9
4. Santander V, Castro J (2002) Deriving use cases from organizational modeling. In: Proceedings IEEE joint international conference on requirements engineering, 2002, pp 32–39
5. Rolland C, Souveyet C, Achour CB (1998) Guiding goal modeling using scenarios. IEEE Trans Softw Eng 24:1055–1071
6. Van Lamsweerde A, Willemet L (1998) Inferring declarative requirements specifications from operational scenarios. IEEE Trans Softw Eng 24:1089–1114
7. Gröner G, Asadi M, Mohabbati B, Gašević D, Silva Parreiras F, Bošković M (2012) Validation of user intentions in process models. Adv Inf Syst Eng 7328:366–381
8. IBM Rational Unified Process. http://en.wikipedia.org/wiki/IBM_Rational_Unified_Process
9. The Business Rules Group: the business motivation model, 2010
10. Cooper K, Abraham SP, Unnithan RS, Chung L, Courtney S (2006) Integrating visual goal models into the rational unified process. J Vis Lang Comput 17:551–583
11. Johnston S (2004) Rational UML profile for business modeling. IBM Rational Software

12. Kaplan RS, Norton DP (1996) Linking the balanced scorecard to strategy. Calif Manage Rev 39:53–79
13. Van Lamsweerde A (2008) Requirements engineering: from craft to discipline. In: Proceedings of the 16th ACM SIGSOFT international symposium on foundations of software engineering, pp 238–249
14. Gigerenzer G, Gaissmaier W (2011) Heuristic decision making. Annu Rev Psychol 62: 451–482
15. Decreus K, Poels G (2011) A goal-oriented requirements engineering method for business processes. Inf Syst Evol 72:29–43

# Chapter 11
# Towards Elimination of Testing Debt in Outsourced IS Development Projects: A View from the Client Side

**Michal Doležel**

**Abstract** Although issues connected with outsourced information systems (IS) projects as well as origins of software project failures have been a part of research literature for several years, software testing is still sometimes viewed as "ugly duckling" in the field of IS research. In this paper software testing is presented as a key part of IS development projects and as a potential source of project failures. From the perspective of client organizations, potential sources of testing debt and consequent project risks in an outsourced project environment are identified. Also, initial findings how to eliminate these problems are presented. Several key principles of recommended client IS development project management practice are proposed, as this paper generally advocates for better test and quality governance from the client side than current observed practices. Additionally, the role of cognitive and psychosocial factors is briefly introduced and some proposals for further research are presented.

**Keywords** Information systems development • Outsourcing • Vendor management • Testing • Test governance • Technical debt • Mum effect • Whistle-blowing

## 11.1 Introduction

Nowadays many organizations, for which software development is not their core business, tend to assign information systems (IS) development activities—partly or wholly—to external partners with the assumption of reaching more efficient and

M. Doležel (✉)
Department of Information Technologies, University of Economics, Prague, Winston Churchill Square 4, Prague, Czech Republic
e-mail: michal.dolezel@vse.cz

M. José Escalona et al. (eds.), *Information System Development: Improving Enterprise Communication*, DOI 10.1007/978-3-319-07215-9_11,

overall cost saving benefits in comparison with maintaining own software development departments. As business reality shows, the consequences of such an approach are not always smooth. New issues connected with proper management of external partnerships arise and new non-formal factors are approaching the scene—to mention only some: trust, commitment and mutual benefits are the ones which are often complicated to codify [1]. Although proper software testing is generally considered as one of key success factors in IS development projects [2], based on several observations it is likely to be one of the first activities, which will be cut down by the supplying vendor as soon as a strict project deadline is approaching fast, or when project economics is jeopardized from the vendor perspective. Although a postulate that vendors try to maximize profit while clients try to minimize costs is generally accepted, clear conclusions on how do to deal with these approaches respecting the "win-win" principle and how they exactly influence software project successes or failures are not yet clear [3].

Recently, the metaphor of "technical debt" has been formally introduced as an interesting issue for software engineering research [4], while in the industry praxis it has already positioned itself in the minds of development managers for some time. Since this fact highlights the technical debt in testing (testing debt) as an important and common problem of IS development projects when outsourcing is involved, this paper focuses on the role of client organization managers in such cases and present a possible approach to the test governance in multi-vendor environment in order to prevent/reduce the testing debt. Therefore, the main aim of this paper is to propose a framework enabling client-side governance over vendor testing activities, as I believe that such approach can effectively provide early warnings on troubled software projects and generally contributes to the prevention of IS development project failures.

## 11.2 Testing in Outsourced IS Development Projects

In the literature one can find an evidence that the choice of development subcontractors based on the lowest proposed price is quite common [5, 6], but from the viewpoint of produced software quality, the consequences of this approach are not always properly managed. Some of main symptoms in such cases are issues coupled with the situations when demanded project deadlines are not met. Consequently, there are already reported stories in research literature that effective management of time pressure and knowledge of factors contributing to pressure situations play a significant role in the success of the whole testing process [7]. Additionally, IS development managers in the industry often perceive many of "pressure situations" to be originated in a form of technical debt. The technical debt was originally understood as a metaphor that describes the situation when "long-term code quality is traded for short-term gain", but currently this metaphor is perceived more broadly, including many real-world "ugly" practises during IS development project lifecycles (architectural debt, testing debt, documentation debt etc.) [4]. For my research, testing debt is the key one.

Testing is divided into several levels with different responsibilities, namely component, integration, system and acceptance (International Software Testing Qualifications [8]). Although testing, especially system testing level in outsourced IS development projects, is in the industry praxis sometimes perceived as a sole responsibility of the vendor without the need of a continual "insider view" from the client managers perspective, I presume that this approach is generally incorrect and such information asymmetry [9] contributes to project escalation [10]. This conforms to the theory that a testing organization entity should be separated from a programming organization entity [11]. Additionally Galin [6] perceives the participation in testing as one of the Software Quality Assurance (SQA) tools when external participants are involved. He states that "character of participation in the testing process is sufficiently comprehensive to enable the contractor's representative to intervene, if necessary, to obtain assurance of the quality demanded of the supplied software and the expected timetable for completion of the testing (and correction) process." Another possible perspective is the one by Black [12] who applies a "due diligence" concept from the area of mergers and acquisitions to the vendor quality evaluation—i.e. looking at technology and process issues from the viewpoint of current vendor practice. However, as the proper vendor (subcontractor) management and the project control during project lifecycle is probably one of the most important factors contributing to software quality in outsourced environment, firstly I briefly focus on the organizational theory and practice in this field.

Ahonen et al. [13] describe interdepartmental model for testing organization, which is used in many non-software intensive businesses. Jain et al. [1] argue that "literature provides limited guidance on CMs' [client managers] project management responsibilities when projects are outsourced, whether vendor is located onshore or offshore." Sanz et al. [14] provide a conceptual proposal of a process model to create a test factory—an independent organization entity providing testing services, but lacking a clear distribution of responsibilities between the client and the vendor. Lu and Käkölä [15] similarly provide a life-cycle model for software testing based on similar initial assumptions as Sanz et al., but with a clear responsibility boundary. They present mainly vendor perspective, but also propose an active client engagement in the process. However, this approach is also focused on independent testing by different organization entity, and I argue that it does not always have to fit client needs when fixed-price IS projects are demanded, and a vendor is responsible for all the delivery aspects.

As testing processes are typically a part of IS development project realization, key findings from this area must be included. Software project management real-word approach (Australia) is described by McBride [16], focusing on the project monitoring, control and coordination. His work in conceptually linked with the classification of different coordination mechanisms adopted from other sources [17]. Unlike real-world praxis reported by McBride, the concept proposed by my research relays heavily on formal technical reviews as a part of project monitoring activities, since I see its role crucial in the outsourced environment. Additionally, to cope with specific nature of quality management activities, a separate role of a quality manager should be established in the IS projects to fulfil specific needs due to nature of this area. These tasks can hardly be performed by a project manager himself or herself, although the overall responsibility for project outputs (including quality) is still held by such a project manager. Additionally, a whole-organization approach is recommended for these activities—ideally with reporting to the chief quality manager/officer outside the project [18].

Based on similar philosophy, in organizations running multiple IS development activities (typically in the form of projects) handled by different teams, implementation of common and coherent principles to oversee testing activities in broader context must be done. I suggest a new term "test governance mechanism" in this context. As the proper definition of this term is not yet established in research literature, understanding of test governance as a high-level process for over sighting of test management activities ranging from the organizational level to the project level [19] is proposed. Implementation of a test governance mechanism in this sense includes establishing of a coherent test policy and test strategy, appropriate organizational structure, escalation mechanism, common reporting standards, testing process descriptions (e.g. test planning, preparation, execution, evaluation, control) and various guidelines as well as other rules of action. I also argue that a shared vision of "testing excellence" must be an essential part of every test governance mechanism. An existence of this mechanism will prove effective especially when a part of or all IS development projects are outsourced and the client organization needs to manage vendors effectively in the field of testing.

## 11.3 Cognitive and Psychosocial Factors

In the context of a proper client test governance process over outsourced IS development projects one can recognize the importance of several people-related aspects. These aspects must be taken into account, especially with an emphasis on the existence of information asymmetry client managers often experience. The importance of people factors for making software projects successful was explored by Chua [2]. I believe that, apart from factors presented in his work, which include stakeholders' involvement and team members' experience, one of the key factors for proper vendor management, linked to project success and information system quality, is a proper process of handling and communicating "bad news" from vendor site, either explicitly or implicitly. Effects of positive psychology based on Beck's cognitive theory were explored by Natovich et al. [20] in project management context—one of the conclusions of their study is confirmation that bad news are withheld by project managers in order to cope with stressful situation ("copying by avoidance"). Thus organizations are encouraged to create an open environment and to eliminate possible "shoot the messenger" syndrome.

Considering a slightly different perspective originated in social psychology, I perceive a huge potential for project failure explanations in cognitive dissonance theory suggested by Leon Festinger. His theory is based on a finding that a person holding two conflicting cognitions causing discomfort and/or frustration tries to eliminate it by changing existing cognitions or adding new ones [21]. Aside from few other forms of cognitive biases, which seem to be important for IS-related research [22], a major factor influencing communication of bad news through the higher levels of organization structure is widely recognized as the mum effect. This phenomena can be characterized as the situation when a person involved in the project, either from the vendor or client side, refuse to transmit unwelcome and negative information concerning project progress [23]. Such information includes, but is not limited to, facts about a project tendency either to completely fail, or to experience major delay, budget overdraw etc. In the opposite case, e.g. when a person decides to provide such information, this action is often referred to as whistle-blowing [24]. Recently, a significant influence of the culture on the mum effect in software outsourcing was described in the work [25], which is especially important for the client management of IS development when interacting with offshore vendors.

As the importance of whistler-blower dilemma for the presented topic should not be undervalued, an attention to Natovich et al. [20] must be paid—they say that study of this dilemma should focus on two different views: people taking a direct part of the project and being directly involved in its success or failure (a project manager, project team members etc.), and "accidental bystanders", referencing to the dilemma of IS auditors described in [24]. Not surprisingly, the relevance of the organizational culture and climate [26] should be also stressed out, especially in terms of the need for developing and communicating "quality culture" of software development [27] and its common principles as well as shared vision of testing excellence. Such principles should be understood as one of the IS development projects success factors, no matter whether development is insourced or outsourced. Moreover, mum effect itself can be viewed as a consequence of specific organization conditions and unhealthy climate heavily predetermining it [24].

## 11.4 Initial Findings

Action research has been used as a research fundamental. Action research has been advocated as a legitimate research method in IS research for several years and it is primary targeted "to solve current practical problems while expanding scientific knowledge" [28]. Control structure [29] of this research project has been informal and has involved collaborative initiation and authority of identity domination—presented interim results are based on author's personal involvement and direct intervention in multiple outsourced IS development projects. The project has been run in a large financial institution in the Czech Republic and involved five different IS development projects/programs. This paper summarizes outputs of Diagnosing and Action Planning phases [30] of this research project, which were carried out over a 6 month period from June 2012 to December 2012.

Following challenges and initial presumptions in a traditional user acceptance testing model performed by client organizations have been identified during diagnosing phase as caveats, which were consequently supported by several informal interviews with client IS development managers being responsible for vendor management in other three client organizations in the same industry segment:

**Challenge No. 1:** Many non-software intensive organizations simply cannot afford building of powerful dedicated testing teams because of their tight IT budgets. They partially or fully rely on their outsourced vendors in terms of software quality concerns, or alternatively, they must involve core-business experts (hereafter "domain testers") into the software testing process during user acceptance testing.

**Challenge No. 2:** Involving of domain testers is usually accompanied with multiple problems, e.g. impossibility to allocate them fulltime because of their regular tasks. This often results in jeopardising the project schedule when commitments and staffing are improperly managed, or cause incomplete user acceptance testing.

**Challenge No. 3:** Domain testers are not well-trained testing experts, their knowledge of testing is strictly based on (and is restricted to) domain expertise—for such reasons they can easily omit some important parts of functional testing. Moreover, this involvement does not assume solution of the need for special types of testing (e.g. performance testing, maintainability testing etc.).

**Challenge No. 4:** Rigorous and extensive software testing at the end of IS development lifecycle is resource inefficient for both involved parties (client and vendor) because it lacks efficient control towards project finalization on time; therefore testing must be handled properly already during system testing phase by the vendor—hand in hand with a sufficient client involvement and a necessary level of supervision from the client side.

**Challenge No. 5:** Omitting of vendor governance principles from the client side can lead to unpredictable results and "quality catastrophic" scenarios in the user acceptance phase of outsourced IS development projects when these projects go awry: some vendors seem to prefer short-term gains and tend to, partially or fully, follow client rules and standards, which may lead to many shortcomings when "nobody is watching". Frankly speaking, this can be understood as an adequate reaction to the pressure associated with the tendency to minimalize price in fixed-price deliveries, but should not be used as a general excuse.

**Challenge No. 6:** Client project managers in non-software intensive organizations often possess more general skills and competencies than their colleagues in core software business who seem to grow up from technical grounds—their (i.e. "non-software" project managers') knowledge of software development and quality management is definitely far less expectable. The tendency to fully rely only on information gained from the vendor seems to be fatal especially in case when a vendor project manager chooses to "keep the silence" about wrong project progress.

### *11.4.1 Sources of Testing Debt*

Testing debt can be generally characterized as a metaphorical expression of inconsistency between planned [31] or rationally expected test activities, and test activities actually performed. According to my research and with respect to generic test process phases (see below), this may especially include: (1) omission of proper test planning phase and stakeholders involvement, causing many serious flaws in concurring test process, (2) improper or inadequate test basis (requirements) analysis and test coverage during test design process, (3) improper or inadequate test execution strategy in the context of concrete IS development project—(a) inadequate number of test runs not respecting development dynamics, (b) ignorance of release management principles during testing, causing test results invalidity, (c) lack of regression testing, (d) defect closure without retesting, (e) undervaluation of severity of identified defects, (f) overall marginalizing of potential impact of lower-severity defects; (4) intentional affirmation of inadequately tested software as suitable for user acceptance testing, with the goal to transfer responsibility and costs of testing to the customer and/or avoid contractual penalties connected with late delivery.

I believe that all of the issues presented above are forming testing debt, with consequences to repay it later, typically during user acceptance testing performed by a client organization. Based on these finding I conclude that the control over sources of testing debt must be actively performed by client organizations in order to eliminate the necessity of later paying for the debt from client own funds. Additionally, negative consequences of testing debt will be furthermore multiplied when the mum effect and forms of cognitive biases take their toll. Such issues can be extremely painful for the clients, especially when dealing with new vendors, e.g. when expectations of both sides are not yet utterly clear.

### *11.4.2 Conceptual Framework*

In order to study and solve presented issues, following conceptual framework (Fig. 11.1) is proposed. Within context of a single project it demonstrates relationships between (1) influencing factors, (2) test process details, (3) test management and governance principles when vendor and client organizations are interacting. Generic test process (2) is based on understanding a test process as being a sequence of several activities (processes). It is conceptually inspired by test process described by Galin [6], Tian [32] and others. The border of responsibility between the client and the vendor is defined (3), hand in hand with typical actions and work items.

### *11.4.3 Guiding Principles*

Based on these findings, presented concept is based on five cornerstones:

1. Reduction of information asymmetry, early client involvement
2. Regular quality management reviews
3. Formal technical reviews serving as quality gates of the project

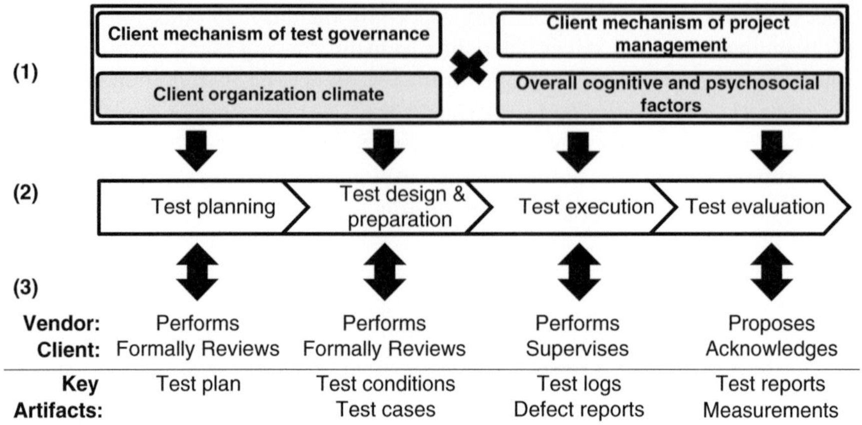


**Fig. 11.1** Conceptual framework (in project context)

4. Assignment of a competent client test/quality manager with test management experience and strong "customer advocacy attitude"
5. Alignment with "soft" factors of the client organization

The demand for reduction of information asymmetry and early client involvement (1) in this concept evocates conformity to consequential principles (2, 3, 4): management reviews (2) consist of organization of vendor-client test/quality status meetings on weekly basis, where key actors are invited: project managers of both client and vendor, vendor test manager, client test/quality manager, and in later phases also client business representatives. A regular set of measurement evaluation based on client submission is presented.

Importance of formal technical reviews (3) aimed at software testing artefacts provided by vendor should be considered as crucial in the whole process of quality governance during outsourced IS development project lifecycle. Formal technical reviews include, but are not limited to, reviewing of test planning documentation, test design documentation (test conditions, test scripts etc.) as well as output artefacts (test logs, test reports) used as a basis for client evaluation of completion of vendor test process and start of client user acceptance testing. The vendor is also expected to implement peer review process internally, as outputs of internal peer reviews should be checked upon request by the client.

With respect to the cornerstone No. 4, this paper promotes the role of an independent test/quality manager from the client organization (hereafter "**client T/Q manager**") (4) acting (apart from other aspects) as a client project manager mentor in the field of software testing and quality assurance. From the project management point of view, client T/Q managers can either (1) participate in project management as a client project co-manager, or (2) act in a more formal position like an SQA auditor. However the role of a client T/Q manager during IS development project greatly differs from standard test management goals and activities during software development projects—client T/Q managers are much more fixed to quality objectives, high-level quality monitoring and overall evaluation, as well as participation in IS development project coordination—in contrast with test planning, evaluation and test team management which are the tasks performed by the vendor test managers. At the same time, the presented approach does not preclude the possibility to make the client T/Q manager responsible for user acceptance test management and coordination—in this case, he or she occupies the "standard" role in the test management process—the only difference is the level/phase of testing. Based on my discoveries I strongly believe that the client T/Q manager who observes multiple IS development projects and who governs multiple vendors work (together with an individual client project manager) can greatly contribute to the solution of the whistler-blower dilemma of these managers, and, at the same time, helps to reduce information asymmetry between client and respective vendors. Typically, there must be also formal support for proposed principles—usually in the form of contract obligations to adhere client rules in the area of software development and testing.

Addressing people-related factors, client T/Q managers should spend some time with the vendor in order to communicate their vision of quality culture, since a shared vision in this area should be perceived as an important trust-supporting factor. Moreover, a new vendor is typically missing the long-term insider view of the client environment. Consequently, the lack of vendor quality culture observed later by client managers during testing or development process can definitely clash the trust between both parties. Both the client T/Q manager and the project manager should focus on possible influence of cognitive and psychosocial factors. This influence must be definitely not underestimated, as reduction of information asymmetry and proper detection of the mum effect symptoms play a crucial role in vendor management.

## 11.5 Discussion and Further Work

Although in recent years coverage of software testing in research literature seems to get better, a certain gap between researchers and industry practice still exists [11]. Historically, testing suffers from general underestimation as a qualified and distinct discipline [13]. Additionally, one may still lack research sources concerning effective testing and quality governance over IS development projects in outsourced environment. This article helps to bridge this gap. My research differs from the prior research in the area of software testing mainly in the way how I understand the need for stronger continual test governance over outsourced IS development projects as a proper solution for the reduction of testing debt. It also provides initial mapping of the importance of cognitive and psychosocial factors greatly influencing the software quality and IS projects success from the viewpoint of possible elimination/reduction by client T/Q managers.

This initial study has clearly some limitations and should be generally viewed as a proof of concept. Basically, only initial phases of action research are captured because the use of presented principles and the conceptual framework for purpose of this validation (after completion with further details) is currently being done in a real test governance process. Therefore, results based on complete action research cycle will be presented in a consequent paper when the validation is complete. Additionally, action research is vulnerable to the researcher bias and generalization of conclusions can be potentially problematic. In order to cope with this issue, discoveries from the action research will be supported by the case-study approach during further research of the phenomena across client organizations in a wider context. Finally, there should be especially noted that other quality assurance activities, such as requirements inspections and code walkthroughs, are not captured in this paper.

This paper addresses issues of both the theory and the praxis. From the viewpoint of the praxis, it allows project and T/Q managers in client organizations to apply similar principles in their work in order to improve vendor management practices and get more confidence in their outsourced IS development projects, taking into account that testing process is run properly and according to best practice in the field. From the viewpoint of the theory, this research contributes to the understanding of how important the governance processes are for a client organization during IS development project realization, because current research is typically addressing problems from the vendor perspective only. The principle of client T/Q manager participation in outsourced IS projects has been strongly advocated as an effective way to deal with some of the factors described in the professional literature as a challenge for IS development project management. I believe that together with the information asymmetry problem and the possible role of cognitive dissonance as well as other cognitive factors, this can be one of the dominant sources of project failures and should be further explored in the presented context. There should also be focus on providing explanations of the fact why client project managers often remain silent when they already have clear data demonstrating a problem, often based on continual poor test results provided by the vendor, but tend to be easily convinced by the vendor project managers that "everything is going fine". Consequently I also encourage to further study the whistle-blower dilemma from the point of view of T/Q managers in client organizations, since this area was only briefly introduced by this study as a conceptual problem, as well as the approach to client T/Q managers positioning and their perception by the project team members (participative versus "quality police" function), which is the dilemma presented likewise by Robey and Keil [24] for IS auditors.

## 11.6 Conclusion

This paper has argued for a better IS development projects coordination and monitoring by the client managers when the projects are outsourced outside the client organization. It has introduced metaphorical concept of testing debt and identified potential sources of this debt as a consequence of wrong vendor practices and insufficient coordination by client managers.

Consequently it clearly divides and describes responsibilities between both clients and vendors from the perspective of governance and project coordination activities performed by clients in the field of quality management. Client T/Q managers in client organizations are promoted as important players in client quality management of outsourced IS development projects. Conceptual framework for test governance is proposed as an interim result of an action research project, since existing research literature does not provide guidelines for activities in this field. This framework also stresses out the role of cognitive and psychosocial factors. I believe that knowledge of the role of these factors can greatly reduce number of failed outsourced IS development projects in client organizations. And finally, I hope that preliminary results presented in this paper may encourage further researchers to study these issues in similar context.

**Acknowledgments** This paper describes the outcome of a research that has been accomplished as a part of the Research program funded by Internal Grant Agency grant number IG406013.

## References

1. Jain RP, Poston RS, Simon JC (2011) An empirical investigation of client managers' responsibilities in managing offshore outsourcing of software-testing projects. IEEE Trans Eng Manag 58(4):743–757
2. Chua A (2009) Exhuming IT projects from their graves: an analysis of eight failure cases and their risk factors. J Comput Inf Syst 49(3):31–39
3. Savolainen P, Ahonen JJ, Richardson I (2012) Software development project success and failure from the supplier's perspective: a systematic literature review. Int J Proj Manag 30(4):458–469
4. Brown N, Ozkaya I, Sangwan R, Seaman C, Sullivan K, Zazworka N, Cai Y et al (2010) Managing technical debt in software-reliant systems. In: Proceedings of the FSE/SDP workshop on future of software engineering research—FoSER'10, p 47
5. Cusumano M (2008) Managing software development in globally distributed teams. Commun ACM 51(2):15–17
6. Galin D (2004) Software quality assurance. From theory to implementation. Pearson Education Limited, Harlow
7. Shah H, Sinha S, Harrold MJ (2011) Outsourced, offshored software-testing practice: vendor-side experiences. In: 2011 IEEE sixth international conference on global software engineering, pp 131–140
8. International Software Testing Qualifications Board (2012) Standard glossary of terms used in Software Testing v. 2.2. http://www.istqb.org/downloads/viewcategory/20.html
9. Keil M, Smith H, Pawlowski S, Jin L (2004) "Why didn't somebody tell me?": climate, information asymmetry, and bad news about troubled projects. Data Base Adv Inf Syst 35(2):65–81
10. Kataja A, Tuunanen T (2006) Information systems development methods and reducing information asymmetry: a way to decrease project escalation in outsourcing? In: Proceedings of the 14th European conference on information systems, ECIS 2006, pp 1–8
11. Mäntylä MV, Itkonen J, Iivonen J (2012) Who tested my software? Testing as an organizationally cross-cutting activity. Softw Qual J 20(1):145–172
12. Black R (2002) Managing the testing process: practical tools and techniques for managing hardware and software testing, 2nd edn. Wiley, Hoboken, NJ
13. Ahonen J, Junttila T, Sakkinen M (2004) Impacts of the organizational model on testing: three industrial cases. Empir Softw Eng 9(4):275–296

14. Sanz A, García J, Saldaña J, Amescua A (2009) A proposal of a process model to create a Test Factory. In: ICSE workshop on software quality 2009, Vancouver, pp 65–70
15. Lu Y, Käkölä T (2011) A dynamic life-cycle model for the provisioning of software testing services: experiences from a case study in the Chinese ICT sourcing market. In: 19th European conference on information systems, ECIS 2011
16. McBride T (2008) The mechanisms of project management of software development. J Syst Softw 81(12):2386–2395
17. Sabherwal R (2003) The evolution of coordination in outsourced software development projects: a comparison of client and vendor perspectives. Inf Organization 13(3):153–202
18. Schulmeyer GG (2008) Handbook of software quality assurance, 4th edn. Artech House, Boston, MA
19. Charankar M (2011) Achieving better test governance. Pract PM J (9). http://practicalpmjournal.com/2011/09/achieving-test-governance/
20. Natovich J, Natovich R, Derzy Z (2011) Withholding bad news in information technology projects: the effect of positive psychology. In: PACIS 2011—15th Pacific Asia conference on information systems: quality research in Pacific
21. Festinger L (1957) A theory of cognitive dissonance. Stanford University Press, Stanford, CA
22. Rouse A, Corbitt B (2007) Understanding information systems outsourcing success and risks through the lens of cognitive biases. In: Proceedings of the 15th European conference on information systems, ECIS 2007, pp 1167–1178
23. Park C, Im G, Keil M (2008) Overcoming the mum effect in IT project reporting: impacts of fault responsibility and time urgency. J Assoc Inf Syst 9(7):409–431
24. Robey D, Keil M (2001) Blowing the whistle on troubled software projects. Commun ACM 44(4):87–93
25. Ramingwong S, Sajeev A (2007) Offshore outsourcing: the risk of keeping mum. Commun ACM 50(8):101–103
26. Schein EH (2010) Organizational culture and leadership, 4th edn. Wiley, Hoboken, NJ
27. Sommerville I (2007) Software engineering, 8th edn. Addison Wesley, Harlow
28. Baskerville R, Myers M (2004) Special issue on action research in information systems: making IS research relevant to practice—foreword. MIS Q 28(3):329–335
29. Avison D, Baskerville R, Myers M (2001) Controlling action research projects. Inf Technol People 14(1):28–45
30. Susman G, Evered R (1978) An assessment of the scientific merits of action research. Adm Sci Q 23(4):582–603
31. Guo Y, Seaman C (2011) A portfolio approach to technical debt management. In: MTD'11 Proceedings of the 2nd workshop on managing technical debt, pp 31–34
32. Tian J (2005) Software quality engineering. Wiley, Hoboken, NJ

# Chapter 12
# From Traditional Practices to a Learning Community: A Knowledge Management Approach to Support Learning in Food Hazard Identification

**Henry Linger, Frada Burstein, and Dora Constanidis**

**Abstract** Active knowledge creation and sharing about food safety hazards can be better fostered within a learning community. This paper proposes an innovative analysis and design framework for supporting knowledge work in a learning community. The framework was informed and derived from an investigation of knowledge creation and sharing practices about food hazard risks specific to the dairy industry. This paper presents a case study illustrating how traditional knowledge sharing practices were transformed through the application of a knowledge management framework. The knowledge management framework enabled the formation of a learning community that in turn facilitated improved integration of disparate instances of knowledge sources and fostered new connections between stakeholders for efficient food safety knowledge flow.

**Keywords** Knowledge management • Learning communities • Knowledge sharing • Food safety management

## 12.1 Introduction: Issues and Insights

Food safety management concerns diverse actors across the industry supply chain including growers/producers, manufacturers, and retailers, as well as health, regulatory, marketing and export authorities. Hazard identification is a critical knowledge capability that facilitates a proactive approach towards safety issues especially in the food industry. In this industry hazards refers to the food-borne health and safety risks. Given the recent food scandals within the industry, predominantly in the EU [1], the early identification of food safety hazards provides stakeholders with an

H. Linger (✉) • F. Burstein • D. Constanidis
Faculty of Information Technology, Monash University, Caaulfield East, VIC, Australia
e-mail: henry.linger@monash.edu

M. José Escalona et al. (eds.), *Information System Development: Improving Enterprise Communication*, DOI 10.1007/978-3-319-07215-9_12,

opportunity to develop procedures that will effectively manage a situation in the event that it should occur. Managing knowledge in a way that maximises the capability to identify potential issues leading to new hazards being discovered will enhance an industry's preparedness for managing future food-borne risks.

This paper describes a Knowledge Management (KM) framework that was used as the lens to study and propose appropriate support of the process of identifying and documenting potential food safety hazards by industry actors. The application of this framework in our case study involved the design of a KM System (KMS) as the basis of a transition from traditional knowledge sharing practices into a learning community (LC). The KMS strengthened existing relationships between industry actors and enhanced the level of trust between all actors. On the basis of these trust relations, the actors were able to transition into a learning community and build the industry's capability for hazard identification.

Our case study was facilitated by the involvement of significant industry actors, including regulators, researchers, producers as well as those involved in information management. The food industry is engaged in building IT-supported systems for managing hazards information. However, our study case extended this effort by facilitated better engagement in knowledge sharing and the formation of a learning community within the industry.

## 12.2 Background: Why a Learning Community?

Originating from organisational learning theories, LCs were proposed to enhance learning as a formal organisation-sponsored approach that engendered a high level of trust amongst the participants [2]. Although the LC concept has been widely adapted [3], a major challenges is to design appropriate mechanism within the LC that enable knowledge creation. Khoo and Cowie [4] propose that the social and emotional ties of members within a LC need to be fostered in order for the aims of the LC to be successful. Exactly how these more tacit aspects of human relationships within a LC are to be fostered still remains a challenge.

In our paper we examines the role knowledge management can play to foster greater understanding and trust between diverse knowledge workers within a LC that may lead to stronger social ties between its members. How members of a LC participate and interact with each other will influence how the degree to which knowledge is created and shared within the LC. Increased knowledge sharing also means that members' expertise is deepened and potentially their knowledge domain changes [5]. In this paper we apply a KM framework to support these knowledge practices.

In the food industry, there is a growing realisation that responsibility for food safety management needs to move from an adversarial system, between the regulator and the producers and manufacturers, into a complex network of co-operative actors, even when stakeholders have competing organisational objectives. Collective responsibility requires knowledge to be shared effectively in order for the actors to

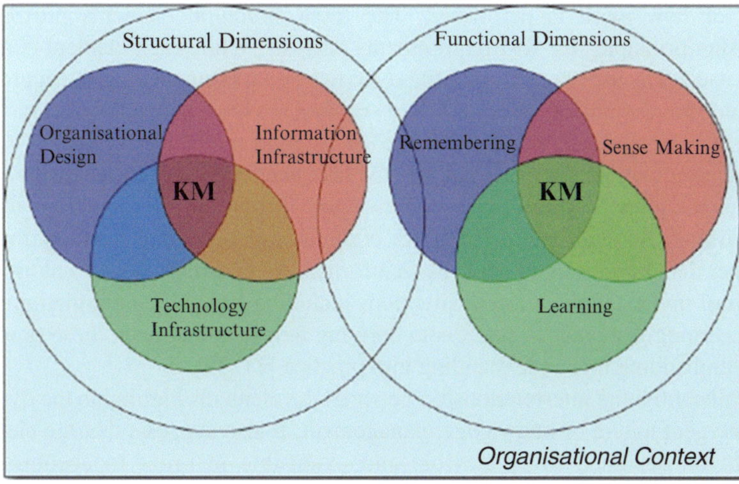


**Fig. 12.1** A framework for KMS: structural-functional co-dependent perspectives of KM

perform their roles and to discharge their organisational responsibilities. When actors create and share knowledge to assist their work tasks there is a need for them to engage in more meaningful and reciprocal ways to forge closer knowledge relationships with each other. This paper argues that there is a need to establish a LC to facilitate active knowledge sharing to identifying food safety risks.

### 12.2.1 Analytical Lens: Task-Based Knowledge Management (TbKM)

The case study adopted the *Task-based Knowledge Management (TbKM)* framework [6] to investigate existing knowledge sharing practices within the Dairy Industry. *TbKM* focuses on articulating knowledge that underlies the ability to perform a specific work task [6]. TbKM adopts a "bottom-up" perspective, defining the work that must be performed at conceptual and pragmatic levels [7]. In this context, a KMS is the implementation of a TbKM.

A KMS requires conceptualisation of the structural elements that define the task and the functions that together relate to the knowledge practices that constitute knowledge work [8]. This conceptualisation of a work task allows stakeholders to articulate some (but not all) of the knowledge and understanding that underlies their ability to perform their work. This conceptualisation of knowledge work is depicted in Fig. 12.1 as a framework for KMS.

The structure of a knowledge management task is conceptualised as the intersection of organisational design and information and technology infrastructures. Organisational design focuses on how work is organised rather than the procedural aspect of how work is performed. The three structural elements provide the organising principles for KM requirements analysis. While each element is important in itself, it is the interaction of these elements that is equally significant to fully determine the design features of KMS to ensure it fits the organisational context and has an architecture that reflects available information and technology.

Concurrent with the structure of the KM task, the functions of KM are conceptualised as the intersection of three cognitive activities; remembering, sense making and learning. Remembering, facilitates complex organisational memory function [9] rather than considering its role as an information repository. Sense making is the process of understanding current situations including reconciling conflicting issues [10]. Learning is a creative process to deriving new meaning from these processes and re-embedding this understanding into practice [11, 12].

Significantly, the interrelationship between the elements highlights the dynamic and emergent nature of knowledge management. It also suggests that the elements need to be designed to be flexible and extensible in order to accommodate changes and evolution while maintaining historical integrity of the tasks supported by the KMS.

### *12.2.2 The Importance of KMS to Learning Communities*

Within a LC, the KMS is a shared resource that supports the actors in their ability to perform their task. The role of a KMS is to support collaboration between the actors in their effort to develop a shared understanding of the task, make sense of new information and learn from past instances of the task. The KMS also aims to meet the needs of the community as a network by addressing the diverse needs of different roles within the community as well as the needs of individual actors. The KMS *emergent*, in that it is not imposed by an outside authority [13], rather it is the LC that defines the resource and is responsible for its evolution in terms of content, organisation and functionality. In our case study, the KMS facilitates the identification, evaluation and communication of food safety issues and the management of hazards raised by these issues. The KMS is an organisational resource that allows actors to structure and document knowledge of what they have learnt from managing a hazard as a record of past instances of the hazard.

## 12.3 The Dairy Industry: An Illustrative Case Study

The case study data was collected through a series of interviews, focus groups and workshops with stakeholders representing various functions in the diary industry. The major source of data appeared to come from Dairy-A (DA), an industry owned service organisation representing dairy producers and manufacturers interests across the country. DA's Technical Issues Group (TIG) was responsible for coordinating the management of food safety issues, acting as the "gatekeeper for the dairy industry issues management capability". This capability contributed significantly to the industry's viability and profitability. As part of its activities, TIG:

- produced an Issues Reference Manual (IRM) which presents a comprehensive documentation of existing issues and the way they need to be managed
- is responsible for collecting and collating relevant information, issuing alerts and generally informing relevant actors
- is implementing an extranet as a tool for monitoring existing issues and improving communication among its constituent members.

## *12.3.1 Structural Elements of the KMS*

### 12.3.1.1 Organisational Design

A proactive approach to hazard identification requires systematic and collaborative activities that are organised as discreet and defined processes. Information gathering and environmental monitoring can be undertaken by the whole community, but a designated group of actors within the community are assigned responsibility to perform this activity on a regular basis as part of their role [14]. In the proposed framework, characterisation and assessment of new information is undertaken regularly by a defined group of actors. Each situation will follow a general approach for that type of issue, but there needs to be flexibility to allow the actors involved to adapt the process to the needs of the situation [15]. Within a LC, such reflection is an explicit process that is done collectively by the community [16] and actors choose to be included rather than being nominated by higher authority [14, 16]. This contrasts with traditional practices where such activity is largely individual and implicit in other activities.

To maintain its issues management capability, TIG established a Reference Group (TIRG) to provide the necessary strategic input for its work program, part of which relates to the food safety issues capability. The main role for the Reference Group was as a consultative forum that allows DA to engage with other industry stakeholders and provide them with an opportunity to strategically input into the issues management process. For DA the forum is an important aspect of the industry network as it allows TIG to tap into the expertise of both the dairy and food industry stakeholders, including government, regulators and technical bodies. The TIG also commissions research and instigates investigations to support its work in maintaining the dairy industry's issues management framework. This is an important form of information gathering and serves to bring researchers into the industry network.

### 12.3.1.2 Information Infrastructure

An information infrastructure situates new information within a relevant context [13, 17]. Hazard identification is essentially the ability to recognise that a particular situation now represents a hazard in terms of food safety. Thus information about a situation is transformed into knowledge that identifies a hazard. This transformation relies on pertinent food safety issues to be articulated and documented.

The IRM contains information regarding existing well-understood food safety issues. It is wide-ranging, standardised documentation of issues that provides the industry with a very concise, informative and useful resource. Internally within DA, the environmental monitoring and authoring of Issues Notes is the responsibility of the Issues Management Group (IMG) under the direction of TIG. The IMG also prioritises issues for inclusion in the IRM based on what they have learned from monitoring the environment and the feedback from IRM holders that evaluate existing issues. Since TIG has sole responsibility for the IRM, it controls the distribution of the IRM to a broad but restricted membership.

#### 12.3.1.3 Technology Infrastructure

The development of a KMS as a shared resource within a LC is dependent on the creative deployment of information and communications technology (ICT) [8, 13, 18]. Much of the discussion of processes and information infrastructure is predicated on a sophisticated ICT infrastructure. Thus, for example, information sharing relies on a central structure that is visible and accessible to all actors. However, technology and functionality needs to be incrementally deployed so as to meet the actual work practices within the community [18].

An extranet commissioned by TIG is designed to complement the IRM processes by providing the industry with better direct access to trusted DA information. It would allow a larger audience to access this information and, by encouraging more interaction, to build a consistent industry view of food safety issues. It is also recognised that the extranet can provides additional functionality, such as access to archival information, and facilitate better interaction between the industry actors and DA with functionality like a discussion forum and access to key experts within DA, the industry and academia. The development of a technology infrastructure, based on the extranet, provides opportunities not available with the existing manual paper-based IRM. The interaction between DA and industry actors is strengthened by the introduction of this new communications channel. But even more significant in the longer term is the intention to record the results of actions, in the first instance providing information about action taken since a previous alert.

### *12.3.2 Functional Elements of the KMS*

#### 12.3.2.1 Remembering

From a KM perspective an organisational memory is a facility that enables actors to make sense and learn from the experience of past episodes of the task [19]. This implies that the memory represent a comprehensive record of that experience [9, 12].

However this only represents aspects of knowledge that can be documented and do not reveal anything of the knowledge required to resolve the situation.

Issues and alerts are treated as episodes of the hazard and, with the extranet, these episodes are being stored in an accessible way. Such a repository can be considered a kernel of an organisational memory. The utility of this repository is somewhat dependent on the technology that supports it. The main issue is the organisation, structure and storage mechanism that facilitate interaction with the content in order to understand and make sense of past episodes by actors who were not directly involved in those episodes. This is the mechanism by which actors use past experience to inform current situations.

#### 12.3.2.2 Learning

Actors must be knowledgeable about their task and learning is an explicit process in the overall performance of the task [6]. To this end, the KMS includes an organisational memory to support reflective practice and single and double loop learning [11]. Reflection-on-practice involves reviewing past episodes of the task with the objective to discover patterns that will change the shared understanding of the task by all actors. Reflective-inpractice applies while performing a task in order to understand the specific circumstances of the episode and adapt knowledge to resolve the situation. Both modes of reflection are facilitated by organisational memory [2].

The organisational processes and the information infrastructure described above provide TIG with a necessary basis to engage in organisational learning. The feedback from IRM, regular revisions of Issues Notes (at least every 2 years), and environmental monitoring triggers learning cycles of issues management that in turn overcomes problems that arise in current practice. The proposed extranet archive provides valuable information to inform and support this learning process. However, the IRM amendment process does not currently seem to explicitly record the context and reasoning that inform revisions, thus omitting an important resource for reviewing existing practices and double loop learning.

#### 12.3.2.3 Sense Making

In order to work effectively, actors need to make sense of the task. This involves gaining an understanding of the task and its context in order to develop a mental model to represent it cognitively [20]. However, with hazard identification there is variety and diversity in the task itself and/or its context. In order to deal with this diversity, each episode of the task needs to be understood and compared against a mental model in order for the actor to act appropriately. Understanding the task, the problem domain and the context is an essential element of knowledge work. From the point of view of the KMS, sense making contributes to the conceptual components of the organisational memory. Aspects of the mental model can be represented as conceptual models that situate an episode of the task within a broad problem domain.

This ability to situate the episode helps the actor decide how to resolve the situation. This also enables actors to recognise situations that are outside their expertise [20, 21].

The process of sense making is generally individualistic as it allows actors to understand the issue from their perspective. This is combined with collaborative processes that construct a shared collective understanding of the issues. In the collaborative environment of the LC, actors are involved in deriving a shared understanding. This process requires actors to articulate aspects of their own mental models and adjust these models to accommodate other viewpoints. This results in a negotiated understanding in which all actors have a vested interest. This shared understanding can also be represented as a conceptual model and stored in the organisational memory. This highlights the need for the KMS to support multiple viewpoints so that actors have access to both the shared community understanding and their own variations [22].

The TIG and other working groups provide forums where individual perspectives can be reconciled, with the resultant view presented in the IRM as the industry view. The TIRG and other working groups allow selected actors to collaborate to make sense of a new situation. These groups address a specific area, have a strong strategic emphasis and use the individual members' expertise and understanding of food safety.

## 12.4 Discussion: Creating a Learning Community for Food Hazard Identification

Food safety management concerns diverse actors across the supply chain but the community can be grouped as three cohorts of stakeholders: (1) **Research:** those involved in research and development; (2) **Practice:** the industry (growers/producers, manufacturers, retailers, etc.); and (3) **Regulatory Authorities:** the regulators, legislators and standards bodies. These cohorts and their interactions are represented schematically in Fig. 12.2a below. In our case study the model is used to represent hazard identification, as the front end of food safety management, as a specialised view of the industry. This emphasises that food safety is the context in which hazard identification is considered. As each cohort represents a variety of actors, there is also a need for these actors to engage with each other. Thus communications need to be supported within and between these cohorts.

The model of the industry, depicted in Fig. 12.2a, represents the structured interaction and feedback between the cohorts. Food safety is one aspect of the industry that relies on the cooperation between actors and stakeholder groups as they share a common interest in maintaining public confidence in the food industry for the industry's long-term future. This cooperation is sustained through various forums, such as committees and workgroups, and the formal and structured communication that is exchanged by the stakeholder groups and actors within each group. However, the ability of the industry to function effectively also relies on informal industry networks. Although very important and significant, these networks are mostly invisible

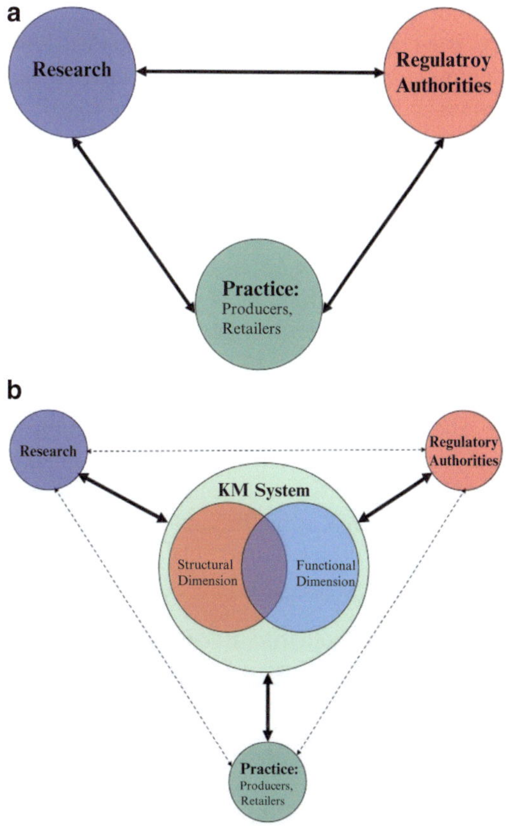


**Fig. 12.2** (**a**) Traditional knowledge sharing practices within a practice-based industry. (**b**) Knowledge management practices within a learning community

in terms of how food safety is documented by the industry. Activities, both structured and unstructured, relate to the broad and diverse areas of interest to the food industry, and include local, national and international forums. By their very nature, such networks are dependent on personalities and the trust that exists between the actors. A striking feature of these trust relationships is the importance of, and pride taken in, these relationships by each actor [23] that extends beyond food safety into other areas of the food industry.

Analysis of current network between actors and stakeholder groups confirmed that a significant proportion of relationships coordinated by DA are informal and are not explicitly documented. The major emphasis of our study was not to overtly change food safety practices, but to make explicit those activities that are central to the effectiveness of this enterprise. The innovation of this study lies in describing how those activities can be organised and documented and how this explicit knowledge can contribute to food safety management when treated as LC.

The proposed innovation was in the form of re-designing the interactions between stakeholder groups through a KMS to support LC needs. The fundamental feature of the LC model is that actors participate in the construction of a shared resource that forms the basis of their informational interaction as a LC. In this approach, the LC is engaged in a collaborative enterprise that is defined by a specific task. The definition of the task identifies the participants, their role within the task, task outcomes and, where appropriate, methods used to perform the task. Figure 12.2 represent the transition of existing knowledge sharing practices (Fig. 12.2a) into a LC with the introduction of an integrated KMS (Fig. 12.2b). The significant difference in Fig. 12.2b is that personal interaction between cohorts is maintained but the formal informational interaction is through the KMS.

The model can be applied across a range of issues and at different levels of detail focused on a specific task. In a broad area like food safety, it is possible for multiple communities to interact through a common KMS. The KMS can be partitioned to support the perspective of each community but the common structures allow knowledge to be shared between communities [24].

The central KMS component shown in Fig. 12.2b is comprised of the structural and functional elements that need to be implemented to support a LC. Interpreting and adopting the six co-dependent elements of Fig. 12.1 ensures that they are appropriate for the task of the LC and hence determine the success of the KMS.

## 12.5 Conclusions: From Traditional Practices to a Learning Community

The initiatives undertaken by TIG within DA illustrate how organisations can initiate changes that could transform their organisations by taking steps towards creating a LC. DA had focused on organisational design and information infrastructure, and has begun to implement a technology infrastructure to support their operations. It is obvious that they have made substantial progress in moving towards a LC although other initiatives would further assist in building on existing changes to progress their transition. An overview of DA's current and evolving practices were discussed above. This discussion provides a practical example of the contrasts between the traditional organisation of knowledge sharing in the food industry and the LC model.

What is apparent from the discussion is that the traditional and LC models are not alternative approaches, but represent transition points on a path toward organisational change. As is shown by the discussion, most sectors of the food industry have developed strategies to deal with their changing environments and the specific requirements of their industry sector. The models can be used to analyse the sector to define the organisational changes required to transform an industry sector into a LC as a means to achieve defined objectives and to see how its operations map onto this transition.

The models are not prescriptive but provide a framework within which current practices can be mapped or changes identified and planned. This allows for a lot of flexibility in terms of how the concepts are applied in practice. In terms of food safety, the major driver for change is the need to take a proactive risk approach in order to predict potential and emerging food safety issues. The LC model as based on a task-based knowledge management approach that focuses on a specific task (hazard identification) and is concerned with revealing the implicit activities that underpin effective performance of the task, making learning an explicit activity within the task and supporting knowledge sharing and collaboration by participants. The knowledge management approach assumes that support for reasoning and reflection will enable participants to be proactive in terms of their ability to recognise and effectively respond to food safety hazards.

**Acknowledgments** This research was supported by a grant from the Australian Food Safety Centre of Excellence under the Commonwealth Government's National Food Industry Strategy program. We gratefully acknowledge Tom Lewis and Doug Eddy for their support and contributions.

## References

1. Dunmore C, Croft A (2013) Horsemeat scandal set to spur tougher EU food tests. REUTERS. http://www.reuters.com/article/2013/02/13/us-eu-horsemeatidUSBRE9 1C0LH20130213. Accessed 15 April 2003
2. Senge PM (1990) The fifth discipline. Doubleday Currency, New York, NY
3. Hord SM (ed) (2004) Learning together, leading together: changing schools through professional learning communities. Teachers College Press, New York, NY
4. Khoo E, Cowie B (2011) A framework for developing and implementing an online learning community. J Open Flexible Distance Learn 15(1):47–59
5. Lave J, Wenger E (1991) Situated learning. Legitimate peripheral participation. Cambridge University Press, Cambridge
6. Burstein F, Linger H (2003) Supporting post-Fordist work practices: a knowledge management framework for supporting knowledge work. J Inf Technol People 16(3):289–305
7. Burstein F, Linger H (2011) Task-based knowledge management approach. In: Schwartz D, Te'eni D (eds) Encyclopedia of knowledge management, 2nd edn. IGI Publishing, Hershey, PA, pp 1479–1489
8. Davenport T (2005) Thinking for a living: how to get better performance and results from knowledge workers. Harvard Business School Press, Boston, MA
9. Walsh JP, Ungson GR (1991) Organizational memory. Acad Manage Rev 16(1):57–91
10. Weick KE (1995) Sensemaking in organizations. Sage Publications, Thousand Oaks, CA
11. Argyris C, Schön DA (1978) Organizational learning: a theory of action perspective. Addison-Wesley, Reading, MA
12. Spender JC (1996) Organizational knowledge, learning and memory: three concepts in search of a theory. J Organ Change Manage 9(1):63–78
13. Tiwana A (2002) The knowledge management toolkit : orchestrating it, strategy, and knowledge platforms, 2nd edn. Prentice Hall PTR, Upper Saddle River, NJ
14. Wenger EC, McDermott R, Snyder WM (2002) Cultivating communities of practice: a guide to managing knowledge. Harvard Business School Press, Boston, MA
15. Dixon NM (2000) Common knowledge; how companies thrive by sharing what they know. Harvard Business School Press, Boston, MA
16. Kimble C, Hildreth PM (2004) Knowledge networks: innovation through communities of practice. Idea Group Pub, Hershey, PA
17. Pfeffer J, Sutton RI (1999) The knowing-doing gap; how smart companies turn knowledge into action. Harvard Business School Press, Boston, MA

18. Standards Australia (2005) AS 5037-2005 knowledge management—a guide, 2nd edn. Standards Australia, Sydney
19. Cross R, Baird L (2000) Technology is not enough: improving performance by building organizational memory. Sloan Manage Rev 41(3):69–78
20. Martin J, Feldman MS, Hatch MJ, Sitkin SB (1983) The Uniqueness paradox in organizational stories. Admin Sci Q 28:438–453
21. Denning S (2001) The springboard; how storytelling ignites action in knowledge-era organisations. Butterworth-Heinemann, Woburn, MA
22. Collins HM (1997) Human machines and the structure of knowledge. In: Ruggles R (ed) Knowledge management tools. Butterworth Heinemann, Newton, MA, pp 145–164
23. Prusak L, Cross R (2002) The people who make organizations go-or stop. Harv Bus Rev 80(6):104–112
24. Alavi M, Leidner DE (1999) Knowledge management systems: issues, challenges and benefits. Commun Assoc Inf Syst 1:Article 7
25. Srinivasan R (2004) Knowledge architectures for cultural narratives. J Knowl Manage 8(4):65–74

# Chapter 13
# Coordinating the Enterprise Modelling Effort by Means of an Entrepreneurial Goal Hierarchy

**Sebastian Bittmann and Oliver Thomas**

**Abstract** Enterprise modelling can be generally used as an approach for developing solutions for specific problems that are faced by an enterprise. However, managing such solutions is an extensive problem as it has to be ensured that a solution will be maintained and eventually excluded from the set of available and applied solutions. Additionally, it has to be ensured that enterprise modelling will be accordingly guided, as it is required to develop solutions that perfectly suit the company. Hence, there is a tremendous need for supporting the lifecycle of an enterprise model. In this paper an enterprise model will be regarded as any kind of model that supports the respective enterprise in achieving its business goals, regardless if it depicts a business process or the organisational structure. During this paper it will be shown, how goal modelling as a special type of an enterprise model can take a central role in supporting the modelling effort and its coordination. Furthermore the concept of an Entrepreneurial Goal Hierarchy will be introduced that will be used for coordinating the whole process of enterprise modelling executed in an enterprise.

**Keywords** Enterprise modelling • Enterprise planning • Goal modelling • Entrepreneurial goal hierarchy • Enterprise model design

## 13.1 Introduction

More specific than conceptual modelling, enterprise modelling is considered with conceptualisations domains that are generally and predominantly of peculiar interest of enterprises. These different domains usually require specific solutions

S. Bittmann (✉) • O. Thomas
University Osnabrueck, Osnabrück, Germany
e-mail: sebastian.bittmann@uni-osnabrueck.de; oliver.thomas@uni-osnabrueck.de

M. José Escalona et al. (eds.), *Information System Development: Improving Enterprise Communication*, DOI 10.1007/978-3-319-07215-9_13,

driven by the capabilities and directions of the different departments, teams and individuals of an enterprise. More specifically, the development of these solutions depends on the different resources an enterprise has and the ultimate difference between enterprises are the people that are a part of this socio-technical system. Hence, any enterprise model should be enterprise-specific. Although between the different enterprises, there will be definitely commonalities or references within the enterprise model, ultimately the differences represent the competitive advantage of an enterprise.

Developing enterprise models for different domains of the enterprise implies the according and proper management of these enterprise models. Although these enterprise model might be directed to different domains of the enterprise, the huge enterprise modelling effort for planning and depicting an enterprise [1], should be coordinated by means of a more holistic perspective. Among probably other aspects, this holistic perspective should be motivated by a common direction of and the appropriate provision of resources for the enterprise models, respectively their enactment. Regarding the management of enterprise models, as shown in other works, e.g. [2], the necessity occurs for identifying commonalities between enterprise models to properly coordinate them. The proper coordination then guides the possibility for integrating the enterprise model. Within this paper, an enterprise model will be regarded as any kind of conceptual model that plans or enables the achievement or contributes to the achievement of the business goals of the respective enterprise. As the depiction of a whole enterprise isn't probably the work of a single modeller and within the specific enterprise might exist domains, in which enterprise modelling is more costly than beneficial, an enterprise model of the classical sense as discussed by [3] or [4] is a composition constituted by a multitude of different enterprise model from different domains of the enterprise. Hence, an enterprise model as a complete depiction of the enterprise may be created after the creation of enterprise model for the specific and different problems the company faces and their integration among different layers [5].

Therefore in this paper a general approach of coordinating the accruing effort of enterprise modelling, which orientates at the goals that the enterprise targets to achieve. In the upcoming Sect. 13.2, the initial basis for such an approach will be introduced, which is goal modelling. In this paper goal modelling, however, will be extended to the concept of an Entrepreneurial Goal Hierarchy, which depicts all the relevant goals that were targeted by the company. Further in this section Enterprise Model Design is introduced that properly guides the development of an enterprise model with respect to the targeted goals, to-be satisfied directions and individual knowledge of the respective modeller. Following, in Sect. 13.3 the concept of an enterprise model lifecycle will be discussed. The enterprise model lifecycle enables the proper management of enterprise models. Further, as this lifecycle is coordinated with the different phases of a specific goal, targeted by the enterprise, the enterprise modelling effort can be coordinated from a more managerial perspective. Finally, the paper ends with a conclusion in Sect. 13.4.

## 13.2 The Concept of an Entrepreneurial Goal Hierarchy

### *13.2.1 Goal Modelling*

Goal modelling was originally derived from goal-oriented requirements engineering [6, 7]. It tries to identify the relevant achievements of a project, in particular a software development project, that are needed to be satisfied in order to regard the project as a success. In an abstract manner, it tries to elucidate the conditions that are needed to be satisfied in order to accomplish a specific goal, which is done by identifying the different goals required to satisfy. Due to its abstract conceptualisation, goal modelling evolved to a manner for strategy setting of an enterprise [8].

A goal may be constituted by multiple further goals that promise the achievement after their own satisfaction. Additionally, goals may depend on further goals that are not directly derived from it. In conclusion, a goal may both be made more concrete by its refinement to more concrete goals and it may be dependent on further goals, whose achievement support the achievement of the initial goal. Within goal modelling two means of an end can be identified. The initial and top design requires the consideration of the enterprise and its business goals [8]. With using a refinement relation, goals are partly justified by this relation with reference to the superior goals. So, based from strategic stated business goals, it is required to refine those goals continuously until the operative goals were derived that represent the desired achievement of actually executed actions. Actions, which are executed by the respective actors, are the other end that can be identified within goal modelling [9]. However, to ultimately derive actions from business goals it is required to refine the business goals and reduce in this manner the respective complexity in a manner that actions are immediately derivable [6].

However, although the complexity of goals may be reducible until an immediate derivation of actions becomes possible. This reduction needs to be coordinated according to the enterprise's needs and therefore should be enterprise-specific. On the one handy, it may be the case that the initial strategic business goals are very generic as they are only represented on an abstract level. On the other hand, the refinement of such strategic business goals depends rigorous on the resources the enterprise owns. An action is not executable, which requires resources that aren't owned by the enterprise. Resources then may refer to tangible or to non-tangible assets such as knowledge or expertise. Therefore the development of a solution that is in charge of satisfying a respective set of goals that were identified by means of a goal model, needs to consider these enterprise-specific circumstances.

### *13.2.2 Enterprise Model Design*

Enterprise Model Design (EMD) may be defined as the constructive and highly creative influenced process of enterprise modelling that aims at the development of solutions achieving set or developed goals under the consideration of the respective

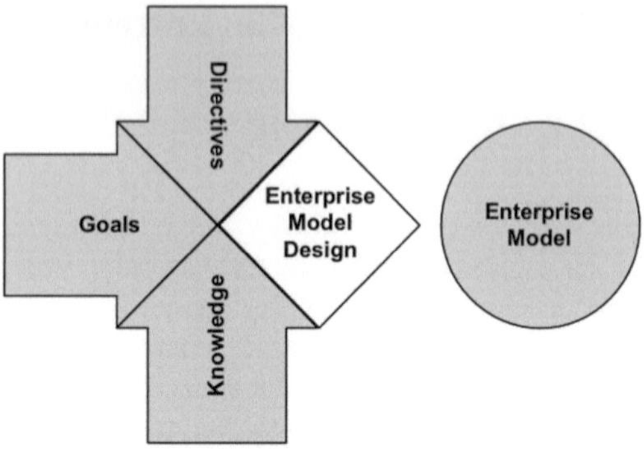


**Fig. 13.1** Conception of enterprise model design

set or identified directives. EMD aims at capturing significant information taken from the actual process of designing an enterprise model, as the significance of the designing process has been identified, next to the actual product or outcome [10, 11]. As not just the model is sufficient for explaining and revealing others indentations and beliefs, it is necessity to capture the different grounds and assumptions with the respective outcome [12] (Fig. 13.1).

With the consideration of goals as an input of this process, the EMD partly guides the required level of abstraction expected from the ultimate outcome of the EMD. The more concrete, a certain goal is, the more concrete the respective enterprise model has to be. However, through the consideration of directives that must be embodied by the enterprise model, the development of a solution that is aligned with the abstraction level of the respective goal, is additionally guided. Directives are either proposed by a superordinate entity that force a certain direction to be followed or specific restriction to be satisfied or by the environment. The last magnitude of influence considered by the EMD is the individual knowledge of the respective modeller that is in charge of creating an enterprise model. Design decisions will be undertaken during the design process of the enterprise model that are neither required with respective to the goals or directives, but as the modeller comes with a certain experience and expertise, these experience and expertise should be used by means of the enterprise model.

The use of EMD is not restricted to any field or domain within the enterprise, moreover it should be suitable for all relevant domains within an enterprise that are considering enterprise modelling prior to direct actions. With the focus on goals, EMD enables to layer different modelling processes within the enterprise, e.g. between the need for modelling for the strategic level and operational level [8]. While the EMD needs to focus on an alignment with the enterprise environment, when developing solutions for the strategical level, e.g. by the specification of a strategy map [13], the EMD applied on a more tactical or even operational level has to consider directives that are directly provided by the strategic level.

**Table 13.1** Influences of goals, direction and individual knowledge

| Goals | Directives | Individual knowledge | Scenario |
|---|---|---|---|
| – | – | + | Requires detailed analyses of the environment and enhanced capabilities for decision making, e.g. strategic decisions. Goals evolve with the EMD in accordance to the analyses |
| + | – | + | Solution is mainly driven by the individual knowledge, captured by the respective modeller. So the modeller can freely design the actual solution, if it is capable of achieving the predefined goals |
| + | + | – | In such a scenario the outcome of the EMD process is basically predefined by the impetus given by the direction and goals. The desired achievement is defined by the respective goals and the approach for achieving is outlined by the direction. Such an EMD is considered with operative choices that are characterised by a low freedom of choice |

The initial set strategic direction has to be decreased by means of the determination of further aspects during the tactical or operational level [14]. However, these determinations shouldn't conflict with those decisions taken on the strategic level. Ultimately, on the operational level, from the operational goals and taken directives, the EMD focuses on the development of solutions that can be directly included by or considered for the actions of the operatives [15]. Within Table 13.1 different scenarios of the application of the EMD are described. The scenarios mainly vary in their explication of the set goals, the directives and the individual knowledge that can be considered for the development of a solution.

As previous explained, the different, exemplary scenarios for applying the EMD vary with the capabilities and the responsibilities of the respective modeller. While in the initial scenario, the solution mainly depends on the individual knowledge the respective modeller has, it rather orientates on a more strategic, direction-taking level. The latter scenarios come with more restrictions that are needed to be respected by the respective developed solution. The level of freeness in creating a solution, or selecting one out of a specific set of solution is more restricted, e.g. due to a strategic direction that has been decided on. Thereby, the latter scenarios, although not absolute, are more orientated at the tactical or operational level. However, although the relation between the strategic, tactical and operational layer can be identified by means of the goals and directives, emergent solutions might occur that were not initially been planned [16]. The EMD explicitly considers such a development, with the explicit consideration of individual knowledge and the possibility for specifying enterprise model, especially strategic enterprise models, on a higher level of abstraction. For example, the more abstract a strategy has been formulated, the higher is the level of freedom of the operative using the EMD.

### 13.2.3 Entrepreneurial Goal Hierarchy

Given the fact that every institution of a company directs its actions to the achievement of the set business goals of the strategy, it can be inferred that every institution may derive its own goals that are derived from the business goals. These goals may either contribute implicitly or explicitly to the achievement of the business goals, but in any case are refined by actions that either contribute to the achievement of the business goals or enable the contribution to the business goals by other actions. Thereby, as a commonality, goals can be identified as a central modelling concept that is dispersed throughout the enterprise.

Moreover, with the identification of a common concept, it is required to align the goals within an enterprise properly. The alignment of the goals within an enterprise needs to consider various aspects. Hence, in order to ensure these types of alignment it makes sense to not just use goal modelling for a specific project, but to use it holistically within the enterprise. In general, it needs to be ensured that the strategic goals are accordingly aligned with the environment of the enterprise by the strategic management and that any further goal within the enterprise is aligned with the strategic goals. So operative goals, from which the actual actions are derived need to be aligned with the strategic goals, preferably by means of a multi layered perspective that is capable of identifying the implications by the operative from the strategic goals [8]. Such considerations are included by the concept of the Entrepreneurial Goal Hierarchy (EGH). An EGH includes all the various goals of an enterprise and aligns them accordingly or identifies contradictions between them. Especially the dependency and the refinement relation of goal modelling support the alignment within the EGH. An EGH helps to identify the contribution of specific operative actions with respect to the superior business goals. Furthermore it ensures the proper derivation of these enterprise actions. Additionally the EGH supports the coordination of various projects that may be located on a rather tactical level and try to enable strategic plans in the operational business. Therefore, the EGH requires a goal proposal to be justified by means of the goals set by the superior. For example, a member of a department shouldn't pursue a goal that will not contribute to the achievement of the general goals of the department.

However, as identified in Sect. 13.2.2 to develop an enterprise model more aspects apart from goals are needed to be considered. If required, directives need to be provided by the respective superiors. Nevertheless, the EGH enables the stepwise refinement of an respective enterprise model according to the refinement of the represented goals. In particular, the execution of a business process might require the preceding development of an information system, which is planned with respect to a superior goal. So a completion of an enterprise model, as a depiction of the actual enterprise, will be possible with the achievement of the respective goals of the EGH. Without the full achievement of the goals, the available enterprise model will only constitute the frame of action provided to the operative business and therefore is on abstract level, as the operative business, although slightly, still is able to contribute its individual knowledge in order to developing a solution for achieving the respective set goals.

## 13.3 Communication of Responsibilities and Conditions Regarding the Enterprise Modelling Effort

### *13.3.1 The Need for an Enterprise Model Lifecycle*

Initially a rather simple, but sufficient lifecycle will be considered, regarding the management of enterprise models. With respect to the BPM Lifecycle [17], the lifecycle of an enterprise model includes the following phases. Initially there is the need for developing a solution. Due to a *diagnosis* phase, the purpose (goals) as well as the different requirements (directives) will be collected. Successively, within the *design* phase an enterprise model will be created with respect to the previous identified relevant aspects. When the design has been approved by all the relevant stakeholders, the enterprise model will be *implemented* within the company. Implementation of an enterprise model then refers either to the introduction of a new business process, the reorganisation by means of an organisational diagram or the implementation of an information system. Succeeding is the *enactment* of the solution provided by the enterprise model. Ultimately there is the need for *diagnosis and/or maintenance* of the enterprise model. Either there is the possibility to adapt the enterprise model to react to new requirements or directions of the enterprise or there is the need to omit the enterprise model (Fig. 13.2).

### *13.3.2 Tightening the Relation Between Enterprise Models and Their Associated Goals*

Regarding the previous introduced enterprise model lifecycle there is a need for indicating the current phase of an enterprise model and to further decide the future direction of the enterprise model. Especially the identification for the necessary adaptation of enterprise models is tremendous, as indicators for required adaptations would increase the agility of the enterprise models and further would decrease the amount of unmonitored, but used enterprise models. Goals are a useful indicator

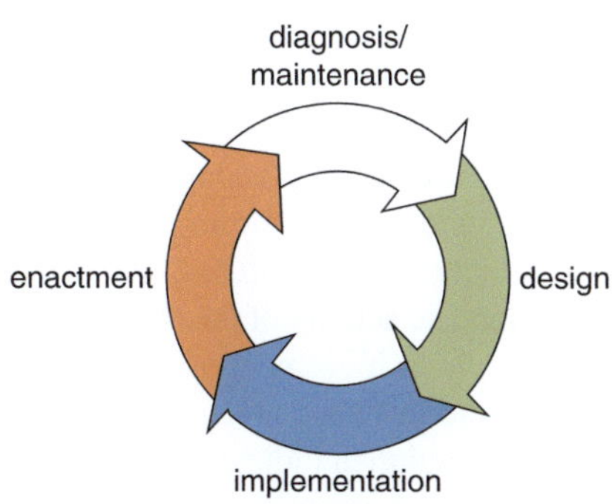


**Fig. 13.2** Lifecycle of enterprise modelling

for both the maintenance of existing as well as the requirement for new enterprise models. The validity and change of a goal, whose achievement is or will be planned by means of an enterprise models, indicates the requirement for creating a new enterprise model or adapt an existing one.

A goal, which was derived from the strategic business goal, will be gain a level of refinement, at which it requires further clarification of how to achieve it, prior its further refinement. It is necessary to decide for a direction for achieving that particular goal, as from the possible set of choices, the particular one that suits the enterprise at most, needs to be selected. As a strategic goal is initially rather abstract, it is not possible to derive immediate actions, but requires further planning, respectively a concretisation, for actions that should be executed in the near future. Thereby, the use of an enterprise model enables the clarification of the specific frame of actions that should be provided for a purposefully execution of the actions of the operative business. As such a clarification has happened, the goal may be refined by means of more concrete goals that orientate towards the clarification established through the EMD. So with new occurring goals the necessity for applying the EMD occurs and the initial phase of the enterprise model lifecycle has been reached.

Regarding the implementation and the enactment of an enterprise model, the related goals especially support the understanding of the different design decision included by the enterprise model [7]. As the enterprise model not just satisfies holistic goals, but further more detailed and refined subordinated goals, these goals can be related to the different design decisions and concepts of the enterprise model, as shown in [15] for business process models. So, the related goals of an enterprise model give an explanation about how the different design decisions and concepts contribute to the superior goal that is associated with the enterprise model. Additionally, the refinement of a goal to an enterprise model or to excerpts of it, lead to objectives regarding the implementation of the enterprise model. These objectives become measurable after the implementation, as the form of the enterprise can be checked with respect to the initial model and which give implications to the achievement of the goal.

Further, these association between a set of goals and the actual enterprise model, enables the identification of a required omission and/or adaptation of the enterprise model. For example, a specific business process model is associated with a goal that is completely achieved by the enactment of the business process. If that is the case, then the exclusion of the respective goal from the EGH implies the suspension of the *business process model* and respectively the business process. Thereby, the need of an enterprise model might be motivated by multiple goals, which are then refined and specifically satisfied by means of the provided concepts of the enterprise model. In a similar manner, the adaptation of the goal requires a modification of the business process model. On a further second level, different goals that are derived from the initial superior goal, will be associated with parts of the enterprise model. For example, a specific activity of the prior discussed *business process model* might be responsible for the achievement of a more specific goal that was derived from the superior goal of the business process model. The evolution of such a particular and more concrete goal then indicates the required adaptation of the enterprise model as well.

However, as the goal is more concrete and associated only partly with the enterprise model, it also indicates which part of the enterprise model needs to evolve.

Additionally, the derivation of the goals that are associated with the enterprise model from the strategic business goals gives an explanation, how the enterprise model contributes to the business goals of the enterprise. So the change of the strategic direction of the enterprise by means of the strategy goals might indicate the required modification of the superior, derived goals. Hence, it is possible to identify necessary changes of the enterprise models by means of the associated goals. The interrelation of the goals from different domains of the enterprise with its strategic goals, was discussed with reference to enterprise architecture in [5]. Moreover, with consideration of the EGH, as concept considered with the enterprise as a whole and holistically, every domain that benefits from enterprise modelling can be included for this strategic alignment.

### *13.3.3 Managing the Enterprise Model Lifecycle*

So with the management of the respective goals of a company, the enterprise models become manageable as well. An occurring goal represent the necessity for the development of a new enterprise model. So with associating the different enterprise models with the respective goals, the adaptation of a goal indicates the necessary adaptation of the enterprise model and the omission of a goal from the EGH and hence, from the targeted goals, indicates the invalidation of an enterprise model. The general conception of the EGH with respect to the use of enterprise models is depicted by Fig. 13.3.

The strategy of the enterprise initially has developed strategic goals that are well aligned with the environment, especially the competitors and collaborators, of the enterprise. So with such a specification the enterprise tries to initially place itself within the market, by means of very abstract strategic business goals. The strategic goals are then further refined, guided by the specification of a strategy map. The process of developing this strategy map was guided by the EMD. The result of the EMD is then used as a basis for the derivation of further more specific goals. For example, initially the strategy map describes a general direction, which should be taken by the company in order to achieve the respective strategic goals. Based on these directions, the derivation of goals for various departments, teams and other subordinates becomes possible, as not just the desired achievements are clarified, but also the direction for achieving them. Hence, the specified strategy map may be used as a *directive* during further and subordinated applications of the EMD. As shown in Fig. 13.3, further enterprise models, e.g. organisational diagrams, business process models and data models, were created with respect to the goals includes by the EGH of Fig. 13.3. So, the application of the EMD by one modeller is purposefully guided by means of the superior goals and directives taken from his superiors. Furthermore, these goals and directives are made explicit by means of the EGH and corresponding enterprise models, which allow a management of the enterprise models.

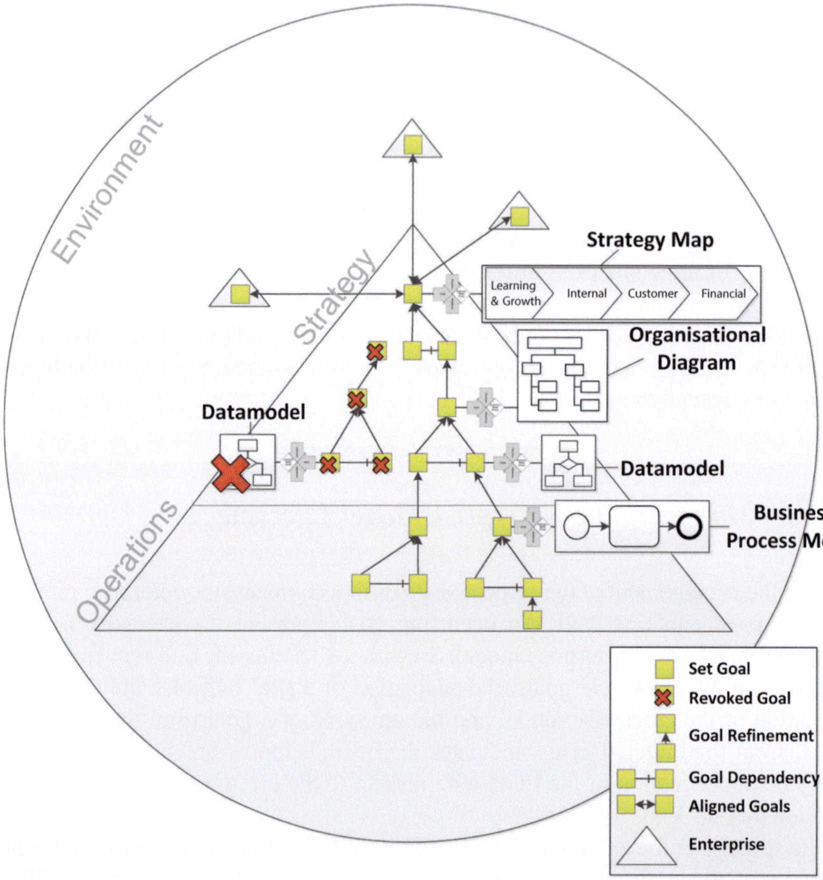


**Fig. 13.3** Conception of the general framework for managing enterprise modelling effort

For example a specific data model has been invalidated within the given example. This data model will be excluded from the EGH, as the associated goals have been revoked. This exclusion can be undertaken from a managerial perspective, as the enterprise models that will need an adaptation or are needed to be omitted can be clearly identified because of the placement and integration of the enterprise model within the EGH.

## 13.4 Conclusion

Within this paper, a general framework for managing different enterprise modelling activities within an enterprise was introduced. Especially the framework tries to abstract from any domain or language specificities and enables a more holistic overview about enterprise modelling in an enterprise. With a more holistic overview, a managerial perspective can be established that is capable of supporting the enterprise model during different lifecycle stages. Furthermore, with identifying a dependency with respect to the different goals, which have to be satisfied by the enterprise models, an integration between those enterprise models is promoted and supported.

In comparison to contemporary enterprise modelling approaches, e.g. MEMO [18], ARIS [19] or SOM [20], the proposed approach relies on a homogenous set of modelling concept for the guidance, coordinating and management of enterprise models. Thereby, it becomes possible to identify whether it is appropriate to integrate certain enterprise models based on their dependency, which can be expressed by means of the goal dependency, or not. Additionally the need for an enterprise model can be justified with respect to the overall strategy of the enterprise. Additional approaches that primarily focus on the management of enterprise models, introduce additional concepts for that particular purpose, e.g. [2]. However, with the increasing expressive power and the claim of depicting the enterprise itself by enterprise model, additional concepts impede the possible benefit of managing enterprise models with familiar and enterprise-related concepts, such as goals. Goal oriented approaches, such as [6] or [21], claim the possibility for deriving actions solely from stated and refined goals. However, these approaches miss to identify enterprise-relevant and additional needed concepts, e.g. legal regulations or available resources, that are offered by enterprise models.

The presented framework enables a coordination of enterprise modelling effort by two different directions. First, the framework clarifies the effort by promoting the identification of the desired target that is needed to be achieved. Second, decisions that have been taken by a higher layer within the EGH are transposed to lower layer by means of directives, explicated due to the application of the EMD. Hence, restriction and prescriptions are not just communicated orally within the enterprise, but are transcribed through the use of a holistic EGH in combination with the EMD. Additionally, the managerial possibilities of the EGH help to maintain the different enterprise model produced by the respective enterprise models. Through the identification of the relevance of goals within the EGH, it can be identified whether to retain, maintain or omit an associated enterprise model. Thereby, development costs of enterprise model are reducible, when they don't contribute to the business goals of an enterprise.

## References

1. Koubarakis M, Plexousakis D (2002) A formal framework for business process modelling and design. Inf Syst 27(5):299–319. doi: 10.1016/S0306-4379(01)00055-2. http://dx.doi.org/10.1016/S0306-4379(01)00055-2
2. Thomas O (2006) Management von Referenzmodellen: Entwurf und Realisierung eines Informationssystems zur Entwicklung und Anwendung von Referenzmodellen. Logos, Berlin
3. Fox MS, Gruninger M (1998) Enterprise modeling. AI Magazine 19(3):109. D10.1609/aimag.v19i3.1399. http://www.aaai.org/ojs/index.php/aimagazine/article/view/1399

4. Frank U (2012) Multi-perspective enterprise modeling: foundational concepts, prospects and future research challenges. Softw Syst Model. doi: 10.1007/s10270-012-0273-9. http://www.springerlink.com/index/10.1007/s10270-012-0273-9
5. Bleistein SJ, Cox K, Verner J (2006) Validating strategic alignment of organizational IT requirements using goal modeling and problem diagrams. J Syst Softw 79(3):362–378. doi: 10.1016/j.jss.2005.04.033. http://dl.acm.org/citation.cfm?id=1140759.1140765
6. Yu E (1999) Strategic modelling for enterprise integration. In: Proceedings of the 14th world congress of the international federation of automatic control
7. Liu L, Yu E (2004) Designing information systems in social context: a goal and scenario modelling approach. Inf Syst 29(2):187–203. doi: 10.1016/S0306-4379(03)00052-8. http://dx.doi.org/10.1016/S0306-4379(03)00052-8
8. Samavi R, Yu E, Topaloglou T (2008) Strategic reasoning about business models: a conceptual modeling approach. Inf Syst e-Bus Manage 7(2):171–198. doi: 10.1007/s10257-008-0079-z. http://www.springerlink.com/index/10.1007/s10257- 008-0079-z
9. Lohrmann M, Reichert M (2012) Modeling business objectives for business process management. In: Stary C (ed) S-BPM ONE scientific research. Lect Notes Bus Inf Process 104:106–126. doi: 10.1007/978-3-642-29133-3 8. http://dx.doi.org/10.1007/978-3-642-29133-3 8
10. March ST, Smith GF (1995) Design and natural science research on information technology. Decis Support Syst 15(4):251–266. doi: 10.1016/0167-9236(94)00041-2. http://dx.doi.org/10.1016/0167-9236(94)00041-2
11. Hevner AR, March ST, Park J, Ram S (2004) Design science in information systems research. MIS Q 28(1):75–105. http://dl.acm.org/citation.cfm?id=2017212.2017217
12. Bittmann S, Thomas O (2013) An argumentative approach of conceptual modelling and model validation through theory building. In: vom Brocke J (ed) DESRIST 2013, LNCS 7939. Springer, Heidelberg, pp 242–257
13. Giannoulis C, Petit M, Zdravkovic J (2011) Modeling business strategy: a meta-model of strategy maps and balanced scorecards. In: 2011 Fifth international conference on research challenges in information science, IEEE, pp 1–6. doi: 10.1109/RCIS.2011.6006868. http://ieeexplore.ieee.org/xpls/abs all.jsp?arnumber=6006868' escapeXml='false'/¿
14. Niebecker K, Eager D, Kubitza K (2008) Improving cross-company project management performance with a collaborative project scorecard. Int J Manag Projects Bus 1(3):368–386. doi: 10.1108/17538370810883828. http://www.emeraldinsight.com/journals.htm?issn=1753-8378&volume=1&issue=3&articleid=1729385&show=html
15. Koliadis G, Ghose A (2006) Relating business process models to goal-oriented requirements models in KAOS. In: Hoffmann A, Kang BH, Richards D, Tsumoto S (eds) Advances in knowledge acquisition and management. Lect Notes Comput Sci 4303:25–39. doi: 10.1007/119612393. http://dx.doi.org/10.1007/11961239 3
16. Mintzberg H, Waters JA (1985) Of strategies, deliberate and emergent. Strateg Manage J 6(3):257–272. doi: 10.1002/smj.4250060306. http://doi.wiley.com/10.1002/smj.4250060306
17. van der Aalst W (2004) Business process management: a personal view. Bus Process Manage J 10(2):135–139. http://citeseerx.ist.psu.edu/viewdoc/summary?doi=10.1.1.58.502 http://archive.itee.uq.edu.au/shazia/Literature/BPM/Papers/Downloads/A04.pdf
18. Frank U (2012) Specialisation in business process modelling: motivation, approaches and limitations. https://opus.zbw-kiel.de/dspace/handle/10419/58159
19. Scheer AW (2000) ARIS-business process modeling, 2nd edn. Springer, New York
20. Ferstl O, Sinz E (2006) Modeling of business systems using SOM. In: Bernus P, Mertins K, Schmidt G (eds) Handbook on architectures of information systems SE-15, international handbooks on information systems. Springer, Berlin, pp 347–367. doi: 10.1007/3-540-26661-5 15. http://dx.doi.org/10.1007/3-540-26661-5 15
21. van Lamsweerde A (2001) Goal-oriented requirements engineering: a guided tour. doi: 10.1109/ISRE.2001.948567

# Chapter 14
# Merger and Acquisition Preparedness Building: An Enterprise Architecture Perspective

**Nilesh Vaniya, Ovidiu Noran, and Peter Bernus**

**Abstract** The increasing rate of Mergers and Acquisitions (M&As) draws attention of researchers and academics alike; however, the significant failure rate of such complex enterprise-wide transformations typically posing 'wicked' problems suggests a significant potential for improvements. Much research has been conducted to investigate the issues and problems in M&As in order to find and propose various ways to increase the percentage of positive outcomes. This abundance of M&A research creates a need to synthesise the existing results in order to propose a collective way to address M&A issues. In addition, due to the complexity, dynamicity and time constraints of M&As it seems that there is little time to implement the recommendation of the research community. This paper proposes a way to address this dilemma through building *preparedness* for M&As. Thus, before the actual M&A implementation, organisations undertake preventative measures to acquire necessary systemic properties (such as flexibility, agility, etc.), in order to be in a better position to address *anticipatable* or emergent issues.

**Keywords** Mergers and Acquisitions (M&As) • Enterprise Architecture • Preparedness building • Management strategy

## 14.1 Introduction

Mergers and Acquisitions (M&As) are practiced in order to achieve a variety of business goals and objectives. This complex and dynamic practice serves a range of stated purposes, such as economies of scale, resource acquisition or market sharing, and governance-related uses such as the restructuring of the banking industry. M&As typically pose complex and 'wicked' [29] problems, where solving one

N. Vaniya (✉) • O. Noran • P. Bernus
School of ICT, Griffith University, Gold Coast, QLD, Australia
e-mail: n.vaniya@griffith.edu.au; o.noran@griffith.edu.au; p.bernus@griffith.edu.au

M. José Escalona et al. (eds.), *Information System Development: Improving Enterprise Communication*, DOI 10.1007/978-3-319-07215-9_14,

aspect in isolation may upset other areas and 'good', rather than 'perfect' solutions exist. This suggests the use of a holistic approach considering all relevant aspects of the M&A project at hand in an integrated rather than isolated manner.

This paper aims to tackle this problem by presenting the principles and use of a 'Mergers and Acquisitions Preparedness Building' (MAPB) approach built on Enterprise Architecture (EA) concepts in order to create and support strategically important transformational activities, in a whole-system life cycle-based manner. The proposed MAPB approach comprises an 'M&A Issues Identification and Categorisation (MAIIC)' Model, as well as an 'M&A Preparedness Building Methodology' (MAPBM). We start by describing the proposed MAPBM which aims to support enterprises in acquiring the necessary systemic properties before merger/acquisition and thus build preparedness for one or more desirable types of M&A, which management considers as future strategic options. Subsequently, we describe a merger case study and using the MAPBM we demonstrate how, with strategic intent, a multifaceted transformation of the participating organisations could have been performed so as to achieve a state of organisational readiness to perform a strategically attractive merger.

## 14.2 Mergers and Acquisitions: Problems and Solutions

Mergers and Acquisitions can be of different types, such as Horizontal, Vertical, Conglomerate [13] or Forced and Voluntary [12, pp. 26–27]. Irrespective of the type, M&As can deliver positive outcomes; Walter [25, pp. 62–77] lists some advantages such as market extension, economies of scale, cost (or revenue) economies of scope and other operating efficiencies.

Current research [3, 16, 19] suggest that while the rate of M&As has increased, the probability of achieving the above mentioned potential benefits has dropped to less than 50 %. In the following discussion we categorise major M&A issues that cause such high failure rate.

The major M&A issues can be grouped into three key categories: (1) Business Management (2) Human Resource (HR) and (3) Information Technology (IT) and Information Systems (IS). These three categories of issues may have interrelationships or interdependencies; for example, as demonstrated by Robbins and Stylianou [21] as well as Baro, Chakrabarti and Deek [5] the success of Post-merger system integration relies on IS and Organisational factors. To consider such interrelationships a so-called 'systems approach' [6, 14, 17, p. 4] would provide a unified framework able to represent the range of problems that arise from the transformation of two systems into a single system. The next questions would be (a) can all issues be addressed in detail before M&A in all circumstances? (b) if not, then what kind of issues can be addressed? (c) at what level of detail? and (d) how to identify those issues? This research is set to identify which type of issues can be addressed, by whom, when and how.

The key requirements for a harmonised solution are discussed in the literature. Examples of such requirements are: maintain business-IS alignment during Post-merger Integration (PMI) [26], consider the level of integration required [24], configure a clear Information and Communication Technology (ICT) vision [14], careful integration planning of IS components such as enterprise applications and platforms (operating systems, communication, security, and database systems) [4]. For PMI, various approaches have been suggested for all three categories of issues. For IS PMI, Giacomazzi, Panella, Pernici and Sansoi [9] suggest a model to list available options (Total Integration, Partial Integration, No Integration and Transition) for given computer and software architectures. For PMI of application systems, recent attention was drawn to the role and utilisation of ERP systems in M&As [e.g., 10, 11], resulting in a methodology to develop an Application Integration Strategy [7, 18]. To address HR PMI issues, Schuler and Jackson [22] suggest a three-stage HR integration model (pre-combination, combination and solidification). Their model covers major HR activities, strategies and planning for successful HR PMI. To explain the PMI process, Mo and Nemes [17] suggest developing the metaphor of an architectural 'DNA' (biological DNA) inheritance. Thus, using a 'DNA EA' concept, they explain post-merger integration as the inheritance of processes, knowledge, control, data, people and assets of the involved organisations into the DNAs of the merged organisation.

During PMI it will be necessary to design an agile and quick responsive operating model which can be tailored to the organisation's needs. Ross, Weill and Robertson [20] suggest that the level of business process integration (sharing data across parts of the organisation and thus requiring business data integration) and the level of business process standardisation (use of uniform business processes across the organisation) can decide an operating model for the organisation.

In summary, although the outlined solutions are able to address individual categories of issues, it is necessary to consider the impact of such solutions in the context of other types of issues. In our view, a systems approach is needed to synthesise the proposed solutions into a workable and comprehensive methodology.

## 14.3 An M&A Preparedness Building Approach

As demonstrated in [1], a potential solution to address the above-mentioned concerns is M&A preparedness building. Figure 14.1 shows the conceptual layout of the proposed MAPB Solution.

Using the conceptual blocks of MAPB approach highlighted in Fig. 14.1, we have previously investigated the *why*, *who*, *how* and *when* of M&A preparedness building [1, 2]. However, one question remains: *how to identify the M&A preparedness building activities?* To answer it, in this paper we propose a way to identify and categorise M&A issues that can occur in PMI and demonstrate how to use these issues to identify the preparedness building activities.

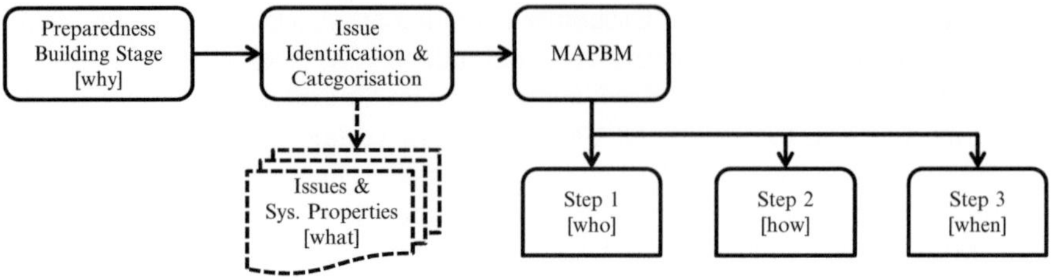


**Fig. 14.1** Proposed M&A Preparedness Building approach

### *14.3.1 Identification and Categorisation of M&A Issues (MAIIC)*

One of the goals of preparedness building is to gain the ability to make informed decisions in time during the merger process. A possible way to achieve such ability is by gaining a comprehensive understanding of the situation at hand (i.e., the merger process and potential issues that the organisation might encounter during the process). To understand the issues and to identify their possible alternatives solutions, we need to consider two aspects of the issues: the ability to control and the ability to anticipate the issues.

Stylianou, Jeffries and Robbins [23] categorise post-merger IS integration success factors as Controllable and Uncontrollable. After identifying the issues pertaining to the merger case, it is necessary to understand the nature of issues (i.e. controllable vs. uncontrollable) in order to tailor the preparedness building exercise case by case.

The other ability can be explained using commonly used terms of medical and pharmaceutical sciences, i.e. preventable and curable. There is a clear difference between these two categories that is the ability to anticipate the disease decides the possible treatments and their timings.

Based on these two abilities of an enterprise any transformational issue can fall into one of the categories shown in Fig. 14.2. This categorisation can:

- *support the planning of potential tasks for a preparedness building exercise,*
- *help determine the objective during the Post-Merger Integration (PMI),*
- *support management to make informed decisions about possible courses of action in case of occurrences of any anticipated or emergent issues.*

As demonstrated in Fig. 14.3, any issue can be categorised as: (a) Controllable and Anticipatable, (b) Uncontrollable and Anticipatable, (c) Controllable and Emergent or (d) Uncontrollable and Emergent. As a consequence, for effective and efficient PMI, organisations should be prepared for all type (a) issues, should avoid type (b) issues, should effectively respond to the type (c) issues, and should survive the effects of type (d) issues. Potential tasks of a preparedness building exercise:

- *prepare for all type (a) issues*
- *convert all the other types of issues into type (a) issues (Arrows 1 to 5, Fig. 14.3) and do the above, or*

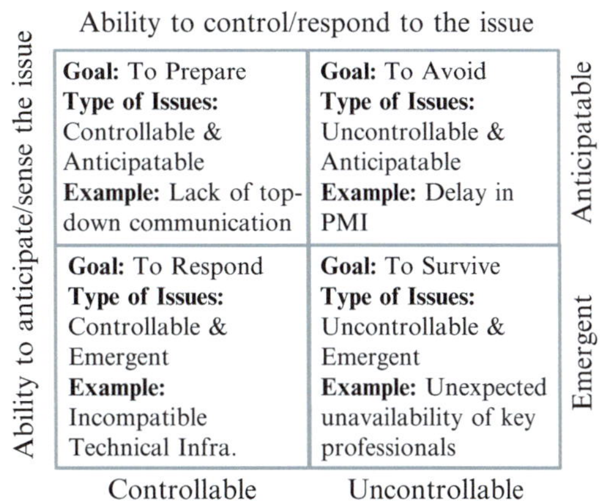


**Fig. 14.2** A model to identify and categorise M&A issues (MAIIC Model)

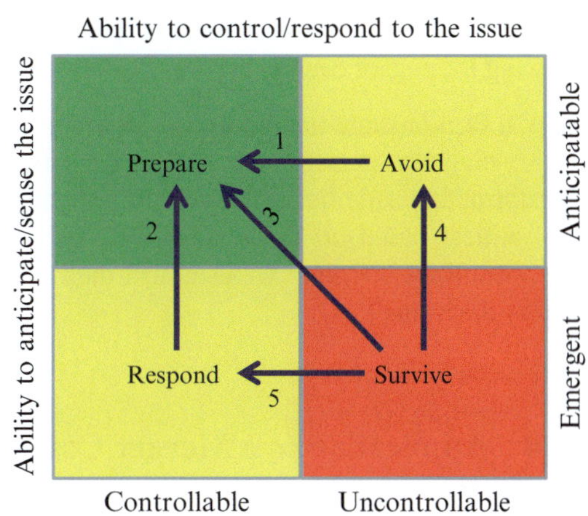


**Fig. 14.3** Potential tasks of MAPBM

- *if the above cannot be achieved then acquire relevant systemic properties and/or architectural design so that the enterprise can meet the respective goals based on the type of issues.*

However, as reported in literature [15], not everything can be planned for at micro level in such dynamic and complex change events as M&A. Therefore, during a preparedness building exercise,

- *Type (d) issues can be responded if their controllable factors can be identified (Path 5), or*
- *Type (d) issues can be avoided if they could be anticipated (Path 4), or*
- *Type (d) issues can be prepared for if both of the above can be done (Path 3)*
- *Type (c) issues can be prepared for if they could be anticipated (Path 2)*
- *Type (b) issues can be prepared for if their controllable factors can be identified (Path 1).*

### 14.3.2 The M&A Preparedness Building Methodology (MAPBM)

The proposed methodology (MAPBM) consists of three main steps (as demonstrated in detail in [2] based on the meta-methodology described in [27]):

**Step 1**: Identify the participating enterprise entities. They can be existing entities (for example existing Management team, business units, affected business processes, IT infrastructure, etc.) contributing to building preparedness, or can be additional entities required to build preparedness (for example Preparedness Building Strategic Program, Gap Analysis Project, Business-, HR- and IS-Preparedness Building Projects, etc., or even strategic partners).

**Step 2**: Show the role of each entity in the preparedness building transformation. Step 2 shows the role of each entity in the preparedness building transformation. Various graphical models can be used for this particular step; we have chosen the so-called 'dynamic business models' proposed by the IFIP-IFAC Task Force [28] showing the role of each entity in other entities' life cycle phases.

**Step 3**: Demonstrate the relative sequence of transformational activities, using life history diagrams (timeline). Step 3 attempts to demonstrate the relative sequence of transformation activities. This step follows the previously identified roles of each of the entities; based on those roles, we first identify activities to match entities' responsibilities and then we establish their relative sequence using so-called 'life history diagrams'.

## 14.4 Application to a Merger Case Study

### 14.4.1 Background

Faculty F within university U contained several schools, with schools A and B having the same profile. School A is based at two campuses situated at locations L1 and L2, while school B is based at a single campus, situated at location L3. Historically, the schools have evolved in an independent manner, reflecting the local educational needs and demographics. This has led to different organisational cultures, HR and financial management approaches. For example, school B enjoyed a large international student intake providing funds that supported heavy reliance on sessional (contract) staff for teaching and wide availability of discretionary funds. In contrast, staff in school A had a larger teaching load and fewer funds available due to a smaller student intake.

Staff profile level between schools was significantly different, with school B featuring fewer senior positions. Course curriculums also evolved separately in the two schools, with similarly named courses containing notably different material.

Thus, although of the same profile, and belonging to the same F, schools A and B were confronted with a lack of consistency in their profiles, policies, services and resources. This situation caused additional costs in student administration and course/program design/maintenance, unnecessary financial losses as well as staff perceptions of unequal academic and professional standing between campuses, all of which were detrimental to the entire faculty.

Therefore, the management of U and F have mandated that the problems previously described must be resolved and have defined the goals of schools A and B becoming consistent in their products and resources strategy, eliminating internal competition for students and being subject to a unique resource management approach. As a solution, it has been proposed that the schools should merge into a single, multi-campus Merged School (MS). The unified MS management and policies would promote consistency in the strategy regarding the products delivered and the resources allocated to its campuses.

After further consultation, the Heads of the participating Schools have set an organisational goal allowing the individual campuses to retain a significant part of their internal decisional and organisational structure after the merger, with an added layer of overall governance structure. This structure was supported by the HR department as the simplest to implement and transition to.

From the point of view of Information Services, the proposed merger presented the opportunity to set the goal to unify and streamline software and hardware deployments across campuses. The business aspect of the merger goals concerned the elimination of internal competition and a unique merged school image.

## *14.4.2 The Results*

The Merged School Project has succeeded, albeit with some difficulties. The decisional, functional, information, resources and organisational models created during the merger have helped significantly to understand the present situation and to select an optimal future state. The use of languages easy to understand and able to manage complexity has resulted in stakeholder buy-in for the project and middle-management consensus on the essential aspect of the future MS.

Unfortunately however, most modelling and mappings (including the modelling of the pre-merger situation) occurred during the merger project itself rather than before; thus, there was insufficient time to achieve appropriate modelling detail. This has led to a 'devil in the detail' situation as the human resources allocated to accomplish the merger and post-merger integration tasks were unable to fully complete the new design before roll-out.

In addition to their inappropriate granularity, the available models were only partially applied. For example, an organisational model showing changes in roles and decisional framework in the transition from the present to the future states was implemented only at the top layer due to the lack of time and proper preparation. As a result, the Head of the newly formed MS had to spend significant amount of time 'putting out fires' (finding short term solutions to re-occurring product/resources imbalances). Thus, unfortunately the interventionist and turbulence issues outlined in the pre-merger (AS-IS) organisational and decisional models were not effectively addressed.

Staff consultation has taken place; however, a significant amount of feedback never translated into changes to the proposed organisational model. This has harmed trust and thereby has reduced the acceptance level of the merger project.

Importantly, the detailed process modelling was never completed and as such the implementation went ahead without detailed models and guidance, in a 'cold turkey' manner (i.e. overnight changeover) resulting in a state of confusion as to 'who does what' lasting several months and affecting staff and students.

On the positive side, the Merged School did achieve a unique image, and in time reached an increased level of integration and consistency across campuses and a more efficient resource administration.

## 14.5 Application of the Preparedness Building Approach

Preparedness can be built for announced and potential M&As. In this case, the merger partners were known, therefore this is a case of preparedness building for an announced merger. Firstly we apply the described model to identify potential issues that can be addressed during preparedness building and then demonstrate how to plan activities based on the objectives identified by MAIIC.

### 14.5.1 *Identifying and Categorising M&A Issues*

From the case study, major issues in PMI can be identified and categorised as shown in Fig. 14.4. These issues could have been addressed for effective PMI and smooth operation of MS, however in this case the merger project failed to achieve that. As a result, there were problems after the merger and pre-merger issues reoccurred even after the merger. Also the PMI resulted in delay in realising the expected synergies and ultimately delaying the delivery of underlying values. According to Epstein [8] often such delays affect the success of M&As.

From the above discussion, the aim of preparedness building can be set as follows:

- *Identify obstacles to the transformation and implement appropriate preventive actions;*
- *Plan for post-merger integration based on the expected outcomes;*
- *Prepare a PMI Plan and an Integration Strategy;*
- *Involve key stakeholders (both schools' management, administration and academic staff) in the preparedness building activities*

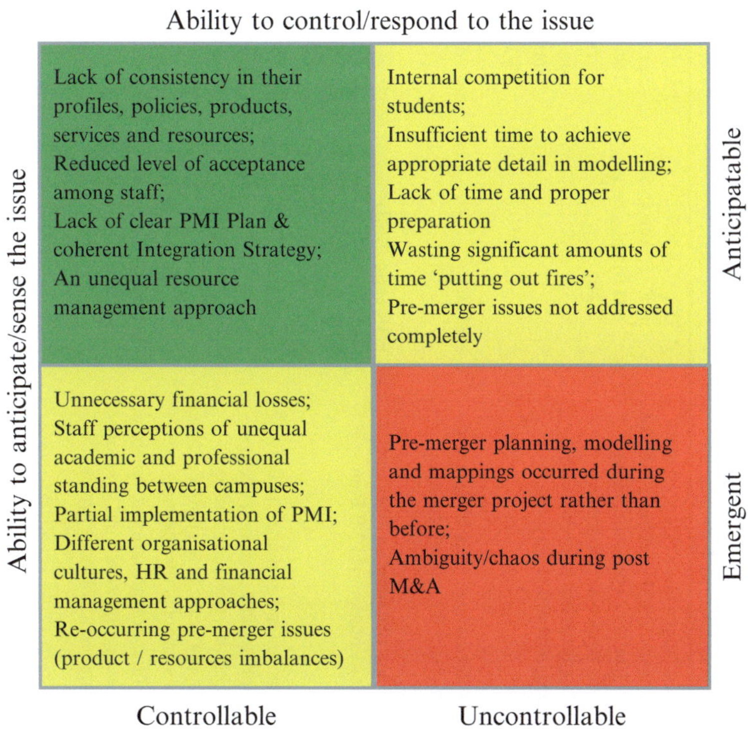


**Fig. 14.4** Identifying and categorising the merger issues in the case study

The identification of issues allows planning for detailed activities which could be implemented during preparedness building. In Table 14.1, based on the MAIIC, we summarise major transformational activities to build preparedness for the merger of the two schools.

## *14.5.2 M&A Preparedness Building Methodology Application*

The previously explained steps of MAPBM can be configured and tailored to plan the detailed activities of needed transformation programs and projects.

In the first step, we identify the entities affected by preparedness building as the management, academic and administration staff, students, services, technical infrastructure and Information Services. In addition, preparedness building requires a strategic program and projects which together could be named Preparedness Building Strategic Program (PBSP), a Business Preparedness Building Project (BPBP) and a HR Preparedness Building Project (HRPBP). The MAIIC model is used to plan the activities for each of these projects and program. Table 14.1 summarises major activities of the preparedness building transformation mandates for the projects and programs involved.

**Table 14.1** Planned activities for Transformational Preparedness Building

| Issue/problem | Nature and objective | Planned activity | Implemented by | Affected entities |
|---|---|---|---|---|
| Lack of consistency in profiles, policies, products, services and resources; unequal resource management approach; differences in requirement specification for technical infrastructure and management of IT resources | Anticipatable and controllable; *prepare* | Identify possible changes into current courses/ programs, organisational structure, policies, reporting mechanisms, communication methods, performance and resource management approaches to manage and maintain them in a unified way | BPBP, HRPBP, ISPBP, Academic and Management Staff | Staff, Products and Services, Technical Infrastructure and Application Services |
| Reduced level of acceptance among staff; internal competition for students | Anticipatable and controllable, *prepare* | Prepare staff for the merger; plan, initiate and continuously foster the culture change | HRPBP | Staff and Students |
| Insufficient time to achieve appropriate detail in modelling<br>Lack of time and proper preparation | Anticipatable and uncontrollable, *avoid* | Effective pre-merger planning for PMI and efficient implementation of PMI Plan | PBSP and ISPBP | |
| Partial implementation of PMI<br>Re-occurring pre-merger issues (product/resources imbalances); wasting significant amounts of time 'putting out fires' | Emergent and controllable, *respond* | Continuously monitor and evaluate the PMI implementation; identify relevant changes at the F and U levels to ensure strategic alignment; support enhanced communication and resources sharing across campuses | PBSP and BPBP | BPBP, HRPBP, ISPBP and Management Staff |
| Staff perceptions of unequal standing between campuses; different organisational cultures; differences in HR management approaches across the campuses | Emergent and controllable, *respond* | Prepare staff for future organisational structure; achieve consistent organisational structure and HR management practices across campuses | HRPBP with the support of BPBP | Management, Academic and Administration Staff |

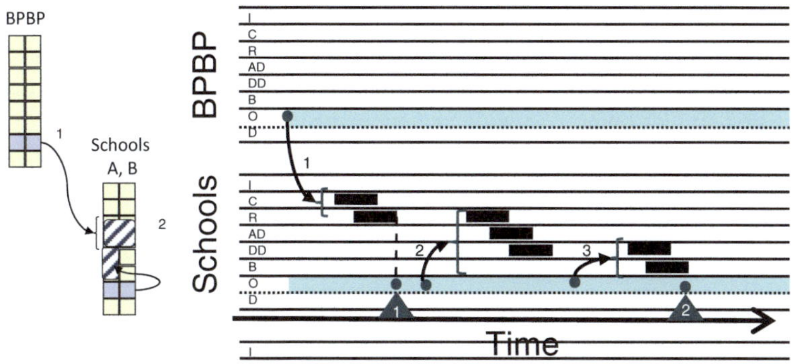


**Fig. 14.5** An example of Dynamic Business Model and life history diagram

The second step is to describe the plan of preparedness building transformation by demonstrating the planned interactions of identified entities. Various graphical representations for such descriptions exist; in this case, we have used the so-called 'dynamic business models' based on the Generalised Enterprise Reference Architecture (GERA) Modelling Framework (MF) [28] which is able to integrate multiple aspects in one representation, in the context of life cycle and history. Each 'relationship' in such models is considered a contribution of an entity to another entity's lifecycle activities. For example, Fig. 14.5 left depicts a relationship (extracted from the models developed for this case study) showing that as a part of its operation, BPBP suggests necessary changes to the structure of Schools A and B (relationship 1) while the Schools' management identifies resulting changes in the architectural and detailed design of products and services offered (relationship 2).

The third step is to determine the activities and their timing. We use so-called 'life history diagrams' to demonstrate *who* (an enterprise entity) will do *what* (detail activity), *how* (by making changes to one or more lifecycle phases of other entity) and *when* (relative sequence). Figure 14.5 shows an extract from the life history models developed for the proposed solution reflecting activities on the time horizon.

## 14.6 Conclusions and Further Work

This paper has proposed a Merger and Acquisition Preparedness Building approach featuring identification and categorisation of M&A issues and a three-step methodology for identifying activities facilitating enterprise-wide transformational preparedness. The approach features EA-specific MF and meta-methodology artefacts in order to achieve an integrated, life cycle-based approach in line with the complexity and 'wickedness' of the problems posed by such endeavours. The paper has also demonstrated the MAPB approach through a case study and has shown what areas could have been addressed by the MAPB components so as to improve the speed, efficiency and facilitate the PMI of the merger project.

The MAPB has the potential to be further evolved into a comprehensive and practical preparedness-building management approach aiming to improve the M&A success rate by addressing the root causes of issues, so that an enterprise is ready for M&As and similar enterprise-wide change endeavours. Therefore, a new research project was started in order to develop a checklist of key M&A issues and their solutions, to define the state of M&A Preparedness in terms of systemic properties and devise a list of optimal M&A Preparedness Building Activities.

## References

1. Vaniya N, Bernus P (2012) Strategic planning to build transformational preparedness: an application of enterprise architecture practice. In: ACIS 2012 proceedings
2. Vaniya N, Bernus P, Noran P (2013) Examining potentials of building M&A preparedness. In: Proceedings of ICEIS 2013 (in press)
3. Alaranta M, Henningsson S (2008) An approach to analysing and planning post-merger IS integration: insights from two field studies. Inf Syst Front 10(3):307–319
4. Bannert V, Tschirky H (2004) Integration planning for technology intensive acquisitions. R&D Manage 34(5):481–494
5. Baro GA, Chakrabarti A, Deek F (2008) Organizational and informational system factors in post-merger technology integration. In: AMCIS 2008 proceedings, paper 359. http://aisel.aisnet.org/amcis2008/359
6. DiGeorgio R (2002) Making mergers and acquisitions work: what we know and don't know—part I. J Change Manage 3(2):134–148
7. Eckert M, Freitag A, Matthes F, Roth S, Schilz C (2012) Decision support for selecting an application landscape integration strategy in mergers and acquisition. In: ECIS 2012 proceedings, paper 88. http://aisel.aisnet.org/ecis2012/88
8. Epstein MJ (2004) The drivers of success in post-merger integration. Organ Dyn 33(2):174–189
9. Giacomazzi F, Panella C, Pernici B, Sansoi M (1997) Information systems integration in mergers and acquisitions: a normative model. Inf Manage 32:289–302
10. Grainger N, McKay J (2007) ERP: more than just a system: the role of ERP in mergers and acquisitions. In: ACIS 2007 proceedings, paper 54. http://aisel.aisnet.org/acis2007/54
11. Hwang M (2004) Integrating enterprise systems in mergers and acquisitions. In: AMCIS 2004 proceedings, paper 12. http://aisel.aisnet.org/amcis2004/12
12. Jayadev M, Sensarma R (2007) Mergers in Indian banking: an analysis. South Asian J Manage 14(4):20–49
13. Kumar R (2009) Post-merger corporate performance: an Indian perspective. Manage Res News 32(2):145–157
14. Larsen MH (2005) ICT integration in an M&A process. In: PACIS 2005 proceedings, paper 95. http://aisel.aisnet.org/pacis2005/95
15. Lauser B (2009) Post-merger integration and change processes from a complexity perspective. Baltic J Manage 5(1):6–27
16. Mehta M, Hirschheim R (2007) Strategic alignment in mergers and acquisitions : theorizing IS integration decision making. J Assoc Inf Syst 8(3):143–174
17. Mo JPT, Nemes L (2009) Issues in using enterprise architecture for mergers and acquisitions. In: Doucet G, Gotze J, Saha P, Bernard S (eds) Coherency management. AuthorHouse, Bloomington, pp 235–262
18. Myers MD (2008) Integration of different ERP systems: the case of mergers and acquisitions. In: PACIS 2008 proceedings, paper 65. http://aisel.aisnet.org/pacis2008/65

19. Rodriguez A (2008) Mergers and acquisitions in the banking industry: the human factor. Organ Dev J 26(2):63–74
20. Ross JW, Weill P, Robertson DC (2006) Enterprise architecture as strategy: creating a foundation for business execution. Harvard Business School Press, Boston, MA
21. Robbins SS, Stylianou AC (1999) Post-merger systems integration: the impact on IS capabilities. Inf Manage 36:205–212
22. Schuler R, Jackson S (2001) HR issues and activities in mergers and acquisitions. Eur Manage J 19(3):239–253
23. Stylianou AC, Jeffries CJ, Robbins SS (1996) Corporate mergers and the problems of IS integration. Inf Manage 31:203–213
24. Vernadat FB (2007) Interoperable enterprise systems: principles, concepts, and methods. Annu Rev Control 31:137–145
25. Walter I (2004) Mergers and acquisitions in banking and finance. Oxford University Press, New York, NY
26. Wijnhoven F, Spil T, Stegwee R, Fa RTA (2006) Post-merger IT integration strategies: an IT alignment perspective. J Strateg Inf Syst 15:5–28
27. Noran O (2008) A meta-methodology for collaborative networked organisations: creating directly applicable methods for enterprise engineering projects. VDM, Saarbrücken
28. ISO/IEC (2005) ISO/IS 15704:2000/Amd1:2005: industrial automation systems: requirements for enterprise-reference architectures and methodologies. Geneva: ISO
29. Rittel H, Webber M (1973) Dilemmas in a general theory of planning. Policy Sci 4:155–169

# Chapter 15
# Starting Building a IT Policy: A Quest to IT Success

**Pedro Neves Rito**

**Abstract** An organization uses information technology (IT), because the IT is consistent with the functional and operational objectives designed by that organization. The investment that is required to place technological innovation in an organization carries significant weight. This weight can be translated into decision making or monetary costs. It is therefore important that after implementation, ensured close monitoring of their use and such monitoring can take many forms, one of which can be explicit through a usage policy. It is the responsibility of the technology manager, to outline the best policy to use the technology that was adopted first by, taking into account the need of the organization and secondly to ensure the construction of monitoring mechanisms and set up conditions to facilitate its successful use. Therefore, the study of the adoption and diffusion of a technological innovation provides these professionals more tools that can assist in the preparation of these strategies, without forgetting that it is necessary to plan at the user level and not only to assist the management in its objectives.

**Keywords** Implementation of technologies • Diffusion of innovations • Resistance to change • Literature review

## 15.1 Introduction

Organizations tend to select the technological innovations that best suit their needs but sometimes they take into account only the objectives of the management. In the literature some cases arise where the selection is made by using

P.N. Rito (✉)
Polytechnic Institute of Viseu, Viseu, Portugal
e-mail: pedronrito@gmail.com

M. José Escalona et al. (eds.), *Information System Development: Improving Enterprise Communication*, DOI 10.1007/978-3-319-07215-9_15,

information systems professionals that outline the strategies necessary for the dissemination and adoption of these innovations. Although they are professionals, sometimes they do not take into account the tools that exist, either in form of models or theories that can assist in this work. Through this study we hope to verify that the delineation and the existence of policy IT strategies have some impact on the user, and the option will be the study of complex technologies for voluntary/optional use.

Explaining the reasons for the use of technologies and processes pertaining, as the adoption process, is crucial for academics, professionals and political, to the extent that technology is one of the key powers of globalization and new opportunities for business.

To exploit the full potential of complex technologies it is necessary for organizations to undertake the encouragement of individuals so that they accept these technologies. The effort made in this direction, by organizations, should consider the complexity of factors influencing the perceptions of individuals about the technologies, as well as their intentions and how the technologies will be used [1]. Only by understanding these factors will the adoption be allowed to proceed in a positive way.

If technology adoption is not done properly, it may be that users never get to explore their full potential, using technology in a superficial way, merely as a means of entry and storage of information rather than as a way to use and analyze data and gain competitive advantages. The adoption of technology varies depending on the technology, users and context.

Thus, it is clear that research into the processes of adoption information technologies has long been important, since it is a prerequisite for the use of technology and the realization of its real value, regardless of technological advances.

When analyzing the issue of existing technologies in organizations, it appears that they relate to one or both of the following guidelines:

i. Those that are essential for use, where the individual is required to use them, not having the ability to deny their use, as is the case of ERP in organizations.
ii. Those that are optional, that is, the individual may or may not use them, but must always bear in mind that it serves to increase his productivity and will facilitate access to information.

These boundaries are not considered airtight, because there are examples where the same technology can be viewed in the two orientations. However the use of technologies (i) and (ii) within organizations are perceived differently by individuals and there is a need to consider whether the existence of an organizational policy in relation to their adoption has an impact on them.

We have no intention at this stage to present results, as they still need to be validated according to the methodology we chose. In this document we will invariably present a literature review on issues related to the use of IT and on the management and IT professional role, during the implementation and post-implementation of a new technology in an organization.

### 15.1.1 *Information Technology*

In most organizations there is pressure to make their operational processes tactical and strategic more efficient and effective [2]. The term IT may refer to an artifact that is based on technology, computer or communication between hardware and software.

The use of IT is considered a facilitator in business environments where organizations gain competitive advantage, and can, therefore, compete. It is necessary that these organizations realize the power of technology and align these technologies with the objectives of the business [3]. One element of this strategy is having the business aligned with the technology implemented, uniform and directly control installed technology and realize that it is necessary to do upgrades, updates or changes.

Most organizations invariably turn to the departments of information systems to help them gain an advantage in the market, through a saving of resources and capacity to be able to respond to the challenges that arise in the environment where they operate [4]. However, organizations can not realize the full benefit of the technology unless they make the necessary changes in the organizational structure, strategies and processes. Some departments begin to implement various technologies (hardware and software) that result in many organizations characterized by systems that do not communicate with each other [5]. Despite the individuals realizing that they could obtain better performance and efficiency by using all information systems, they are incompatible. The information in this case is usually shared at the local level and not at a global level, creating islands that have little impact on the productivity of an organization. It is important to explain how new technology can assist the individual in their performance of tasks [6].

## 15.2 IT Implementation

Organizations can maximize the benefits of investment in IT but need to understand and manage the implementation process [2]. IT represents a substantial investment for most organizations and is a significant asset, but this value is only realized when the information systems are used by their respective users in a way that contributes to the strategic and operational objectives of the organization [7].

Sometimes members of an organization avoid changing technologies because they are used to using a particular system for a long period of time, and that barrier is usually observable in the implementation of a new technology [4].

Implementors are usually the organizations managers, functional managers or IT professionals. Implementors of information technologies are, thereby, those who are responsible for the introduction of technologies to potential users [8]. They are also those responsible for the success of using the system which has been implemented. This includes implementing it on time and within the budget and still matching the desired benefits [9].

The change in management and user involvement are the most important factors determining the success/failure of information systems projects [10].

### 15.2.1 *IT Implementation Failure*

Among the main factors associated with failures in the implementation of information systems projects is the resistance of users [11]. Resistance is a behavioral reaction to this or any situation that occurs and is perceived as negative, unfair, threatening or stressful feeling [12].

Understanding the users and how they interact with IT is critical to the successful implementation of new systems within organizations [13]. It is important to know the characteristics of individual decision makers, since people do not process information or reach decisions in the same way [6].

## 15.3 The Changing Process

It is necessary for organizations to realize that part of the change process is to understand the "why" factor of change. Organizations are advised to increase the emphasis in management practices during the process of new implementation. This is necessary because the way change is introduced will affect the response of the employees [14].

Organizations are systems defined by shared standards, will that provide a space for individuals and sections to pursue their interests. These systems are social in that they are based on interdependence between the individuals within the organization. Organizations must understand the information systems and that individuals use signals to communicate and to perform intentional actions [15]. The organization must, therefore, be aware of these signs, which most of the time conceal resistance.

## 15.4 Resistance to Organizational Change

Organizational change is an empirical observation about an organizational entity that has had variations in form, quality or state over time, after having been deliberately introduced to new ways to thinking, acting and operating [16].

The problem of resistance has been presented and discussed as one of the most common reasons for not using innovations. If a new IT project is implemented in an organization for reasons related to economic sustainability, the staff first have to be prepared for these changes, since there is a risk of protest and refusal to use innovation, that can lead to unintended results [17]. On the other side, change begins with the perception of its necessity, and a wrong perception is the first barrier to change [16].

The organizational change to occur has to take into account the possible resistance from those who are part of this organization [18]. The change in the organization that leads to symptoms of resistance does not arise from the working group by a minority. There are responsible individuals in organizations, particularly executives, managers, project leaders, leaders of teams and the teams that want to improve organizations who have to take into account the resistance of those who will be affected by these changes [19]. If changes do not succeed, it is usually not due to technical reasons but human reasons, since the promoters of these changes do not value the reactions of those to whose daily routines will be influence by changes [20].

Most organizational change is managed from a technical standpoint without recognizing or realizing how human factors influence the success or failure of this change [21]. It is easier for management to focus its attention and to be concerned with technical factors that are quantifiable and predictive than to develop strategies and action plans, calculate profits and streamline resources which take into account the individual aspects of individuals. Management has a tendency to neglect and ignore the size of the human factor when implementing changes. For these authors there is too little investment in communication, and the training and follow-up necessary for the successful implementation of change. There are sometimes attempts by managers, by way of manipulation and coercion to cause success. However, these practices tend to result in distrust and resentment, making it even harder to implementation.

Researchers have attempted to explain the terminology of resistance to change, it usually results in explanations of why certain efforts to introduce large-scale changes in technology, production methods, management practices or to compensate for the failures by information systems fail [22].

There are differences between organizations in their capabilities to manage complex processes of implementation technologies, as is the case of complex technologies. It is necessary to understand and manage individual resistance and maintain the support and acceptance by the members of the organization affected by these implementations [23]. The change sometimes does not work because organizations underestimate the importance of the responses by individuals at every level: cognitive, affective and behavioral.

The literature on change typically indicates that employees follow their head (aspects of cognition) and heart (affections) when they have to deal with responses to changing events. Employees use a combination processes of cognition and affection for the change to take effect and have an answer. Cognitive processes are related to believe about the need for change, realizing the significance of the change and the favorable results regarding this change, i.e., extend the change will be beneficial both personally and organizationally [24].

## 15.5 Management Actions

The decisions about a technology are fraught with complexity and uncertainty. The impact of technology can take years to be realized, and the benefits of adoption are often uncertain [25]. Sometimes there are situations in organizations that require procedures for the management of these organizations. These actions can be triggered by different situations related with the implementation of ideas developed by staff in the organization or with the use of terminology suitable for non-specialists.

Resistance should not be treated as a black box, this is like having a single solution. Managers must have the ability to manage via different actions [12].

Conversely, regarding actions supported by the management and the involvement of innovation being implemented, communication and training are important activities that management has to realize [26]. The management of organizations must, therefore, learn effective strategies for managing the different stages of change and not have a only one approach to resolving the situations. On the other hand studies have found evidence that resistance to change on the part of employers is a major reason for non-use of innovations [11]. Without the existence of an appropriate procedure to change, the initial attitudes of employees regarding the change will probably continue [14, 23].

### *15.5.1 Strategies to Manage Resistance*

Some of the problems that arise related to resistance to change are attitudes that individuals created by themselves about their work and about their own ideas. If an innovation arises and is supported by the management of the organization, because it is seen as an event favoring the development of the organization, it is expected that this innovation is subsequently shared with the other elements organization [18]. The management in organizations can influence these attitudes of resistance and deal with problems directly from their origin.

The critical factors management must take into account regard the costs related to changing the resistance of individuals. Management can try to reduce these costs by improving favorable opinions about the changing technology and enhancing self- efficacy in individuals for change. This change can be accomplished through advertising about the need for new technology and persuading individuals who are identified in the organization as opinion leaders so that they are the first to accept the change. In addition to changing the views, management can provide training for individuals to improve their skills and confidence in the new system. They should also show the advantages of the new system in terms of these individuals. There must be organizational support for change through training, mentoring and learning resources, always using a clear communication strategy, before implementation [9, 27, 28].

It is necessary to understand the user and how they interact with IT for there to be success in the implementation of the new system. Systems are becoming larger and more complex. They have an impact on both the organization and individuals at all levels. One factor of success is use, but resistance is something that is normal when a new system is introduced and thus resistance may contribute to the failure of the new implementation [29].

### 15.5.2 User Resistance to IT

The organizational actors involved in technological change are not a homogeneous group but rather a group of users that belong to different groups, each of which has its own organizational agendas, personal agendas and social point of view. Resistance is diverse in its nature [30]. For most individuals they use a system only if they believe that the use will support the current position of power, otherwise they resist [31]. The individual does not resist change per se, he resists the uncertainty and the possible outcomes that change can cause [32].

Researchers acknowledge that people have a natural tendency to keep things that are familiar to them as opposed to accepting an innovation that unknown [22]. The success of any change depends on the people and their ability to perform the activities that are necessary [27]. Users of information technologies may react differently to a new technology. They may reject the technology, partially, actively resist, gradually accept or embrace it in full [7].

Resistance can take various forms, including [18]: persistent reduction in individual production, increasing the number of dropouts, increasing the number of transfer requests, emergence of chronic quarrels, obstinate hostilities, individuals consistently angry, individuals likely to participate in strikes and daily use of explanations to justify why the change does not result. These reasons may lead to the need for a review by the management of the organization's budget for the implementation of new technology [33].

Although literature acknowledges resistance to change as, it is a complex phenomenon which can not be easily described [34].

### 15.5.3 Strategies for Dealing with User Resistance

Develop a unique recipe to overcome resistance is problematic for agents of change. The existence of the user with heterogeneous resistance must be taken into account by the agents of change [30]. The generic recipe to overcome the resistance does not take into account the diversity in resistance strategies used by various organizational actors. Attention should be paid to the heterogeneity of the user or group and their answers, regarding the development of the information system as it will contribute to enrich the knowledge about the phenomenon of resistance.

#### 15.5.3.1 User Participation

The participation of the user may be defined as a set of behaviors or activities performed by the user during system development [35]. This participation enhances self- efficacy and has a positive impact on the usefulness and ease of use [36]. The participation by the users create a sense of appreciation, empowerment and ownership, by creating opportunities to influence decisions about the new system. Empowerment is one factor that helps create the feeling of control, and promotes the enthusiasm of the user and reduces resistance when the technology changes [29].

#### 15.5.3.2 User Training

Training is a process of transferring knowledge and skills to the operational user of technology. It is used to provide users with skills so they can use the system. The training of the user of the technology increases their understanding of the new system and can be used as a tool to increase feelings of self-efficacy. On the other hand, training was identified as a tool for anxiety [37, 38]. Training helps you gain confidence in the handling of the new technology without fear of losing important information [39].

#### 15.5.3.3 User Support

User support can be defined as a variety of modes that are available to help the end user to resolve their problems with the technologies or applications [40]. The office help desk is one way to provide support to users. The help desk provides help upon request including: training, consulting on systems, technical assistance, documentation about the software, consulting and support on hardware and development support [41]. If it is an organization where the users are laggards the experience of a department of an IT department is important [39].

#### 15.5.3.4 Communication

Communication is one of the most important strategies to prevent resistance by the user. It is a tool that helps organizations during a process of change in technology make users feel important, and increases their levels of confidence in the changes that will arise [11]. Communication can be oral, written or non-verbal.

#### 15.5.3.5 Consultants

The use of consultants may be beneficial during deployment since they have experience and knowledge. These are external agents to the organization that are not involved in the organization's policies and decisions are related to what they think is best for the organization's business [42]. They are generally identified as change agents who are responsible for managing, both social and technological change within organizations [43]. They also reduce anxiety about the risks involved, technical problems, job change and lack of adequacy [39].

#### 15.5.3.6 Make Use of Champions

Champions are more than leaders. They are transformational leaders who will inspire other individuals in the organization to transcend [44]. These leaders promote a vision of IT and surpass the obstacles on the authorization and implementation. They manage to overcome the bureaucratic barriers and drive change in the organization [45].

Some authors [46] suggest the use of subject matter experts (SME), who are individuals with knowledge of the business in the organization and about the processes that are critical in analyzing the gaps and also configurations of the systems. These individuals are able to configure the new system that is being implemented. Project managers plan carefully and safely. They ask those individuals within the organization that are considered the best and brightest at each functional or operational department level to help. After implementation, these individuals assume the role of functional units with great ability to influence future decisions about improvements and process improvements.

#### 15.5.3.7 Mix of Personalities

Individuals with a resisting personality, generally have a negative attitude towards the use of information technologies. Information systems in organizations are only beneficial if used by relevant staff. If there are many people with this personality type it is not beneficial to the process of implementation and use of new technology. Thus, the departments of human resources and recruitment should try to encourage a healthy mix between individuals with IT profiles and other technical departments [38]. There is evidence that individuals have different attitudes when adopting new technology [47].

#### 15.5.3.8 User Friendly System

If the system is user friendly, users will feel they have more control over the system, which increases conformity and skills, i.e. self-efficacy. Complex systems and technical problems increase the resistance of the user, so the design and the characteristics are significant [48, 49, 50]. Computer systems should be designed in a way that is easy to use, simple to learn and flexible in interaction [51].

#### 15.5.3.9 Policy

An IT policy is built to handle issues related to the dissemination of information, use of information and dissemination and use of a technology [52]. The policy may further clarify the duties of stakeholder's responsibilities and rights and specify what is efficient and acceptable. Such policies should be aligned and take into account the needs of the community. Politics is, therefore, a set of guidelines which is addressed to the stakeholders of the organization [3]. For the current policy to be appropriate it should involve all stakeholders during its development. The policy of one organization may not be compatible with other organizations [52].

## 15.6 Outlook and Conclusions

This review of literature emerged by the need to understand the concepts that served to build an initial framework and to help contextualize what was found in the case studies. As we progressed in research, in which was used qualitative methodology with interpretative analysis of the data, we verified that it was essential to deepen this review, as the case studies were revealing information that had not been foreseen in that initial framework. It has been required, then, to clarify the initial concepts and the "new" concepts that came with cases.

In the work that we have been developing, four case studies have been analyzed. The four cases are universities without any institutional grating. In each of the schools they identified and analyzed the different realities with regard to the implementation process and diffusion and acceptance of Management Learning Systems. In each organization they found there are different definitions of successful use of technology. They have been implemented in all usage policies but only one has accomplished the verification of politics. Thus, the empirical work we have done at this stage is to validate a framework that results from the cross-reference between the literature and four case studies. Our goal is to build a tool to support the work of IS professionals for the purpose of implementation and post-implementation of a new technology, which can be achieved by developing a proper policy on the use of technology.

## References

1. Youngberg E, Olsen D, Hauser K (2009) Determinants of professionally autonomous end user acceptance in an enterprise resource planning system environment. Int J Inf Manage 29: 138–144
2. Cooper RB, Zmud RW (1990) Information technology implementation research: a technological diffusion approach. Manage Sci 36:123–139
3. Alinaghian R, Rahman AA, Ibrahim R (2011) Information and communication technology (ICT) policy; significances, challenges, issues and future research framework. Aust J Basic Appl Sci 5:963–969
4. Rajagopal P (2002) An innovation–diffusion view of implementation of enterprise resource planning (ERP) systems and development of a research model* 1. Inf Manage 40:87–114
5. Jarvenpaa SL, Ives B (1993) Organizing for global competition. Decis Sci 24:547–580
6. Chakraborty I, Hu PJ-H, Cui D (2008) Examining the effects of cognitive style in individuals' technology use decision making. Decis Support Syst 45:228–241

7. Agarwal R, Karahanna E (2000) Time flies when you're having fun: cognitive absorption and beliefs about information technology usage 1. Mis Q 24:665–694
8. Rivard S, Lapointe L (2012) Information technology implementers' responses to user resistance: nature and effects. Manage Inf Syst Q 36:897–920
9. Martinko MJ, Zmud RW, Henry JW (1996) An attributional explanation of individual resistance to the introduction of information technologies in the workplace. Behav Inf Technol 15:313–330
10. Elie-Dit-Cosaque C, Pallud J (2010) User adaptation and IS success: an empirical investigation among French workers. In: ICIS 2010 proceedings
11. Jiang JJ, Muhanna WA, Klein G (2000) User resistance and strategies for promoting acceptance across system types. Inf Manage 37:25–36
12. Lapointe L, Rivard S (2005) A multilevel model of resistance to information technology implementation. Mis Q 29:461–491
13. Russell DM, Hoag AM (2004) People and information technology in the supply chain: social and organizational influences on adoption. Int J Phys Distrib Logist Manag 34:102–122
14. Calvert C (2006) A Change-Management Model for the Implementation and Upgrade of ERP Systems
15. Al-Rajhi M, Liu K, Nakata K (2010) A Conceptual Model for Acceptance of Information Systems: An Organizational Semiotic Perspective
16. Del Val MP, Fuentes CM (2003) Resistance to change: a literature review and empirical study. Manage Decis 41:148–155
17. Laumer S, Maier C, Eckhardt A (2010) Why do they resist?—an empirical analysis of an individual's personality trait resistance regarding the adoption of new information systems. In: ECIS 2010 proceedings
18. Lawrence PR (1969) How to deal with resistance to change. Har Bus Rev 32(3):49–57
19. Palmer B (2004) Change management: overcoming resistance to change. Qual Prog 37:35–39
20. Sultan F, Chan L (2000) The adoption of new technology: the case of object-oriented computing in software companies. Eng Manage IEEE Trans 47:106–126
21. Bovey WH, Hede A (2001) Resistance to organisational change: the role of defence mechanisms. J Manage Psychol 16:534–548
22. Klaus T, Blanton JE (2010) User resistance determinants and the psychological contract in enterprise system implementations. Eur J Inf Syst 19:625–636
23. Cohen J (2010) Cognitive, affective and behavioural responses to an ERP implementation: a dual perspective of technology acceptance and organisational change. In: ACIS 2010 Proceedings
24. Smollan RK (2006) Minds, hearts and deeds: cognitive, affective and behavioural responses to change. J Change Manage 6:143–158
25. Bhattacherjee A (2001) Understanding information systems continuance: an expectation-confirmation model. Manage Inf Syst Q(25)
26. Shang S, Su T (2004) Managing user resistance in enterprise systems implementation
27. Garaca Z (2011) Factors related to the intended use of ERP systems. Manage J Contemp Manage Issues 16:23–42
28. Wanchai P, Andrade A, Techatassanasoontorn A (2012) Adjusting to mandatory information systems: understanding individual adaptation to ERP systems. In: Conf-Irm 2012 proceedings
29. Tavassoli L, Toland J (2008) Strategies for Dealing with End-User Resistance
30. Griffiths M, Light B (2006) User resistance strategies and the problems of blanket prescriptions: a case study of resistance successes. In: ECIS 2006 proceedings
31. Markus ML (1983) Power, politics, and MIS implementation. Commun ACM 26:430–444
32. Waddell D, Sohal AS (1998) Resistance: a constructive tool for change management. Manage Decis 36:543–548
33. Kim H-W, Kankanhalli A (2009) Investigating user resistance to information systems implementation: a status quo bias perspective. Mis Q 33:567–582

34. Laumer S (2011) Why do people reject technologies—a literature-based discussion of the phenomena resistance to change in information systems and anagerial psychology research. In: ECIS 2011 proceedings
35. Barki H, Hartwick J (1989) Rethinking the concept of user involvement. Manage Inf Syst Q(13)
36. Venkatesh V, Davis FD (2000) A theoretical extension of the technology acceptance model: four longitudinal field studies. Manage Sci 46:186–204
37. Huang AH (2002) A three-tier technology training strategy in a dynamic business environment. Adv Top End User Comput 2:263
38. Eckhardt A, Laumer S, Nguyen NT (2010) Social influence in technology adoption research—a scientometric study over two decades behavior. In: Digit 2010 proceedings
39. Järveläinen J (2012) "I wouldn't go back to the old system": a technology laggard organization, resistant users and system implementation. In: Bled 2012 proceedings
40. Yager SE, Schrage JF, Berry RL (2002) Preparing end user support specialists. J Educ Bus 78:92–96
41. Marcella R, Middleton I (1996) The role of the help desk in the strategic management of information systems. Oclc Syst Serv 12:4–19
42. Adams B, Berner ES, Wyatt JR (2004) Applying strategies to overcome user resistance in a group of clinical managers to a business software application. J Organ End User Comput 16:55–64
43. Winston ER (2002) IS consultants and the change agent role. SIGCPR Comput Pers 20:55–74
44. Beath CM (1991) Supporting the information technology champion. Mis Q 15:355
45. Bradley J (2005) Are All Critical Sucess Factors in ERP Implementation Created Equal? Amcis 2005 Proc
46. Gallagher KP, Gallagher VC (2012) Organizing for post-implementation ERP: a contingency theory perspective. J Enterp Inf Manag 25:170–185
47. Rogers EM (2003) Diffusion of innovations, 5th edn. Simon & Schuster, New York, NY
48. Ellen PS, Bearden WO, Sharma S (1991) Resistance to technological innovations: an examination of the role of self-efficacy and performance satisfaction. J Acad Mark Sci 19:297–307
49. Compeau DR, Higgins CA (1995) Computer self-efficacy: development of a measure and initial test. Mis Q 19:189
50. Venkatesh V, Morris MG, Davis GB, Davis FD (2003) User acceptance of information technology: toward a unified view. Mis Q 27:425–478
51. Al-Jabri IM, Al-Hadab A (2008) End user adoption of ERP systems: investigation of four beliefs. In: Americas Conference on Information Systems (AMCIS), pp 1–8
52. Labelle R (2005) ICT policy formulation and e-strategy development: a comprehensive guidebook. Elsevier, New Delhi

# Chapter 16
# Design Within Complex Environments: Collaborative Engineering in the Aerospace Industry

**Fernando Mas, José Luis Menéndez, Manuel Oliva, Javier Servan, Rebeca Arista, and Carmelo del Valle**

**Abstract** The design and the industrialization of an aircraft, a major component, or an aerostructure is a complex process. An aircraft like the Airbus A400M is composed of about 700,000 parts (excluding standard parts). The parts are assembled into aerostructures and major components, which are designed and manufactured in several countries all over the world. The introduction of new Product Lifecycle Management (PLM) methodologies, procedures and tools, and the need to reduce time-to-market, led Airbus Military to pursue new working methods to deal with complexity. Collaborative Engineering promotes teamwork to develop product, processes and resources from the conceptual phase to the start of the serial production. This paper introduces the main concepts of Collaborative Engineering as a new methodology, procedures and tools to design and develop an aircraft, as Airbus Military is implementing. To make a Proof of Concept (PoC), a pilot project, CALIPSOneo, was launched to support the functional and industrial design process of a medium size aerostructure. The aim is to implement the industrial Digital Mock-Up (iDMU) concept and its exploitation to create shop floor documentation.

**Keywords** Collaborative engineering • iDMU (industrial Digital Mock-Up) • PLM (Product Lifecycle Management) systems

---

F. Mas (✉) • J.L. Menéndez • M. Oliva • J. Servan • R. Arista
AIRBUS, Av. García Morato s/n, 41011 Sevilla, Spain
e-mail: fernando.mas@airbus.com

C. del Valle
Sevilla University, Reina Mercedes s/n, 42012 Sevilla, Spain
e-mail: carmelo@us.es

M. José Escalona et al. (eds.), *Information System Development: Improving Enterprise Communication*, DOI 10.1007/978-3-319-07215-9_16,

## 16.1 Introduction

The design and the industrialization of aircrafts is a complex process. An aircraft like the Airbus A400M is composed of about 700,000 parts (excluding standard parts). Parts are assembled into aerostructures and major components, which are designed and manufactured in several countries all over the world [1]. Airbus design started in 1969 [2] with the launch of the Airbus A300B.

At that time drawings were created on paper. By the mid-eighties, the Airbus A320 and the CASA CN235 were designed using a Computer Aided Design (CAD) tool, for the 3D surfaces and drawings, and a Computer Aided Manufacturing (CAM) application, for numerical control programming.

In the mid-nineties, Product Data Management (PDM) systems and 3D solid modeling let Airbus to start building a product Digital Mock-Up (DMU), mainly to check functional design interferences. Then Concurrent Engineering was introduced and a huge project, named ACE (Airbus Concurrent Engineering) [3], started to develop and deploy methods, process and tools along all the functional design disciplines. A brief summary of the product Digital Mock Up (DMU) at Airbus is presented in [4].

The complexity of an aircraft as a product is not only at functional design level. The complexity is also related to the industrial design of the aircraft and the generation of the manufacturing documentation. The lifecycle of a typical aircraft could be more than 50 years. The number of versions, variants, customer customizations, modifications due to flight security, improvements, etc. and the need to develop and implement them is another important source of complexity.

During the last years different methodologies and techniques have been applied to deal with this complexity. The core idea of the Collaborative Engineering is to start with functional and industrial design from the beginning of the lifecycle building an iDMU. The iDMU lets influence Product, Processes and Resources each other and perform virtual manufacturing. PLM methodologies and tools are used intensively to help in the implementation and deployment.

The remainder of this paper is organized as follows: the next section explains the role of PLM in managing the process complexity in the aerospace industry. Follows two sections with an analysis of the 'As Is—To Be' situations and the functional model developed. Finally the paper ends with an introduction to CALIPSOneo Project as a proof of concept and some conclusions about the work.

## 16.2 The Role of PLM to Cope with This Complexity

PLM methods and tools are targeted to manage processes and engineering process in the first place. Therefore PLM is being introduced in the aeronautical industry to manage process complexity within functional and industrial aircraft design. The introduction of PLM methods and tools are part of the evolution of the engineering process already taking place in the aircraft industry. This evolution is due mainly to the technological evolution of software tools, the need to short time-to-market, to reduce cost and to increase quality and maturity of the development. PLM is also a main enabler of this evolution.

This evolution can be resumed in three engineering paradigms: traditional, concurrent and collaborative.

## 16.3 Traditional Engineering

The Traditional Engineering approach to design a product is the implementation of design tasks in sequence (Fig. 16.1). Main disadvantages of this approach are: only focused on product functionality, supported in drawings, different teams with lack of communication between them for which is often referred to as the "over-the-wall" approach [5], problems pushed down to the end of the lifecycle, and long time-to-market.

## 16.4 Engineering

About 20 years ago, and with the introduction of the emergent PLM tools, a wide-company project, ACE (Airbus Concurrent Engineering) [3] was launched. Framed within the Airbus A340-500/600 development program, it aimed to develop and deploy concurrent engineering practices and the associated PLM tools.

Concurrent Engineering diminished the disadvantages of the traditional paradigm (Fig. 16.2). The wall described in Traditional Engineering still exits but it is not so high. The industrialization tasks are not as advanced as functional tasks in terms of PLM tools usage. The current deliverable is again the product DMU, interoperability and working practice issues cause that compact disks or memory sticks flies over

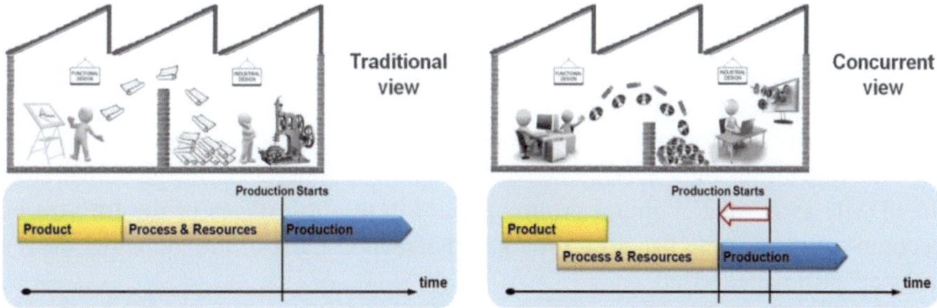


**Fig. 16.1** Traditional view vs. concurrent view

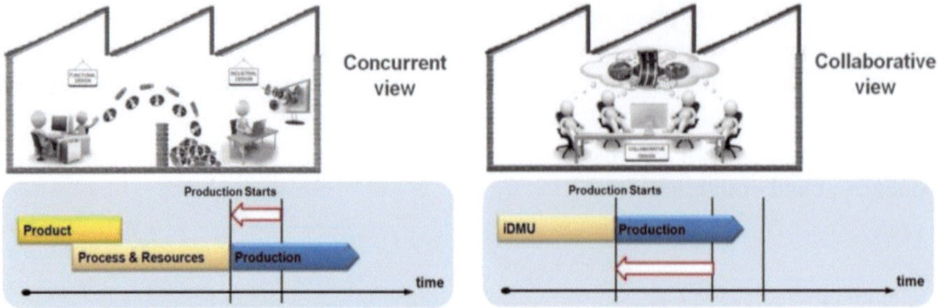


**Fig. 16.2** Concurrent view vs. collaborative view

the wall instead of paper based drawings. Industrial design is not fully integrated with functional design, and it has little influence over the latter. They comprise two separate teams with dissimilar skills.

In fact, one of the most important reasons to have the wall is the certification process by the aeronautical authorities. Traditionally certification was made using the product definition, drawings, and marks the end of the aircraft design process. Concurrent Engineering still holds this idea and considers the aircraft design only as the functional design, enriched with manufacturing constrains and needs.

## 16.5 Collaborative Engineering

Nowadays the aim is a design process with a single team that creates a single deliverable including both the functional design and industrial design (Fig. 16.2). The main advantage is a further time-to-market reduction applying virtual validation by means of virtual manufacturing techniques. It is a new methodology that needs new working procedures and new PLM tools.

## 16.6 Analysis As-Is To-Be

Current or "As Is" situation (Fig. 16.3) shows an optimized functional design area with a clear deliverable: the product DMU. The concurrent process closes the gap between functional design and industrial design, and is intended to promote "Design for Manufacturing" and "Design for Assembly." The functional design deliverable, the "DMU as master," being a profitable item in the first stages of the life- cycle, becomes of decreasing use with time and meanwhile most of the industrial design tasks are still paper based.

Future or "To Be" situation shows an optimized functional and industrial design area with a clear deliverable: the iDMU [6]. The previous gap is eliminated by the

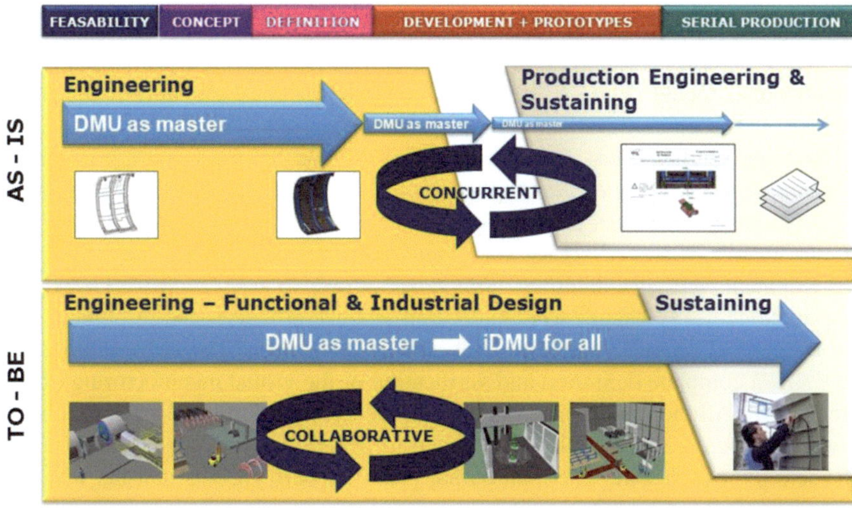


**Fig. 16.3** Analysis "As Is - To Be"

collaborative way of building the iDMU and the virtual validation of it. The design (functional and industrial) deliverable, the "iDMU for all," is a valuable item along the whole lifecycle. The information contained in the iDMU can now be exploited by Sustaining, to produce shopfloor documentation in a wide variety of formats.

This new collaborative design methodology requires also new procedures. In this work we can only discuss the core of the procedures change, and this can be done within this analysis As-Is To-Be.

Current As-Is Concurrent Design procedures are based in a well defined DMU lifecycle made of exhaustively defined maturity states. DMU maturity states are controls for the industrial design tasks. The different Industrial design tasks cannot be started until the DMU reaches the corresponding maturity state. There are also procedures controlling the evolution from a maturity state to the following based on approval of all the stakeholders. This way, the industrial design only can progress on the tracks of the functional design, and processes and resources life- cycles are less exhaustively defined.

In the To-Be Collaborative Procedures, the maturity of product, processes and resources must evolve at the same time. The product DMU is not delivered at defined states to the industrial design. Instead, product, processes and resources are available at any time to all the development stakeholders working in a common environment. Maturity states are superseded by control points or maturity gates which are common to product, processes and resources and are passed by agreement between the stakeholders.

This new collaborative design methodology requires also a rearrangement of functions as proposed below.

## 16.7 Functional Model

A functional model (Fig. 16.4) shows the main functions and information flow involved in the development and production of an aircraft.

Management activities are represented by the box "*Manage*," comprise "*program management, cost & planning*" and influence all the downstream functions.

Development activities are represented by the box "Engineer," controlled by the output from "*Manage*," "*Customer requirements*" and "*Industrial strategy*". Development activities include "*Functional Design*" and "*Industrial Design*," working together as a single team to develop product, processes and resources from the conceptual phase to the start of the serial production. The deliverable is an "*iDMU*," a complete definition and verification of the virtual manufacturing of the product [6]. All the deviations coming from the shopfloor, in terms of "*Deviations* (*non conformances, concessions*), are inputs to "*Engineer*," included in the "*iDMU*" and sent to "*Operation*." The final output is an "*As built*" iDMU that fits with the real product launch by "*Operation*".

Production activities are represented by the box "Operation," controlled by the output from "*Manage*" and by the output from "*Engineer*" "*iDMU*." Operation activities include "Sustaining," which is in charge of exploit the iDMU, with the help of MES (Manufacturing Execution Systems), to launch "*Shopfloor Documents*" to "*Serial production*." The "*Manufacturing Problems*" that can be managed without

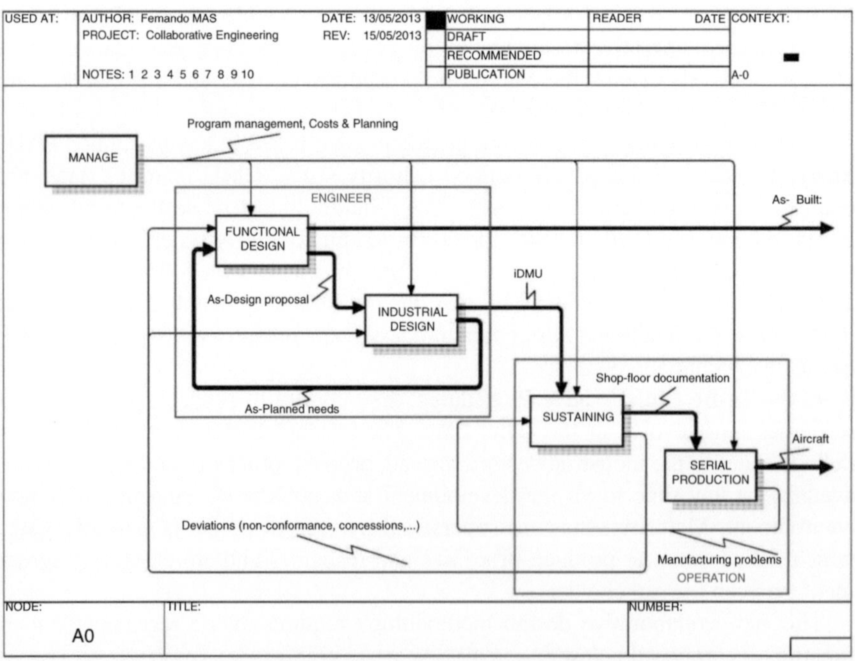


**Fig. 16.4** Collaborative functional model

modifying the iDMU, are managed by "*Sustaining*." Any other item affecting the iDMU is derived to "*Engineer*" as deviation. The output from "*Operation*" is the final physical product that fits 100% with the "*As built*."

## 16.8 Digital Mock Up (DMU) and Industrial Digital Mock Up (iDMU)

Collaborative Engineering involves a lot of changes: organizational, teams, relationships, skills, methods, procedures, standards, processes, tools and interfaces. It is really a business transformation process [7]. One of the key changes is the engineering deliverable: from the "DMU as master" to the "iDMU for all" (Fig. 16.3).

"DMU as master" is a standard inside Airbus [4]. All the information related to the functional aspects of the product is included in the product DMU, e.g. aspects like "design in context" and clashes-free product are fully deployed. The DMU is the reference for the product functional definition, and it is built in concurrent engineering taken into account manufacturing constraints.

The "iDMU for all" is a new concept. It is the main enabler of the Collaborative Engineering approach and provides a common virtual environment for all the aircraft development stakeholders. Functional design and industrial design are part of a single design process where they progress together and influence each other.

The iDMU collects the information related to functional design plus all the information related to industrial design: manufacturing and assembly process, associated resources, industrial means and human resources [6]. All is defined in an integrated environment, where complete and partial simulations are done continuously, and at the end of the design phase, they guarantee a validated solution. Figure 16.5 shows an example of a 3D view of an iDMU.

## 16.9 Proof of Concept Project: CALIPSOneo

To make a proof of concept (POC) of implementing the Collaborative Engineering paradigm using PLM tools, an R+D+i project called CALIPSOneo was launched. CALIPSOneo is a joined effort that involves Engineering Companies, IT companies, PLM Vendors, Research Centers and Universities.

CALIPSOneo uses the newest PLM tools. It takes as input developments from previous projects, related to digital manufacturing techniques implementation [6, 8] and aircraft conceptual design modeling [9] and other initiatives reported in literature [10–12]. The project aims to demonstrate the capabilities of newest PLM methodologies and tools to:

- To implement the Collaborative Engineering concepts described above. To implement a common 3D work environment for all the stakeholders. To implement an iDMU and virtual validation by using it.

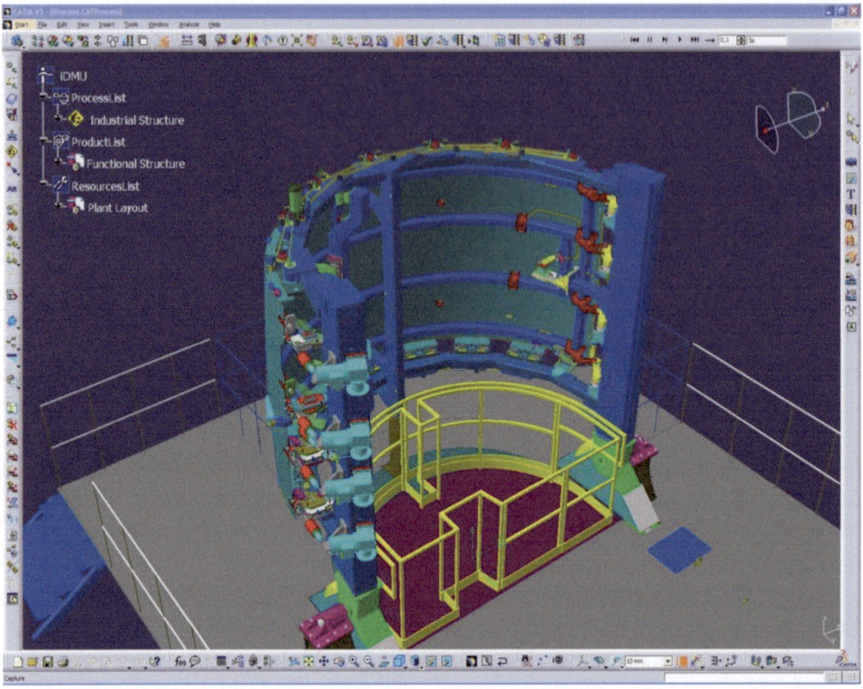


**Fig. 16.5** Fan cowl industrial digital mockup (iDMU) (not a real iDMU due to confidentiality)

- To implement configuration based on individual specimen (single aircrafts).
- To assess the benefits of the iDMU concept when compared to the usual practices for the industrial design.
- To assess the availability of PLM tools to develop and deploy new capabilities. To assess the capability to exploit the iDMU to produce advanced shopfloor documentation.

To use of the latest PLM tools for a proof of concept of the new design collaborative paradigm needs a sound technical support, for this reason the PLM tools manufacturer Dassault Systèmes were incorporated as a partner.

Also a sound project development methodology was necessary; the NDT (Navigational Development Techniques) methodology was selected. NDT was developed by the IWT2 research group (University of Sevilla) [13].

## 16.10 Conclusions

Collaborative Engineering is a broader approach derived from the previous Concurrent Engineering experiences. The availability of new PLM systems and the maturity of the teams are the key elements in the success of its implementation. During the development of the CALIPSOneo project, which is still running, several issues were identified. One of them is them is the complexity of managing a real pilot case in parallel with the development and deployment of the associated PLM tools.

The PLM tools are the key enabler in creating and managing the industrial DMU (iDMU) and the iDMU is the key enabler of the collaborative approach. For this issue, the adopted solution was to set up a multidisciplinary working team model, where engineers, experts on the industrial design tasks, and PLM experts work altogether conducting industrial and CALIPSOneo R + D + i tasks.

Engineers were trained in understanding how PLM tools could help in the industrialization design process and PLM experts were focused on customizing and using the PLM tools to create the iDMU.

**Acknowledgements** Authors wish to express their sincere gratitude to colleagues from Airbus Military, Universidad Politécnica de Madrid, Universidad de Sevilla and partners of CALIPSOneo project for their collaboration and contribution to the project.

## References

1. Aviation News Magazine—AIRBUS A400M. http://www.aviation-news.co.uk/archive/a400m.html. Accessed Jan 2013
2. Airbus. The timeline. Airbus, a leading aircraft manufacturer. http://www.airbus.com/company/history/the-timeline. Accessed Jan 2013
3. Pardessus T (2004) Concurrent engineering development and practices for aircraft design at Airbus [conference]. In: Proceedings of the 24th ICAS meeting, Yokohama
4. Garbade R, Dolezal WR (2007) DMU at Airbus—evolution of the digital mock-up (DMU) at airbus to the Centre of Aircraft Development. In: The future of product development. Proceedings of the 17th CIRP design conference, 3–12. Springer, Berlin
5. Prasad B (1996) Concurrent engineering fundamentals. Prentice-Hall PTR, New Jersey
6. Menendez JL, Mas Morate F, Servan J, Rios J (2012) Virtual verification of an aircraft Final Assembly Line industrialization: an industrial case. Key Eng Mater 502:139–144. ISSN: 1662-9795
7. Lu SC-Y, Elmaraghy W, Schuh G, Wilhelm R (2007) A scientific foundation of collaborative engineering. CIRP Ann Manuf Technol 56(2):605–634
8. Mas Morate F, Menéndez JL, Serván J, Gómez A, Ríos J (2012) Aerospace industrial digital mock-up exploitation to generate assembly shop floor documentation. In: Proceedings of 29 international manufacturing congress, Belfast
9. Mas Morate F, Rios J, Menendez JL, Gomez A (2013) A process-oriented approach to modeling the conceptual design of aircraft assembly lines. Int J Adv Manuf Tech. 67(1): 771–784, ISSN: 0268-3768. doi: 10.1007/s00170-012-4521-5
10. Butterfield J et al (2007) Optimization of aircraft fuselage assembly process using digital manufacturing. J Comput Inf Sci Eng 7(3):269–275
11. Van Wijk D, et al (2009) Integrated design and PLM applications in aeronautics product development. In: Proceedings of the 19th CIRP design conference competitive design, Cranfield University
12. Rouchon C (2012) V6 deployment in new business Jet. Top 5 PLM breakthroughs. In: Proceedings of 3DEXPERIENCE FORUM Europe 2012, Brussels
13. García-García JA, Alba M, García L, Escalona Mª J (2012) NDT-suite: a model-based suite for the application of NDT. Lecture Notes in Computer Science (The 12th International Conference on Web Engineering, ICWE), vol 7387, pp. 469–472

# Chapter 17
# Understanding Contradictions in Enterprise System Implementations: A Case for Stakeholder Theory

**Stig Nordheim, Kai R. Moseid-Vårhus, and Arnfinn Min Bærø**

**Abstract** Enterprise Systems (ES) implementation is challenging, and handling conflicting interests may be vital for success. Previous research has established how ES implementation involves dialectics, often related to multiple stakeholders. Involved stakeholders have in previous studies been analyzed in a power perspective, through the lens of organizational influence processes. Stakeholder theory (ST) takes a wider perspective, by including legitimacy and urgency in addition to power. An interesting perspective is therefore a suggested combination of ST and dialectics. This paper presents an ES implementation case where the explicit combination of ST and dialectics was tried out in the data analysis. In this case, two types of contradictions surfaced in implementation process. The case demonstrates that stakeholder theory contributes to a richer understanding of these contradictions than a focus on power only. Based on previous research, power would be expected to be decisive for the outcome of contradictions. However, in this case urgency and legitimacy compensated for lack of power. This suggests that the combination of ST and dialectics is a useful theoretical perspective on ES implementation processes, to better understand contradictions.

**Keywords** Enterprise systems • Implementation • Dialectics • Stakeholder theory • Contradictions

S. Nordheim (✉)
University of Agder, Kristiansand, Norway
e-mail: stig.nordheim@uia.no

K.R. Moseid-Vårhus • A.M. Bærø
Norwegian Tax Administration, Oslo, Norway

M. José Escalona et al. (eds.), *Information System Development: Improving Enterprise Communication*, DOI 10.1007/978-3-319-07215-9_17,

## 17.1 Introduction

Implementing an Enterprise System (ES) in an organization is both challenging and expensive [1]. In ES implementations value conflicts occur between stakeholders [2], and a dialectic perspective thus explains important aspects of the ES implementation process [3].

In a dialectic perspective the stakeholders are important, since multiple stakeholders with divergent interests can play a vital role in contradictions occurring in ES implementation, thus affecting ES success [4, 5]. Without going deep into the body of general stakeholder theory (ST) (e.g. [6–8]), ES implementation has been be perceived as a negotiation process where various parties try to use the project to defend or to advance their individual or group interests [4]. Therefore value conflicts occur between stakeholders in ES implementation processes [2]. An ES entails many stakeholders who typically have multiple and often conflicting objectives and priorities, and rarely agree on a set of common aims [9].

Previous research has established how ES implementation involves dialectics, related to multiple groups of stakeholders [10]. These contradictions have previously been analyzed in a power perspective, through the lens of organizational influence processes [11]. Here we try out another perspective, to see if it contributes to understand contradictions. A combination of ST and dialectics, which is being suggested and applied in another context [12], is a promising perspective for analyzing ES implementation. We are not aware that this combined theoretical perspective has been applied to ES research, and this paper therefore raises the following research question:

*Faced with the challenges of an ES implementation, how may we understand contradictions through the perspective of stakeholder theory?*

After a brief presentation of dialectics and ST in the following, the research method and case study is presented. The case findings are presented and discussed, showing how the contradictions were understood in view of ST.

## 17.2 Theoretical Background

ES implementation in general is a large research area [13], and beyond the scope of this paper. We limit the ES background literature to dialectics in ES implementation, after a brief description of dialectics. Stakeholder theory (ST) is presented briefly, before summarizing the suggested combination of ST with dialectics.

### *17.2.1 Dialectics*

The dialectic perspective on organizational change and development emphasizes a pluralistic world of colliding events, forces, or contradictory values that compete with each other for domination and control [14]. In dialectic process theory, stability and change are explained by reference to the balance of power between the

**Table 17.1** Different types of contradictions in the ES implementation literature

| Contradiction | Literature |
|---|---|
| A dialectic of learning: old vs. new knowledge | [16] |
| Job and governance conflicts in the shakedown phase | [15] |
| Misfit: ERP package and organizational requirements | [3,17] |
| A dialectic of adaptation: features of the ES packages vs. organizational requirements | [10] |

opposing entities. Change occurs when these opposing values, forces, or events gain sufficient power to confront and engage the status quo. The resolution of the contradiction becomes a synthesis, and such a synthesis can be a novel construction. This synthesis, in turn, becomes a new thesis as the dialectic process continues. However, a contradiction does not necessarily result in a synthesis. An observed contradiction may continue in the organization(s), maintaining the pluralist or conflicting status quo, or it may result in survival of the thesis or antithesis alone [14].

### *17.2.2 Dialectics and Enterprise Systems*

Dialectics has been applied for some time to analyze ES implementation, and a dialectic perspective has proven useful to characterize important aspects of ES implementation processes [3, 10, 15, 16]. Table 17.1 gives an overview of the types of contradictions that this previous ES research has established.

Understanding contradictions in the ES implementation process is necessary to be able to deal with them in a constructive way. Previous research on understanding contradictions in ES implementation suggests a number of different perspectives, from vendor challenges to stakeholder interests in the project phase [18]. Contradictory stakeholder interests in the project phase have previously been analyzed in view of organizational influence processes [11]. Their study concluded that power, operationalized as organizational influence processes, provides important insights on key stakeholders involved in ES dialectics. An adequate formal position was decisive for the outcome of the project phase dialectics. Based on this insight, we here take one step further and suggest a wider perspective on dialectics, i.e. ST.

### *17.2.3 Stakeholder Theory*

Stakeholder theory (ST) development owes a lot to Freeman [6]. His purpose was to outline an alternative way of strategic management, by acknowledging that an organizations' stakeholders need to be actively managed. ST consists of three mutually supportive elements: normative assumptions, descriptive aspects, and instrumental aspects [7]. Key aspects of descriptive ST involve definition of stakeholders as well as tools to identify them (e.g. stakeholder analysis) and concepts that represent stakeholder salience towards managers. Salience refers to why some stakeholder claims are attended to while others are not. According to Mitchell et al. [8], salience is composed of the attributes of power, legitimacy, and urgency.

Power refers to a relationship in which one social actor, A, can get another social actor, B, to do something that B would not have otherwise done. Legitimacy is defined as a generalized perception or assumption that the actions of an entity are desirable, proper, or appropriate within some socially constructed system of norms, values, beliefs or definitions. Urgency is understood as the degree to which stakeholder claims call for immediate attention, and is based on time sensitivity and criticality ([8], p. 869).

Stakeholders possessing all three attributes are more salient towards managers than other stakeholders and are thus termed definitive stakeholders in a stakeholder typology. ST has spread to different disciplines, including information systems [19, 20].

### *17.2.4 Stakeholder Theory and Dialectics Combined*

ST has been referred to a few times in relation to ES, such as using ST as a lens to look at stakeholder knowledge sharing during ES implementations [21]. Although ST implicitly is dialectic in nature [22], an interesting and explicit combination of the two was suggested by Flak et al. [12]. They present an analytical approach for investigating contradictory stakeholder interests by combining descriptive ST and dialectic process theory. ST is concerned with why some stakeholder claims are attended to, and dialectic process theory views organizational change and development as the result of contradictory values competing for domination. The suggested combined theory was applied to study a G2G project [12].

None of the reviewed ES literature on dialectics (Table 17.1) has included ST. Nor are we aware that this combined theoretical perspective [12] has been applied to ES research. We therefore explore how such a theoretical perspective possibly applies to an ES implementation case, by attempting to understand contradictions through ST.

## 17.3 Research Method

As the knowledge interest here is understanding, an interpretive case study is considered an appropriate research method [23]. We used a qualitative research approach [24], to be able to gain an in- depth understanding of the conflicting stakeholder perspectives and their influence on the ES implementation project. A qualitative approach is useful when the phenomenon to be studied is little known in advance [25]. The research has been based on a single case, and was carried out as interpretive research [26]. The principal data collection method was in-depth interviews with key stakeholders, conducted during the spring 2012.

**Table 17.2** Overview of informants

| Informant | Company | Company role |
|---|---|---|
| Main project manager | 'P' | Implementation partner |
| Project owner | 'I' | Case company |
| Super user | 'I' | Case company |
| Two project managers | 'I' | Case company |
| Implementation manager/responsible | 'I' | Case company |
| Two department managers | 'I' | Case company |
| Project manager | 'O' | Vendor |

### *17.3.1 Interviews*

We used stakeholder analysis to select stakeholders [8], and to map the different stakeholders according to power, legitimacy and urgency. We chose three types of data collection, in-depth interview, document analysis and analysis of audio-visual material. The semi-structured interviews were dialogues based on open questions, and lasted between 60 and 110 min. Interesting occurring themes were followed up in subsequent interviews. A total of 9 informants were interviewed, as presented in the case description (Table 17.2). The informants themselves were asked to classify key stakeholders according to the salience attributes of power, legitimacy and urgency.

### *17.3.2 Data Analysis*

All the interviews were recorded and later transcribed, then annotated and categorized and compared. This is in line with Miles and Huberman [24]. After connecting and comparing the data, we compared them with previous research. We started with observations and gradually reached conclusions based on patterns and interrelationships in the data. We also used content analysis of interviews and documents, as a technique to split text in thematic categories to study frequency of themes.

We used the combination of ST and dialectics as suggested by Flak et al. [12] as the theoretical basis for our data analysis, using dialectics to uncover contradictions and ST to analyze the stakeholders involved in the contradictions. The informants' classification of the stakeholders' salience was analyzed in order to understand why the contradictions did result in the particular syntheses.

### *17.3.3 Validation*

The data analysis has been validated by following the fundamental principle of the hermeneutic circle [27]. This principle was applied at different levels, considering the research questions, the interviews and project documents as wholes.

Statements were the typical parts of interviews and documents. We iterated between the independent meanings between the interviews and the project documents. Through the interaction between us as researchers and the informants, our assumptions were constantly questioned and modified. We also contacted key informants after the interviews to clarify and verify our understanding.

## 17.4 Case Background and Description

This research is based on a single case study, involving three companies: the case company, enterprise 'I', which used implementation partner 'P' to implement an ES from vendor 'O'. The case enterprise 'I', is a major Nordic actor within IT Service, with more than 75 service offices and almost 2,500 employees in Norway, Sweden, Denmark and Finland. We got access to 'I' through 'P', which was chosen by 'I' to be the implementation partner, to manage their implementation of a SaaS (Software-as-a-Service) ES.

The implemented system was a planning system from a third party we label 'O', with main office in Sweden. 'O' develops, sells and supports software for planning and optimizing customers' field services. The main focus of the interviews was the ES implementation project in Norway.

The implementation partner 'P' is an IT consultancy company located in southern Norway. They have about 50 employees. 'P' offers consultancy in project management, procurement, systems development, integration and management.

We considered this case well suited for the research question, because the enterprise 'I' has implemented an ES in numerous locations in several countries. With an initial access to the project manager from P, and their interest in this study, we had good access to data for this research.

We interviewed key stakeholders who were selected based on their degree of involvement in the project. An overview of the nine informants is presented in Table 17.2.

Seven informants were from 'I', the case company implementing the ES, one informant was from the implementation partner 'P', and one informant was from the vendor 'O'. They had mainly a management perspective. In addition to the informants, about 400 pages of internal and external project documents were analyzed.

The system implemented is called Laps, and schedules service assignments and tracks for all technicians, based on parameters such as competence, home address and Service Level Agreement (SLA). Laps may eventually replace the manual planning done by a centralized dispatch group in each country. Dispatch is a management function that decides which people to send on the different jobs. Norway did not have a centralized dispatch group, and the centralization was done as Laps was implemented. Almost every Norwegian location used to have a local dispatch function. The centralization was to be located in Oslo. A pilot project was run at the Oslo office in May 2011, and system rollout started late October 2011. In January 2012 the project was formally terminated, leaving just a few locations to get the system rolled out.

## 17.5 Case Study

The stakeholder analysis led to an overview of 18 key stakeholders, categorized by the informants according to the stakeholder typology of Mitchell et al. [8]. The space does not allow a detailed overview here, but a summary of stakeholder groups is presented in Table 17.3, where the 3 salience attributes of 14 stakeholders within project management and 4 within specialist staff are summarized.

After an analysis of the salience attributes of the stakeholders, we performed a dialectic analysis to identify potential contradictions between the different stakeholders. Already early in the interview phase two contradictions emerged as important for the project outcome. These are presented next.

### *17.5.1 Contradiction 1: Solution Roll-Out*

Whether to follow the roll-out plan for 'I' was an important issue in this contradiction, which is summarized in Table 17.4.

The thesis implied following the plan for solution roll-out (Table 17.4), to profit from savings already in the current year. The antithesis stressed that a high quality implementation required an adjusted solution roll-out. The synthesis emerged as several modifications of the plan, and the deadline was extended twice. As expressed by the Nordic project manager:

> We could not have such an aggressive roll-out as initially planned.

And as expressed by the Norwegian project manager:

> ... with this type of customers... I know it is very critical at Christmas time.

Instability related to the existing solutions amplified this:

**Table 17.3** Attributes of the stakeholder groups

| Stakeholders | Power | Legitimacy | Urgency |
|---|---|---|---|
| Project management (stakeholders 1–14) | High | High | Average |
| Specialist staff (stakeholders 15–18) | Low | High | High |

**Table 17.4** Contradiction 1: planned versus realistic roll-out

| Entity | Description |
|---|---|
| Thesis | Follow budget and timeframe for solution roll-out |
| Antithesis | High quality through adjusted solution roll-out |
| Synthesis | Adjusted solution roll-out with extended deadline |
| Involved stakeholders | Project management (thesis) Super user (Norway) and Project manager (Norway) (antithesis) |
| Decisive stakeholder typology | Urgency |

**Table 17.5** Contradiction 2: centralization vs. decentralization of dispatch

| Entity | Description |
|---|---|
| Thesis | Centralization, for efficiency and saved costs |
| Antithesis | Decentralization, for efficiency based on local knowledge |
| Synthesis | Centralized solution with some local adaptations |
| Involved stakeholders | Project management (thesis) Department manager (antithesis) |
| Decisive stakeholder typology | Legitimacy (although combined with urgency and power) |

> Instability regarding IT infrastructure. I think that was the greatest challenge, and the fact that we started with Laps and centralized dispatch simultaneously. (Super user Norway).

Delays also led to the antithesis:

> …the systems went down and this created of course a lot of noise and insecurity whether this was a good solution for 'I' (Project manager Nordic, 'P').

Despite a pressure from the project management, the plan was partially postponed:

> …in a way we had some all-round pressure on us. The Nordic project manager and project owner I who constantly wanted us to speed up. But we chose to take the different divisions into consideration, to avoid any chaos before moving on to the next division (Project manager Norway).

The decisive stakeholder typology involved in contradiction 1 was urgency. The urgency that decided the synthesis was especially present in the different divisions of the enterprise. Two bases for this urgency were decisive: time-sensitivity and criticality. This urgency led to the synthesis of an adjusted solution roll-out with extended deadline (Table 17.4).

### *17.5.2 Contradiction 2: Centralization of Dispatch*

Centralization of the dispatch function in 'I' was an important issue in this contradiction, which is summarized in Table 17.5.

The thesis was to centralize the dispatch function, to save costs and resources (Table 17.5). This would also harmonize the Norwegian branch with the other countries. As expressed by the Norwegian project manager:

> I am saying that these dispatchers in Oslo are top motivated for the job and ready for the task, and there are some people sitting out there that have to let go of this job. It is a conflict of interests.

The antithesis was to keep a decentralized dispatch function, for efficiency and applying local knowledge for better customer service. To acquire and keep satisfied customers, efficiency was considered important.

> To become efficient, is a question of local knowledge (Department manager 2).

The synthesis became a compromise, mainly a centralized solution, but with some local adaptations (Table 17.5). One local dispatcher in question would work as before, but organized under the central dispatch function.

> But now we managed to get our local dispatcher to be part of the organisation of the centralized dispatch function. So we have in a way been able to convince the project, and after a lot of pressure the local dispatcher we had here, will sit here most of the time, so we will not notice much change on our location (Department manager 2).

The project management had the official power, but the antithesis prevailed because department manager No. 2 could show that his location would have no cost savings through centralization. We found that the synthesis emerged as a result of urgency, combined with power and legitimacy. Legitimacy was acquired by talking to the right people in the organization.

## 17.6 Discussion

We heeded the advice to apply ST in analyzing an ES implementation [21], combining ST with dialectics as suggested by Flak et al. [12]. The two contradictions presented in this paper are not to be found in the existing ES literature (Table 17.1), although centralization versus decentralization (contradiction 2) is known from other domains such as car manufacturing [28]. Contradiction 1 may in a general sense be viewed as management of dialectic tensions, such as discussed by Vlaar et al. [29]. To deal with contradictions constitute a challenge for project management.

We raised the initial question: *Faced with the challenges of an ES implementation, how may we understand contradictions through the perspective of stakeholder theory?*

According to previous research by Nordheim and Nielsen [11], certain ES implementation contradictions may be understood by analyzing how power has been applied. Organizational influence processes accounted for important differences that explained outcomes of certain contradictions [11]. We applied a wider perspective than power, the other ST concepts of urgency and legitimacy were included in the analysis. The salience attributes of the stakeholder groups were explicitly identified by the informants. The lack of power would indicate that specialist staff (Table 17.3) would be unable to persist in these contradictions. But specialist staff persisted in this case, due to legitimacy and urgency. Power alone was therefore not decisive; power was compensated for by urgency and legitimacy. The outcome of contradiction No. 1 can be explained by urgency, i.e. time-sensitivity and criticality. The outcome of contradiction No. 2 can be explained by all three salience attributes combined, and in particular by legitimacy. The contribution to ES research is that ST with its salience concept in this case provided a richer basis for understanding the syntheses. Salience is composed of three attributes, where power is one of them. The other two, legitimacy and urgency, were important for the syntheses in this case. Especially the concept of urgency is an important ST contribution to understand the contradictions in this case, as the synthesis of contradiction 1 can be understood as a direct result of time- sensitivity and criticality. This means the degree to which stakeholder claims call for immediate attention [8]. Based on the findings of this case, we would therefore argue that time-sensitivity in particular is important to explain outcomes of contradictions, in addition to organizational influence processes as found by Nordheim and Nielsen [11]. We thus find the combined perspective of ST and dialectics presented by Flak et al. [12] to be useful in analyzing this ES implementation. Understanding dialectics based on power issues, would have been insufficient in this case.

To what extent urgency and legitimacy may compensate for power in other contradictions, remains to be explored by future research. This is a limited case with nine informants and raises a question whether previous attempts to understand dialectics could have benefited from a wider theoretical basis such as suggested by Flak et al. [12]. Our findings indicate that future research on understanding contradictions in ES should include all facets of the theoretical ST concept of salience.

## 17.7 Conclusion and Implications

ST with its salience concept gives a promising perspective for a richer understanding of ES contradictions involving stakeholders. Of the stakeholder salience attributes, power is challenging to apply in a case analysis. Previous research has shown that this may be done through organizational influence processes. However, findings presented in this paper imply that urgency with its notion of time- sensitivity and criticality also needs to be analyzed, to understand ES implementation contradictions involving stakeholders. So urgency and legitimacy are important for this understanding, together with a focus on how power is applied, e.g. through organizational influence processes.

This study also has practical implications for projects. An initial mapping of stakeholder typology, combined with an awareness of emerging contradictions, may be useful for project management. The salience concept has implications for research on organizational implementation of ES. To what extent urgency and legitimacy may compensate for power in other contradictions, remains to be explored by future research.

## References

1. Seddon PB (2005) Are ERP systems a source of competitive advantage? Strat Change 14:283–293
2. Allen JP (2005) Value conflicts in enterprise systems. Inform Tech People 18(1):33–49
3. Sia SK, Soh C (2007) An assessment of package–organisation misalignment: institutional and ontological structures. Eur J Inform Syst 16(568–5):83
4. Boonstra A (2006) Interpreting an ERP-implementation project from a stakeholder perspective. Int J Proj Manag 24(3):8–52

5. Sarker S, Lee AS (2003) Using a case study to test the role of three key social enablers in ERP implementation. Inform Manag 40(8):13–829
6. Freeman RE (1984) Strategic management: a stakeholder approach. Pitman, Boston
7. Donaldson T, Preston LE (1995) The stakeholder theory of the corporation—concepts, evidence, and implications. Acad Manage Rev 20(1):65–91
8. Mitchell RK, Agle BR, Wood DJ (1997) Toward a theory of stakeholder identification and salience: defining the principle of who and what really counts. Acad Manage Rev 22(4):853–886
9. Sedera D, Gable GG, Chan T (2004) Measuring enterprise systems success: the importance of a multiple stakeholder perspective. In: Proceedings of the 13th European conference on information systems, Turku, Finland
10. Nordheim S, Päivärinta T (2006) Implementing enterprise content management: from evolution through strategy to contradictions out-of-the-box. Eur J Inform Syst 15(6):648–662
11. Nordheim S, Nielsen PA (2008) Enterprise system implementations: organizational influence processes for corporate user representatives. In: Golden W, et al. (eds) Proceedings of the 16th European conference on information systems, Galway, Ireland, Paper 187
12. Flak LS, Nordheim S, Munkvold BE (2008) Analyzing stakeholder diversity in G2G efforts: combining descriptive stakeholder theory and dialectic process theory. e-Serv J 6(2):3–23
13. Esteves J, Bohorquez V (2007) An updated ERP systems annotated bibliography: 2001–2005. Instituto de Empresa Business School Working Paper No. WP 07-04
14. Van De Ven AH, Poole MS (1995) Explaining development and change in organizations. Acad Manage Rev 20(3):510–540
15. Besson P, Rowe F (2001) ERP project dynamics and enacted dialogue: perceived understanding, perceived leeway, and the nature of task-related conflicts. SIGMIS Database 32(4):47–66
16. Robey D, Ross JW, Boudreau M-C (2002) Learning to implement enterprise systems: an exploratory study of the dialectics of change. J Manag Inform Syst 19(1):17–46
17. Soh C, Sia SK (2005) The challenges of implementing vanilla versions of enterprise systems. MIS Quart Exec 4(3):373–384
18. Nordheim S (2011) Towards understanding contradictions in enterprise system implementations—insights from a case study. In: Pkorny J et al (eds) Information systems development. Business systems and services: modelling and development. Springer, New York, pp 563–574
19. Pouloudi A, Whitley EA (1997) Stakeholder identification in inter-organizational systems: gaining insights for drug use management systems. Eur J Inform Syst 6(1):1–14
20. Vidgen R (1997) Stakeholders, soft systems and technology: separation and mediation in the analysis of information system requirements. Inform Syst J 7(1):21–46
21. Sathish S, Pan S, Raman K (2004) Using stakeholder theory to analyze knowledge sharing during enterprise systems implementations. In: PACIS 2004 proceedings, Paper 26. http://aisel.aisnet.org/pacis2004/26
22. Friedman AL, Miles S (2002) Developing stakeholder theory. J Manag Stud 39:1–21
23. Braa K, Vidgen R (1999) Interpretation, intervention, and reduction in the organizational laboratory: a framework for in-context information system research. Account Manag Inform Tech 9(1):25–47
24. Miles MB, Huberman AM (1994) Qualitative data analysis: an expanded sourcebook, 2nd edn. SAGE Publications, Thousand Oaks
25. Creswell JW (2009) Research design: qualitative, quantitative, and mixed methods approaches, 3rd edn. SAGE Publications, Thousand Oaks
26. Walsham G (2006) Doing interpretive research. Eur J Inform Syst 15:320–330
27. Klein HK, Myers MD (1999) A set of principles for conducting and evaluating interpretive field studies in information systems. MIS Quart 23(1):67–93
28. Henfridsson O, Yoo Y, Svahn F (2009) Path creation in digital innovation: a multi-layered dialectics perspective. Sprouts: working papers on information systems, vol 9, article 20. http://sprouts.aisnet.org/9-20
29. Vlaar PWL, Van Den Bosch FAJ, Volberda HW (2007) Towards a dialectic perspective on formalization in interorganizational relationships: how alliance managers capitalize on the duality inherent in contracts, rules and procedures. Organ Stud 28(4):437–466

# Chapter 18
# Company Process Support by Software Systems: Research in Small Software Companies in the Czech Republic

**Jan Mittner and Alena Buchalcevova**

**Abstract** The aim of this paper is to introduce the results of a survey which focuses on the state of company process support by software systems in small software companies in the Czech Republic. The survey includes various processes from software development to other areas of company activities, e.g. human resources, finance, quality management, or utilities. The survey results show that two thirds of respondents are not satisfied with the current state of company process support.

**Keywords** Company processes • Small software company • Process support by software systems

## 18.1 Introduction

Although much attention has recently been devoted to the topic of software engineering, the results of published surveys show that a significant number of software projects are not successful. According to [2] the ratio of successful software projects ranges to 60 % while the rest is categorized as challenged or failed. Overall, researchers' effort worldwide concentrates rather on software development methodologies and practices than on other areas of company processes. There is a lack of comprehensive endeavor to look at an operation of a company as a whole. The Annual State of Agile Development Survey [9] together with other published surveys [1, 3] focus also on the causes of software projects failures among which are often mentioned broader organizational or communication problems. In addition, this survey brought up a desire of more frequent use of management tools in small software companies as an important method for future

J. Mittner (✉) • A. Buchalcevova
Department of Information Technology, University of Economics,
Prague W. Churchill Sq. 4, Prague, Czech Republic
e-mail: xmitj05@vse.cz; buchalc@vse.cz

M. José Escalona et al. (eds.), *Information System Development: Improving Enterprise Communication*, DOI 10.1007/978-3-319-07215-9_18,

development (especially project, release, requirements and ideas management). Considering all of the facts stated above, we identified a potential for improvement, as the topic of software systems completely supporting the operation of software companies is not widely examined in literature. Firstly, we determined a current situation in this field and also a demand level set by software companies. These findings are the main objective of this paper and serve as an identifier of a possible opportunity to e.g. design a new information system specifically for software companies.

## 18.2 Related Works

To gain a general overview of this topic, we researched literature in this field, particularly EBSCO, ProQuest, ACM-DL databases and Google. However, most of the resources focus only on the particular topic of software development, not on the operation of a software company as a whole. There is no single model that would cover all key processes in a software company. Nevertheless, the concept of Application Lifecycle Management described e.g. in [7], and process improvement frameworks as e.g. CMMI-DEV or People-CMM [8] try to approach a software company as an inseparable piece. Nalbant goes further and focuses on designing an information system to streamline software development process in software companies [6]. However, the depth of his paper is not sufficient and thus can rather be used as a basis for further work.

We believe that it is important to concentrate not only on software development itself, but also on the overall company functioning. Many software development methodologies and practices exist, but in our opinion they do not solve the problems as a matter of fact for several reasons. In particular:

- SW development methodologies focus in most cases only on the area of software development itself; they do not provide any complex view of the overall operation within a company. However, this approach does not fulfill the needs of a software company which also requires linking its development with invoicing, remuneration of employees, training, customer relationship management, etc.
- There is a lack of suitable software systems that properly facilitate an introduction and usage of SW development methodology while allowing a complete control of major areas of a company's life.

Software development processes influence a complex company operation, as they are naturally its primary concern. Therefore, they are closely related to the areas such as project management, human resources, finance and quality management. Thus, to make an overall improvement in software development processes, we also need to focus on enhancing processes in the areas mentioned above. Information system should include such an approach and automate and/or improve these processes. The relationship of an information system and software processes is shown in Fig. 18.1.

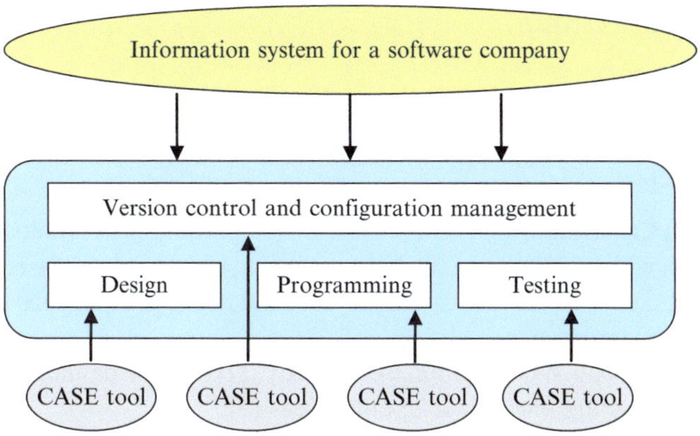


**Fig. 18.1** Relationship of IS and software development processes, based on [6]

CASE (Computer Aided Software Engineering) tools allow an effective implementation of partial development processes, such as programming, testing, version control and configuration management, etc., while the information system integrates these processes into a single environment with shared information.

However, we did not encounter any surveys focused on the support for software company processes by software systems. Nevertheless, several surveys contain questions regarding the usage of software development tools. The 7th Annual State of Agile Development Survey [9] states that agile project management tools are used in 60 % of companies but does not include their level of satisfaction. Thirty-seven percent of companies wish to implement management tools (project, requirements, release or ideas management) in future. According to [4] companies in the Czech Republic perceive a lack of suitable software tools for agile development.

To find out the state of company process support by software systems in small software companies in the Czech Republic we conducted a questionnaire survey at the conference WebExpo 2011.

## 18.3 Research Characteristics

Our research objective was to determine to which extent company processes are supported by software systems in software companies, which problems software companies have to address and whether they perceive that a suitable software system would help to improve their software processes. We carried out the research for the whole segment and further analyzed the results according to company size. This article focuses solely on small software companies and their survey results.

### 18.3.1 Target Group

As mentioned above, this paper concentrates on small software companies. When defining the target group of small companies, we considered several approaches applied in practice:

- Pursuant to Czech Act No. 47/2002 Coll. on the promotion of small and medium enterprises, which is based on the EU legislation, a small company is understood as the one that has fewer than 50 employees, and micro company the one that has fewer than 10 employees.
- The working group WG24 within the ISO/IEC JTC 1 SC7 defines small companies as those having fewer than 25 employees [5].
- According to the Czech Social Security Administration, small organizations are those with up to 25 employees.

We believe that the definition of small companies according to the EU is not applicable for software development, as it is a technology and knowledge intensive sector, where 50 employees mean quite a high number at least in the case of the Czech Republic. On the contrary, micro-companies up to 10 employees would be far too limiting. Thus, we decided to use the WG24 working group's definition and for the purposes of this paper we consider small companies as those with fewer than 25 employees.

### 18.3.2 Research Model

As stated above, every software company needs more than a pure software development for its overall operation. It is bound to deal with planning and managing projects, organizing work, recruiting, managing and rewarding employees, managing its finances and invoicing, ensuring knowledge transfer between employees, managing work quality, etc. To design our survey properly, we identified key process areas crucial for a software company. The model of the process areas is shown in Fig. 18.2.

As we haven't found any single model covering all key process areas of a software company, we took into account multiple sources when designing this model—proposal in [6], process areas in CMMI-DEV, the concept of ALM and also our experience from real businesses. We identified five areas of processes that are crucial for even the smallest software companies—Project Management, Human Resources, Finance, Quality Management, and Utilities. Although there could be named more areas of processes, these are the most important ones that should be present in every software company. Software development processes are inherently affected by all these other areas of company processes.

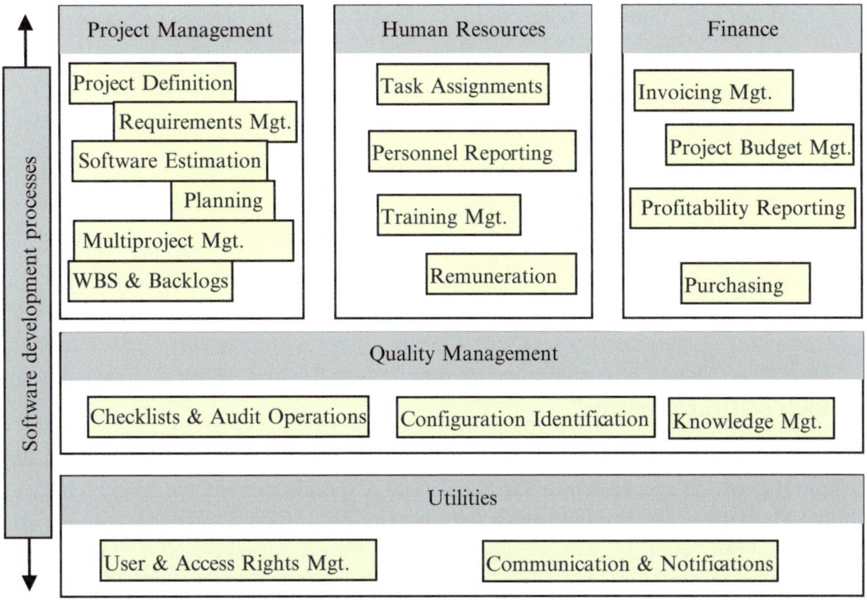


**Fig. 18.2** Key company process areas, based on [6]

### *18.3.3 Research Method*

To perform our research, we applied the method of a questionnaire survey. The questionnaire consisted of three main parts:

- Segmentation questions,
- A section focused on software development methodologies and practices,
- Questions concerning the support of company processes by software systems.

Since this paper concentrates solely on the support of company processes by software systems, methodologies and practices are not further analyzed. The questionnaire was carried out in a printed as well as electronic form. Thus, the respondents could choose their preferred method of response. The online form in original Czech version is available at http://spi.vse.cz. Questions that are relevant for this paper are listed below.

**Question: Are you or your company involved in software development?**

The respondents picked "yes" or "no" answer. This question was designed to filter out those respondents who are completely irrelevant for our questionnaire survey.

**Question: How large is your company?**

The respondents selected the size of their company from the predefined intervals listed in Table 18.1. The aim of this question was to create relevant segments of respondents depending on the size of their companies.

**Table 18.1** Company size

| Company size | Absolute frequency | Relative frequency (%) |
|---|---|---|
| Freelancer | 14 | 13 |
| 2–5 Employees | 18 | 17 |
| 6–10 Employees | 16 | 15 |
| 11–25 Employees | 16 | 15 |
| 26–50 Employees | 7 | 7 |
| 51–250 Employees | 25 | 24 |
| More than 250 employees | 9 | 9 |
| **Total** | **105** | **100** |

**Question: Does your company support the following areas of processes by suitable software systems?**

Regarding this question, we defined areas of company processes based on the article [6], which are shown in Fig. 18.2. The respondents selected in each area to what extent they are satisfied with the support by software systems within their company. Four possible answers were provided for every process area: "Yes, to the full satisfaction," "Yes, but we would like to achieve an improvement," "No, but we feel the need," "No, we do not feel any need."

**Question: What are the problems and shortcomings in functioning of your company in terms of software development processes and their support by suitable software systems?**

We designed this question as an open one to gain insight into our respondents' view of possible problems their company might be facing and relevant to our work.

**Question: Do you think that it is possible to improve the processes of software development in your company by using suitable software development methodologies, practices and software systems?**

The respondents selected from "yes" or "no" answer. This question was included to verify our presumption that the improvement can be achieved using suitable methodologies, practices and software system.

### *18.3.4 Sample Structure*

We conducted the questionnaire survey at the conference WebExpo 2011. This conference took place from the 22nd till the 24th of September, 2011 in Prague and was held by the University of Economics. According to [10], such a conference is the biggest event for web enthusiasts aimed at business aspects, design and development of web applications in the region of Central Europe. Overall, 1,100 participants signed up mainly from small and medium-sized companies that are engaged in software development primarily within the internet environment. This conference is traditionally the most important professional event of the year for many web developers not only from the Czech Republic.

Conference participants received the questionnaire in a paper form at registration as well as electronically in e-mail newsletters. One hundred and eight participants responded to our questionnaire survey. Thus, the response rate was 9.8 %. The paper form was preferred by 37 respondents, 71 respondents submitted the online form. 105 of 108 respondents are involved in software development, i.e. 97 %. Therefore, our selected group (conference participants) represents a suitable target group for this survey. The answers of irrelevant respondents are not included in the results.

Regarding the company size of our respondents, all 105 respondents replied to this question. Table 18.1 summarizes the frequencies for each category. Based on the results, we identified the segment of small companies, which includes 64 respondents according to the definition in Sect. 18.3.1.

## 18.4 Research Results

This section analyzes those survey questions related to the support of company processes by software systems. Further, the problems in this area as well as possible opportunities for a future improvement listed by the respondents are discussed.

### *18.4.1 Company Process Support by Software Systems*

Concerning the support of company processes among software companies, 98 respondents participated in this question, 61 of them from the segment of small companies. Figure 18.3 presents the overall scope of support of company processes by software systems. To interpret the graph in Fig. 18.3, it is meaningful to sum up the values "No, but we feel the need" and "Yes, but we would like to achieve an improvement," since these values represent the ratio of respondents who wish to make an improvement in supporting their company processes. Therefore, the response options are listed in the following order.

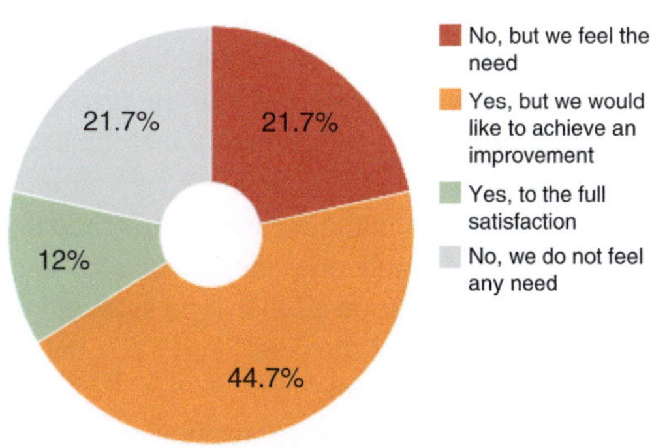


**Fig. 18.3** Support of company processes by software systems

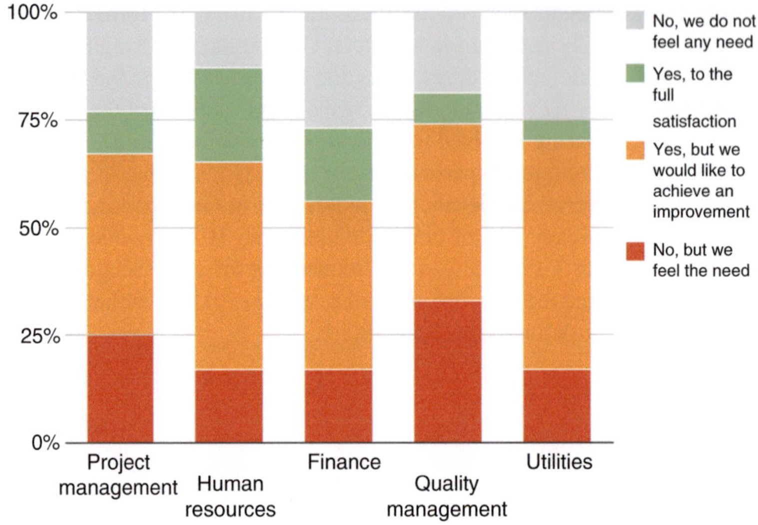


**Fig. 18.4** Support of company processes by software systems in individual process areas

**Table 18.2** Support of company processes by software systems

| | Small companies (61) | | | | | | | |
|---|---|---|---|---|---|---|---|---|
| | No, but we feel the need | | Yes, but we would like to achieve an improvement | | Yes, to the full satisfaction | | No, we do not feel any need | |
| Process area | Abs. | Rel. (%) | Abs. | Rel. (%) | Abs. | Rel. (%) | Abs. | Rel. (%) |
| Project management | 15 | 25 | 25 | 42 | 6 | 10 | 14 | 23 |
| Human resources | 10 | 17 | 29 | 48 | 13 | 22 | 8 | 13 |
| Finance | 10 | 17 | 23 | 39 | 10 | 17 | 16 | 27 |
| Quality management | 20 | 33 | 25 | 41 | 4 | 7 | 12 | 20 |
| Utilities | 10 | 17 | 32 | 53 | 3 | 5 | 15 | 25 |
| ø | | **22** | | **45** | | **12** | | **22** |

It is evident from Fig. 18.3 that two thirds of the respondents are not satisfied with the current situation within their company and would like to achieve an improvement. More than a fifth of the respondents do not have their processes supported by software systems but do feel the need to do so.

Figure 18.4 depicts the detailed view in individual areas of company processes. This figure shows that there are not any fundamental differences between the areas, but the most desirable ones include Quality and Project Management.

Frequency of responses that express an interest in improving the current situation (i.e. "No, but we feel the need" and "Yes, but we would like to achieve an improvement") is shown in Table 18.2.

The segment of small companies compared to the medium and large organizations differs primarily in the lower number of respondents that are fully satisfied

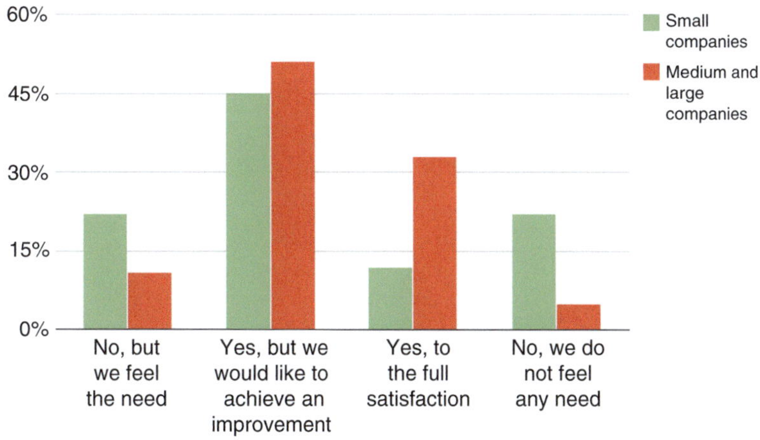


**Fig. 18.5** Support of company processes by software systems in small, medium and large companies

with the current situation of software system support in their company, as seen in Fig. 18.5. Moreover, much fewer respondents feel the need to have their processes supported.

### *18.4.2 Other Issues Within the Frame of Software Development Process Support by Software Systems*

Regarding the shortcomings in company operation among our target group, only ten respondents replied, nine being from small companies.

Three respondents perceive the problems in functioning of their company in poorly designed processes, or even in their complete absence. Three respondents experience difficulties in communication among team members (both in terms of process and communication tools). In addition, missing automated deployment, helpdesk, and poor documentation were described as further shortcomings.

Unfortunately, these results do not provide us with a new and significant problem area that would not be included in defined process areas already.

### *18.4.3 Improvement Possibilities of Software Development Processes*

The vast majority of respondents think that software processes in their company can be improved by using suitable software development methodologies, practices and software systems, as seen in Fig. 18.6. The results were quite identical with the segment of small companies.

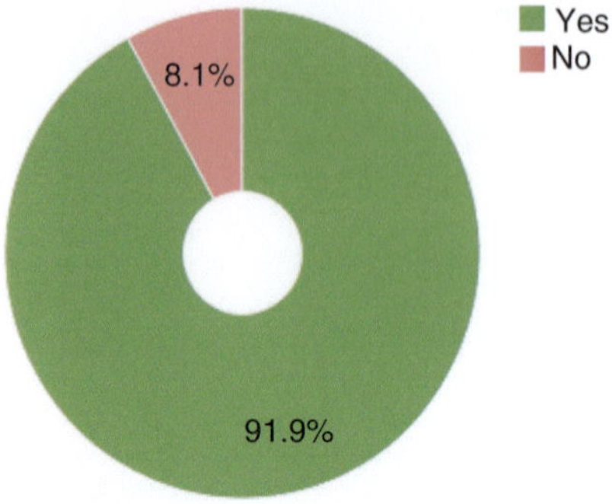


**Fig. 18.6** Possibilities of improvement of software development processes

We believe that it is crucial for an upcoming change of company's approach to their processes being supported by software systems that companies are convinced that using a good methodology, practices and information system can lead to an overall improvement of their functioning.

## 18.5 Conclusion and Future Work

The questionnaire survey conducted at the conference WebExpo 2011 fulfilled the given objective to discover how software company processes are supported by software systems. Sixty-six percent of the respondents either do not support their processes at all but feel the need to do so, or support their company processes, but wish to achieve an improvement. Moreover, 92 % of respondents believe that software processes in their company could be improved by using suitable methodologies, practices and software systems.

These results clearly indicate that the issue of software development process support by software systems has a great potential for further work. We identify an opportunity for a detailed requirements analysis of information system aimed at small software companies and optimally supporting their processes. Based on this requirements analysis results, two possible solutions could be performed; either to implement an integrated currently available software system or to implement a prototype of a new system.

**Acknowledgment** This paper describes the outcome of a research that has been accomplished as a part of the Research program funded by Internal Grant Agency grant No. IG406013.

## References

1. Ambler SW (2009) Agile practices survey results: July 2009. Ambysoft [online]. http://www.ambysoft.com/surveys/practices2009.html
2. Ambler SW (2011) 2011 IT project success rates survey results. Ambysoft [online]. http://www.ambysoft.com/surveys/success2011.html

3. Ambler SW (2012) Surveys exploring the current state of information technology practices. Ambysoft [online]. http://www.ambysoft.com/surveys/
4. Buchalcevova A (2009) Research of the use of agile methodologies in the Czech Republic. In: Barry C, Conboy K, Lang M et al (eds) Information systems development: challenges in practice, theory and education. Springer, New York, pp 51–64. doi:10.1007/978-0-387-68772-8_5
5. Laporte C (2007) Applying international software engineering standards in very small enterprises. CrossTalk - J Defense Software Eng 2007:29–30
6. Nalbant S (2004) An information system for streamlining software development process. Turk J Electr Eng Comput Sci 12:97–106
7. Rossman B (2010) Application lifecycle management—activities, methodologies, disciplines, tools, benefits, ALM tools and products. Newstead, Australia. Tebbo. ISBN: 978-1742446646
8. Software Engineering Institute (2011) CMMI Version 1.3 Information Center. Carnegie Mellon University [online]. http://www.sei.cmu.edu/cmmi/index.cfm
9. VersionOne (2013) 7th Annual State of Agile Development Survey. VersionOne.com. http://www.versionone.com/pdf/7th-Annual-State-of-Agile-Development-Survey.pdf
10. WebExpo (2011) Conference WebExpo Prague 2011. WebExpo [online]. http://webexpo.net/prague2011/

# Chapter 19
# An Integrated Information Systems Success Model: A Case Study of an Australian Hospital

**Tian Yu Goh, Morgan Priestnall, Sedigheh Khademi, and Christopher Bain**

**Abstract** Currently, the majority of Information Systems (IS) theories focus on either the technological side of impact analysis or attempts to rigidly define impact into discrete categories that wind up ignoring a part of the bigger picture. In this paper we will examine the UTAUT and ISS model as representations of these theories. In our opinion, the main shortcomings of these theories are that while they overlap slightly, the UTAUT model downplays the role of technological aspects involved in impact and that the ISS model oversimplifies the impact factor of system use by its users, thus not covering the necessary breadth needed to properly examine impact on a system's users. To address these issues, we propose a new theoretical model, the Integrated Information System Success (IISS) model that integrates the two models. We have conducted a case study utilizing the model with a major Australian Hospital (Hospital Y), analyzing the impact of a recently implemented Patient Referral System (PRS), on users at the individual level. Our findings through the use of IISS indicate that there was an increase in workflow processes for the administrative staff as well as a sense of guilt and frustration that some of the front-line users experienced. These findings would not have been possible with the use of either the ISS or UTAUT models alone.

**Keywords** Systems success • User acceptance • Dashboard • Case study • Impact

T. Yu Goh (✉) • M. Priestnall • S. Khademi
Caulfield School of Information Technology, Monash University,
PO Box 197, Caulfield East, VIC 3145, Australia
e-mail: Tian.Goh@monash.edu; morgan.priestnall@monash.edu; skha59@student.monash.edu

C. Bain
Alfred Health, The Alfred, PO Box 315, Prahran, VIC 3181, Australia
e-mail: christopher.bain-informatics@alfred.org.au

M. José Escalona et al. (eds.), *Information System Development: Improving Enterprise Communication*, DOI 10.1007/978-3-319-07215-9_19,

## 19.1 Introduction

In the last five decades, IS has grown into a multidisciplinary, multifaceted field of research in the use of technology and systems in organizations [1, 2]. Given that the field is still growing, the current theoretical models cannot encompass all the practical aspects. One particular aspect that we will investigate in this paper is the lack of depth needed to investigate the impact of a new system on organizations at an individual level. Specifically, we will examine the drawbacks and strengths of two of the most prominent models that evaluate systems based on user acceptance and system success. These models are the Unified Theory of Acceptance and Use of Technology (UTAUT) and the Information Systems Success (ISS) model, which we consider to be representative of two different approaches—focused on user acceptance and system success, respectively—to the theoretical evaluation of impact in organizations at an individual level. This is critical to our case study as, in healthcare, both facets—system success and its acceptance—are of great importance, given that there have been many cases of implemented systems that have either been unused or rejected by users for not having one or both of these facets [3–5]. The purpose of our research is to further this process of theory evolution by integrating and revising each of these models into our proposed model to further the evaluation of system impact at an individual level. This paper will be comprised of the following four sections: literature review and an outline of our proposed model, research methodology, case study discussion and finally, our conclusions and avenues for future research.

## 19.2 Literature Review

An important first step to understanding how the issue of system impact will be handled in this paper, is to properly define what impact means in the field of IS. DeLone & McLean define impact as having "improved… decision making productivity, a change in user activity, or has changed the decision maker's perception of the importance or usefulness of the information system" [1, p. 69]. In this paper, we are focusing on evaluating the effect of a system on the users' workflows, practices and on themselves. This effect, or 'impact', is often viewed through the lens of 'success' or 'acceptance' in IS research. Two of the main theories, UTAUT and the IS Success Model, will be outlined and discussed in the following sections. From these theories, we derived a combined model that enables us to investigate the impact at the individual level whilst allowing for breadth in the analysis of the impact measures.

### *19.2.1 Unified Theory of Acceptance and Use of Technology Model*

Over the years, research in the Information Technology field has been rapidly evolving. Researchers have developed many models to explain the measures that affect Information Technology adoption, e.g. Technology Acceptance Model [6], the Theory of Reasoned Action [7], and the Theory of Planned Behavior [8]. Venkatesh et al. [9] stated that the eight predominantly used theories had many similarities and so it was appropriate to create a unified theoretical model, which they named the Unified Theory of Acceptance and Use of Technology model (UTAUT). They theorized that there were various motivating or influencing factors that would lead to an 'Intention to Use' (Behavioral Intention/Acceptance of Technology) by an individual and which ultimately affect the actual usage of technology (Use Behavior).

UTAUT has subsequently been validated using actual workplace data and has been proven to outperform the eight individual models it encompasses [10]. UTAUT uses four core determinants for examining the acceptance of technology: performance expectancy, effort expectancy, social influence, and facilitating conditions. The UTAUT model views the acceptance of a new system as such: as an individual accepts the challenge of adopting a new technology, the ideal outcome is the actual usage of the technology, but infrastructure must be in place for this to occur, thus facilitating conditions are important for use behavior independent of whatever effect they might have on behavioral intention. We believe that UTAUT is a good fit for this evaluation. It enables comprehensive measurement of the various influences on the Intention to Use a system and for the purposes of this study, we need to have a clear understanding of what the individual users expect to gain from using the system, a priori that motivates them to engage with it or to reject it. UTAUT has also previously been used in case study research [11–14]. However, UTAUT does not examine the impact of system usage on the individuals and their organizations. Moreover it does not study the influence of technology characteristics on the use of the system. Therefore we looked for an additional theoretical model that evaluated impacts, which led us to DeLone & McLean's IS Success Model [2]

### 19.2.2 Information Systems Success Model

The IS Success model was originally proposed in 1992 by DeLone and McLean, and was based on previous theoretical and empirical IS research [1]. This model was further improved upon by DeLone and McLean, culminating in their 2003 paper, "The DeLone and McLean Model of Information Systems Success: A Ten Year Update." [2]. The original model described six constructs that can be used to evaluate information systems: system quality, information quality, system use, user satisfaction, impact on the individual user and impact on the organization. We utilized the IS Success model for its comprehensiveness and the distinct evaluation measures, furthermore it has been extensively validated and has been previously used in evaluating systems within health care services [15–17]. However its measurements are slanted differently to the UTAUT system—which emphasizes an individual's motivation and intentions in response to new technology, whereas the IS Success model examines the technology itself, and quantifies its usage by and impact upon individuals and organizations. In a previous study, Petter, DeLone & McLean [18] concluded that the measure of System Use was oversimplified and needed to be expanded in order to sufficiently provide the necessary breadth to the analysis of impact itself. In this paper, we are focusing on impact affecting the success of a system; hence it is necessary to consider the other measures affecting System Use.

### *19.2.3 UTAUT Extended Research Model*

There had been prior attempts to combine the two models together such as in Cody-Allen & Kishore's UTAUT Extended Research Model [10] where they expand the UTAUT model with Quality, Satisfaction & Trust constructs. Their focus was on an individual user's perception of an e-business system, however as they were not focused on examining the impact of a system at the individual level within an organization, their model was not suitable for analyzing the system success and user acceptance together.

## 19.3 A Model for Impact at the Individual Level in Relation to IS Systems Success and User Acceptance

The apparent shortcomings of these two influential models have led us to consider integrating the two together so that we could utilize the strengths of both whilst negating the downsides as much as possible. Thus, we propose the Integrated IS Success (IISS) Model as an improvement on the current theoretical models available; to better address the needs of examining the impact of a new system on the individual. One of the major boons of the IISS Model is that it will resolve UTAUT's inability to examine the system impact and measures of technological factors of the system, and DeLone & McLean's IS Success Model's inability to reconcile the human factors involved in use and intention to use of a system. The IISS Model which we propose is represented diagrammatically in Fig. 19.1. In the proposed model, we have taken out the Service Quality measure originating from the IS Success model, while adding in the UTAUT core determinants in order to increase the depth of analysis available with regards to the dimension of Use and Intention to Use. This allows us to better focus on the system impact on the individual user and not on the organization. The arrows in the model represent the core determinants' influence on the Use, Intention to Use and User Satisfaction factors. The vertical arrows represent how the factors motivate and affect each other e.g. if a user is satisfied with the system, they are motivated to use it (Intention to Use) and are more likely to actually use the system (Use), which in turn makes them more satisfied with the system as it meets their needs (User Satisfaction). These factors contribute together towards how an individual user feels towards a system.

The core determinants in the proposed model relate to the Use and Intention to use, which in turn affects the User Satisfaction factor, which is also influenced by the Information and System Quality factors. The Use and Intention to use combine

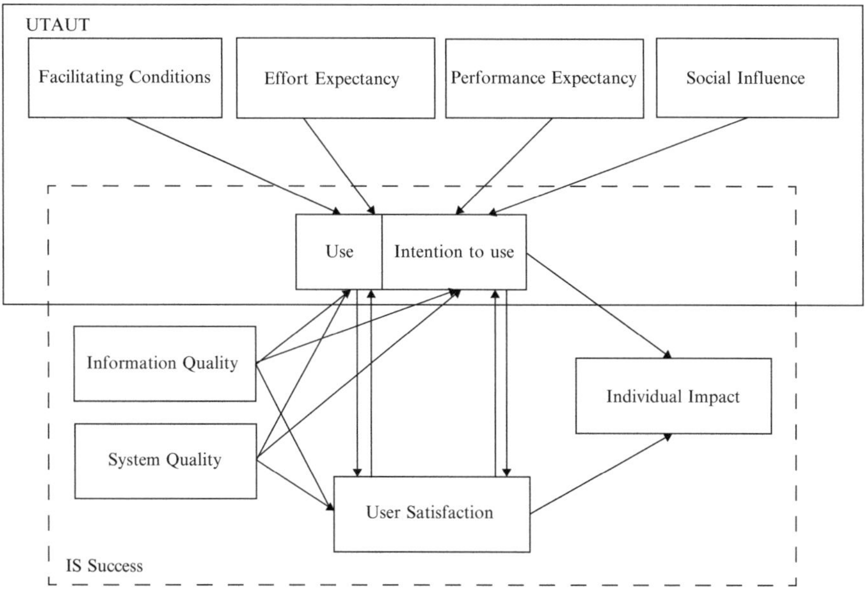


**Fig. 19.1** Integrated information systems success (IISS) model

with User Satisfaction which results in the overall Individual Impact, which in turn has an effect on the success and user acceptance of a given system. Unlike the original models which were validated quantitatively, the proposed model identifies the major factors but not in a causal relationship. The two boxes (solid and dashed lined) illustrate the original models that the proposed model have been derived from. To demonstrate the applicability of our proposed model in practice, we have applied it in a case study which will be discussed further in the following sections.

## 19.4 Methodology

As our case study research is based on investigating impact at an individual level as well as validating our proposed model, we have chosen to use qualitative research methods, specifically semi-structured interviews in a case study setting. As part of fulfilling the ethics requirements for human research, an application for ethics was completed and submitted.

### *19.4.1 Research Methods*

We utilized a semi-structured interview process to collect data from participants to analyze the impact of the Patient Referral System (PRS) on the participants' workflows and on themselves, this was based on prior case study research done within the IS field [19]. The participants for this study were selected via a voluntary, anonymous opt-in method and on the basis that they use or interact with the system, or had a hand in its design or development. We selected the semi-structured interview approach as it enabled detailed information collection about how users' day to day work has been impacted by the introduction of PRS. This approach enabled the chance to identify unanticipated outcomes [20], and allowed a degree of flexibility in adjusting the interview questions on the fly to further probe and engage participants. Overall, we have selected 10 participants from two distinct groups—users that were involved in the development of PRS, and those who were not. By providing participants freedom in answering the questions, the chance of bias in the responses are lowered, which is ideal. Similarly, we could prompt them to explore their responses in greater depth, which may not have been possible in a structured interview or a survey [21]. Interview questions were based on the themes derived from our proposed model and were compared to questions from several studies which have used the same method for analyzing user acceptance and systems success in relation to impact [11]. We analyzed the data gathered from the case study by utilizing an Adapted Conceptually Clustered Matrix (ACCM) adapted from Miles & Huberman [22, p. 128] with five themes and sub-themes as coded segments, to identify relationships between the segments in the matrix. We have chosen the ACCM over other thematic analysis techniques as it allows for a better fit with both the data we collected and the outcomes we were examining. Through examining the statements arranged in the ACCM, interpretations could be drawn regarding the impact on the participants, allowing us to demonstrate the validity and robustness of the IISS model.

## 19.5 Case Study

Hospital Y is a major trauma hospital in Australia, with a strong focus on acute medical care. A large part of their workflow is to ensure that the patients' get the correct medical care at the right time. As Hospital Y focuses on acute care, they transfer patients that are suitable for other treatment streams such as sub-acute and aged care streams to other sister hospitals. Therefore, it is pertinent that there be a degree of transparency and clarity in information exchange between departments, and similarly—other hospitals where patient transferals are to be made. In attempting to optimize that, Hospital Y has implemented a Patient Referral System (PRS) [23, 24]. Its users are able to view up-to-date information about the referrals and transferal status of their patients' to other treatment streams and hospitals. The PRS updates automatically four times a day and uses data entered by the administrative role type users to generate its view. In an attempt to validate the IISS model, we have used it to analyze the impact on the individual users' of the PRS using the ACCM. In our dataset, we have interviewed participants from different usage roles in the PRS. The participant role breakdown is presented in Table 19.1.

**Table 19.1** Participant role breakdown

| Role | Description | Amount |
|---|---|---|
| Manager | Strategic decision making | 3 |
| Operations | Day to day workflow optimisation | 4 |
| Administrative | Data quality, data entry, timely data dissemination | 3 |

**Table 19.2** Summarized table of impacts by user role

| User role | Summarized Impact |
|---|---|
| Manager | Reduced time taken in gathering information for strategic overview, decision support—instant view of the current state anytime during the day. |
| Operations | Increased perceived ability in streamlining daily work processes. Reduced time taken in gathering information on the current state—available anytime during the day. |
| Administrative | Increased perceived pressure to keep data up to date in the system |

The case study discussion is broken up into two sections, the derived impact on the users' workflow and on themselves at a personal level. The division of the users' workflow and personal areas of impact allows a more objective view into the different kinds of impact factors, and simplifies the usage of the ACCM.

## 19.5.1 *Impact on Workflow*

A major element affecting the Intention to Use, Use and User Satisfaction factors is the impact a system can have on an individual user's workflow as it heavily drives whether or not the use of a system is feasible to an individual—as if it negatively affects their workflow, it will be seen as a detriment or a burden to their work. Similarly, if it positively affects their workflow, the user would perceive the system more favourably because of this. Table 19.2 illustrates an overview of changes identified in the users' workflow as a result of the implementation of the PRS. It can be seen that there is an overall improvement in workflow efficiency. We identified a negative impact on the administrative user role types where their primary interaction with the PRS was to provide accurate, good quality data input in a timely manner to the PRS. These users were compelled to try and keep the information available to PRS as up to date as possible, which was influenced by the knowledge that the PRS updated at certain times of the day. This meant that they spent more time moving around between patient and their office in an attempt to put as much data in the moment they get it. On the other hand, there was a positive impact on the rest of the user role types, as it was shown to reduce the amount of time needed to request information about what was going on, as it was available in a single location on the PRS. This also meant that they could spend the time to concentrate on clearing up

**Table 19.3** Selected summarized impacts from an adapted conceptually clustered matrix (ACCM) showing impact categorized by factors in the IISS model

| Factor | Impact |
|---|---|
| Information quality | Information is up to date (perceived) on the PRS; the data is clear and sufficient. Still able to enquire for clarification when needed |
| System quality | System was live, accurate and updated four times a day based off data entered in the back end. Accessible at any time from a networked Hospital Y computer to users with access to it. Simple to use and sortable |
| Facilitating Conditions | Sufficient training was conducted for all staff interested in using the system. Support provided by IT Services |
| Effort expectancy | System is easy to use, especially useful when checking for patient referral status. It is still faster to check the old existing back-end systems when needing to look up general patient information. |
| Use | System in use by all relevant parties such as upper management (for strategic overview), operational management (direction of focus for daily workflow), and administrative users who mostly provide the data into the system |
| Performance expectancy | Reduced the amount of time spent enquiring about patient referral status. Information about patient referral available at a glance increased. Great improvement in information transparency and visibility in regards to patient referral status data |
| Social influence | Compelled to keep the data available in the PRS up to date as possible by updating information once it is obtained. Administrative role felt guilt in taking resources away from other staff—felt compelled to travel back to their own office to enter the data instead |
| Intention to use | Intention to use driven by interest, users expressed great interest in using system to improve their own daily workflows |
| User satisfaction | System is simple and friendly to use. Has been described as being organic and straightforward to use |

items undone, and to provide better care to the patients. For the user role type manager, it was noted that the PRS became a great tool to have an overview of the situation, and provide a clear basis to make strategic decisions on.

### *19.5.2 Impact on Users*

Building upon our workflow analysis, we utilized our ACCM for thematic analysis; we generated a consolidated list of impacts on the user base broken up into themes based on the IISS model's factors. We have briefly summarized these impacts in Table 19.3. From the data, it can be seen that the amount of enquiries and time spent on such activities has been reduced. Users across the board are able to use the system easily, and information available on the system is clear and sufficient. Users are able to better focus their daily workflow based on the current status of the patient referrals, being able to address outstanding issues identified on the PRS.

One interesting point of note was that the analysis conducted using our model identified that the administrative users felt a sense of guilt in taking resources (computers) away from the other staff, and were compelled to return to their office—which increased time spent travelling around—in order to enter the data into the system. It can be argued that it is both a good and bad impact, since it improves the efficiency by providing a source of encouragement to put in data as soon as it is obtained, and on the other hand, it subconsciously pressures these users to do so.

Overall, our analysis indicates that there has been a positive impact on the users interviewed. There were some negative impacts that have affected the administrative users to some extent, but when considering the benefits on an individual level—encouraging users to not put off work for later, it resulted in a mixed impact to the PRS as the users accepted the change. The other user roles experienced a positive impact as the increase of transparency and information flow assisted their work and optimized the users' daily work tasks. This is also reflected at the management level as it increased the efficiency in obtaining the information required to make strategic and operational decisions. The users were not forced to use the PRS, but were encouraged to do so by their peers, and their core motivation was that they wanted to improve their own workflow efficiency, which was critical factor in system acceptance. The IISS model was developed to bridge this gap between system acceptance and system success when investigating impact on an individual level within an organization. The Hospital Y case study results validate our model's theoretical raison d'être and highlight the importance of examining system success and user acceptance together in order to determine the overall impact on individual users of a system. Thus, our model has been shown to fill this particular theoretical niche of evaluation of a practical system.

## 19.6 Conclusion

Following our use of the IISS model in the hospital case study, we have reached the following findings; our model has successfully examined the impact on the individual users in this case. The analysis of the data derived from the case study indicated that, there was a positive overall impact on the individual users since the implementation of PRS; however a minor negative impact was identified regarding the workflow processes of staff needing to provide input that was utilized by PRS. This strengthens the validity of our model, given that had we only utilized the IS Success model or the UTAUT model, this would have remained unidentified. The inclusion of UTAUT's four core determinants [9] to DeLone & McLean's IS Success model [2] allows our proposed model to overcome the shortcomings that both models have in isolation. Avenues of future research could include research into utilizing our proposed model in other scenarios, both inside and outside of the health industry as well as expanding the scope from an individual level to an organizational one. Similarly, research could be done to replicate the use of our model in a similar scenario to verify the veracity and robustness of our model. The maturation of IS theory over the last five decades have seen these two models evolve and grow into what they are today. However, they are by no means perfect or complete. On the other hand, without such models, improvements in the field of IS research could have been delayed significantly, which would have been made a large impact to the quality and efficiency of IS as a whole. Our model seeks to continue this trend of continual evolution and growth in IS research and methods, thus contributing a new way of examining impacts on the individual within an organizational setting.

**Acknowledgements** We would like to acknowledge Monash University and Alfred Hospital for funding this research, Professor Frada Burstein for her support throughout this endeavour, and Mohamed Nasher for his help with this research.

## References

1. DeLone W, McLean E (1992) Information systems success: the quest for the dependent variable. Inform Syst Res 3(1):60–95
2. DeLone W, McLean E (2003) The DeLone and McLean model of information systems success: a ten year update. J Manag Inform Syst 19(4):9–30
3. Beynon-Davies P (1995) Information systems 'failure': the case of the London ambulance service's computer aided despatch project. Eur J Inform Syst 4(3):171–184
4. Zheng K, Padman R, Johnson MP, Diamond HS (2005) Understanding technology adoption in clinical care: clinician adoption behavior of a point-of-care reminder system. Int J Med Inform 74(7–8):535–543
5. Trafton J, Martins S, Michel M, Lewis E, Wang D, Combs A, Scates N, Tu S, Goldstein MK (2010) Evaluation of the acceptability and usability of a decision support system to encourage safe and effective use of opioid therapy for chronic, noncancer pain by primary care providers. Pain Med 1(4):575–585
6. Davis F (1989) Perceived usefulness, perceived ease of use, and user acceptance of information technology. MIS Quart 13:319–340
7. Fishbein M, Ajzen I (1975) Belief, attitude, intention and behavior: An introduction to theory and research
8. Ajzen I (1991) The theory of planned behavior. Organ Behav Hum Decis Process 50(2): 179–211
9. Venkatesh V, Morris M, Davis G, Davis F (2003) User acceptance of information technology: toward a unified view. MIS Quart 27:425–478
10. Cody-Allen E, Kishore R (2006) An extension of the UTAUT model with equality, trust, and satisfaction constructs. In: Proceedings of the 2006 ACM SIGMIS CPR conference on computer personnel research: forty four years of computer personnel research: achievements, challenges & the future, ACM, pp 82–89
11. BenMessaoud C, Kharrazi H, MacDorman K (2011) Facilitators and barriers to adopting robotic-assisted surgery: contextualizing the unified theory of acceptance and use of technology. PLoS One 6(1):e16395
12. Hennington A, Janz B (2007) Information systems and healthcare XVI: physician adoption of electronic medical records: applying the UTAUT model in a healthcare context. Comm Assoc Inform Syst 19(5):60–80
13. Kijsanayotin B, Pannarunothai S, Speedie S (2009) Factors influencing health information technology adoption in Thailand's community health centers: applying the UTAUT model. Int J Med Inform 78(6):404–416

14. Wills M, El-Gayar O, Bennett D (2008) Examining healthcare professionals' acceptance of electronic medical records using UTAUT. Issues Inf Syst 9(2):396–401
15. Jen W, Chao C (2008) Measuring mobile patient safety information system success: an empirical study. Int J Med Inform 77(10):689–697
16. Otieno G, Hinako T, Motohiro A, Daisuke K, Keiko N (2008) Measuring effectiveness of electronic medical records systems: towards building a composite index for benchmarking hospitals. Int J Med Inform 77(10):657–669
17. Yusof M, Paul R, Stergioulas L (2006) Towards a framework for health information systems evaluation. In: System sciences, 2006. HICSS'06. Proceedings of the 39th annual Hawaii International Conference on, IEEE, vol 5, pp 95a–95a
18. Petter SME, DeLone W (2008) Measuring information systems success: models, dimensions, measures, and interrelationships. Eur J Inform Syst 17(3):236–263
19. Yin RK (2003) Case study research: design and methods (applied social research methods). Sage Publications, Thousand Oaks
20. Creswell JW (2008) Research design: qualitative, quantitative, and mixed methods approaches. Sage Publications, Thousand Oaks
21. Ovaska S, Aula A, Majaranta P (2005) Käytettävyystutkimuksen menetelmät. Tampereen yliopisto, Tietojenkäsittelytieteiden laitos, Tampere
22. Miles M, Huberman A (1994) Qualitative data analysis: an expanded source book. Sage Publications, Thousand Oaks
23. Kim-Hwang JE, Chen AH, Bell DS, Guzman D, Yee HF, Kushel MB (2010) Evaluating electronic referrals for specialty care at a public hospital. J Gen Intern Med 25(10):1123–1128
24. Warren J, Gu Y, Day K, White S, Pollock M (2012) Electronic referrals: what matters to the users. Stud Health Technol Inform 178:235–241

# Chapter 20
# Identifying Essential and Optional Decision Constructs in On-line Transactional Processes

**Chris Barry, Mairéad Hogan, and Ann M. Torres**

**Abstract** Decision-making for users during an on-line transactional process has become fragmentary and 'start-stop'. Much of this discontinuance arises from decision points presented to users or consumers. Recent studies into Web development and user interface design practices amongst some airlines concluded that many optional extras are not presented as opt-in decisions, and are therefore in breach of European Union regulation. Furthermore, some airlines are using 'imaginative', sometimes unusual, decision constructs such as a 'must-opt'. This construct has been the source of the research question in this paper; what are the nature and types of decision constructs that users encounter throughout on-line transactional processes? The findings presented herein make an incremental contribution in identifying and categorizing some new decision constructs alongside established ones.

**Keywords** Information systems • Website design • Regulation • Opt-in/opt-out • Must-opt • Decision constructs

## 20.1 Introduction

The whole area of decision-making during an on-line transactional process is far more fragmentary and 'start-stop' than one might anticipate. Much of this discontinuance arises from decision points presented to users or consumers. Recent studies [1, 2] into Web development and user interface design practices amongst some airlines, concluded that all optional extras are not presented as opt-in decisions to users, and are therefore in breach of European Union regulation [3]. A further key finding in the 2011 study is a novel decision construct that forces users to make

C. Barry (✉) • M. Hogan • A.M. Torres
National University of Ireland, Galway, Ireland
e-mail: chris.barry@nuigalway.ie; mairead.hogan@nuigalway.ie; ann.torres@nuigalway.ie

M. José Escalona et al. (eds.), *Information System Development: Improving Enterprise Communication*, DOI 10.1007/978-3-319-07215-9_20,

choices on optional extras (coined a 'must-opt' decision), has been developed. This must-opt appears to assist firms in circumventing the regulations. This construct has been the source of the research question in this paper; what are the nature and types of decision constructs that users encounter throughout on-line transactional processes?

## 20.2 A Brief Exploration of Optionality

An initial consideration of optionality proffers the notion that an option presented to a user is a straightforward choice—you either wish to secure the option or not. The reality is that optionality is far more complex. While the European Union directive dealing with optional charges by European airlines states that "all optional price supplements should only be accepted by the consumer on an 'opt-in' basis", it does not define optionality or opt-in. Thus, our starting point must go back a little further. In seeking to define the notion of optionality, the following were identified:

- Merriam Webster define optional as 'involving an option: not compulsory'
- Geddes and Grosset define to 'opt' as 'to choose or exercise an option'
- Merriam Webster have no definition for opt-in but define opt-out as 'to choose not to participate in something'
- The Oxford English Dictionary define opt-in as to 'choose to participate in something' and opt-out to 'choose not to participate in something'

A more nuanced consideration is found on http://en.wiktionary.org/wiki/opt-in where the following distinction is made between opt-in and opt-out.

- To opt-in—of a selection, the property of having to choose explicitly to join or permit something; a decision having the default option being exclusion or avoidance.
- To opt-out—of a selection, the property of having to choose explicitly to avoid or forbid something; a decision having the default option being inclusion or permission.

Another dimension of decision constructs is question framing. Questions may be framed in terms of acceptance (e.g., I would like to receive email) or rejection (e.g., I would not like to receive email). Alternatively, Lai and Hui [4] described them as 'choice' and 'rejection' frames, where they suggest positive phra sing corresponds with choice and negative phrasing corresponds with rejection of an option.

## 20.3 Previous Research

Barry, Hogan and Torres [1] explored user views on whether two Irish airlines were acting in good faith in their compliance with European Union consumer protection legislation [3]. Rather than the more usual opt-in/opt-out mechanisms used to offer

**Table 20.1** Ancillary services categorised

| | Airline | | | |
|---|---|---|---|---|
| | Ryanair | | Aer Lingus | |
| Option presented as | Option | No. | Option | No. |
| **Opt-in** | Baggage<br>Sports equipment<br>Special assistance | 3 | Flex fare (1)<br>Flex fare (2)<br>SMS confirmation<br>Special assistance<br>Voucher<br>Baggage<br>Extra baggage weight<br>Sports equipment<br>Lounge | 10 |
| **Opt-out** | – | 0 | Mailing list | 1 |
| **Must-opt** | Priority boarding<br>Travel insurance (1)<br>SMS confirmation<br>Ryanair approved cabin bag<br>Terms and conditions<br>Travel insurance (2)<br>Hertz Rent-a-car | 7 | Terms and conditions<br>Travel insurance<br>Parking | 3 |

ancillary services, it was found the airlines were using a new approach, referred to here as a must-opt selection. What appears to be an opt-in option is presented to the user who may choose to overlook it in the normal course of events. However, when they attempt to move to the next webpage, they are informed, generally via a pop-up, they must go back and make a selection. Elsewhere the construct has been referred to as a 'no default' [5] and a 'neutral condition' [6]. The difficulty with these definitions is they do not adequately reveal the more subtle distinctions within the constructs.

The findings of the analysis are reproduced in Table 20.1. They identify the nature of the optional decision constructs encountered during a flight reservation process. One airline presented ten different decision points (i.e., three opt-in and seven must-opts), while the other presented 14 (i.e., 10 opt-in, 1 opt-out and 3 must-opts) to be negotiated before a flight was booked. Dictionary definitions do not fully capture the way in which must-opt optional extras are presented. It can be argued they are both opt-in and opt-out as the user must explicitly choose or decline to participate. However, the must-opt can also be viewed as neither opt-in nor opt-out, since the default option is to prevent the user from continuing until they either choose or decline the option. The net effect is users cannot overlook this type of decision and must give it their full consideration.

Although EU regulations specifically state optional charges be accepted on an opt-in basis, the airlines seem to have found a technical mechanism to by-pass the regulations—the must-opt construct.

## 20.4 Decision Constructs and Their Impact

The lack of clarity in the definition of optional price supplements has resulted in a case being taken to the European Court of Justice [7]. Article 23(1) of Regulation No. 1008/2008 [3] states: "Optional price supplements shall be communicated in a clear, transparent and unambiguous way at the start of any booking process and their acceptance by the customer shall be on an 'opt-in' basis". The judgement in relation to this regulation has clarified the issue somewhat. It states optional price supplements are not unavoidable and:

> "In particular, the last sentence of Article 23(1) of Regulation No 1008/2008 refers to 'optional price supplements', which are not unavoidable, in contrast to air fares or air rates and other items making up the final price of the flight, referred to in the second sentence of Article 23(1) of that regulation. Those optional price supplements therefore relate to services which, supplementing the air service itself, are neither compulsory nor necessary for the carriage of passengers or cargo, with the result that the customer chooses either to accept or refuse them. It is precisely because a customer is in a position to make that choice that such price supplements must be communicated in a clear, transparent and unambiguous way at the start of any booking process, and that their acceptance by the customer must be on an opt-in basis, as laid down in the last sentence of Article 23(1) of Regulation No 1008/2008.
>
> That specific requirement in relation to optional price supplements, within the meaning of the last sentence of Article 23(1) of Regulation No 1008/2008, is designed to prevent a customer of air services from being induced, during the process of booking a flight, to purchase services additional to the flight proper which are not unavoidable and necessary for the purposes of that flight, unless he chooses expressly to purchase those additional services and to pay the corresponding price supplement."

While the regulation only applies to airlines, the definition above relating to optional price supplements is clear and could be used to define optional price supplements on other e-commerce sites.

The European Union has recognised the need for regulation in relation to other forms of distance and off-premises contracts, which would include e-commerce transactions. They have introduced a new directive on consumer rights [8] whose intent is to protect the consumer in distance contracts. In this directive it is stated that additional payments above and beyond the minimum cost of the transaction require the explicit consent of the consumer. The directive states in Article 22, with respect to additional payments, that:

> "Before the consumer is bound by the contract or offer, the trader shall seek the express consent of the consumer to any extra payment in addition to the remuneration agreed upon for the trader's main contractual obligation. If the trader has not obtained the consumer's express consent but has inferred it by using default options which the consumer is required to reject in order to avoid the additional payment, the consumer shall be entitled to reimbursement of this payment."

The European Union recognise consumers need to be protected against unscrupulous practices that may result in an inadvertent purchase that is not a necessary part of the transaction. For airlines, they state the additional options may only be purchased on an 'opt-in' basis while for all other distance contracts, the consumer's express consent is required and the vendor may not use default options that require the consumer to reject the option. However, neither piece of legislation defines what is meant by an 'opt-in' or what type of constructs are allowed where the consumer must make a decision on an optional extra. There is, however, a definition of 'consent' in the Data Protection Directive [9] relating to the use of an individual's data. Consent is defined as:

> "any freely given specific and informed indication of his wishes by which the data subject signifies his agreement to personal data relating to him being processed."

While no definition of this nature is included in the Directive on Consumer Rights, it is possible the European Court would deem it an acceptable definition for this directive. However, there is still no indication of what are considered acceptable ways of obtaining consent other than stating the use of default options that the consumer must reject are unacceptable. It is therefore at the discretion of the vendor to determine the most suitable method of obtaining the consent.

The Office of Fair Trading in the UK [10] carried out a study on the impact of pricing practices on consumer behaviour, where they described a process referred to as "drip pricing". This practice presents the user with an element of the price up front and then presents additional components as "drips" throughout the buying process. The drips can be either compulsory, where they are inherent to the prices of the product (e.g., shipping cost) or optional, where they are generally add-ons (e.g., an optional warranty). These "drips" can be presented as opt-ins, opt-outs or must-opts. Their review of the available literature indicated consumers tend to retain the default option presented, even if it is detrimental to them. This retention may be due to inertia and an inherent believe the default is a recommendation by the vendor. Consumers may also choose the default in order to avoid the cognitive effort required to make a decision. Therefore, where the vendor uses an opt-out policy, the consumer may accept options that are detrimental to them or make purchases they do not need or want.

Much research has been carried out to determine whether users are more likely to participate when an option is framed as an opt-out rather than an opt-in [6, 11–13]. They generally conclude an individual is more likely to retain the default option than to change it. That is, they are more likely to participate if an option is presented as an opt-out, rather than an opt-in. Johnson and Goldstein [6] also found there was little difference in acceptance rates between an opt-out and a must-opt (referred to as a neutral condition in their paper). The main reasons identified for this negligible difference are participant inertia and a perception that the presentation of a default is a recommendation. McKenzie et al. [13] take that conclusion further and state those presenting the choice are more likely to present it in a way that indicates their beliefs or attitudes towards the choice. They also state those choosing an option are less likely to accept the default if they are educated about the issues in question. Although no study was found that examined this question, perhaps the vendor could also influence the consumers' decision by providing additional information that is biased in favour of the vendor's preferred choice, even if they are required to present the decision as an opt-in.

Lai and Hui [4] have carried out additional research into the impact of the question framing on user decisions. Their study indicates the way in which the option is described as well as the selection mechanism has an impact on user choice. They found for opt-in decisions using check boxes, users are more likely to accept an un-selected opt-in over a pre-selected opt-in. They posit the language of acceptance (i.e., referred to here as acceptance format) inherent in an unselected opt-in is likely to influence the users' decision (e.g., 'Please send me newsletters' with the checkbox un-ticked versus 'Please do not send me newsletters" with the check box ticked). However, for the opt-out mechanism, they did not find a significantly different acceptance rate between the pre-selected opt-out and the un-selected opt-out. Belman et al. [5] found similar results for opt-in using radio buttons, where an option in an acceptance format was more likely to be accepted than one in a rejection format (e.g., 'Notify me about more health surveys' with the No button pre-selected versus 'Do NOT notify me about more health surveys' with the Yes button pre-selected). However, Belman et al. [5] did identify a difference between the acceptance rates for pre-selected opt-outs using radio buttons. They found users were more likely to accept an option when the language was phrased in an acceptance format, rather than a rejection format (e.g., 'Notify me about more health surveys' with the Yes button preselected, versus 'Do NOT notify me about more health surveys' with the No option pre-selected). They also considered a must-opt format that forced users to choose an option. In this case, users were more likely to choose the option when it was framed in an acceptance format rather than a rejection format (e.g., 'Notify me about more health surveys' versus 'Do NOT notify me about more health surveys'). This point concurs with Lai and Hui's [4] finding that users perceive the way in which the selection is presented as guidance rather than a neutral choice.

## 20.5 Research Approach

The researchers have constructed a research plan to investigate how decision constructs are presented to users engaged in business-to-consumer (B2C) electronic commercial transactions. The authors are planning an extensive and systematic study to examine how this presentation is made. Before this research can be conducted it is necessary, in as far as possible, to identify an exhaustive list of the various decision constructs users encounter when purchasing a product or service whilst on-line. The constructs are not core to the actual product or service and are for the most part options. Thus, an exploratory study was conducted that examined typical e-commerce transactions consumers would carry out. These transactions included:

- Buying a book
- Buying a bus ticket
- Taking out an insurance policy (home, health, motor)

- Buying clothes
- Buying DVDs and gadgets
- Booking flights, ferries and trains
- Hiring a car
- Selecting a package for a soccer match
- Booking a hotel room
- Choosing a phone package

More than 20 on-line retailers' Websites were explored and on some, several products or services were studied. The transactional process on each website is made up of a number of sequential webpages that end, ultimately, in a payments page where the exploration terminated. During the process, normally after the core product or service has been selected, various decisions are presented to users. These constructs are the subject of this study.

## 20.6 Findings of the Exploratory Study

The researchers confirmed the identification of the must-opt construct [1], made distinctions between essential and optional decisions and identified more elaborate and complex constructs.

### *20.6.1 Decision Types: Essential and Optional*

Most decisions, other than those relating to the core product or service, are real 'options' that may or may not be chosen. However, there are also common decisions that must be made which involve various options. Such decisions are 'essential' to obtaining the product or service. Thus, the first meta-category of decisions is whether they are essential or truly optional.

#### 20.6.1.1 Definition of an Essential Decision

An essential decision is where the customer must choose between variants of a necessary and fundamental aspect of the product or service. The customer will not be able to complete the purchase without choosing one of the variants. For example, choosing between different delivery methods or choosing between different payment methods. It is a non-intrinsic aspect of the product or service. Thus, it is not the garment size or colour decisions; nor is it the dates or destination decisions for a flight.

**Table 20.2** Decision construct descriptions

| Decision construct | Description |
|---|---|
| Un-selected opt-in | This decision structure has a default option of not receiving the option. It is generally presented as an un-ticked check box or a radio button set to off, where the question is framed in an acceptance format. Thus, the terminology states the customer wants the option. |
| Pre-selected opt-in | This decision structure has a default option of not receiving the option. It is generally presented as a ticked check box or a radio button set to on where the terminology states the customer does not want the option. |
| Un-selected opt-out | This decision structure has a default option of receiving the option. It is generally presented as an un-ticked check box or an radio button set to off where the terminology states the customer does not want the option. |
| Pre-selected opt-out | This decision structure has a default option of receiving the option. It is generally presented as a ticked check box or a radio button set to on where the terminology states the customer wants the option. |
| Must-opt | A must-opt decision occurs when an optional extra is presented to a customer as un-selected. It is not possible to proceed to the next webpage without having made a selection. It is generally presented as radio buttons, command buttons or a drop down list. |
| Un-selected essential decision | An un-selected essential decision is where none of the variants has been pre-selected for the customer. For example, the customer chooses a payment method. |
| Pre-selected essential decision | A pre-selected essential decision is where one of the variants has been pre-selected for the customer. It may be in either the customer's or the vendor's favour—or it may be neutral. For example, fast delivery for a surcharge may be pre-selected. |

#### 20.6.1.2 Definition of an Optional Decision

An optional decision is where the customer may choose an optional extra. It is not a necessary or fundamental aspect of the product or service. The customer will be able to complete the purchase without choosing this option. For example, choosing an extended warranty or receiving a n SMS message. There is normally an extra charge for the optional extra. It is an ancillary aspect of the product or service.

### *20.6.2 Decision Constructs*

To assist the following discussion, Tables 20.2 and 20.3 present descriptions and illustrations of each decision construct. What emerged through the transactional process was that optional decisions are not always presented in a manner users might

**Table 20.3** Illustrations of decision constructs

| Decision construct | Description | Framing |
|---|---|---|
| Un-selected opt-in | ☐ I want an extended warranty | Acceptance |
| Pre-selected opt-in | ☑ I do not want an extended warranty | Rejection |
| Un-selected opt-out | ☐ I do not want an extended warranty | Rejection |
| Pre-selected opt-out | ☑ I want an extended warranty | Acceptance<br>Rejection |
| Must-opt | ○ I want an extended warranty<br>○ I do not want an extended warranty | Acceptance |
| Un-selected essential decision | ○ Express delivery in 2 days (€5.00)<br>○ Fast delivery in 3-4 days (€2.00)<br>○ Free delivery in 5-7 days (free) | Neutral |
| Pre-selected essential decision | ✔ Express delivery in 2 days (€5.00)<br>○ Fast delivery in 3-4 days (€2.00)<br>○ Free delivery in 5-7 days (free) | Neutral |

anticipate. Opt-in decisions are normally just that—you choose or do something in order to receive the option. This construct would involve ticking a check box or choosing an item from a drop down list—thus an un-selected opt-in. However, a pre-selected opt-in involves much more uncertainty. If a check box on an option is ticked, it is suggestive of something having been pre-selected for the user. Instinctively users are likely to quickly de-select an option for, say, opting-out of receiving an e-mail newsletter. However, the construct can be used with rejection framing such as "I do not want an extended warranty". To opt-in you un-tick the box. The construct is counter-indicative and uses un-ticking of the check box to opt-in, while the text uses rejection framing to NOT receive the option. Undoubtedly, this phrasing would be extremely confusing for users.

It is usual that an opt-out decision appears as a pre-selected tick in a check box with acceptance framing—for example, 'I wish to receive email'. Drawing attention to the option in this manner may result in the user giving the option more consideration than they would otherwise. While it may be questionable whether firms should force users to opt-out of a pre-selected option, at least the pre-selection may be logically interpreted to mean the user will receive that option. However, one opt-out construct was found to be actually un-selected. This instance is a most extraordinary, counter-intuitive means of designing an opt-out structure. Essentially, the option appears like a 'normal' opt-in decision. If conventionally used, a user might safely overlook an un-selected option, assuming it to be opt-in. However, the un-selected opt-out construct is designed so that a user must tick a box to reverse out of an opt-out decision. The decision framing is rejection—a negation of the decision. In this case—'I do not want Collision Damage Waiver' as part of insuring a hired car.

A must-opt decision occurs when an optional extra is presented to a user as unselected. However, it is not an opt-in since the user is prevented from progressing onto the next webpage unless they explicitly accept or reject the option—thus they 'must-opt'. Typically, a user is only alerted to this construct when they click a 'Continue', or similar button, at the bottom of the webpage. Instead of progressing to the next webpage, the user must go back and read the option variants and choose one. For example, some airlines force customers to explicitly indicate they do or do not want travel insurance.

The final constructs relate to essential decisions, which may also be pre-selected (e.g., a fast delivery method) or more usually un-selected (e.g., choice of a payment method). Table 20.2 summarises the seven decision constructs identified in the study while illustrations are shown in Table 20.3.

The more comprehensive analysis of decision constructs carried out here is clearly merited in light of the European Union's recognition of the requirement for consumer consent on optional price supplements and other provisions. What is clear is the basis of their directives and judgments needs to be more finessed so they account for all the decision constructs identified in this study.

## 20.7 Conclusions

This study set out to identify all possible ways in which essential and optional decision constructs can be presented to a user in an on-line transactional process. The genesis for the research question was to explore whether sectors, other than the airline sector, were acting in good faith in relation to consumer protection regulations. As noted earlier, the European Union has recognised programming constructs are being used to nudge consumers to behave in a way that airlines wish. It would appear that these constructs are also being used in many other sectors. Furthermore, with the must-opt and other ambiguously presented decisions, it is clear that European Union regulations deal with the notion of optionality inadequately. The authors believe they have captured all decision constructs presently in use. However, it is likely firms will continue to behave inventively as they seek ways of attracting users attention to various ancillary products and services. The theory of cultural lag identified by Ogburn [14] is a resilient one—in this case, firms are using new technologies to shape user behaviour in their favour—researchers and regulators take note.

## 20.8 Further Research

From this study, a number of issues emerged that will contribute to the analysis in the next phase of research. Some are objective observations whilst others will require a more subjective interpretation. These are:

- The decision construct control type (e.g., radio buttons, check boxes, command buttons, drop-down menus)
- The method used to present the decision (e.g., vertically or horizontally)
- How additional information is presented (e.g., hyperlink, on screen or rollover) and how easy is it to comprehend
- Price visibility and clarity
- The clarity of the decision construct (e.g., is it clearly a must-opt?)
- The clarity of the optionality of the decision
- How the option is posed (persuasive or dissuasive)
- The framing of the question (acceptance, rejection or neutral)
- The ease of reversing a decision
- The ease of use for the decision construct

Thus, a more elaborated study investigating how decision constructs are presented to users engaged in business-to-consumer (B2C) commercial transactions is planned. It will be framed with the decision constructs identified in this paper and the emergent issues specified here.

## References

1. Barry C, Hogan M, Torres A (2011) Perceptions of low cost carriers' compliance with EU legislation on optional extras. In: 20th international conference on information systems development, Edinburgh, Scotland, 24–26 August 2011
2. Torres A, Barry C, Hogan M (2009) Opaque web practices among low-cost carriers. J Air Transport Manage 15:299–307
3. European Union 1008/2008 (2008) Regulation of the European Parliament and of the Council on Common rules for the Operation of Air Services in the Community (recast)
4. Lai Y, Hui K (2006) Internet opt-in and opt-out: investigating the roles of frames, defaults and privacy concerns. In: Proceedings of SIGMIS-CPR'06, Claremount, CA, 13–15 April
5. Belman S, Johnson E, Lohse G (2001) To opt-in or opt-out? It depends on the question. Comm ACM 44(2):25–27
6. Johnson E, Goldstein D (2003) Do defaults save lives? Science 302:1338–1339
7. ebookers.com Germany v BVV (2012) Case C-1 12/11, European Court of Justice Judgment.
8. European Union 2011/83/EU (2011) Directive on consumer rights
9. European Union 1995/46/EC. Data Protection Directive. 1995
10. Ahmetoglu G, Fried S, Dawes J, Furnham A (2010) Pricing practices: their effects on consumer behaviour and welfare. Prepared for the Office of Fair Trading by Mountain Learning. http://www.oft.gov.uk/shared oft/business leaflets/659703/Advertising-of-prices/Pricing Practices.pdf. Accessed 9 May 2013
11. Madrian B, Shea D (2001) The power of suggestion: inertia in 401(k) participation and savings behavior. Quart J Econ 116(4):1149–1187

12. Junghans C, Feder G, Hemingway H, Timmis A, Jones M (2005) Recruiting patients to medical research: double blind randomised trial of "opt-in" versus "opt-out" strategies. Br Med J 331:940
13. McKenzie C, Liersch M, Finkelstein S (2006) Recommendations implicit in policy defaults. Psychol Sci 17(5):414–420
14. Ogburn W (1957) Cultural lag as theory. Sociol Soc Res 41(3):167–174

# Chapter 21
# Ontology and SOA Based Data Mining to Business Process Optimization

**Aleksander Pivk, Olegas Vasilecas, Diana Kalibatiene, and Rok Rupnik**

**Abstract** The need to improve business process efficiency, to react quickly to changes and to meet regulatory compliance is the main driver for using *Business Process Intelligence* (BPI). BPI refers to the application of Business Intelligence techniques, like data warehousing, data analysis, and data mining, to find correlations between different workflow aspects and performance metrics, to identify the causes of bottlenecks, and to find opportunities for business process prediction and optimization, e.g. elimination not necessary steps. In this paper we propose an ontology and Service Oriented Architecture (SOA) based approach for data mining process implementation for business processes optimization. The proposed approach was implemented in eight commercial companies, covering different industries, such as telecommunications, banking and retail. The experiment achieved shows that companies having data warehouse had a significant advantage, e.g. it allows us to eliminate not necessary operations and optimise business process.

## 21.1 Introduction

Nowadays organizations aim to automate their business processes with the objectives of improving operational efficiency, reducing costs, improving the quality of service offered to customers and reducing human error. From a business process

A. Pivk
Department of Intelligent Systems, Jozef Stefan Institute, Ljubljana, Slovenia
e-mail: aleksander.pivk@ijs.si

O. Vasilecas • D. Kalibatiene (✉)
Information Systems Department, Vilnius Gediminas Technical University, Vilnius, Lithuania
e-mail: olegas.vasilecas@vgtu.lt; diana.kalibatiene@vgtu.lt

R. Rupnik
Faculty of Computer and Information Science, University of Ljubljana, Ljubljana, Slovenia
e-mail: rok.rupnik@fri.uni-lj.si

M. José Escalona et al. (eds.), *Information System Development: Improving Enterprise Communication*, DOI 10.1007/978-3-319-07215-9_21,

automation perspective: a) business processes should be correctly designed, b) their execution should be supported by a system that can meet the workload requirements, and c) the (human or automated) process resources should be able to perform their work items in a timely fashion. Research in the business process automation area has been mostly focusing on developing new process models and process automation techniques; little work has been done in the areas of process analysis, prediction, and optimization.

*Business process intelligence* (BPI) aims to apply data warehousing, data analysis, and data mining techniques to process execution data, thus enabling the analysis, interpretation, and optimization of business processes. In this paper we focus on using data mining for optimization, e.g. elimination not necessary operations, of business processes in commercial organisations. The second force, which helps to be faster in implementing new business strategies and products, is *service oriented architectures* (SOA). As data mining becomes more and more an integral part of executing a business, data mining functionality needs to be integrated into SOA in the context of existing applications [1]. And the third, but not the last, force, which helps to represent and capture knowledge of a business domain, is *ontologies* [2].

However, the review of the related works presented in Sect. 21.2 shows that there is lack of approaches and methods for the use of data mining for business process optimization. Therefore, according to the three forces we propose an ontology and SOA based data mining process implementation into business process approach in Sect. 21.3. The last part of the paper is structured as follows. Section 21.4 presents a case study in eight companies on the use of data mining to support direct marketing according to the presented approach. Section 21.5 concludes and summarises the paper.

## 21.2 Related Works

According to Davenport and Short [3], a business process is *a set of logically related tasks performed to achieve a defined business outcome*. In today's business processes, such as order processing, account replenishment, etc., are typically *integrated* and *automated* (mostly partially). Therefore, they are of long duration, involve coordination across many manual and automated tasks, require access to several different databases and the invocation of several application systems (e.g., ERP systems), and are governed by complex business rules [4]. A typical business process may consist of up to 100 IT transactions.

Nowadays IT offers very good solutions for implementing business process renovation. The contributions of IT in business process renovation could be categorized in two different ways [5]: as a facilitator to the process of renovation and as an enabler to master the new process in the most effective way. In our research we are interesting in the second contribution.

The use of data mining[1] in business processes is motivated by the following reasons [6]: a) *business process analysis*, which refers to the analysis of past and/or current process executions. It is helpful to find correlations between different workflow aspects and performance metrics, to identify the causes of bottlenecks, and to find opportunities for business process optimization [7]. b) *Business process discovery*, which refers to the analysis of business events recorded in event logs to discover process, control, data, organizational, and social structures [8] and to identify malfunctions or bottlenecks. c) *Business process monitoring*, which refers to the monitoring of running process instances (e.g., their progress, bottlenecks and times spent in each known patterns and relationships hidden in data, where the data can be stored in databases, data warehouses, or other information repositories [9, 10]. activity) and their analysis results (e.g., percentage of instances not completing successfully) to inform users about unusual or undesired situations (i.e., alerts) [11]. And d) *conformance checking*, which can be applied to analyse whether a log conforms to a process model and to identify undesired behaviour [12].

**CRISP-DM** (CRoss-Industry Standard Process for Data Mining) [13] is a standard process model for data mining that depicts corresponding phases of a project, their tasks, and relationships between these tasks. Even though CRISP-DM describes the general procedure of performing data mining projects; it does not suffice, as this approach is too general. The CRISP-DM model lacks in the deployment phase [14] and misses phases important for engineering projects [15]. There is no specification or support for standards on how to deploy the data mining results into the business process.

Kohavi and Provost [16] argue that it is important to enable the use of data mining in business processes through automated solutions. However, it can be a rife with pitfalls. Authors also argue that one must deal with social issues when deploying automated solutions to previously manual processes.

Gray [17] discusses data mining as an option of knowledge sharing within the enterprise. Ciflikli and Ozjirmidokuz [18] discusses the use of data mining in manufacturing processes and state that the knowledge acquired through data mining improves processes. Kurgan and Musilek [19] states that the use of knowledge acquired through data mining can improve work of business users through semi-automating or automating activities. He argues that semi-automating is more realistic than automating and declares this as important new trend in knowledge discovery. Wegener and Ruping [20] propose a pattern-based approach for integrating data mining in business process. They approach is based on CRISP-DM and includes the definition of data mining patterns and a meta-process for applying patterns to business processes. Details on the definition, architecture and implementation of the underlying services remain to future work. In addition, should be decided how to model the data within the process.

[1] Here Data mining is understood as extracting or "mining" knowledge from large amounts of data in order to discover implicit, but potentially useful information.

Another promising solution for just-in-time, distributed and privacy-protected data mining process is using SOA, like presented in [21, 22]. According to *The Open Group*,[2] SOA is an architectural style that supports service-orientation—a way of thinking in terms of services, service-based development and the outcomes of services. Adopting SOA for data mining (DDM) has at least three advantages [21]: 1) implementing data mining services without having to deal with interfacing details such as the messaging protocol, 2) extending and modifying DDM applications by creating or discovering new services, and 3) focusing on business or science problems without having to worry about data mining implementations.

Authors of [23] propose *Anteater service-oriented architecture for data mining* that relies on Web services to achieve extensibility and interoperability, offers abstractions for users, and supports computationally intensive processing on large amounts of data through massive parallelism. Anteater is designed as a set of distributed components that offer their services through well-defined interfaces, so that users can employ them as needed.

The use of *ontologies* for the implementation of data mining into business processes is beneficial in the following issues [24]: a) sharing common understanding among people or software agents [25], b) enabling reuse of domain knowledge, c) making explicit domain assumptions, d) separating the domain knowledge from the operational knowledge, and e) analysing domain knowledge. This allows us to choose proper data mining method to do the knowledge discovery, explore similar situations or problems and unify the data mining domain and the demand for formalized representation of outcomes of data mining process. In this context *Ontology* defines the basic concepts, their definitions and relationships comprising the vocabulary of a domain and the axioms for constraining interpretation of concepts and expressing complex relationships between concepts [26, 27].

Authors of [28] propose heavy-weight ontology, named *OntoDM*, based on a general framework for data mining. It represents entities such as data, data mining tasks and algorithms, and generalizations. Authors of [29] use a data mining ontology for selecting data mining algorithms and for semantic searching. As they state, this method is helpful for data mining beginners who don't know well about which kind of algorithm can perform their tasks.

Ankolekar et al. [30] present DAML-S, a DAML+OIL ontology for describing the properties and capabilities of Web Services. They describe three aspects of the proposed ontology: the service profile, the process model, and the service grounding. Their proposed the *Process Ontology* describes a service in terms of its inputs, outputs, preconditions, effects, and, where appropriate, its component subprocesses. The *Process Control Ontology* describes each process in terms of its state, including initial activation, execution, and completion. Pinto and Guarda [31] introduce process oriented ontology for database marketing knowledge based on Data Mining system architecture. They proposed the ontology and the knowledge extraction process based solution is used for end user assistance in the entire process development.

[2] http://www3.opengroup.org

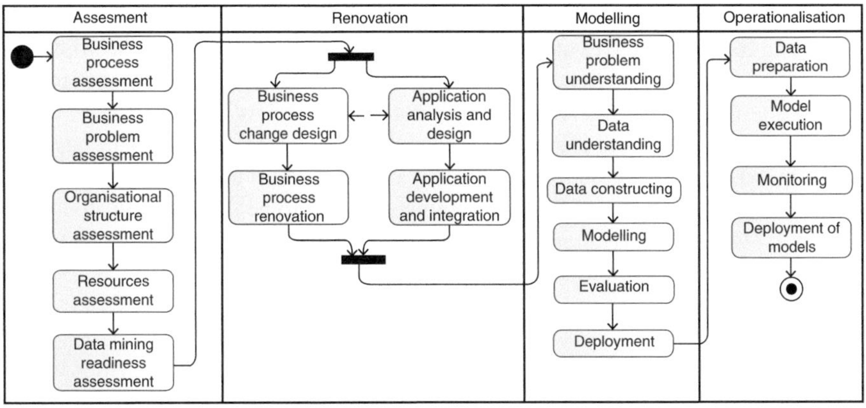


**Fig. 21.1** The proposed approach of implementation of data mining into business processes

However, there is lack of a combined approach or a method of using ontology, SOA and data mining for business process optimization. Therefore, in the future sections we present our ontology and SOA based data mining implementation to business processes optimization.

## 21.3 Implementation of Data Mining to Business Processes

We propose an approach for the implementation of data mining into business processes which is based on the following contributions: 1) approaches to BPI implementations (e.g. [13, 32, 33]); 2) a methodological framework to business process renovation and IS development [13]; 3) CRISP-DM framework, and 4) our experience with the implementation of data mining in analytical business process. The main elements of the proposed methodology are (Fig. 21.1):

1. ***Assessment of organizational processes and their structure***. The purpose of this phase is to determine readiness of an organization to incorporate data mining into its business processes. The assessment of each activity must consider the following evaluation elements: *Risks* (examination of diverse factors that can bring about different types of risks, e.g. strategic, compliance, financial, operational), *Challenges* (strategic, legal and compliance, financial, operational, portfolio), *Opportunities* (potential improvements on all levels of the organization), and *Potential value* (the potential impact on the organization's performance in terms of money or market position).

Identification, description and generation of the potential improvement scenarios of the affected business processes are created and evaluated for all evaluation elements at *Business process assessment* activity. Then according to the obtained results, the originated business problems are identified, described and evaluated according four perspectives at *Business problems assessment* activity. The impact of the new or renovated business processes on the existing organizational structure is assessed at *Organizational structure assessment* activity. At *Resources assessment* activity two main parts of issues are analysed: 1) availability and skills of business analysts, business users, and support teams is analysed, 2) availability, sophistication, and interoperability of the data mining platform to be used is evaluated. And finally, at *Data mining readiness assessment*, the level of readiness for incorporating data mining operations into business processes is determined.

2. ***Renovation***

For successful implementation of data mining there are two key activities which must be executed more or less simultaneously: *process change design* and *applications analysis and design*. The aim of the first is to define and design the changes that happen in business process in order that the use of data mining brings added value. The aim of the second is to make analysis and design of the application that will use data mining and should be developed. There is not only new application that should be developed, there are also existing operational and analytical applications that should be adapted to the use of data mining models and support changes in business process. Both activities are very dependent on each other. A particular application must provide the functionalities suitable for changes designed in activity *process change design*. After the activity *applications analysis and design* is finished the development of application is initiated through activity *applications development and integration*. The aim of this activity is not only to develop application, but also to integrate it into information system of the company, i.e. integrate it with other applications. When the activity *business process change design* is finished the activity *business process renovation* is initiated. The aim of this activity is to renovate the business process according to the changes defined and designed in previous activity.

3. ***Extending CRISP-DM—Modelling.*** The purpose of the modelling phase is to create a stable data mining model that needs to tackle the identified business problem.

Understanding the business problem and transforming it into a data mining problem, using domain ontology (DO), is performed at *Business problem understanding* activity. An initial data collection and understanding, to discover first insights into the data and to identify data quality problems, using DO, is performed at *Data understanding* activity. *Data constructing* activity is required to construct the final data set from the initial raw data including data integration, data selection, data cleaning, data transformation, data imputation. The *Modelling* activity starts with the selection of data mining methods, proceeds with the creation of data mining models and finishes with the assessment of models. Data mining ontology (DMO) is useful here for selecting various data mining algorithms and models. *Evaluation* of the data mining models created in the modelling phase and confirmation that the models are well performing and stable are performed here. At *Deployment* activity knowledge gained through data mining models are organized and presented for users.

4. ***Operationalization of data mining results***, i.e. the use of data mining models in business processes.

*Data preparation*—data mining models being developed, evaluated, and deployed in the previous phase, are being executed in this phase as part of the *Model execution* activity. The *Monitoring* activity tracks the quality, stability, and performance of deployed data mining models through time. In the activity *deployment of models* to various applications data mining results (scores) are deployed to various applications which are adapted to use them.

In the context of using data mining for business processes, ontology is useful from three perspectives: *business process ontology* (B PO), which describes business process properties, like inputs, outputs, participants, effects, preconditions, constraints and components (activities); *domain ontology* (DO), which describes domain in terms of main concepts, their properties, relationships among concepts and axioms; and *data mining ontology* (DMO), which describes data mining process and used for selecting data mining algorithms and for semantic searching

In the *assessment* phase, BPO is used to assess business process and to generate potential improvement scenarios, BPO and DO is used to identify business problems to be solved using data mining approach. After generating potential improvement scenarios, potential organizational structures can be generated from DO and BPO. DMO is used in *Data mining readiness assessment* to match and to combine data mining activities with business process activities.

In the *Business process renovation* phase three ontologies together with generated process improvement scenarios are used for process change design.

During *Operationalization* phase ontologies are useful as a documentation of processes, domain and data mining for users.

The architecture for ontology and SOA based data mining implementation to business process (Fig. 21.2) is based on the research of [23]. In our architecture model we add ontology server. The functionality of the *data server* (DS), which responsible for the access interface to data sets and all associated metadata, the *mining server* (MS), which executes the data mining algorithms, the *visualization server* (VS), which creates the visual metaphors the users see, and the *application server* (AS), which is used to ensure user's interaction with all servers, are the same as in [23, 34]. The *ontology server* (OS) is used to process three ontologies: DMO, used for selecting data mining algorithms and for semantic searching, BPO, used to model business process and to select optimal algorithm for processing tasks of the previously modelled business process, and DO, used for queering data from data sources.

Figure 21.3 presents typical data flow architecture for incorporating data mining into business processes.

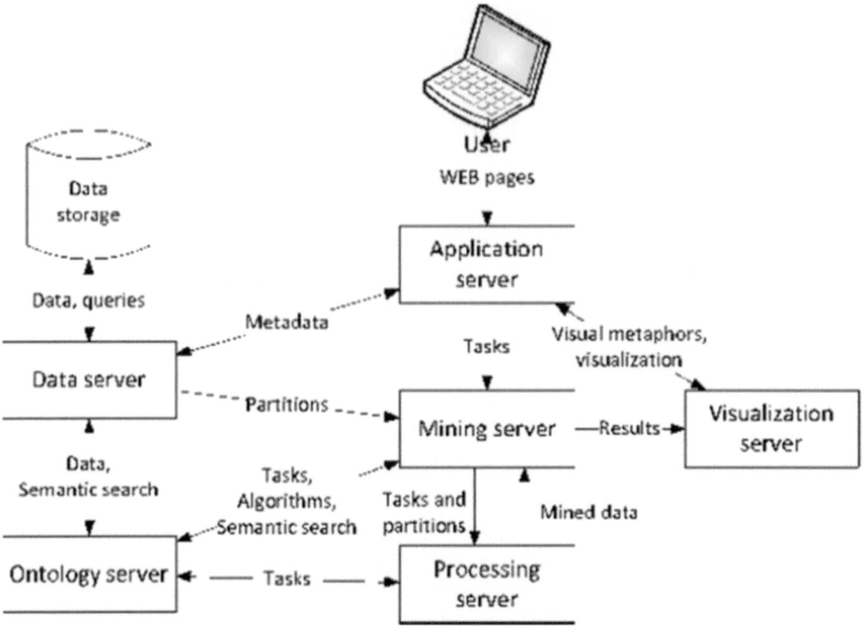


**Fig. 21.2** The architecture of the ontology and SOA based data mining into hacia business process

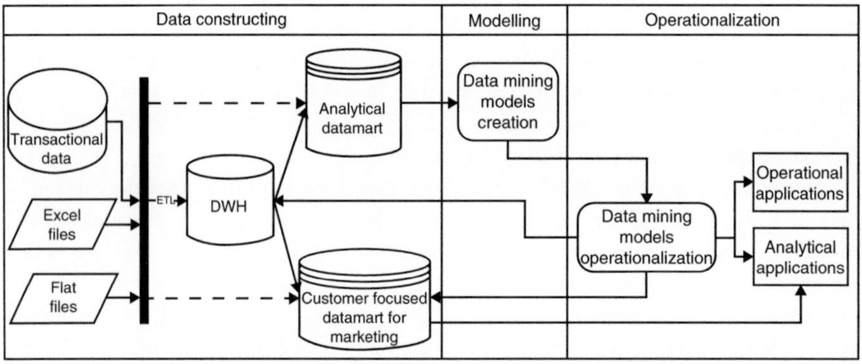


**Fig. 21.3** Data flow architecture for the proposed approach

In order for an organization to be able to fully support the development of data mining models, typically a dedicated analytical data mart is developed which is fed either from a DWH or directly from transactional systems. For marketing purposes typically a separate customer focused datamart is created, which incorporates all relevant information regarding each customer: customer data, account data, analytical scores, contact history, response history, etc.

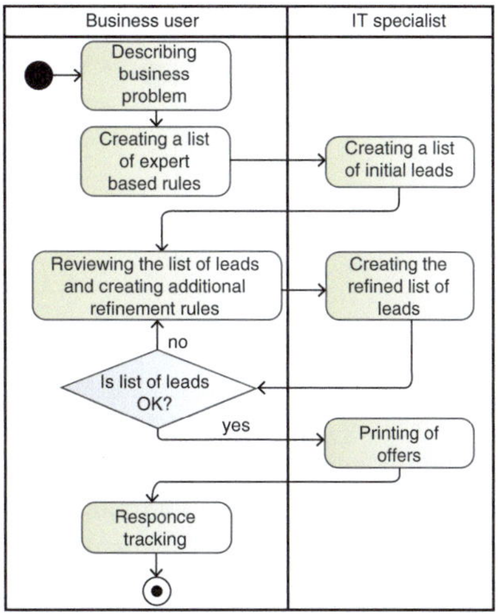


**Fig. 21.4** BEFORE data mining implementation

The *Data mining model creation* and the *Data mining models operationalization* activities are covered by the Modelling activity in our approach. During the second activity created models are written to DWH and other operational or analytical applications, which are used by Business users and front-line personnel (e.g. clerks in the bank) for various purposes, like direct marketing (e.g. to derive the best leads and accompanying offers), credit scoring (e.g. approving a loan to a customer), front-line services (e.g. up-sell a new product or service), etc.

## 21.4 The Case Study in Eight Companies

In this section we present a case study in which we analyse the impact of the implementation of data mining within direct marketing business processes in eight organizations: four banks, three telecommunication companies and one retail company.

### *21.4.1 BEFORE Data Mining Implementation*

Summarized *BEFORE data mining implementation process model* in eight companies includes the following activities (Fig. 21.4):

1. *Business problem*: Business user decides to make a marketing campaign to potential customers to achieve a particular business goal.
2. *Creating a list of expert based rules*: A list of expert based rules, typically expressed in natural language, is created by business user.

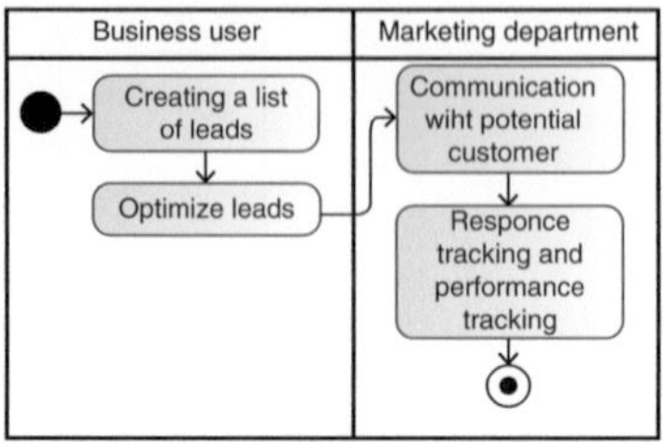


**Fig. 21.5** AFTER data mining implementation

3. *Creating a list of initial leads*: IT specialist resolves human-language rules and builds queries to create an initial list of leads and returns the list to the business user accompanied with requested data fields.
4. *Reviewing the list of leads and creating additional refinement rules*: a business user reviews the list of initial leads and potentially further explores the target groups by means of provided data fields. As part of the review business user defines additional refinement rules.
5. *Creating the refined list of leads*: IT specialist creates a new/refined list of leads based on additional refinement rules defined by the business user.
6. *Printing of offers*: After the lists of target groups are finalized and approved, the marketing offer is prepared and the lists with corresponding offers are sent to printing office. Offers are printed and mailed to customers.
7. *Response tracking:* As customers can response to offers over different channels (e.g. in person, over telephone, or e-channel) the tracking of responses is typically very limited mostly due to the fact that data appears in multiple transactional systems.

### 21.4.2 AFTER Data Mining Implementation

Summarized *AFTER data mining implementation process model* (Fig. 21.5) in eight companies includes the following activities:

1. *Creating a list of leads*: Business user has access to an operational application, which uses the dedicated customer focused datamart. Through this data mart business user analyses the customer base, does selections of target groups based on expertize, tests several scenarios and performs required refinements. *Note* that IT experts are no longer required to create target lists.
2. *Optimization step*: The dedicated customer focused datamart includes also results (scores) of data mining models for particular business problems, where these results are typically populated on regular basis. Additionally, the operational application allows the users to run the models on demand providing the most up-to-date information regarding the customers. Business users can use this data to further refine and optimize the final target groups, in example to calculate the expected value or profit for accepting the offer, or to determine the propensity to respond to the offer.

**Table 21.1** Data mining level of sophistication

| Numerical value | Descriptive value | Description for data mining level | Description for data level |
|---|---|---|---|
| 1 | Low | No data mining | No DWH |
| 2 | Mid | Some data mining modelling in place with manual work to operationalize | DWH in place with some dedicated datamarts |
| 3 | High | Data mining modelling regularly done, operational applications adapted to the use of data mining results, automated monitoring processes, etc. | DWH in place, dedicated datamarts, automated data flows |

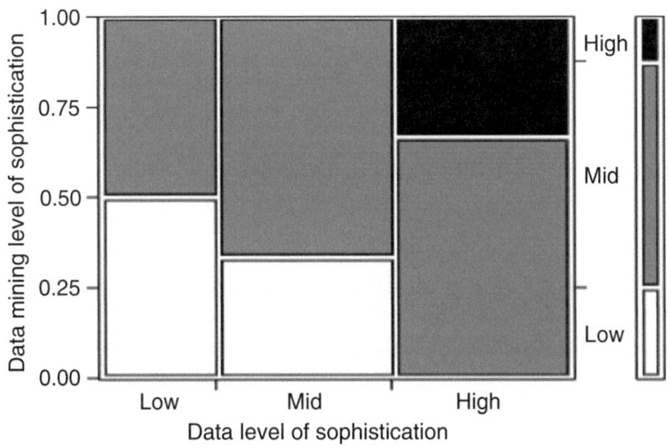


**Fig. 21.6** Relation between data level of sophistication and data mining level of sophistication

3. *Communicating with potential customers*: After the lists of target groups are finalized and approved and the offers created, the commutations are executed through appropriate channels (e.g. mail, e-mail, SMS, retail branch, shop).
4. *Response tracking and performance tracking*: All the responses from all associated systems are gathered into dedicated customer focused datamart. The automation of this process enables automated response tracking and performance measurement.

## 21.4.3 Discussion of the Obtained Results

Data mining has been implemented in eight companies. The progress of implementation projects from beginning to end has revealed that companies having DWH had a significant advantage. Those companies implemented data mining in two levels of sophistication: *data mining level* and *data level* (Table 21.1).

Through project monitoring we have created the snapshot of the autumn 2010 projects status an present the results obtained in Fig. 21.6. It shows relationship between data level and data mining level of sophistication: *higher data sophistication level implies higher value of data mining sophistication level.*

The analysis of data mining implementation in eight companies revealed the following benefits of AFTER data mining implementation process model:

- Significantly less involvement of IT department—the IT involvement was reduced by 60–90 %. It leads to the independence of business users.
- The redefined and optimized business processes led to the following improvements on a monthly basis: Number of executed campaigns raised by 3–12 times; Reduced marketing cost by 50–75 %; Number of contacted leads reduced by 30–87.5 %; Return on investment growth by 50–400 %; Time to market dropped by 45–85 %; Conversion rate improvement by 20–100 %; Improved effectiveness of business users by 20–30 %.
- Better control of the process.
- Consistent tracking and reporting (the ability to learn and improve).

## 21.5 Summary, Conclusions and Future Work

Nowadays, there are few approaches for integrating data mining into business processes, since it is not trivial to implement them in real situations, like e-commerce, fraud detection, banking, etc. The proposed approach on integrating data mining into business processes is based on the CRISP-DM model and advantages of using ontology and SOA.

The carried out experiment of using data mining to support marketing shows that companies having data warehouse had a significant advantage. It allows us to optimise business process. Moreover, the presence of data warehouse indicates a higher level of data integration and thus a much better basis for data mining.

The advantages of using data mining in marketing business process are: 1) Business users are more independent from IT users; 2) Marketing process is better controlled and more efficient.

According to the survey results, we propose using ontologies and SOA for data mining implementation into business processes on higher level of maturity. We suggest using three ontologies—domain ontology (DO), business process ontology (BPO) and data mining ontology (DMO)—to make a possibility of defining and modelling characteristics of business process, business domain and business rules within domain, how data mining will be used in those business processes, and store those definitions and models in an independent place and enable the use of models to every authorized user and application system. SOA ensures data exchange, distribution and protection during data mining process.

## References


1. Hornick MF, Marcade E, Venkayala S (2006) Java data mining: strategy, standard, and practice. Morgan Kaufmann, San Francisco, CA
2. Guarino N (1998) Formal ontology and information systems. In: Guarino N (ed) FOIS'98 formal ontology in information systems. IOS Press, Amsterdam, pp 3–15

3. Davenport TH, Short JE (1990) The new industrial engineering: information technology and business process redesign. Sloan Manage Rev 31(4):11–27
4. Dayal U, Hsu M, Ladin R (2001) Business process coordination: state of the art, trends, and open issues. In: Proceedings of the 27th VLDB conference, Roma, Italy, 2001
5. Chang SL (2000) Information technology in business processes. Bus Process Manage J 6(3): 224–237
6. Cardoso J, Aalst W Hershey, (2009) Handbook of research on business process modeling. Information Science Reference, Hershey, PA: IGI Global
7. Castellanos M, Casati F, Sayal M, Dayal U (2005) Challenges in business process analysis and optimization. In: Bussler C, Shan MC (eds) Technologies for E-services, 6th international workshop, TES 2005, Trondheim, Norway, 2–3 September 2005. Revised Selected Papers, LNCS. Springer-Verlag Berlin Heidelberg, 2006, vol 3811, pp 1–10
8. van der Aalst WMP, Reijers HA, Weijters AJMM, van Dongen BF, Alves AK, de Medeiros SM, Verbeek HMW (2007) Business process mining: an industrial application. Inform Syst 32(1):713–732
9. Han J, Kamber M (2006) Data mining: concepts and techniques, 2nd edn. Elsevier, San Francisco
10. Witten IH, Frank E (2005) Data mining: practical machine learning tools and techniques. Morgan-Kaufmann, San Francisco
11. Grigori D, Casati F, Castellanos M, Dayal U, Sayal M, Shan M (2004) Business process intelligence. Comput Indus 53(3):321–343
12. Rozinat A, van der Aalst WMP (2008) Conformance checking of processes based on monitoring real behavior. Inform Syst 33(1):64–95
13. Shearer C (2000) The CRISP-DM model: the new blueprint for data mining. Journal for Data Warehousing 5(4):13–22
14. Rupnik R, Jaklic J (2009) The deployment of data mining into operational business processes. In: Ponce J, Karahoca A (eds) Data mining and knowledge discovery in real life applications. I-Tech, Vienna
15. Marban O, Segovia J, Menasalvas E, Fernndez-Baizn C (2009) Toward data mining engineering: a software engineering approach. Inform Syst 34(1):87–107
16. Kohavi R, Provost F (2001) Applications of data mining to electronic commerce. Data Min Knowl Discov 5(1–2):5–10
17. Gray P (2005) New thinking about the enterprise. Inform Syst Manage 11(1):91–95
18. Ciflikli C, Ozjirmidokuz EK (2010) Implementing a data mining solution for enhancing carpet manufacturing productivity. Knowl Based Syst 23(8):783–788
19. Kurgan LA, Musilek P (2006) A survey of knowledge discovery and data mining process models. Knowl Eng Rev 21(1):1–24
20. Wegener D, Ruping S (2011) Integration and reuse of data mining in business processes? A pattern-based approach. Int J Bus Process Integration Manage 5(3):218–228
21. Cheung WK, Zhang X-F, Wong H-F, Liu J, Luo Z-W, Tong FCH (2006) Service-oriented distributed data mining. IEEE Internet Computing 10(4):44–54
22. Birant D (2011) Service-oriented data mining. In: Funatsu K (ed) New fundamental technologies in data mining. InTech, pp 1–18. http://www.intechopen.com/books/new-fundamentaltechnologies-in-data-mining/service-oriented-data-mining. Accessed 30 May 2014
23. Guedes D, Meira JW, Ferreira R (2006) Anteater: a service-oriented architecture for high-performance data mining. IEEE Internet Computing 10(4):36–43
24. Noy NF, McGuinness DL (2003) Ontology development 101: a guide to creating your first ontology. Technical Report, Stanford University
25. Gruber TR (1993) A translation approach to portable ontology specifications. Knowl Acquis 5:199–220
26. Vasilecas O, Kalibatiene D, Guizzardi G (2009) Towards a formal method for transforming ontology axioms to application domain rules. Inform Technol Control 38(4):271–282

27. Kalibatiene D, Vasilecas O (2012) Application of the ontology axioms for the development of OCL constraints from PAL constraints. Informatica 23(3):369–390
28. Panov P, Dzeroski S, Soldatova LN (2010) Representing entities in the OntoDM data mining ontology. In: Dzeroski S, Goethals B, Panov P (eds) Inductive databases and constraint-based data mining, Part 1. Springer, New York, NY, pp 27–58
29. Gong X, Zhang T, Zhao F, Dong L, Yu H (2009) On service discovery for online data mining trails. In: The second international workshop on computer science and engineering, IEEE Computer Science, pp 478–482
30. Ankolekar A, Burstein M, Hobbs JR, Lassila O, McDermott D, Martin D, McIlraith SA, Narayanan S, Paolucci M, Payne T, Sycara K (2002) DAML-S: web service description for the semantic web. In: Horrocks I, Hendler J (eds) Proceedings of the 1st international semantic web conference Sardinia, Italy, 9–12 June 2002, vol 2342, LNCS. Springer, Heidelberg, pp 348–363
31. Pinto FM, Guarda T (2011) Database marketing process supported by ontologies: a data mining system architecture proposal. In: Funatsu K (ed) New fundamental technologies in data mining. InTech, pp 19–42
32. Moss LT, Atre S (2003) Business intelligence roadmap: the complete project lifecycle for decision-support applications. Addison-Wesley Professional, 2003
33. Williams S, Williams N (2007) The profit impact of business intelligence. Morgan Kaufmann, San Francisco
34. Pivk A, Vasilecas O, Kalibatiene D, Rupnik R (2013) On approach for the implementation of data mining to business process optimization in commercial companies. Technol Econ Dev Econ 19(2):237–256

# Chapter 22
# Data-Aware Conformance Checking for Declarative Business Process Models

**Diana Borrego, Irene Barba, and Pedro Abad**

**Abstract** Since the accurate management of business processes is receiving increasing attention, conformance checking, i.e., verifying whether the observed behaviour matches a modelled behaviour, is becoming more and more critical. Moreover, declarative languages are more frequently used to provide an increased flexibility. However, little work has been conducted to deal with conformance checking of declarative models. Furthermore, only control-flow perspective is usually considered although other perspectives (e.g., data) are crucial. In addition, most approaches exclusively check the conformance without providing any diagnostics. To enhance the accurate management of flexible business processes, this work presents a constraint-based approach for conformance checking over declarative models (including control-flow and data perspectives) and for providing related diagnosis.

**Keywords** Business process management • Process mining • Conformance checking and diagnosis • Declarative business process models • Constraint programming

---

D. Borrego (✉) • I. Barba
Departamento Lenguajes y Sistemas Informáticos, Universidad de Sevilla, Sevilla, Spain
e-mail: dianabn@us.es; irenebr@us.es

P. Abad
Departamento Tecnologías de la Información, Universidad de Huelva, Huelva, Spain
e-mail: pedro.abad@dti.uhu.es

M. José Escalona et al. (eds.), *Information System Development: Improving Enterprise Communication*, DOI 10.1007/978-3-319-07215-9_22,

## 22.1 Introduction

Nowadays, there exists a growing interest in aligning information systems in a process-oriented way [1] as well as in the accurate management of Business Processes (BPs). In such a setting, conformance checking[1] [2] (which consists of verifying whether the observed behaviour recorded in an event log matches a modelled behaviour) is critical in many domains, e.g., process auditing or risk analysis [3].

Moreover, flexible Process-Aware Information Systems (PAISs) [4] are required to allow companies to rapidly adjust their BPs to changes. Typically, processes are specified in an imperative way. However, declarative process models are increasingly used. In a declarative model (e.g., constraint-based model) everything that is not forbidden is allowed. Therefore, the declarative specification of processes is an important step towards the flexible management of PAISs [5].

Declarative languages based on LTL (Linear Temporal Logic), e.g., ConDec [6], can be fruitfully applied in the context of compliance checking [7]. However, whereas there exist solid conformance checking techniques for imperative models, little work has been conducted for declarative models [3].

In existing proposals, only control-flow perspective of BPs is typically considered when performing conformance checking. In fact, the data-flow of a process is usually only considered for determining the control-flow (e.g., which option of a XOR gateway is selected for execution). However, other perspectives (e.g., data) are also crucial when modelling and executing a BP. In this respect, we propose incorporating to the process specification business data rules to describe the data semantics for the representation of the relations between data values in a process model. The fact of including business data rules in a process allows to decrease the cost related to the modification of its business logic, to shorten the development time, to externalize and easily share rules among multiple applications, and to make changes faster and with less risk [8].

Moreover, most existing approaches exclusively focus on determining whether a given process instance conforms with a given process model or not without providing any detailed diagnostics [3, 7].

To enhance the accurate management of flexible processes, this work presents an approach for performing conformance checking over declarative models which include business data rules and for providing related diagnostics. With this aim, efficient constraint-based approaches are proposed to allow for an increased effectiveness when performing conformance checking as well as when dealing with diagnosis. Moreover, to validate the proposed approach, the analysis of different performance measures has been performed. The proposed approach is focused on addressing realistic problems from different domains which present flexible nature and where the accurate management of BPs is crucial (e.g., medical guidelines, processes from automotive industry and flight planning).

---

[1]Note that conformance checking is one of the three main application scenarios of process mining.

There is some related work on conformance checking of declarative BP models (e.g., [3, 7, 9, 10]). However, the approach presented in this paper differs from existing approaches in various ways: (1) besides the control-flow constraints, our approach is able to deal with business data rules; and (2) the proposed approach is based on constraint programming, which: (a) allows for efficient conformance checking and diagnosis processes, and (b) due to its versatility, facilitates the management of many aspects related to BPs which are not usually considered in previous work (e.g., dealing with non-atomic activities).

The rest of the paper is organised as follows: Sect. 22.2 gives backgrounds on related areas, Sect. 22.3 presents the proposed approach, Sect. 22.4 deals with the evaluation, and Sect. 22.5 concludes the paper.

## 22.2 Background

In this work, the declarative language ConDec [6] is used as basis for specifying constraint-based process models.

**Definition 1** A **constraint-based process** model $S=(A, _{CBP})$ consists of a set of activities $A$ and a set of constraints $CBP$ limiting execution behaviours.

The activities of a constraint-based process model can be executed arbitrarily often if not restricted by any constraints. In a ConDec model control-flow constraints are specified trough templates, which are grouped into:

1. Existence templates: unary relations stating the number of times one activity is executed, e.g., Exactly(N,A) specifies that A should be executed exactly N times.
2. Relation templates: positive binary relations used to establish what should be executed, e.g., Precedence(A,B) specifies that before any execution of B at least one execution of A must be executed.
3. Negation templates: negative relations used to forbid the execution of activities in specific situations, e.g., NotCoexistence(A,B) specifies that if B is executed, then A cannot be executed, and vice versa.

In addition to control-flow relations, we consider business data constraints to describe the data semantics for the representation of relations between data values in a process model, resulting in a data constraint-based process model.[2] These constraints can be used for describing both the behaviour of each single activity independently and the behaviour of the overall process in a global way.

[2] In a similar way, [11] deals with modelling data in ConDec.

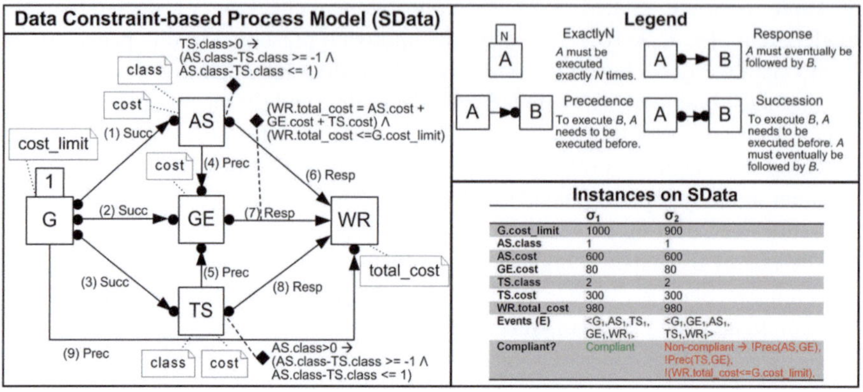


**Fig. 22.1** Running example: data constraint-based process model and instances

**Definition 2** A **data constraint-based process model** $SData = (A, Data, C_{BP}, C_{Data})$ is a constraint-based model $S = (A, C_{BP})$ (cf. Def. 1) which also includes a set of data variables *Data*, and a set of business data constraints $C_{Data}$ relating the variables included in *Data*.

Business data constraints are linear or polynomial equations or inequations over data-flow variables, related by a boolean combination, according to a BNF grammar.

**Definition 3** Let *v* E *Data* be a variable, *int val* be an integer value, *nat val* be a natural value, and *float val* be a float value. The set of business data constraints which can be included in $C_{Data}$ is generated by the following grammar in BNF:

$$\begin{aligned}
Constraint &::= Atomic_Constraint\ BOOL_OP\ Constraint \\
&\quad \mid Atomic_Constraint \mid \neg Constraint \mid (Constraint) \\
BOOL_OP &::= \wedge \mid \vee \mid \rightarrow \\
Atomic_Constraint &::= Function\ \ PREDICATE\ \ Function \\
Function &::= v\ \ FUNCTION\ \ Function \\
&\quad \mid v \mid int_val \mid nat_val \mid float_val \\
PREDICATE &::= < \mid \leq \mid = \mid > \mid \geq \\
\underline{FUNCTION} &::= + \mid - \mid *
\end{aligned}$$

Figure 22.1 shows a data constraint-based process model which represents a hypothetical travel agency.[3] This agency manages holiday bookings by offering: transport (*TS*), accommodation (*AS*), and guided excursions (*GE*). After the client request (*G*) is

[3] This model is an adaptation of the model presented in [12].

carried out, the agency must write a report (*WR*) containing the information in answer to the request, which will then be sent to the client. This process is specified as the following data constraint-based process model (cf. Def. 2): *SData*=({*G,AS, GE, TS, WR*}, {*G.cost limit, AS.class, AS.cost, GE.cost, TS.class, TS.cost, WR.total_cost*}, {*Exactly*( 1, *G*), *Succ*(*G, AS*), *Succ*(*G, GE*), *Succ*(*G, TS*), *Prec*(*AS, GE*), *Prec*(*TS, GE*), *Resp*(*AS, WR*), *Resp*(*GE,WR*), *Resp*(*TS, WR*), *Prec*(*G, WR*) }, {*AS.class*— *TS. class*≥—1, *AS.class*— *TS.class*≤1, *WR.total cost*=*AS. cost*+*GE.cost*+*TS.cost*, *WR. total_cost*≤*G.cost_limit*}). Note that in the graphical notation, business data constraints are depicted linked to activities and relations, as proposed by [11].

As the execution of a data constraint-based model proceeds, related information is recorded in an execution trace.

**Definition 4** Let *SData*=(*A,Data*, $C_{BP}$, $C_{Data}$) be a data constraint-based process model (cf. Def. 2). Then: a **trace** σ=(*Values,E*) is composed of: (1) *Values*= {($Data_i$, $Value_i$), $Data_i$ ϵ *Data*}, which is a set of pairs ($Data_i$, $Value_i$) stating the $Value_i$ assigned to $Data_i$ in the trace; and (2) *E*, which is a sequence of starting and completing events $<e_1, e_2, \ldots e_n>$ regarding activity executions $a_i$, *a* E *A*, i.e., events can be:

1. *start*($a_i$, *T*), meaning that the i-th execution of activity *a* is started at time *T*.
2. *comp*($a_i$, *T*), meaning that the i-th execution of activity *a* is completed at time *T*.

A process instance represents a concrete execution of a data constraint-based process model and its execution state is reflected by the execution trace.

**Definition 5** Let *SData*=(*A,Data*, $C_{BP}$, $C_{Data}$) be a data constraint-based process model. Then: a **process instance** *I*=(*SData*, a) is defined by *SData* and a corresponding trace a.

A running process instance *I*=(*SData*, a) is **compliant** with *SData*=(*A, Data*, $C_{BP}$, $C_{Data}$) if a satisfies all constraints stated in $C_{BP} \cup C_{Data}$. In this way, when checking the conformance of a process instance, both the control-flow and the data-flow conformances are checked. As examples, Fig. 22.1 includes one compliant instance and one non-compliant instance[4] due to the constraints {*Prec*(*AS, GE*), *Prec*(*TS, GE*), *WR.total_cost*<= *G.cost_limit*} are not satisfied.

There are few proposals for checking the conformance of constraint-based models (e.g., [3, 7, 9]). However, we are not aware of any constraint-based approach for checking conformance in constraint-based models. To solve a problem through Constraint Programming (CP) [13], it needs to be modelled as a constraint satisfaction problem (CSP).

[4]For the sake of clarity, only completed events for activity executions are included in the trace representation.

**Definition 6** A **CSP** $P=(V, D, C_{CSP})$ is composed of a set of variables $V$, a domain of values $D$ for each variable in $V$, and a set of constraints $C_{CSP}$ between variables, so that each constraint represents a relation between a subset of variables and specifies the allowed combinations of values for these variables.

A solution for a CSP consists of assigning values to all the CSP variables. A solution is feasible when the assignment of values to variables satisfies all the constraints. In a similar way, a CSP is feasible if there exists at least one related feasible solution, overconstrained otherwise. To solve the latter, Max-CSPs, which are CSPs which allow the violation of some constraints, together with reified constraints (constraints associated to CSP boolean variables which denote its truth value) can be considered. Specifically, the identification of the constraints to relax (or remove) to make the overconstrained model feasible can be performed by maximizing the number of reified constraints whose truth value is equal to true. This gives rise to a constraint optimization problem (COP).

**Definition 7** A **COP** $POF=(V, D, C_{CSP}, OF)$ is a CSP which also includes an objective function $OF$ to be optimized.

Several mechanisms are available for the solution of CSPs and COPs (e.g., complete search algorithms). Regardless of the search which is used, global constraints can be defined to improve the modelling of the problems. The global constraints can be implemented through filtering rules (rules responsible for removing values which do not belong to any solution) to efficiently handle the constraints.

## 22.3 Our Proposal

Although a declarative process model is verified correct from the control-flow and data-flow perspectives at build-time, the process may not behave correctly at run-time. This anomaly can be detected by performing conformance checking by means of the comparison between the instances collected in the event log, which reflect the actual behaviour of the process, and the declarative model indicating the expected behaviour. The discrepancies detected after the execution of process instances can be derived from: (i) a declarative model that has not been correctly designed in accordance with the business goals; or (ii) an abnormal execution of the process.

In this work, a method for checking the conformance of declarative models which include business data constraints and for diagnosing the parts of the model that were not executed as they were modelled is proposed. The method starts from a data constraint-based process model (cf. Def. 2) defined by business analysts. During the process enactment, the events related to the start and the completion of activity executions and to data are stored in an event log, collecting the process instances (cf. Def. 5).

To detect discrepancies, conformance checking is performed over the input model and the instances recorded in the event log. As a result, the instances are classified as *compliant* or *non-compliant* if they comply or fail to comply with the model, respectively. Finally, using the *non-compliant* traces, the model-based fault diagnosis of the process model is performed. This diagnosis determines the parts of the model that were either incorrectly designed or abnormally executed by users. Both conformance checking and diagnosis phases are detailed below.

**(1) Conformance checking**

To check the conformance of a process instance, a constraint-based approach is proposed. To deal with the fact that in ConDec BP activities can be executed arbitrary often if not restricted by any constraint, process activities are modelled as repeatable activities (cf. Def. 8).

**Definition 8** A **repeatable activity** $ra=(nt, a, sActs)$ is an activity $a$ which can be executed several times, where $nt$ is a CSP variable specifying the number of times the activity is executed (which is the number of activity occurrences related to $ra$), and $sActs$ is a sequence of its related activity occurrences ($sActs=[a_1,\ldots,a_{nt}]$).

For sake of brevity we write $nt(ra)$, $sActs(ra)$, etc, when referring to attributes $nt$, $sActs$, etc, of the repeatable activity $ra$. For the data constraint-based model of Fig. 22.1 there are five repeatable activities (e.g., $(nt(G), G, [G_1, \ldots, G_{nt(G)}])$).

For each repeatable activity, $nt$ activity occurrences exist (cf. Def. 9).

**Definition 9** An **activity occurrence** $a_i=(st, et, ra)$ represents i-th execution of a repeatable activity $ra$ ($sActs(ra)$ $[i]=a_i$), where $st$ and $et$ are CSP variables which refer to the start and the end times of the activity occurrence, respectively.[5]

As an example, related to the activity $G$ of the data constraint-based model of Fig. 22.1 there are $nt(G)$ activity occurrences (e.g., $(st(G_1),et(G_1),G)$).

Furthermore, to improve the modelling of the problems and to efficiently handle the constraints when checking the satisfiability of a process instance, the proposed constraint-based approach includes a global constraint implemented through a filtering rule for each ConDec template [14].

Moreover, to model the problem as a CSP, a CSP-Conformance problem (cf. Def. 10) related to a data constraint-based model is created.

[5] Note that the CSP variables st and et allows dealing with non-atomic activities.

**Definition 10** Let $SData=(A,Data, C_{BP}, C_{Data})$ be a data constraint-based process model (cf. Def. 2), and *RActs* be the set of repeatable activities related to *SData* ($RActs=\{ra=(nt, a, sActs)$ , $a$ E $A\}$. Then: a **CSP-Conformance problem** related to *SData* and *RActs* is a CSP $P=(V,D,C_{CSP})$ (cf. Def. 6), where:

- The setofvariables $V=\{nt(ra)$ ,$ra$ E$RActs\}$U$\{st(ai)$ ,$et(a_i)$ ,$a_i$ E $sActs(ra)$ ,$ra$ E $RA\}$ U $\{Data_i,Data_i$ E $Data\}$.
- The set of domains *D* is composed of the domains for each variable from *V*.
- The set of constraints $C_{CSP}$ is composed of: (1) the global constraints (implemented by the filtering rules) related to the ConDec constraints included in $C_{BP}$ and (2) the constraints related to the business data constraints included in $C_{Data}$.

For the data constraint-based model in Fig. 22.1, $V=\{nt(G), nt(AS), nt(GE), nt(TS), nt(WR), st(G_1), et(G_1), \ldots, st(WR_{nt(WR)}), et(WR_{nt(WR)}), G\_cost\_limit, AS\_class, AS\_cost, GE\_cost, TS\_class, TS\_cost, WR\_total\_cost\}$, *D* is composed of the domains of each variable in *V*, and $C_{CSP}=\{Exactly(1,G), Succ(G, AS), Succ(G, GE), Succ(G, TS), Prec(AS, GE), Prec(TS, GE), Resp(AS, WR), Resp(GE, WR), Resp(TS, WR), Prec(G, WR), |AS\_class - TS\_class|<1, WR\_total\_cost=AS\_cost+GE\_cost+TS\_cost, WR\_total\_cost<G\_cost\_limit\}$.

Let $I=(SData, \sigma)$ be a process instance (cf. Def. 5) defined by the data constraint-based process model $SData=(A, Data, C_{BP}, C_{Data})$ (cf. Def. 2) and the trace $\sigma=(Values, E)$ (cf. Def. 4). Moreover, let $P=(V, D, C_{CSP})$ be the CSPConformance problem related to *SData* (cf. Def. 10). Then, the problem of checking the conformance of *I* regarding *SData* is equivalent to check the feasibility of *P* when instantiating its CSP variables as follows:

- $\forall a \in A$, $nt(a)=argmaxi(comp(a_i, T) \in E)$, stating that the number of times a repeatable activity *a* is executed is instantiated to the index of the last execution of *a* which was completed in the trace,
- $\forall a \in A$, $i \in [1..nt(a)]$, $st(a_i)=T\,|(start(a_i, T) \in E)$, stating that the start time of each activity occurrence is instantiated to the time when this occurrence started in the trace,
- $\forall a \in A, i \in [1..nt(a)]$, $et(a_i)=T\,|(comp(a_i,T) \in E)$, stating that the end time of each activity occurrence is instantiated to the time when this occurrence was completed in the trace,
- $\forall Data_i \in Data$, $Data_i=Value_i\,|\,(Data_i, Value_i) \in Values$, stating that all the
- variables related to data are instantiated with their actual values in the trace.

In this way, the process instance *I* is compliant if the CSP-Conformance problem *P* instantiated as mentioned is feasible, non-compliant otherwise.

**(2) Model-based fault diagnosis process**

For each non-compliant instance, a model-based fault diagnosis process [15] is applied to determine which parts of the model are potentially the source of the problem.

The proposed diagnosis process has the goal of identifying the minimum set of constraints of the data constraint-based model (cf. Def. 2) which need to be relaxed/removed to make such model feasible according to the instance *I* (which is equivalent to solve the related Max-CSP, cf. Sect. 22.2). To this end, the related CSPConformance problem $P=(V, D, C_{CSP})$ (cf. Def. 10) is modified (by including reified constraints) and translated into a COP $POF=(V',D', C'_{CSP}, OF)$ [16] where:

- $V'=V$ U *RefVars* U *OF*, where *RefVars* is the set of the boolean CSP variables which hold the truth values of the reified constraints (note that there is the same number of boolean variables in *RefVars* as the number of reified constraints),
- $D'=D$ U *Dom(RefVars)* U *Dom(OF)*, where *Dom(RefVars)* is the set of domains of each variable in *RefVars*,
- All the constraints included in $_{CCSP}$ are transformed into reified constraints, in such a way that $C'_{CSP}$ is composed of: (1) the reified constraints related to the global constraints which are included in $C_{BP}$ and (2) the reified constraints related to the business data constraints which are included in $C_{Data}$ (cf. Def. 10).
- $OF=max|TrueRefVars|$, where $TrueRefVars=\{var$ E $RefVars, var=true\}$.

The fact of transforming a constraint into a reified constraint makes the satisfaction of such constraint optional since a solution to a CSP which violates any reified constraints is a feasible solution. Most of the variables of this COP are instantiated in the same way than when checking the conformance of *I* (i.e., all the CSP variables except the reified variables and *OF* are instantiated according to *I*). However, after transforming the constraints into reified constraints, the instantiation of the CSPConformance problem according to *I* becomes feasible. Moreover, since the goal consists of diagnosing the source of the problem, the maximization of the number of reified constraints is included as the objective function to be optimized.

Solving COPs takes exponential time due to the dependency of their complexity on the number of values each variable can take [17]. The proposed constraint-based approach is based on the first-fail principle [18] which orders the variables by increasing set cardinality, breaking ties by choosing the variable with the smallest domain size, and reducing the average depth of branches in the search tree. Therefore, as shown in next section, the time it takes to check the conformance of declarative models with the presented approach is acceptable.

After solving this COP, the minimum set of constraints to be relaxed/removed to obtain a model compliant with the instance *I* is equal to the constraints related to reified variables which are false in the solution to the COP.

As an example, for the CSP-Conformance problem of the model in Fig. 22.1, *V* is extended, including {*exactlyG*, *succG_AS*, *succG_GE*, *succG_TS*, *precAS_GE*, *precTS_GE*, *respAS_WR*, *respGE_WR*, *respTS_WR*, *precG_WR*, *dc*1, *dc*2,*dc*3, *OF*}. *D* is also extended with the domain of the new variables, and $C_{CSP}$ is modified by transforming each constraint *c* ϵ *CSP* into *rVarC* == (*c*), being *rVarC* the reified variable related to *c* (e.g., *exactlyG* == (*Exactly*(1, *G*)) and *dc*1 == ($|AS\_class - TS\_class| \leq 1$)), and by including the constraint related to *OF*.

## 22.4 Empirical Evaluation

The purpose of this section consists of analyzing the effectiveness and efficiency of the proposed approach over a diversified set of models.

### Experimental Design

Different data constraint-based process models are generated considering: (i) correctness, i.e., the models must represent feasible problems without conflicts and without any dead activities, and (ii) representativeness, i.e., the models must represent problems which are similar to actual BPs. Consequently, we require the test models to be of medium-size and comprise all three types of ConDec templates (cf. Sect. 22.2). Moreover, the considered models must include business data constraints. Overall, one template model *testCase* which contains ten repeatable activities (cf. Def. 8), nine control-flow constraints, and eight data-flow constraints is considered (cf. Fig. 22.2). Note that although this model includes ten activities the related instances generally contain much more activity executions (i.e., activity occurrences, cf. Def. 9) depending on the constraints which are included.

During the experiments this generic model is specified by instantiating the relations with concrete constraints. Since the filtering rules for the different ConDec constraints significantly vary in their computational complexity [14], four models (testCaseA, testCaseB, testCaseC, testCaseD) are generated by substituting the labels Relation and Negation by: (A) Response and Negation Response, (B) Response and Negation Alternate Response, (C) Alternate Response and Negation Response, and (D) Alternate Response and Negation Alternate Response, respectively. Moreover, a value for label *N* must be established (*N* E { 10,20, 30} is considered). Therefore, 12 different specific models (*Model* E {*testCaseA10, testCaseA20, testCaseA30, testCaseB10, testCaseB20, testCaseB30, testCaseC10, testCaseC20, testCaseC30, testCaseD10, testCaseD20, testCaseD30*}) are considered. Furthermore, activity durations are randomly varied between 1 and 10, resulting in 750 games of durations.

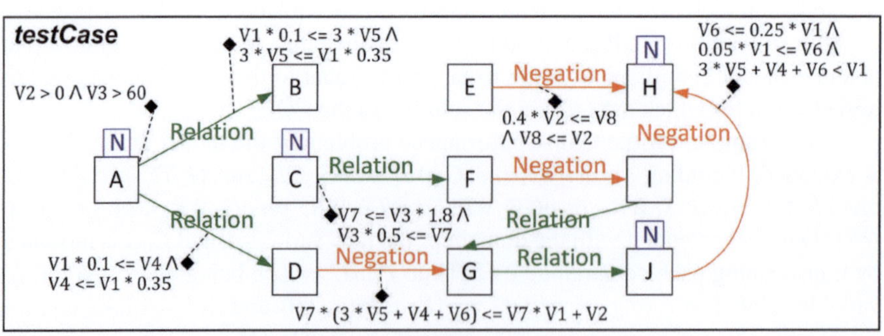


**Fig. 22.2** Generic data constraint-based process model

For each specific model, compliant and non-compliant instances are randomly generated to create a diversified synthetic log. As mentioned, related to a specific declarative model there may exist multiple executions which meet the constraints (i.e., several feasible execution plans). In previous approaches (cf. [12, 19]), feasible optimized enactment plans from ConDec specifications were generated. To this end, activity durations and resource availabilities were considered. In the current work, the proposal presented in [12, 19] is used for generating the flow of both compliant and non-compliant instances related to the 12 different models. Since resource perspective is out of the scope of this paper, this aspect is not considered when generating the instances.

1. Generation of compliant instances: The approach presented in [12, 19] is used to generate an enactment plan related to each combination of specific model plus a specific sample of activity durations (12×750 compliant instances are generated). Regarding data, 750 combinations of values which are compliant with the business data rules (cf. Fig. 22.2) are randomly generated by using a simple constraint-based approach.
2. Generation of potentially non-compliant instances: To this end, the approach presented in [12, 19] is also used. Specifically, 750 potentially non-compliant enactment plans are generated by randomly removing one of the ConDec relations or negations of the considered models. Regarding data, 750 combinations of values which are not compliant with the business data rules (cf. Fig. 22.2) are randomly generated by using a simple constraint-based approach.

From this, four groups of tests are generated: (1) compliant instances; (2) instances non-compliant in data; (3) instances potentially non-compliant in control-flow; and (4) instances potentially non-compliant in control-flow and non-compliant in data. Note that the way in which the instances non-compliant in data are generated always ensures the non-compliance. However, the way in which the instances noncompliant in control-flow are generated may lead to compliance. In fact, the percentage of compliant instances in the complete set of instances is 38.9 %.[6]

The suitability of our approach is tested regarding: (1) *checkTime*: average time for checking the conformance, (2) *diagTime*: average time for diagnosing, and (3) $val_{OF}$: average value which is reached for the objective function to be minimized (which is the number of constraints which should be relaxed/removed to make the model compliant). Note that the variables *diagTime* and $val_{OF}$ only make sense when diagnosing is required (that is, for non-compliant instances).

To solve the constraint-based problems, a complete search algorithm based on the first-fail heuristic is integrated in the COMET™ system [20]. This algorithm is run on a Intel Core I7 processor, 3.4 GHz, 8 GB RAM, running Windows 7.

[6] All the instances which are generated for the empirical evaluation can be accessed at http://www.lsi.us.es/-quivir/index.php/Main/Downloads.

**Table 22.1** Experimental results

| Tested model | *#inst* | *#actOcc* | *checkTime* (ms) | *diagTime* (ms) | $val_{OF}$ |
|---|---|---|---|---|---|
| testCaseA10 | 1,170 | 43 | 267.91 | – | – |
| testCaseB10 | 1,165 | 43 | 268.82 | – | – |
| testCaseC10 | 1,168 | 70 | 268.59 | – | – |
| testCaseD10 | 1,167 | 70 | 268.59 | – | – |
| testCaseA20 | 1,167 | 83 | 316.18 | – | – |
| testCaseB20 | 1,165 | 83 | 316.17 | – | – |
| testCaseC20 | 1,165 | 140 | 319.15 | – | – |
| testCaseD20 | 1,167 | 140 | 319.44 | – | – |
| testCaseA30 | 1,167 | 123 | 317.47 | – | – |
| testCaseB30 | 1,165 | 123 | 318.35 | – | – |
| testCaseC30 | 1,165 | 210 | 327.49 | – | – |
| testCaseD30 | 1,167 | 210 | 328.52 | – | – |
| testCaseA10 | 1,831 | 43 | 267.58 | 272.61 | 2.49 |
| testCaseB10 | 1,834 | 43 | 267.80 | 272.53 | 2.49 |
| testCaseC10 | 1,831 | 70 | 269.29 | 272.39 | 2.49 |
| testCaseD10 | 1,834 | 70 | 268.91 | 271.78 | 2.61 |
| testCaseA20 | 1,834 | 83 | 311.54 | 318.98 | 2.49 |
| testCaseB20 | 1,834 | 83 | 312.50 | 318.52 | 2.49 |
| testCaseC20 | 1,834 | 140 | 315.80 | 321.00 | 2.59 |
| testCaseD20 | 1,834 | 140 | 317.07 | 321.16 | 2.61 |
| testCaseA30 | 1,834 | 123 | 313.61 | 319.79 | 2.49 |
| testCaseB30 | 1,834 | 123 | 314.59 | 318.82 | 2.49 |
| testCaseC30 | 1,834 | 210 | 318.63 | 323.00 | 2.59 |
| testCaseD30 | 1,834 | 210 | 319.25 | 323.59 | 2.60 |

### Experimental Results and Data Analysis

The response variables are calculated by considering the average values for the instances of each model (*Model*) differentiating between compliant and non-compliant instances, as shown in Table 22.1. In this table, besides the identifier of the model and the values for the response variables, each row also includes additional information: *#inst*, which is the number of compliant/non-compliant instances related to that model which are included in the complete set of instances (i.e., the number of instances used for calculating the averages) and *#actOcc*, which is the number of activity occurrences (cf. Def. 9) of each instance (note that each activity occurrence leads to two events (start and end)).

As can be observed in Table 22.1, the execution times which are obtained for both conformance checking and diagnosis are rather low (less than 324 ms for all cases) for problems of medium-size (number of activity occurrences varying between 43 and 210). Furthermore, the execution time remains quite stable regardless of the relations which are given between the activities and of the number of constraints which should be relaxed/removed ($val_{OF}$ column). However, as the number of activity occurrences increases, the execution times slightly increase as well. Moreover, it can be seen that, as expected, the diagnosis takes longer than the conformance checking since reaching an optimal solution in a COP (cf. Def. 7) is usually more time-consuming than checking the feasibility of a CSP (cf. Def. 6).

## 22.5 Conclusion and Future Work

This work presents a constraint-based approach for conformance checking over declarative models (including control-flow and data perspectives) and for providing diagnostics. This differs from existing approaches in various ways (e.g., dealing with business data rules). It has been validated using models of varying complexity, concluding that the execution times which are obtained for both conformance checking and diagnosis are rather low for all the tested cases. As for future work, it is intended to apply our proposal to actual processes which present flexible nature and where the accurate management of BPs is crucial.

## References


1. Weske M (2007) Business process management: concepts, languages, architectures. Springer, Berlin
2. Rozinat A, van der Aalst WMP (2008) Conformance checking of processes based on monitoring real behavior. Inform Syst 33(1):64–95
3. De Leoni M, Maggi F, van der Aalst WMP (2012) Aligning event logs and declarative process models for conformance checking. In: BPM 2012. LNCS, vol 7481. pp 82–97
4. Reichert M, Weber B (2012) Enabling flexibility in process-aware information systems. Springer, Berlin
5. van der Aalst W, Pesic M, Schonenberg H (2009) Declarative workflows: balancing between flexibility and support. Comput Sci Res Dev 23(2):99–113
6. Pesic M, van der Aalst WMP (2006) A declarative approach for flexible business processes management. In: Proceedings of BPM workshops. pp 169–180
7. Burattin A, Maggi F, van der Aalst W, Sperduti A (2012) Techniques for a posteriori analysis of declarative processes. In: EDOC 2012. pp 41–50
8. Weber B, Sadiq SW, Reichert M (2009) Beyond rigidity – dynamic process lifecycle support. Comput Sci Res Dev 23(2):47–65
9. Grando MA, van der Aalst WMP, Mans RS (2012) Reusing a declarative specification to check the conformance of different CIGs. In: BPM 2011 workshops, Part II. LNBIP, vol 100. pp 188–199
10. Montali M, Pesic M, van der Aalst WMP, Chesani F, Mello P, Storari S (2010) Declarative specification and verification of service choreographies. ACM Trans Web 4(1):art. no. 3
11. Montali M (2009) Specification and verification of declarative open interaction models–a logic- based framework. PhD Thesis, University of Bologna
12. Barba I, Weber B, Del Valle C, Jiménez-Ramirez A (2013) User recommendations for the optimized execution of business processes. Data Knowl Eng 86:61–84
13. Rossi F, van Beek P, Walsh T (eds) (2006) Handbook of constraint programming. Elsevier Science Inc., New York, NY, USA
14. Barba I, Del Valle C (2013) Filtering rules for ConDec templates – pseudocode and complexity. http://regula.lsi.us.es/MOPlanner/FilteringRules.pdf

15. de Kleer J, Kurien J (2003) Fundamentals of model-based diagnosis. Safeprocess 3:25–36
16. Borrego D, Gasca RM, Gomez-Lopez MT, Parody L (2010) Contract-based diagnosis for business process instances using business compliance rules. In: Proceedings of the 21st international workshop on principles of diagnosis (DX-10). pp 169–176
17. Kumar V (1992) Algorithms for constraint satisfaction problems: a survey. AI Magazine 13(1): 32–44
18. Van Hentenryck P (1989) Constraint satisfaction in logic programming. MIT Press, Cambridge, MA
19. Barba I, Del Valle C, Weber B, Jiménez-Ramirez A (2013) Automatic generation of optimized business process models from constraint-based specifications. Int J Cooper Inform Syst 22
20. Dynadec. Comet Downloads (2010) http://dynadec.com/support/downloads/ [Online; accessed 13 Mar 2013]

# Chapter 23
# Taxonomy of Anomalies in Business Process Models

**Tomislav Vidacic and Vjeran Strahonja**

**Abstract** Anomalies in business process models refer to deviations from their expected structure, functionality, behavior, semantics, use of concepts and their expression, among others. The research in this paper focuses on anomalies in business process models with the aim to propose taxonomy of anomalies and devise ways to avoid them or to eliminate them when they occur. Anomalies are divided into basic categories and subcategories to the level of granularity that allows their differentiation, description of their specific causes and workarounds as well as their expression in the pattern format.

**Keywords** Business process models • Anomalies detection

## 23.1 Introduction

Over the past decades there have been extensive developments in process modeling in terms of modeling methods on the one hand and validation and verification methods on the other. Along with modeling methods, model anomalies and methods of their detection and prevention have emerged as a novel area of research. In general, anomalies refer to deviations from the expected structure, functionality behavior and use of concepts, among others. A model that contains anomalies is considered invalid. Whereas some authors characterize anomaly as a form that does not meet the definition of normal behavior [1], others describe it as a noise in data [2]. Anomalies can occur at various stages of model design. The most common causes

T. Vidacic (✉)
HPB d.d., Kapucinski trg 5, Varaždin, Croatia
e-mail: tomislav.vidacic@hpb.hr

V. Strahonja
Faculty of Organization and Informatics, University of Zagreb, Pavlinska 2, Varaždin, Croatia
e-mail: vjeran.strahonja@foi.hr

M. José Escalona et al. (eds.), *Information System Development: Improving Enterprise Communication*, DOI 10.1007/978-3-319-07215-9_23,

of anomalies are the human factor or the transformation mechanism in conversion from one abstract level to another. For proper consideration of possible consequences of anomalies it is necessary to first provide methods of their recognition and resolution. Modeling is a creative human activity, which means that the probability of errors will be fairly high. The causes of these errors can be diverse, including lack of notation knowledge, vastness and accidents, among others. Consequently, anomaly detection is an integral part of model validation and verification (V&V). This paper presents the results of our research into the causes and types of business process models anomalies. The research was aimed at devising a taxonomy of anomalies as a tool for determining ways to avoid anomalies or to eliminate them when occur.

## 23.2 Present Status and Approaches to Study of Anomalies in Business Process Models

There are several papers that deal with classification of process models anomalies. In the research by Solti-Rogge et al. anomalies are investigated with the aim to design a business process configuration wizard and consistency checker for BPMN 2 [3]. They divide anomalies into three basic categories: lack of data, redundant data and conflicting data. According to Solti-Rogge et al., anomalies caused by lack of data occur when probability exists that the data that have been read by an activity will never be written down. Redundant data occur when data objects are stored but are never read by certain activities. Conflicts arise when data are written by two or more competing activities. On the basis of the above, we can conclude that anomalies associated with the data generally do not incorporate improper use of concepts, semantic inconsistency or misleading execution. The categorization of anomalies proposed by Awad, Decker and Lohmann contains: anomalies caused by over-restrictive conditions, the ones caused by implicit routing and those caused by implicit constraint on the execution order. it happens when there are two competing activities that share a precondition and the anomaly occurs when an activity changes the state of data [4]. Kim, Lee and Soon distinguish two basic categories of anomalies in business processes: syntactic and structural anomalies [5]. Whereas syntactic anomalies are related to the misuse of concepts and rules of modeling, structural anomalies occur when business processes do not control the interaction of tokens due to the misuse of gateways. According to Kim, Lee and Soon, most model verification tools deal exclusively with AND and OR gates. Finally, in the study by Dohring and Heublein, the following three basic categories of anomalies are mentioned: control flow anomalies, rule based anomalies and data flow anomalies [6]. The analysis of the aforementioned existing research indicates three different approaches to the study of process models anomalies:

1. Theoretical approach, which seeks to build a set of consistent and comprehensive rules, methods, techniques and other aspects of the business process modeling theory. If disciplined application of such rules or other parts of the modeling theory results in anomaly, it is necessary to change or amend them.

2. Empirical approach, which is based on the application of analytical diagnostic methods on real projects and models. This approach seeks to identify the causes of anomalies and discrepancies between the model and reality observed in the application of the model. The ultimate goal is to determine best practices that result in models with no anomalies.
3. Design approach, which is based on the theory of design patterns and seeks to define unacceptable design patterns or anti-patterns, the presence of which indicates anomalies. Patterns become design elements, while anti-patterns are identified as shapes that are either prohibited or can be used with care.

## 23.3 Proposed Taxonomy of Anomalies in Business Process Models

Building a taxonomy is a demanding job that includes the establishment of a classification scheme of concepts (hierarchical arrangement of terms), creation of a thesaurus (expression of semantic relationships of hierarchy, equivalence and related terms) and the definition of metadata. Considering that all of the aforementioned research and existing classifications of anomalies in business process models overlap in some detail, we attempted to consolidate them into a single taxonomy. Ensuring a proper granularity of terms presented one of the key issues related to building the taxonomy. Anomalies that have so far been identified in the existing studies into business process models are shown in Fig. 23.1, where they are divided into three basic categories: structural, semantic and syntactic. Each of the anomalies included has their own specific causes and workarounds. Since the presented anomalies are related to the BPMN notation, it needs to be noted that anomalies concerning EPC, UML activity diagram or other functional models are somewhat specific.

### *23.3.1 Structural Anomalies*

Structural anomalies are divided into flow control anomalies, anomalies in rules and data flow anomalies. **Flow control** anomalies include control flow redundancy, undefined gate, conditions, special issues with dynamic flows and violation of soundness. **Anomalies of rules** are divided into anomalies of consistency and anomalies of coverage. Examples of anomalies comprised within anomalies in rules include rule redundancy, conflicting rules and circularity.

**Unnecessary control flow** refers to the occurrence of more semantically identical connections between any two elements in a flow, as in the case when there are more sequence flows going from one activity to another. Since they do not bring any added value to the model, such sequence flows are redundant [7]. **Modeling redundancy** is common in flexible notations like BPMN and results from one branch being different from another by only a few activities. Each of these branches

| Anomaly taxonomy | | | | Anomalies | Error type | | | | | | | | Solution | | | | |
|---|---|---|---|---|---|---|---|---|---|---|---|---|---|---|---|---|---|
| | | | | | Redundancy | Undef. Con. | Inconsisten. | Asynchron. | Livelocks | Deadlocks | BPMN | Other | Ruleverify. | OCL | BPMN | Simulations | Other |
| Structural anomalies | Flow control | | Control flow redundancy | Unnecessary control flow | • | | | | | | | | | • | | | |
| | | | | Modeling redundancy | • | | | | | | | | | | • | | |
| | | | Gateway condition issues | Undefined gateway conditions | | • | | | | | | | | • | | | |
| | | | | Contradictive gateway conditions | | • | | | | | | | • | | | | |
| | | | | Deterministic lack of synchroniz. | | | | • | | | | | | • | | | |
| | | | | Undeterministic lack of synchroniz. | | | | • | | | | | | • | | | |
| | | | Special issues with dynamic workflows | Dangling states | | | | • | | | | | | | • | | |
| | | | | No tolerance for change of order | | | | • | | | | | | | • | | |
| | | | | Inconsist. when ins. paral. branch | | | | • | | | | | | | • | | |
| | | | | Inconsistent change of past | | | | • | | | | | | | • | | |
| | | | | Misssing loop tolerance | | | | • | | | | | | | • | | |
| | | | Violation of soundness | Flow livelock | | | | | • | | | | • | • | | | |
| | | | | Flow deadlock | | | | | • | | | | • | • | | | |
| | Rules | Consistency | Rule redundancy | Redundant rules | • | | | | | | | | • | • | | | |
| | | | | Subsumed rules | • | | | | | | | | • | • | | | |
| | | | Conflicting rules | Direct contradiction | | | • | | | | | | | | • | | |
| | | | | Contradiction in input | | | • | | | | | | • | | | | |
| | | | | Contradiction in conclusion | | | • | | | | | | • | • | | | |
| | | | Circularity | Rule base-livelock\| | | | | | • | | | | | | • | | |
| | | | | Rule base-deadlock\| | | | | | | • | | | | | • | | |
| | | | Coverage | Isolated rules | • | | | | | | | | | | • | | |
| | | | | Useless conclusions | | | | | | | | | | | | | |
| | | | | Dangling conditions\| | | • | | | | | | | • | | | | |
| | | | | Missing rules | | • | | | | | | | • | • | | | |
| | Data flow | | | Redundant data | • | | | | | | | | | | | • | |
| | | | | Lost data | | | | • | | | | | | | | • | |
| | | | | Missing data | | | • | | | | | | | | | • | |
| Semantic ano. | Element names | | | Non-uniqueness of element names | | | | | | | | • | | | | | • |
| | | | | No element names | | | | | | | | • | | | | | • |
| | Invalid attribute values | | | Invalid data length | | | | | | | | • | | | | | • |
| | | | | Invalid data type | | | | | | | | • | | | | | • |
| | | | | Invalid characters of attrib values | | | | | | | | • | | | | | • |
| Syntactic anomalies | Invalid use of flow | | Events | Invalid use of start event | | | | | | | • | | | | • | | |
| | | | | Invalid use of end event | | | | | | | • | | | | • | | |
| | | | | Invalid use of intermediate event | | | | | | | • | | | | • | | |
| | | | Activity | Activities without activation | | | | | | | • | | | | • | | |
| | | | | Activities without termination | | | | | | | • | | | | • | | |
| | | | | Invalid use of receiving task | | | | | | | • | | | | • | | |
| | | | Gateway | Invalid use of data-base XOR gat. | | | | | | | • | | | | • | | |
| | | | | Invalid use of event-base XOR gat. | | | | | | | • | | | | • | | |
| | Invalid use of connection objects | | | Input sequence with a start event | | | | | | | • | | | | • | | |
| | | | | Output seq. with an end event | | | | | | | • | | | | • | | |
| | | | | Input sequence with intermediate event in a boundary activity | | | | | | | • | | | | • | | |
| | | | | Multiple input sequence wiht interm. event in boundary activity | | | | | | | • | | | | • | | |
| | | | | Invalid use of conditional seq.flow | | | | | | | • | | | | • | | |
| | | | | Invalid use of message flow with start event and gateway | | | | | | | • | | | | • | | |
| | Invalid use of containers | | | Invalid use of pool | | | | | | | • | | | | • | | |
| | | | | Invalid use of lane | | | | | | | • | | | | • | | |

**Fig. 23.1** Taxonomy of anomalies

must be defined separately, which is a form of redundancy owing to which workflow maintenance and modification are further complicated [8]. **Undefined gateway conditions** occur due to the possibility of defining a XOR gate without any conditions, which is a feature allowed by the BPMN standard that can cause anomalies [9]. Contradictive gateway conditions occur when conditions on a XOR gate partially or completely overlap, which entails that any further course of execution does not depend on those conditions [9]. **Lack of synchronization** is a flow control anomaly that occurs when the flow structure is incidentally executed a few times [10]. In other words, it refers to unintentional activation of multiple activities that follow a XOR gateway connection. Lack of synchronization can be further subdivided into Deterministic and Nondeterministic lack of synchronization. **Deterministic lack of synchronization** is defined as one of the competitive activities connected with a XOR gateway that always results in unintended multiple execution [5]. **Nondeterministic lack of synchronization** is defined by a XOR gateway in which one of the competing activities may result in unintentional multiple execution [5]. The **Dangling states** anomaly occurs when no distinction between activated and initiated activities is possible. Such a state often leads to loss of work, because it prevents the deletion of activated or activity or disables delete initiated activities [11]. **No tolerance for change of order** is an anomaly that concerns correct replacement order of application operations such as parallelism, traceability or replacement activities [11]. **Inconsistency when inserting parallel branches** occurs during the insertion of new parallel branches. Inconsistency when inserting parallel branches and No tolerance for change of order anomalies are related to dynamic changes of the model [12]. **Inconsistent change of past** is related with the "Do not change the past instances" rule. Ignoring this rule can lead to inconsistencies in the state of instances or missing input data for next activities to be executed [12]. **Missing loop tolerance** is a special type of anomaly that occurs due to the definition of terms that emerge from the loop. This anomaly is an issue that concerns proper and reasonable determination of changes in the loop structure. Namely, in some cases it is not necessary to exclude instances of changes merely based on the appropriateness of the changes in the loop [12]. The **Livelock flow** anomaly is commonly denoted as an infinite loop. It occurs due to poorly modeled loop conditions that prevent the exit from the loop [13]. **Deadlock flow** describes a situation where the flow is stuck in its path and cannot be completed. A deadlock occurs when activating all output flows required for passage through a parallel gateway is not possible. In such a case, the flow cannot be continued beyond that gateway [13]. The **Redundant rules** anomaly refers to a set of rules that are executed in the same context and have the same effect [14]. **General rules** refer to a set of rules that have identical conclusion and conditions, where one state is a generalization of another state. The state of one set of rules applies to another context of another set of rules [14]. The **Contradiction in input** anomaly refers to rules that contain contradictory terms, such as those which apply if one variable has two different values at the same time (e.g. IF x=10 and x=20 THEN B) [14]. **Direct contradiction** involves simultaneous application of mutually exclusive rules, predicates or arguments [14]. **Contradiction in conclusion** denotes a set of rules whose conclusions do not work when they are implemented at the same time although they have the same conditions (e.g. when each conclusion assigns a different value to the same external variable) [14]. The **Rule-base livelock** anomaly concerns problems with the derivation of rules that are interdependent even though they should not be so. Livelock anomalies occur when the output loop rule contains overlapping conclusions, as in: If A then B and if B then A [14]. **Rule-base deadlock** occurs when two or more rules prevent one another from execution. In other words, a certain rule cannot be executed because its conditions represent the conclusion for other rules, whose conclusions in turn depend on the conclusions of the first one, as in: If A and C then B, If B and C then A [14]. In the **Isolated rules** anomaly a rule is considered to be isolated if and only if [14]: all proposals within the premises are complex and cannot be explained in terms of the basic concept of U; conclusion R is not a basic concept and there is no rule R for statement α, where α ϵ Ψ. In **Useless conclusions** the main concept of the fundamental rule consists of a number of attributes resulting from that rule. Considering the entry, if the derived conclusion is not identical with the main concept, or if some or all of the attributes cannot be concluded from the basic rules, a set of rules applied to this input is incomplete. The main concept is therefore dependent on the domain and requires explicit definition [14]. For example, in ($p1 \wedge p2 \rightarrow R$) R is not a fundamental concept, and there is no rule. **Dangling conditions** is the reverse case of the Useless conclusions anomaly in which conditions can probably be fulfilled in the basic context, but the conclusions are modeled in a way that can never yield any effect. In Dangling conditions the situation is reversed since conclusions do have some effect but the conditions can never be met [6]. The **Missing rules** anomaly occurs if a business process modeler defines a rule that is performed in a particular context, but does not define a rule that will apply if the context is not performed [6]. **Redundant data** occurs if the flow sends some data that is not necessary for it to be performed. In other words, redundant data refers to the data that is created in the process but is never used afterwards [6]. The **Lost data** anomaly is created when some data is overwritten before it has been read. Therefore when activity-created data is changed before it has been read by other activities, it actually represents lost content [6]. The **Missing data** anomaly occurs when a certain activity cannot read the information that it needs for its execution, for instance, when some data is not provided by one activity to another [6]. Since methods of modeling business processes such as BPMN use graphic notation, it is evident that cases of anomalies can also be presented graphically. In our research we aimed to devise anomaly patterns (i.e. anti-patterns) shown in a comprehensive taxonomy. Due to space limitations, however, only a few examples of such patterns are presented in Fig. 23.2, while other examples can be found in [5, 6, 9, 11–14].

### *23.3.2 Semantic Anomalies*

In modeling theory, semantics concerns the meaning of the modeling language and concepts that are used as well as their mapping to the real world. Semantic anomalies, which deal with violation of meaning, can be divided into two basic types:

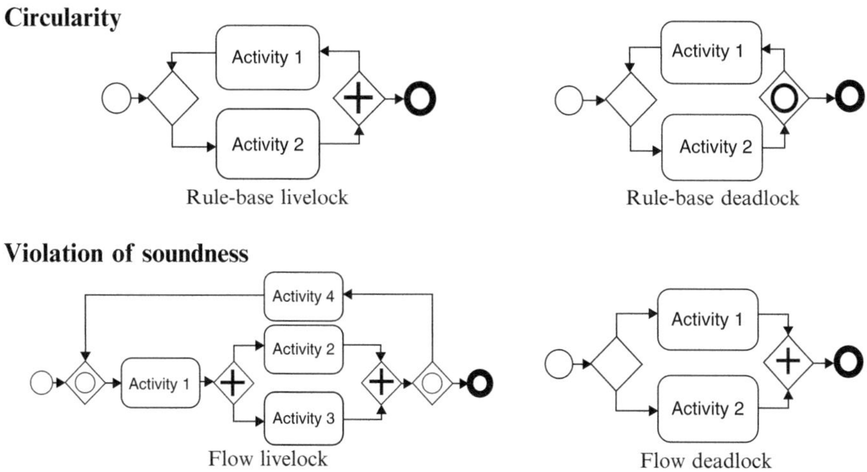


**Fig. 23.2** Example of structural anomalies in BPMN notation

incorrect element names and invalid attribute values. Incorrect element names, such as non-defined, ambiguous and non-unique names, which include synonyms and homonyms, are the most common causes of semantic anomalies. Synonyms are lexemes that belong to the same word type but have different expressions to which the content is completely or partially matched. Homonyms are lexemes that have identical expression with different meanings. Semantic anomaly can also be the result of impossibility of mapping individual objects to a class. Invalid length, type, character or data values represent invalid attribute values.

## *23.3.3 Syntactic Anomalies*

In the case of models, modeling syntax comprises the set of allowed modeling concepts, reserved words and their parameters and the correct way in which modeling concepts are used [15]. Syntactic anomalies of graphical models primarily concern incorrect use of graphic elements and notation. They can be divided into three basic categories that are related with the type of elements, namely: **Invalid use of flow**, **Invalid use of connections objects** and **Invalid use of containers**. These anomalies are elaborated in detail by Kim, Lee and Soon [5]. **Invalid use of flow** comprises invalid use of events, activities and gateways. **Invalid use of events** appears as [5]: Invalid use of start event, Invalid use of end event and Invalid use of intermediate event. Common causes of anomalies in the invalid use of the start event are: more than one start event in the model, disconnection between start events and other elements in the model, and lack of the end event. Common causes of anomalies in the improper use of the end event are: disconnection between end events and other elements in the model, and lack of the start event. Most common problems concerning intermediate events can be further classified into three main groups: Invalid use of trigger in intermediate event, Invalid use of trigger types in boundary activity and Invalid use of intermediate event destinations. Problems related to the Invalid type of trigger of intermediate event occur due to strictly defined intermediate event types that include: nothing, message, timer, compensation, condition and connection or signal, among others. Trigger types related with other events include abandonment, error or multiple trigger. Invalid use of trigger types in boundary activity arises when one or more intermediate events can be connected to boundary activities, with the event falling into one of the following trigger types: message, timer, error, cancel, compensation, condition, signal or multiple trigger. Error is created when a boundary intermediate event coincides with trigger types like nothing or connection. Finally, invalid use of intermediate event destinations is created when the destination is not an activity or a sub process, but another event or gateway. The **Invalid use of activity** [5] anomaly occurs when:

- There are activities in a process, but no external or internal activation of the process;
- There are activities in a process but no end event;
- Use of the receive task is invalid. (The receive task is completed when the message is received, so it is often used as the initial assignment process.)

**Invalid use of gateways** can be divided into [5]: Invalid use of a XOR gateway based on data and Invalid use of a XOR gateway based on events. Invalid use of a XOR gateway based on data usually occurs when the destination XOR gateway is not a data object. An event object should not be used with a XOR gateway based on data. If we use a XOR gate based on the events, the receiving task or an intermediate event with trigger-type messages, timer and signal must be set as the destination. If we use a gateway based on events for process initialization the following conditions must be fulfilled [5]:

- The process does not have start event and gateway has no input sequence flow.
- The input sequences flow gateway has its source in the start event.
- The destination of the gateway must not be an intermediate event with the timer shutter type.

**Invalid use of connection** is usually manifested in cases when a model contains one of the following [5]:

- Input sequence with a start event,
- Output sequence with an end event,
- Input sequence with an intermediate event in a boundary activity,
- Multiple input sequence with an intermediate event in a boundary activity,
- Invalid use of the conditional sequence flow,
- Invalid use of the message flow with a start event and gateway, i.e. the message flow is connected with the start event or a gateway inside the container.

**Invalid use of containers** can be divided into two basic categories [5]: Invalid use of pools and invalid use of lanes. Invalid use of pool occurs when the flow passes the sequence borders. Pool borders should typically include all flow objects and sequences. The sequence flow must not exceed the borders, for which message flows should be used instead. Invalid use of lanes occurs when lanes are used before pools, because lanes are used to categorize activities within the pool.

## 23.4 Resolution of Anomalies in Business Process Models

The taxonomy proposed in this paper indicates that each type of anomalies (i.e. structural, semantic and syntactic) should be treated differently. Structural anomalies require the use of complex methods of recognition and mechanisms that prevent their formation during execution. The basis for the resolution of structural anomalies in business process models is metamodeling. A metamodel is a "model of models" that contains modeling rules and grammar and defines concepts, relationships, and semantics for the exchange of user models between different modeling tools. Reasons for metamodeling in the process of model verification were elaborated by Eessaar [16]. Emerson and Sztipanovits described the role and use of MOF (Meta Object Facility) and GME (Generic Modeling Environment) [17]. Breton and Bézivin compared different concepts, such as PIF, NIST, PSL, UPM and WfMC, with UML and MOF [18], while Kretschmer compared JWT and BPMN [19] at the metamodel level. Such comparisons and mappings are the foundation for the transformation of concepts and model structures. Awad, Decker and Lohmann investigated the detection and correction of data in business processes models [4]. They used BPMN notation mapping and Petri nets to provide an overview of transformation patterns for these. The basic elements required for BPMN model verification are: metamodel of notation, additional metamodel details derived in OCL and notation rules. Since the metamodel of graphical notation does not allow a complete expression of details, rules and restrictions of models that will subsequently be derived from the metamodel, it is necessary to develop further specifications by using some formal language, for example OCL as a part of UML. OCL enables more detailed description of certain elements of the MOF model in a way that is not possible to achieve using graphics only. The Business Process Definition Metamodel (BPDM) is a standard definition of concepts used to express business process models by means of a metamodel, which was adopted by OMG. Similar to the MOF model, BPDM aims to unify diverse business process definition notations that exist in the industry based on comprehensive semantics. These theoretical approaches need to be complemented by an automated mechanism in the form of a rules verifier that is capable of recognizing a specific anomaly and correcting it, or notifying the user. The foundation of any rule verifier is a Business Rule Management System (BRMS). Every BRMS system relies on a rule engine that is based on the declarative definition of business rules needed for a certain part of the task to be performed.

A mechanism for resolving semantic anomalies depends on their types. Basic activities in verifying irregular elements names include: checking the length of elements names and their uniqueness on the one hand and invalid characters on the other. These are followed by a more complex verification of synonyms in elements names, for which dictionaries of synonyms are commonly used. Syntactic anomalies can be resolved on the basis of a series of rules related to a particular notation. Rules for the BPMN notation were set by OMG [18, 20]. Based on a comprehensive analysis of more than 1,000 BPMN models, Rozman, Polancic and Vajde Horvat presented the 15 most common BPMN anti-patterns and proposed their resolution [21]. As the opposite of design patterns [22, 23], anti-patterns are commonly repeated examples of bad practice or misunderstanding. According to the design approach, the use of patterns and anti-avoidance samples should prevent the occurrence of anomalies in models. The experimental approach to identifying and preventing anomalies is based on a large number of samples and statistical methods. According to Mendling [24], most approaches use some kind of decomposition technique, or a combination of reduction and reachability analysis. Mendling also introduced the following correctness criteria for verifying industry-scale EPC process models: soundness, relaxed soundness, interactive verification approach, EPC soundness, single-entry-singleexit decomposition and structuredness. To avoid the appearance of semantic anomalies, manual checks are usually applied. In theory, semantic anomalies of business processes models could be addressed with the aid of computational linguistics and language technologies. However, such an approach is not applicable to graphical models. The first step in automated proofing of semantic anomalies should therefore be the translation of graphical models into formal models that provide input for some automated-reasoning tool. So far disciplined implementation of regulatory standards pertaining to naming conventions, data formats, approved domain of attributes and similar seems to have been the best protection from semantic anomalies. It is very difficult to implement standardization through a number of applications which are distributed throughout the organization. There are several basic approaches to resolving the occurrence of synonyms, such as dictionaries of namespaces. Larson, Navathe and Elmasri [25] suggested a mechanism of comparison of attributes in databases that can be applied to business items in functional models.

## 23.5 Conclusion

Anomalies in business process models refer to deviations from their expected structure, functionality, behavior, semantics, use of concepts and their expression, among others. The research in this paper focuses on anomalies in business process models with the aim to propose a taxonomy of anomalies and devise ways to avoid them or to eliminate them when they occur. Anomalies are divided into basic categories and subcategories to the level of granularity that allows their differentiation, description of specific causes and workarounds, as well as their expression in the pattern format.

A taxonomy of anomalies in functional models is one of prerequisites for their avoidance and resolution, for which automated methods should preferably be used. However, such methods of avoiding and resolving anomalies are only possible with structural anomalies, using design patterns and anti-patterns. Syntactic anomalies, on the other hand, are largely handled by tools that are based on metamodels and modeling rules prescribed by a particular methodology. Finally, dealing with semantic anomalies necessitates human intervention, since mechanisms for identifying and resolving such anomalies are not yet completely reliable and require experts to select a particular potential solution. Our research into business process models anomalies has so far yielded a taxonomy proposed in this paper. The following directions of our future research have also emerged. Firstly, since the presented taxonomy is primarily a diagnostic tool, it is necessary to complement it with methods of avoiding and eliminating anomalies. Secondly, the faceted taxonomy should be upscaled to the level of ontology by developing strict and richer specification rules and semantic relationships among terms.

**Acknowledgments** The research presented in this paper, conducted within the doctoral dissertation of T. Vidacic, is part of the project "Modeling of procedural regulation" funded by the Ministry of Science, Education and Sports of the Republic of Croatia (Nr. 016-0161217-0870), led by Prof. V. Strahonja, Ph.D.

## References

1. Chandola V, Banerjee A, Kumar V (2009) Anomaly detection: a survey. ACM Comput Surv 41(3):1–58
2. Teng H, Chen K, Lu S (1990) Adaptive real-time anomaly detection using inductively generated sequential patterns. In: Proceedings of IEEE computer society symposium on research in security and privacy. IEEE Computer Society, pp 278–284
3. Solti-Rogge A, Kunze M, Awad A, Weske M (2011) Business process configuration wizard and consistency checker for bpmn 2.0. In: 12th international conference, BPMDS 2011 and 16th international conference, EMMSAD 2011 held at CAiSE 2011, London, UK, June 2011, Proceedings. Springer, Berlin, pp 231–245
4. Awad A, Decker G, Lohmann N (2010) Diagnosing and repairing data anomalies in process models. In: Business process management workshops, BPM 2009 international workshops, Ulm, Germany, September 2009. Springer, Berlin, pp 5–16
5. Kim GW, Lee JH, Soon JH (2009) Classification and analyses of business process anomalies. In: Communication software and networks. ICCSN'09. International. pp 433–437
6. Dohring M, Heublein S (2012) Anomalies in rule-adopted workflows – a taxonomy and solutuions for vBPMN. In: 6th European conference on software maintenance and reengineering. IEEE, Szeged, Hungary, pp 117–126
7. Olkhovich L (2006) Semi-automatic business process performance optimization based on redundant control flow detection. In: AICT-ICIW'06. p 146
8. Dohring M, Zimmermann B (2011) vBPMN: event-aware work-flow variants by weaving BPMN2 and business rules. In: EMM-SAD. Springer, London, pp 332–341
9. Business Process Model and Notation (BPMN) 2.0 (2011) Object management group, Technical Report
10. Sadiq W, Orlowska ME (1999) Applying graph reduction techniques for identifying structural conflicts in process models. In: CAiSE'99. Springer, London

11. Rinderle S, Reichert M, Dadam P (2004) Correctness criteria for dynamic changes in workflow systems: a survey. DKE 50(1):9–34
12. Sun SX, Zhao JL, Nunamaker JF, Sheng ORL (2006) Formulating the data-flow perspective for business process management. Inform Syst Res 17(4):374–391
13. Laue R, Awad A (2010) Visualization of business process modeling anti patterns. Proceeding of ECEASST. Chapter 25. pp 1–12
14. Zaidi AK, Levis AH (1997) Validation and verification of decision making rules. Automatica 33(2):155–169
15. Strahonja V (2006) Modeling legislation by using Uml state machine diagrams. In: Conference proceedings of the Canadian conference on electrical and computer engineering (IEEE CCECE 2006). pp 624–627
16. Eessaar E (2007) Using metamodeling in order to evaluate data models. In: AIKED'07 Proceedings of the 6th conference on 6th WSEAS international conference on artificial intelligence, knowledge engineering and data bases – vol 6. World Scientific and Engineering Academy and Society (WSEAS), Stevens Point, WI
17. Emerson MJ, Sztipanovits J (2004) Implementing a MOF-based metamodeling environment using graph transformations. In: 4th OOPSLA workshop on domain-specific modeling. pp 83–92
18. Breton E, Bézivin J (2000) An overview of industrial process meta-models. In: 13th international conference software and system engineering and their applications, ICSSEA 2000–14, Paris, France
19. Kretschmer P (2011) JWT metamodel compared to BPMN metamodel. http://wiki.eclipse.org/JWT_Metamodel. Last accessed on June 2014
20. The Rules of BPMN. http://brsilver.com/the-rules-of-bpmn/. Last accessed on June 2014
21. Rozman T, Polancic G, Vajde Horvat R (2008) Analysis of most common process modeling mistakes in BPMN process models. In: 2008 BPM and workflow handbook. Future Strategies, pp 293–306
22. Aalst W, Hofstede A, Kiepuszewski B, Barros A (2003) Workflow patterns. Distributed and Parallel Databases 14:5–51
23. Aalst W (2000) Workflow verification: finding control-flow errors using petri net-based techniques. In: van der Aalst W, Desel J, Oberweis A (eds) Proceedings of business process management: models, techniques and empirical studies, vol 1806, Lecture notes in computer science. Springer, Heidelberg, pp 161–183
24. Mendling J (2009) Empirical studies in process model verification. In: van der Aalst J (ed) Transactions on petri nets and other models of concurrency II, vol 5460, Lecture notes in computer science. Springer, Heidelberg, pp 208–224
25. Larson AJ, Navathe BS, Elmasri R (1989) A theory of attribute equivalence in database with application to shema integration. Trans Softw Eng 15:449–463

# Chapter 24
# An Automated Approach for Architectural Model Transformations

**Grzegorz Loniewsli, Etienne Borde, Dominique Blouin, and Emilio Insfran**

**Abstract** Software architectures are frequently represented as large models where many competing quality attributes have to be taken into account. In this context, there may be a large number of possible alternative architectural transformations that the architecture designer has to deal with. The complexity and dimensions of the solution space make that finding the most appropriate architecture considering several quality attributes is a challenging and time-consuming task. In this paper, we present a model transformation framework designed to automate the selection and composition of competing architectural model transformations. We also introduce a case study showing that this framework is useful for rapid prototyping through model transformations.

**Keywords** Model Transformations • Architecture Refinement • NFR

G. Loniewsli (✉) • E. Borde
Institute Telecom, TELECOM ParisTech,
LTCI—UMR 514, 46, rue Barrault, 75013 Paris, France
e-mail: grzegorz.loniewski@telecomparistech.fr; etienne.borde@telecomparistech.fr

D. Blouin
Lab-STICC, Universite de Bretagne-Sud, Centre de Recherche,
BP 92116, 56321 Lorient Cedex, France
e-mail: dominique.blouin@univ-ubs.fr

E. Insfran
Department of Computer Science and Computation, Universitat Politècnica de València,
Camino de Vera, s/n, 46022 Valencia, Spain
e-mail: einsfran@dsic.upv.es

M. José Escalona et al. (eds.), *Information System Development: Improving Enterprise Communication*, DOI 10.1007/978-3-319-07215-9_24,

## 24.1 Introduction

Since software systems are constructed to satisfy business goals, the design activities and especially architecture design must also be responsive to those business goals [1]. Business goals are described in [2] as *high-level objectives of the business, organization, or system that capture the reasons why a system is needed and guide decisions at various levels within the software development life cycle*. According to [3], the view that the rationale for a Non-Functional Requirement (NFR)[1] can be found in business goals gives software and system architects a new lens through which to examine and realize the NFRs of software systems.

The dimensions of the architectural models along with the number of alternative architectural strategies to be applied lead to a situation, where architects have to be given support in the candidate architectures exploration. Otherwise, architects manual work is time consuming, and essential design decisions are not always well motivated.

Model-driven Development (MDD) emphasizes the use of models and models transformations as the primary artefacts for automating the software production process. However, in a previous study [4], we showed that there is still a lack of MDD approaches, which starting from requirements, benefit from this automation. Most of the MDD approaches only consider system functional requirements, without properly integrating NFRs into their development processes [2]. Also, approaches which make use of the knowledge about architectural strategies implemented as transformations in the context of design space explorations are lacking.

There exist several partial solutions to deal with design space evaluation with respect to NFRs [5–7] and NFRs in design decisions [3, 8, 9]. However, approaches which consider these two dimensions in an integrated MDD environment are not common.

In this paper, we introduce a MDD approach, tailored for embedded software architecture design, where architecture refinements are implemented as model transformations driven by NFRs and goals. The approach is fully automated making use of: (1) *higher-order transformations* (HOT) in ATL with which to generate model refinements for specific design model; (2) extended goals specification from an extension of the Requirements Definition and Analysis Language (RDAL) [10]; (3) and finally the knowledge about the impact of particular model transformations of different quality properties.

The remainder of this paper is structured as follows. Section 24.2 introduces the problem statement, assumptions and challenges. Section 24.3 introduces our approach giving details about the use of MDD knowledge to select the architectural transformation, and the composition of architecture refinements. Section 24.4 presents a case study using a theoretic example to show the feasibility of the approach. Finally, Sect. 24.5 presents the conclusions and further work.

---

[1] Some authors use different names, remarkably "quality attribute requirement" or "quality requirement" as a synonymous of NFR.

## 24.2 Problem Statement and Challenges

This section discusses in details the problem that we address in this paper. First, we present the inputs and assumptions of our contribution. Then we enumerate the challenges raised by current limitations in interpreting this knowledge.

To our understanding, MDE processes for embedded systems are keen on providing the level of information considered as the main inputs of the method:

- The model of a system architecture that represents software and hardware components, along with the bindings between them. Such model constitutes a candidate solution that could be refined or improved through applying tailored architectural model transformations.
- System NFR and goals that must be met by the system. For example, constraints related to: end-to-end data flow latency; services response time; availability; and power consumption are part of usual NFR of embedded systems. In addition, the definition of quality goals refines some of the NFR into an objective of minimizing or maximizing a quality attribute. For example, minimizing power consumption. The main difference between a quality goal and a NFR is that the latter gives a strict definition when to approve or discard a solution while the former aims at improving the characteristic of a solution (as long as all the requirements are met).
- A set of model transformations that implement refinements or improvements of the source model, each of them focusing on increasing the satisfaction of a quality objective (reducing power consumption, increasing availability, etc.).

In the following, we explain the challenges raised by merging this information in order to automate design decisions.

As explained in the Introduction section, our objective is to take advantage of the formalization of (1) requirements, (2) sensitivities, and (3) model transformations in order to automate the design space exploration towards an architecture that satisfies requirements and reach goals at best.

The first challenge raised by this objective comes from the necessity of solving tradeoffs between competing goals. For example, availability can be increased by redundancy patterns to the price of increasing the system's power consumption. To the best of our knowledge, there are no existing approaches that would describe how to solve the tradeoff between quality attributes expressed as design pattern (as a model transformation). When a design pattern is implemented as a model transformation improving a quality goal, solving the competition between these goals leads to selecting and composing these model transformations. Indeed, transformations are usually defined to be generic (like design patterns) whereas they will compete with other transformations when used in practice. As a consequence, a combination of competing model transformations should be used during the design of an embedded system.

The second challenge comes from the size of the design space to explore in order to study such tradeoffs. To answer this challenge, it is thus of prime importance to automate the selection and composition of model transformations. Of course, the quality of the exploration algorithm is very important in this context, as well as the implementation of a framework to enable the selection and composition of model transformations. The evaluation and optimization of the exploration algorithm is left as a future work of this contribution, which focuses first on automating the selection and composition of model transformations.

Finally, the last main challenge of this approach is the management of dependencies between model transformations, as model transformations can be used sequentially and potentially with an imposed order. In this paper, we consider model transformations that apply to a given element as competing and independent transformations: only one of them can be selected and applied to this element independently of other transformations.

## 24.3 Method Overview

At coarse grain, model transformations are grouped into two main categories of model transformation languages: (1) rule-based transformation languages, or (2) imperative transformation languages [5]. Rule-based transformation languages, such as ATL and QVT-R, are of great interest in our context since they express model transformation logics in a way that is easy to interpret and adapt. Indeed, modifying the pattern matching part of a transformation rule is sufficient to modify the set of elements this rule applies to. In addition, superimposition helps to combine different sets of rules [11]. Contrasting with rule-based formalisms, the adaptation of imperative transformation languages requires a deep understanding of the control flow graph that leads to the execution of one or another transformation. Model transformations we deal with in this paper are endogenous (AADL [12] to AADL model transformations) in order to refine or improve the architecture of an embedded system while representing the result of this refinement in an analyzable model (using AADL as an output language).

To facilitate aforementioned pattern matching modifications, thus creating new model transformation rules in an automated manner, the artifacts described hereafter have been designed.

The Transformation Rules Catalog (TRC) was designed to gather MDD knowledge, describing not only available model transformations, but also their impact on quality attributes. Figure 24.1a presents defined metamodel for the TRC artifact. Quality attributes are defined in the RDAL specification. The impact of a model transformation on a given quality attribute is specified by an integer value ranging from −5 (very negative effect on the quality attribute) to +5 (very positive impact on the quality attribute). It is thus of the responsibility of the transformations specifier to define the impact of each of the transformations on the defined quality attributes.

The Transformations Iteration Plan (TIP) is an artifact with which to describe which transformation rules should be applied and how to produce the final architecture refinement rules. Its metamodel is presented on the Fig. 24.1b.

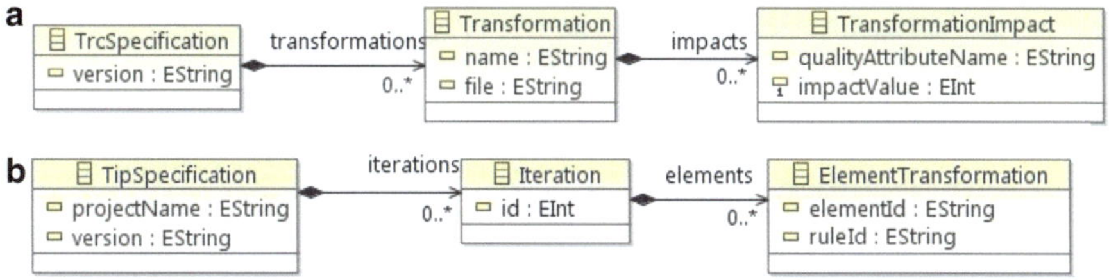


**Fig. 24.1** Metamodels (**a**) transformation rules catalog, (**b**) transformations iteration plan

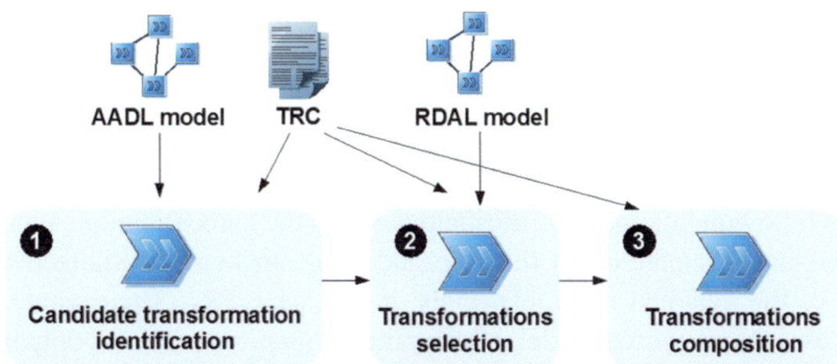


**Fig. 24.2** Process schema

In the following subsections, we present the overall process that we have implemented in order to automate the specific architecture refinements generation. Figure 24.2 illustrates this process and its three main activities.

## 24.3.1 *Candidate Transformation Identification*

In order to identify the transformations that can be applied to a particular element of the input model, we introduce a method based on the steps presented hereafter.

**Step 1.** Application of the HOT that takes as input the generic model transformations referenced in the TRC and generates new model transformations that after their application on the architecture model produce lists of transformations that can be applied to each element of the input model.

**Step 2.** Execution of transformations generated in step 1. Data collected after executing all of the newly created transformations allow creating a set of tuples <*element, list of transformations*>. It constitutes the output of the pattern-matching phase of the transformations execution engine;

Figure 24.3 illustrates the ATL implementation of the HOT where rule *TR1* is transformed into a new rule *TR1'*. Rule *TR1* originally transforms an input element of type *A* (that conforms to *SimpleMetamodel*) into an output element of type that conforms to the same metamodel. On the right side of Fig. 24.3 the output of the

```
rule TR1 {
    from
      in : SimpleMetamodel!A
    to
      out : SimpleMetamodel!A (
        key <- in.qualifiedName )
}
```

HOT →

```
rule TR1' {
    from
      in : SimpleMetamodel!A
    to
      out : PatternMatching!Result (
        key <- in.name )
    do {
        out.value.add('TR1');
    }
```

**Fig. 24.3** Schema of the higher-order transformation for pattern matching

HOT is a rule *TR1'* that transforms all the elements of type *A* (same pattern matching clause as the original rule *TR1*) into an output element *out* of type *Result* from the *PatternMatching* metamodel (defined by us). The output element is then initialized in the following manner: *key* attribute receives the name (a unique identifier) of the element that was matched by this rule, and *value* attribute is initialized with a list of names of the matched rules that apply, in this case *TR1*. The execution of the rule *TR1*' on a given source model conforming to *SimpleMetamodel* outputs a XMI file containing tuples of all possible rules (among those referenced in TRC) that can be applied to each source model element.

Each of the transformations declared in the TRC, when applied the aforementioned steps 1 and 2, outputs the result of the pattern matching condition checking applied on the model, giving the set of candidate transformations for a given design element.

### 24.3.2 *Transformations Selection*

We distinguish the following inputs to the selection process: (1) Set of tuples *<element, list of transformations>* resulting from the candidate transformations identification; (2) RDAL specification—information about the quality attributes to optimize along with responsible for them architecture elements (sensitivities); (3) TRC artifact—information on the judged impact that each transformation has with respect to all the quality attributes which are important for the system's architecture.

The output of the selection process is the Transformations Iteration Plan (TIP), which defines which model transformations should be used in an iteration of the architecture refinement. It consists of the actual tuples *<element, transformation>* to be applied. Figure 24.1b shows the ecore metamodel of the TIP artifact. *Element7ransformation* meta-class defines aforementioned tuples describing which transformation (attribute: *ruleld*) is going to be applied on which model element (attribute: *elementld* is the element's qualified name which identifies every element in deterministic manner). After the iteration planning, the architecture refinement composition is to be performed.

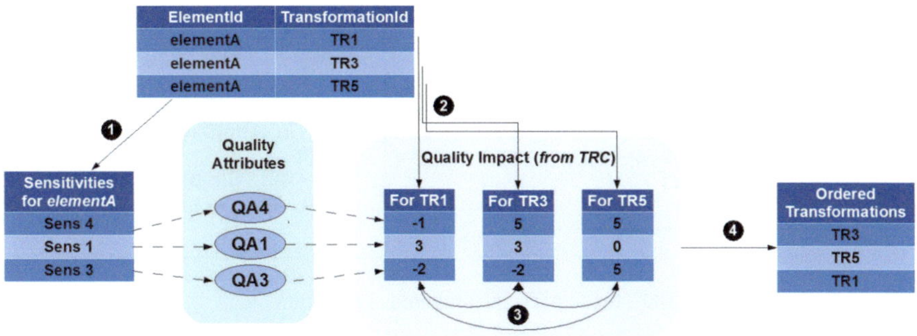


**Fig. 24.4** Transformations selection process

The inputs presented above (set of tuples, RDAL specification, and TRC) combines several factors to be considered while analyzing possible architectural refinements. These factors can not be analyzed in isolation, as their complex relationships have to be considered all together. For this purpose, different multi- criteria decision making methods are known (e.g. AHP [7]). They can be easily integrated to our approach to analyze provided input data from the goals' specification and MDD knowledge. We propose to use simple rules with which to select transformations of the best possible impact on targeted quality goals.

The selection process is illustrated on Fig. 24.4. In Step 1, we iterate on the elements for which some applicable transformations were found (candidate tuples). For each of these elements, we retrieve from the RDAL specification a list of sensitivities. Sensitivity is simply a reference to a quality attribute from an element of the architecture. In the retrieved list, sensitivities are ordered with respect to their priority in RDAL metamodel. On Fig. 24.4, *elementA* is sensitive, by order of importance, to quality attributes *QA4*, *QA1* and *QA3*. In Step 2, for each candidate transformation to be applied on *elementA*, the list of quality impact values is retrieved from the TRC specification. E.g. *TR1* has negative impact (−1) on quality attribute *QA4*, a positive (+3) impact on *QA1*, and a negative impact (−2) on *QA3*. In Step 3, values obtained in step 2 are then used for the selection of rules to be applied. Our tool performs pair-wise comparisons considering the priority of the quality attribute and the impact value. However, different multi criteria decision making methods can be applied here. In Step 4, the output of the selection process is a list of transformations to be applied on particular model element. They are ordered from the best to the worst with respect to the quality properties that should be taken into account. Optimal refinement selection is a very challenging task. However, our selection algorithm along with the refinements generation provides means to review the refinements space in an efficient manner. Identified solutions should be evaluated in the proposed in TIP order, validating the correctness of the identified refinements if the already search solutions are not satisfactory, and alternatives have to be explored.

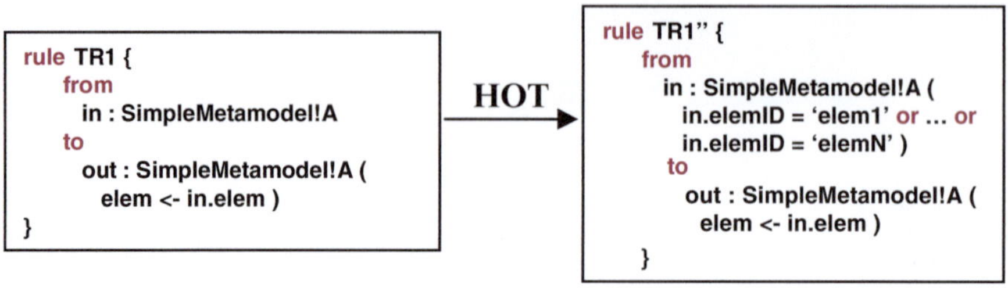


**Fig. 24.5** Schema of the HOT for actual transformations generation

### *24.3.3 Transformations Composition*

The composition of architecture refinement is performed by using the HOT to transform original rules into a new set of rules which are going to be applied on particular model elements. This process takes as input the rules definitions (TRC) and also the refinements configurations (TIP).

Figure 24.5 illustrates the realization of mentioned above refinements generation with an ATL implementation of the HOT where rule *TR1* is transformed into a new rule *TR1"*. Rule *TR1* was originally specified to transform all the elements of type A in source metamodel *SimpleMetamodel. TR1"* uses the transformation mechanism of the original rule but with a limited set of elements for which this rule has been selected: as depicted on Fig. 24.5 (changed pattern matching condition in the output transformation *TR1"*).

Finally, the framework implemented to support presented process executes generated model transformations on the input AADL model to produce a refined AADL model. If the evaluation of the output model is not satisfactory, the architect has the possibility to continue the design space exploration from step 2 (i.e. transformations selection), to produce output models with different results in terms of quality attribute.

Next section details designed intermediate artifacts and shows their use in the automation of the complete process introduced in this section.

## 24.4 Case Study

Architectures of embedded systems are frequently very large models where many competing quality attributes have to be taken into account. At the same time, the amount of possible architectural refinements increases as developers are becoming familiar with model-driven techniques. However, the complexity and the dimensions of the solution space make that finding the most correct solution is a challenging, time-consuming task.

In this section, we present a case study where given candidate architecture is refined in order to ensure the satisfaction of two quality attributes: *MemoryFootprint*

a
```
thread producer
features
  p_out: out event data port Int
{Output_Time => ([Time => Deadline; Offset => 0 ns .. 0 ns;]);};
end producer;

thread consumer
features
  p_in: in event data port Int {dequeue_protocol => AllItems;};
end consumer;

process implementation proc.impl
subcomponents
  T1: thread producer.impl {Period=>7 ms; Deadline=>7 ms; Priority=>3; };
  T2: thread producer.impl {Period=>10 ms; Deadline=>10 ms; Priority=>2; };
  T3: thread consumer.impl {Period=>10 ms; Deadline=>10 ms; Priority=>1; };
  T4: thread producer.impl {Period=>10 ms; Deadline=>10 ms; Priority=>2; };
connections
  cnx1: port T1.p_out -> T3.p_in {Timing=>Delayed;};
  cnx2: port T2.p_out -> T3.p_in {Timing=>Delayed;};
  cnx3: port T4.p_out -> T3.p_in {Timing=>Delayed;};
end proc.impl;
```

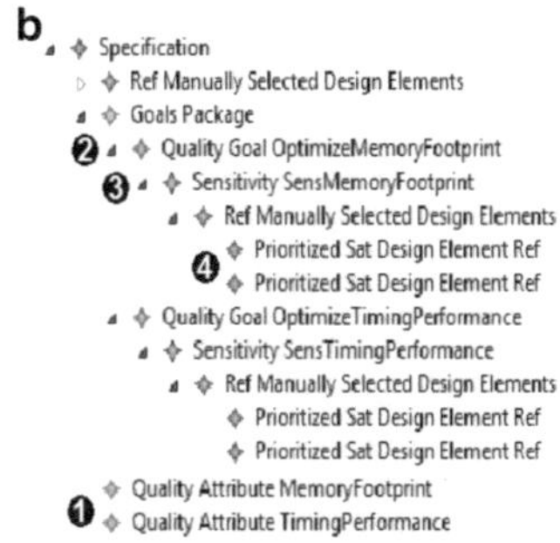


**Fig. 24.6** Case study artifacts: (**a**) AADL specification, (**b**) RDAL specification

and *TimingPerformance*. ***Memory footprint*** refers to the amount of memory resources that the running system needs for proper execution. It is one of the characteristics of embedded systems which possess limited memory resources. Optimization of this quality property is often a challenging task. ***Timing performance*** is one of the most important quality properties of embedded real-time systems. At design time, architects have to ensure that it is possible to satisfy the imposed on the system timing constraints. The application of the process presented in this paper is automated by JAVA/ATL implementation.

This case study bases on a theoretical example to show how our approach deals with the complexity of embedded systems' architectures design. In this example, where input model consists of 4 elements, 2 transformations apply to 3 of them, and another 2 transformations apply to the fourth element. Thus we get ten alternative possible architectures. Even in this very simple example, the automation of architectural refinement proves beneficial, as the different refinement combinations are not always easy to be configured manually.

Figure 24.6a presents a simple system specification where three threads of type *Producer* (*T1*, *T2*, *T4*) interact with one *Consumer* thread (*T3*) through event data ports. Moreover, the specification consists of the definitions of a system, one processor with one process running.

In the requirements specification (RDAL), important quality attributes are assigned to elements of the architecture. This assignment is called *sensitivity*. It indicates the quality attributes to be considered while transforming sensitive elements. In our example, connections *cnx1* and *cnx2*, as well as input event data port *p_in* are sensitive to *MemoryFootprint* while for *cnx3* the priority is assigned to *TimingPerformance* (see Fig. 24.6b).

In our example, TRC consists of four rules. Rules *TR2* and *TR4* are specified to transform an *in event data port* and rules *TR1* and *TR3* apply to a *connection*. These transformations can be used to implement lock-free queues for managing delayed communications between periodic tasks [13]. The underlying theory is

**Table 24.1** Quality impact from transformation rules catalog

| | | Transformation Rule | | | |
|---|---|---|---|---|---|
| | | TR1 | TR2 | TR3 | TR4 |
| Quality property | Memory Footprint | –4 | –4 | 2 | 2 |
| | Timing Performance | 5 | 5 | –3 | –3 |

```
<iteration id=”1”>
   <elements elementId=”root_impl_Instance.p.T3.p_in” ruleId=”TR4”>
   <elements elementId=”root_impl_Instance.p.T1.p_out->T3.p_in” ruleId=”TR3”>
   <elements elementId=”root_impl_Instance.p.T2.p_out->T3.p_in” ruleId=”TR3”>
   <elements elementId=”root_impl_Instance.p.T4.p_out->T3.p_in” ruleId=”TR1”>
</iteration>
<iteration id=”2”>
   <elements elementId=”root_impl_Instance.p.T3.p_in” ruleId=”TR4”>
   <elements elementId=”root_impl_Instance.p.T1.p_out->T3.p_in” ruleId=”TR3”>
   <elements elementId=”root_impl_Instance.p.T2.p_out->T3.p_in” ruleId=”TR1”>
   <elements elementId=”root_impl_Instance.p.T4.p_out->T3.p_in” ruleId=”TR1”>
</iteration>
```

**Fig. 24.7** Identified architectural refinements (TIP artifact)

presented in [14]. Transformations *TR1* and *TR2* implement queues accesses with lookup tables that contain indexes of the queue which a task can access for each of its activation (over a hyper-period). Transformations *TR3* and *TR4* implement queues accesses by computing indexes at runtime. Thus, *TR3* and *TR4* consume more memory but less CPU time in comparison to *TR1* and *TR2*. Details about their influence on particular quality attributes can be found in Table 24.1.

In order to validate the different steps described on Fig.24.3 Sect. (24.3), we have executed the process on the use-case described in previous subsection. Several configurations of the architectural refinements have been automatically identified and necessary tailored transformations have been generated. Figure 24.7 shows the output of that process for two successive iterations of the selection algorithm.

- For iteration 1, the result shows that the architecture refinement will contain different transformations applied to different model elements, since three elements of the input model are sensitive to *MemoryFootprint* quality attribute, they are transformed by the low memory footprint version of available transformations (*TR3*). One input element is transformed by the transformation resulting in lower CPU consumption, as this element is primarily sensitive to *TimingPerformance*.
- For iteration 2, the identified architecture refinement sets the transformation that is preferable for improving the *TimingPerformance* of the architecture (*TR1*) on one more element of the input model in comparison to the iteration 1 (element highlighted by a dashed line on Fig. 24.7).

The resulting architecture refinements produced in each of the iterations are next executed in RAMSES execution framework [13], producing candidate architectures that need to be validated by the analysis tools and domain experts.

The main outcome of this contribution is to automate the design space exploration, based on the selection and composition of legacy transformations in architectural refinement and improvement processes. The performed case study shows design space exploration by the automated production of different candidate architectures. Moreover, it proves to be useful in rapid prototyping and evaluation of embedded systems architectures, with respect to specified quality properties.

## 24.5 Conclusions and Further Work

It is well accepted in the software architecture community that goals and quality attribute requirements are the most important drivers of architecture design [15]. However, dealing with several competing goals and quality attribute requirements to perform well-informed architectural design decisions is not an easy task. MDD techniques, based on explicit modeling of these goals and quality attribute requirements, and design decisions (this latter as model transformations) will allow not only to perform automated and documented design decisions but to preserve important architectural knowledge. In this paper, we introduced a specific MDD approach tailored for embedded software architecture design with the aim of helping software architects to perform these informed design decisions based on automating the selection and composition of competing model transformations [16]. The main contribution of the paper is to bring together quality attribute requirements to design decisions (represented as model transformations) and to use this information to automate the selection of architectural refinements based on these relationships.

Currently, some improvements are being planned in the architecture refinement framework, among them: integrating an evaluation process for the resulting architecture to quantitatively measure the resulting architecture (regarding the goals and quality requirements stated); reviewing the RDAL metamodel to be able to capture more expressive requirements; and enriching our implementation tool by adding a user-friendly interface to allow architects to easily define and test different architectural transformations. Finally, we are developing a tool framework to automate all this process and we are planning to use this tool to conduct an industrial case study in the context of a medium-size embedded system development project.

**Acknowledgments** This research work is partially funded by the "Chaire Ingénierie des Systèmes Complexes" and by the MULTIPLE project (MICINN TIN2009-13838).

## References

1. Ozkaya I, Bass L, Sangwan R, Nord R (2008) Making practical use of quality attribute information. IEEE Software 25(2):25–33
2. Anton A, McCracken W, Potts C (1994) Goal decomposition and scenario analysis in business process reengineering. In: 6th Conference on advanced information systems engineering (CAiSE'94), Utrecht, Holland

3. Clements P, Bass L (2010) Using business goals to inform a software architecture. In: 18th IEEE Int. Requirements engineering conference (RE'10. IEEE CS), pp 69–78
4. Loniewski G, Insfran E, Abrahão S (2010) A systematic review of the use of requirements engineering techniques in model-driven development. In: 13th MODELS Conf. Springer, Berlin, pp 213–227
5. Elahi G, Yu E (2007) A goal oriented approach for modeling and analyzing security tradeoffs. In: 26th Int. Conf. on conceptual modeling (ER'07). Springer, Berlin, pp 375–390
6. Letier E, van Lamsweerde A (2004) Reasoning about partial goal satisfaction for requirements and design engineering. In: 12th ACM SIGSOFT Int. Symp. on Foundations of Software Eng (SIGSOFT '04/FSE-12, ACM, New York), pp 53–62
7. Svahnberg M, Wohlin C, Lundberg L, Mattsson M (2003) A quality-driven decision support method for identifying software architecture candidates. Int Journal of Software Engineering and Knowledge Management 13(5):547–573
8. Ameller D, Franch X (2012) Linking quality attributes and constraints with architectural decisions. In: CoRR abs/1206.5166
9. Sterritt A, Cahill V (2008) Customisable model transformations based on non-functional requirements. In: IEEE congress on services, CS, Washington, pp 329–336
10. The Requirements Definition and Analysis Language Annex of AADL, https://wiki.sei.cmu.edu/aadl/index.php/Standardization
11. Wagelaar D, Tisi M, Cabot J, Jouault F (2011) Towards a general composition semantics for rule-based model transformation. In: 14th Int. Conf. on model driven engineering languages and systems (MoDELS 2011), LNCS 6981, pp 623–637
12. The Architecture Analysis & Design Language (AADL), version 2, Jan 2010, http://standards.sae.org/as5506a/
13. Cadoret F, Robert T, Borde E, Pautet L, Singhoff F (2013) Deterministic implementation of periodic-delayed communications and experimentation in AADL. In: 17th International symposium on object/component/service-oriented real-time distributed computing, June 19–21, Paderborn
14. Cadoret F, Borde E, Gardoll S, Pautet L (2012) Design patterns for rule-based refinement of safety critical embedded systems models. In: 17th Int. Conf. on Eng. of complex computer systems (ICECCS'12). IEEE CS, Washington, DC, pp 67–76
15. Ameller D, Franch X, Cabot J (2010) Dealing with non-functional requirements in model-driven development. In: 18th IEEE Int. Requirements Engineering Conf. (RE'10). IEEE Computer Society, Washington, DC, USA, pp 189–198
16. Czarnecki K, Helsen S (2003) Classification of model transformation approaches. In: Workshop on generative techniques in the context of model-driven architecture (OOPSLA'03), Anaheim, CA

# Chapter 25
# A Discrete-Event Simulation Metamodel for Obtaining Simulation Models from Business Process Models

**M. Teresa García, M.A. Barcelona, M. Ruiz, L. García-Borgoñón, and I. Ramos**

**Abstract** Organizations need to be agile and flexible to meet the continuous changes. Business Process Management (BPM) is harnessing the continuous changes suffered by organizations in the value chain and, therefore, in their processes. Simulation models offer the ability to experience different decisions and analyze their results in systems where the cost or risk of actual experimentation are prohibitive. BPMN models are not directly executable nor is it possible to simulate their behavior in various input parameters. This paper proposes the application of model-driven engineering (MDE) to integrate the definition of business processes with Discrete- Event Simulation (DES) as a tool to support decision-making. We propose a platform independent DES metamodel and a set of rules, to automatically generate the simulation model from BPMN 2.0 based business process in accordance with previous metamodel.

**Keywords** Model Driven Engineering • Business Process • Discrete-Event Simulation • Decision-Making

---

M.T. García (✉) • M. Ruiz
University of Cádiz, C/Chile, S/N, 11003 Cádiz, Spain
e-mail: mayte.garcia@uca.es; mercedes.ruiz@uca.es

M.A. Barcelona • L. García-Borgoñón
Aragon Institute of Technology, c/María de Luna 7, 50018 Zaragoza, Spain
e-mail: mabarcelona@ita.es; laurag@ita.es

I. Ramos
University of Seville, Av. Reina Mercedes S/N, 41012 Sevilla, Spain
e-mail: iramos@us.es

M. José Escalona et al. (eds.), *Information System Development: Improving Enterprise Communication*, DOI 10.1007/978-3-319-07215-9_25,

## 25.1 Introduction

Due to the changing and competitive environment facing organizations today, it is necessary to keep alive the concept of continuous improvement towards excellence.

Organizations need to be agile and flexible to meet the continuous changes that are subjected (e.g. change in customer needs, in applicable law or technology available at the organization's staff) be effective and efficient if they want to build their business in the medium and long term. The business and technical layers must be able to combine their efforts.

In the business context, Business Process Management (BPM) is harnessing the continuous changes suffered by organizations in the value chain and, therefore, in their processes. BPM techniques enable analysts to handle all aspects of these processes. However, today, there remains a significant gap between the analysts and software engineers who must develop applications that support (all or part) the business process.

Simulation models offer the ability to experience different decisions and analyze their results in systems where the cost or risk of actual experimentation are prohibitive. At company level, the simulation of processes associated with business can be very helpful for making strategic decisions, tactical and operational.

The business process modeling is the representation of the activities of the business processes of an organization to be analyzed and improved [13]. Currently, BPMN 2.0 [10] is used to model business processes, which is an evolved version of BPMN 1.0 [9] which is considered the de facto standard for this. However, BPMN models are not directly executable nor is it possible to simulate their behavior in various input parameters. Thus, for these models resulting from greater utility, particularly during the decision-making process, it would be interesting to transform them into others who can simulate and analyze.

Today there are several tools for modeling business processes, but they do not allow their simulation in the broad sense of the word. Similarly, it is possible to find numerous simulation tools, although not specifically designed to simulate business processes.

This paper proposes the application of model-driven engineering (MDE) to integrate the definition of business processes with Discrete-Event Simulation (DES) as a tool to support decision-making. We propose a platform independent DES metamodel and a set of rules, to automatically generate the simulation model from BPMN 2.0 based business process in accordance with previous metamodel.

The rest of the paper is organized as follows. First, we present the state of the art and related work in Sect. 25.2. Then Sect. 25.3 exposes the problem statement. Basic concepts related to DES are introduced in Sect. 25.4, which precedes the metamodel showed in Sect. 25.5. A case study is presented in Sect. 25.6. Finally, Sect. 25.7 shows the conclusions and directions for future research.

## 25.2 Related Work

In this section we present related work on the subject. The term Software Process Simulation Modeling (SPSM) comes from the first ProSim workshop in 1998 in which Kellner, Madachy and Raffo (KMR) discussed the "why, what and how" of process simulation [7]. They identified the questions and issues that simulation could be used to address, the scope and variables that could be usefully simulated and the modeling approaches that could be employed.

Ten years later Zhang et al. [14] performed a systematic literature review of software process simulation papers fom the ProSim series[1] and they analized the facts, trends and future research directions. They identified as major trends: (1) system dynamics (SD) and discrete-event simulation form the main stream of SPSM paradigms; (2) new simulation paradigms continue to be introduced; (3) continuous modeling gradually lost its dominant position in SPSM research in comparison with discrete approached during the decade; (4) most of newly introduced paradigms enhanced the research capability at the micro-process level; (5) in recent years, micro- processes have been attracting more simulation researchs. They identified the following future research directions: (1) more recent modern software development processes need to be further investigated (e.g. agile, open-source and global development; (2) more new simulation paradigms need to be experimented and introduced; (3) more attempts are needed to effectively tackle the uncertainty of software process in practice; (4) hybrid simulation models should address more than SD and DES in vertical integration; (5) process simulation models should become more reusable, which makes them easier to build.

Raffo et al. [12] described in detail how Process Simulation Modeling supports each of the various CMMI Process Areas from level 2 through level 5. They pointed out some of the tangible benefits that PSIM can provide: (1) selection of the best possible development process, quality assurance strategy, or set of tools for a given project, situation or circumstance; (2) improved project planning through the use of an objective and quantitative basis for decision making; (3) enhanced project execution and control through PSIM's ability to quickly evaluate alternative responses to unplanned events; (4) a deeper understanding of the many factors that influence success for complex software development projects; (5) enhanced ability to communicate process choices and alternatives; (6) an objective way for project managers to answer difficult questions such as "Will adding resources early in the project really reduce overall project cost?".

van der Aalst et al. [1] focused on the behavior of resources and propose a new way of characterizing their availability. Experiments show that it is possible to capture human behavior in the business processes of a much better way. By incorporating best resource characterization tools, business process simulation can finally fulfill his promise slope.

---

[1] Proceedings of International Workshop on Software Process Simulation Modeling (ProSim) and International Conference on Software Process (ICSP), and the special journal issues.

Guizzardi et al. [5] present a Discrete-Event Simulation Ontology that defines the basic concepts that any general purpose DES language should support in order to evaluate DES languages. Then they make a study of how they adapt the basic elements of BPMN models to represent Agent Based Simulation to analyze the following properties: consistency, completeness, clarity and conciseness [6].

Onggo et al. [11] propose the use of BPMN to represent a conceptual model of Agent-Based Simulation. BPMN is widely used by organizations and can serve for business users adopt Agent Based Simulation to design this visual modeling tool called ABMN Designer (Agent-Based Model and Notation Designer).

Cetinkaya et al. [3] proposed a Model Driven Development work environment for Modeling and Simulation (MDD4MS). This environment includes: (1) the life cycle of Modeling and Simulation; (2) the definitions for conceptual metamodel for BPMN what you have selected; (3) the stages of modeling, specification, for which they have chosen DEVS (Discrete-Event System Specification), and implementation; (4) the transformation of metamodels; (5) and a tool architecture for the overall process.

## 25.3 Problem Description

In this section we describe the context of the problem we want to solve. As discussed in Sect. 25.1, today is very important for organizations to have business processes that are flexible and can be adapted to the dynamic conditions of a global market increasingly competitive. The modeling and simulation of business processes have become an essential tool for scenario analysis and the results would anticipate changes before they occur to support decision-making.

Currently, those responsible for modeling business processes need to have experts in simulation to obtain the corresponding models and carry out the tests because, they do not know simulation tools. This implies that the BP modeler must define any process change, the variables to be adjusted and the elements for analyzing, then the simulation team creates and executes the simulation model. After that, the BP modeler is responsible for interpreting the results to decide how to improve the BP defined. This process is shown in Fig. 25.1a.

This gap between the BP modeler and simulation experts may create inconsistencies between what the BP modeler wants to analyze and how it is interpreted by the technical team, it makes more difficult to create new scenarios for analysis and interpretation of simulation results can be subjective regarding the BP domain.

Ideally, BP modeler could also perform simulation. It is essential to develop tools that enable BP designers: (1) automatically obtain simulation models from BPMN models of their business processes, (2) complete such models with initial values of the simulation parameters and (3) generate the input file for execution on a particular simulation tool.

To this end, we propose a MDE approach [4] to address this problem by following the flow defined in Fig. 25.1b, which we explain in more detail below:

- First, the BP designer must create a model of your business process. This model must conform to the BPMN 2.0 metamodel.

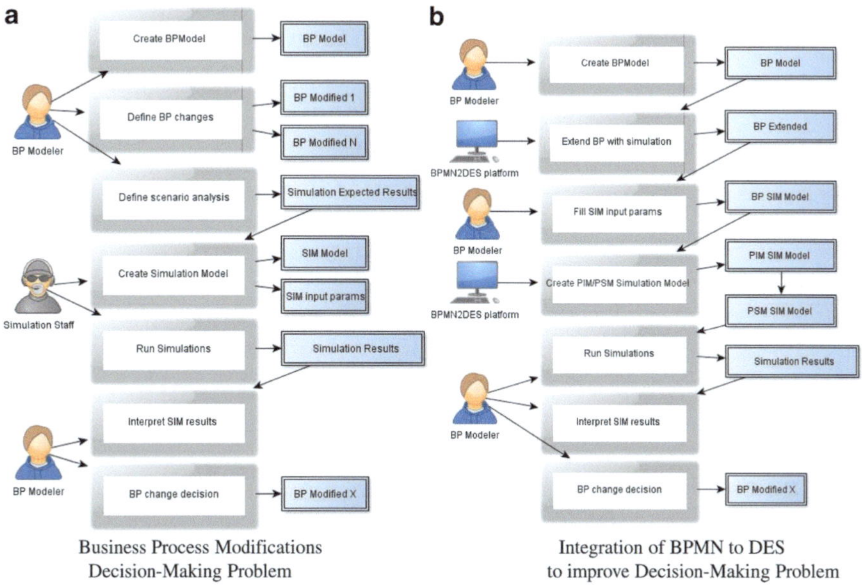


**Fig. 25.1** Business process modifications decision-making problem and BPMN to DES integration

- Then the model developed in the previous step will become an extended BPMN model. This model contains all the information of the original model and a set of parameters, initially empty, necessary for process simulation. To automate this task we implement both the Extended BPMN metamodel and the corresponding M2M transformation.
- At this point, the designer must set (manually) the specific values for the parameters of the model generated in the previous step for those who want to carry out the simulation.
- Once the extended BPMN model was completed, it will be transformed into a simulation model independent of the concrete platform on which finally will perform the simulation. To automate this task we implement both the metamodel and the corresponding PIM Simulation M2M transformation.
- Then, using M2T transformations (to be defined once for each considered simulation platform), we will obtain the input files that can be executed in the corresponding simulation tools, for example, Vensim[2] or AnyLogic.[3]
- The simulation results will be interpreted by the BP Modeler to take the appropriate decisions.

In summary, the MDE proposed approach contains a number of metamodels and transformations between them, as shown in Fig. 25.2.

[2] http://www.vensim.com

[3] http://www.xjtek.com

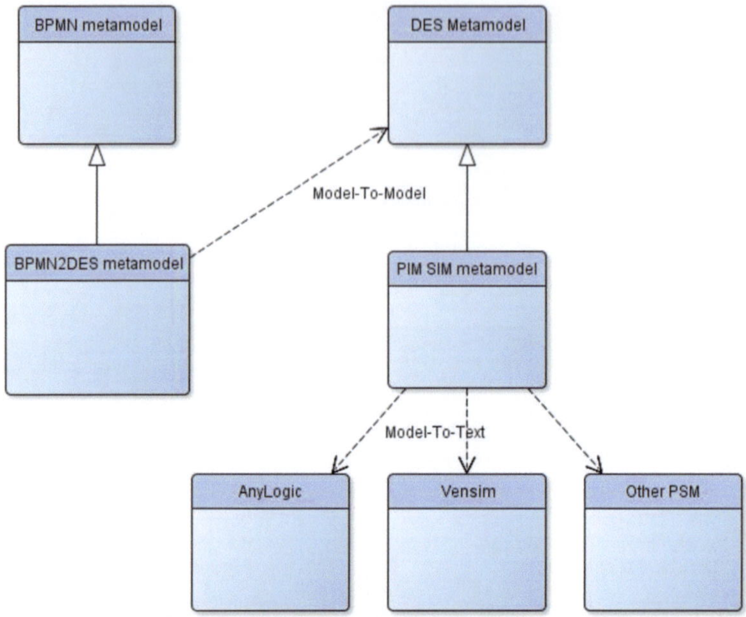


**Fig. 25.2** MDE approach from BPMN to DES

## 25.4 Discrete-Event Simulation

A *model* is an abstraction or a simplified representation of a complex system that can be real or conceptual. A model is designed to show certain characteristics of the system that you want to study, predict, modify or control. Therefore, a model includes some aspects, but not all, of the system to be analyzed.

A *simulation model* is a computational model that has the characteristics cited above and which represents a dynamic system. Simulation models offer as main advantage the possibility of experiencing different decisions and analyze their results in systems where the cost or risk of actual experimentation are prohibitive. Furthermore, the simulation allows the analysis of systems of such high complexity that are impossible to represent by static models.

The common goal of simulation models is to provide mechanisms for experimentation, prediction of behavior, resolution of questions of the type *What would it happen if…?* and learning of the system represented, among others [3].

A *Discrete-Event Simulation Model* is one in which the model state changes occur at discrete points in time, which are the events [8]. Measuring the time of the simulation is given in units of time, appropriate for the system in question.

The main components involved in DES are [2]:

- *Entities*: they are the objects that move, change state and interact with others through the system. Some are permanent, they are found always in the system, and temporary staying for a limited time.

- *Attributes*: they identify entities, they include a number indicating the time of the next event.
- *Variables*: they reflect the characteristics of the system as a whole. They are not associated with entities, although they may be modified by them. they can be predefined by the simulation software or be set by the user.
- *Resources*: they are a special type of entity used by other entities to perform an action. An entity requests a resource. When assigned uses and releases it later.
- *Queues*: they are the place where entities expect when they can move, for example, because the resources needed are not available.
- *Activities*: they are functions that resources perform over entities. Any activity can have a defined duration, empirical or stochastic, but can also be defined with a fictional length. In each often be quoted an entity with one or more resources. The tasks performed in each process would be our system activities.
- *Events*: they are facts that occur in a given time, and which rise to changes in the state of a system entity. They may be endogenous, if given by conditions of the model, or exogenous, if the causes are external. Keep in mind that during an activity can not be given any event.

## 25.5 DES Simulation Metamodel

In order to develop the proposed MDE approach, as we have shown in Fig. 25.2, it is necessary to have a Platform Independent DES metamodel. In this paper we propose the metamodel presented in Fig. 25.3.

A model consists of *Parameters*, *Variables*, *Events*, *Institutions*, *Activities* and *Transitions*. On one hand *Entities* are related to *Events* that can cause a change of state in the same as a trigger condition, and the *Activities* that will act on them performing a particular action (i.e. create, add to the queue or destroy) expressed through the action link attribute. The *Event* may occur when given a certain condition and activation value, which are its attributes.

On the other hand, *Activities* that act on the *Entities* can be of three types: *initial*, *final* and *intermediate*. The *initial* is what starts the process simulation using an activation function and a value, the *final* contains an attribute that collects information in order to end the simulation and the *intermediate* activities are those that allow the process to move forward and keep your life in that attribute. *Transition* class allows passage from one *Activity* to another, and subclass represents the alternative decision between two *Activities*, depending on the evaluation of a condition. Any *Transition* has an activity of *origin* and a *destination*.

Figure 25.4a shows the graphical representation of the elements which form a pattern of discrete-event simulation for modeling according to the defined metamodel. Once we have defined the metamodel, it is necessary to establish a model-to-text transformation scheme in order to obtain the files that serve as input to a simulation tool. Figure 25.4b shows the relationship between our DES metamodel elements and basic elements of AnyLogicTM Enterprise Library tool for discrete-event simulation.

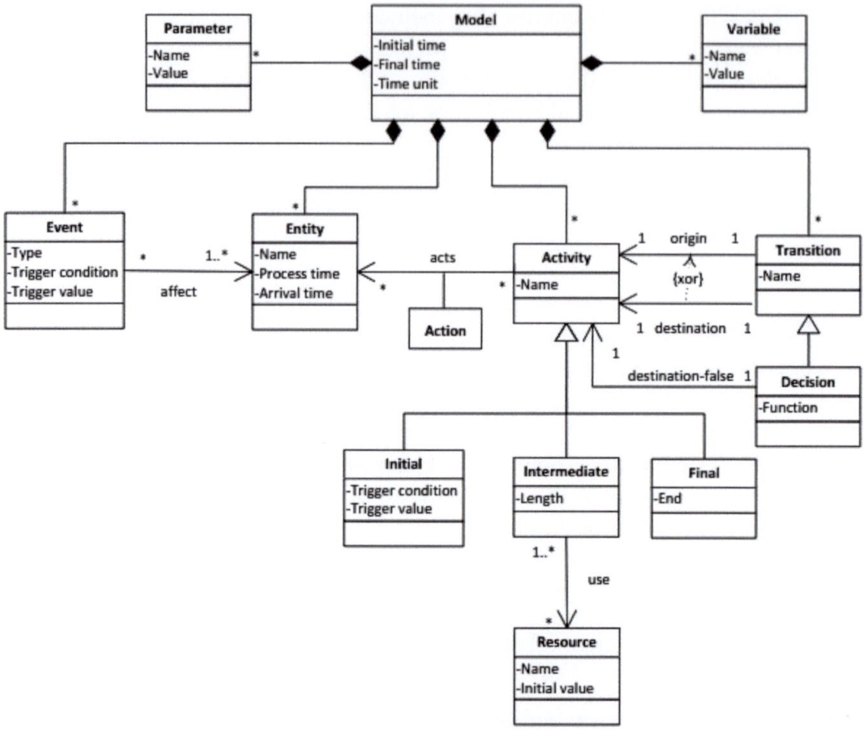


**Fig. 25.3** DES metamodel. (**a**) DES elements model graphic representation. (**b**) Metamodel transformation from DES elements to AnyLogic

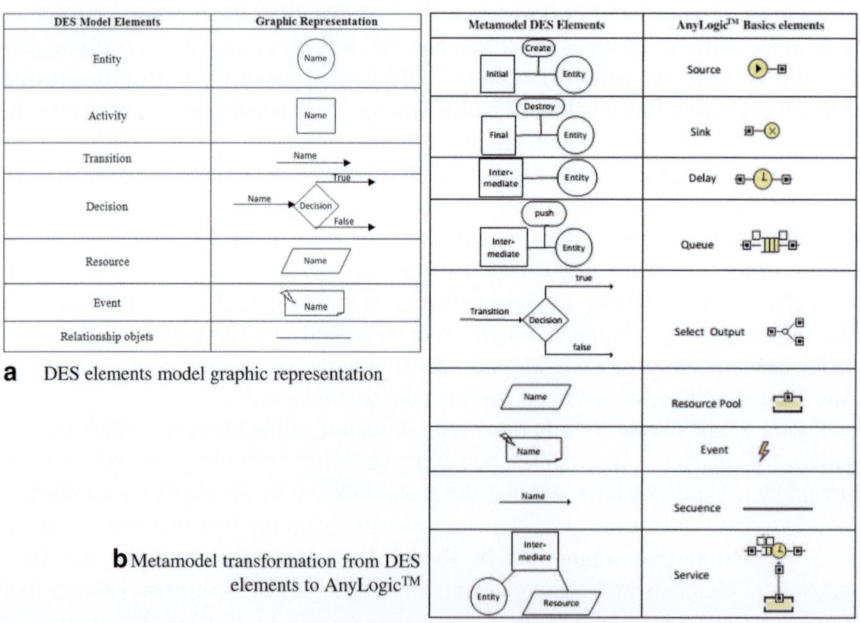


**a** DES elements model graphic representation

**b** Metamodel transformation from DES elements to AnyLogic™

**Fig. 25.4** DES model graphic representation and M2T transformation to AnyLogic

## 25.6 Case Study

To validate our proposal, we propose a simulation model for customer service process in a bank branch. Customers are coming to the branch and, on arrival, they decide if they want to perform the operation that requires the intervention of an employee of the branch or can do it themselves in the Automated Teller Machine (ATM). If they decide to use the ATM, there may already be another client using it, so they must wait in line. Conversely, if required attention window, head to one of the windows that are free or, if all are busy, wait in line for their turn.

Having a simulation model of a system like this allows management of the branch test different decisions regarding the allocation of staff to the windows depending on the demand at different times of the working day and analyze the effects of these decisions toward implementing a solution that optimizes the results.

The input parameters for the simulation of this process are set to the following values:

- Arrival rate of people in the office: 1.7 persons/minute.
- Average shooting time cashier: Triangular (0.5, 1, 1.5).
- Probability of using ATM: 50%.
- Number of windows: 4.

The discrete-event simulation model for this example, compliant with our DES metamodel, is shown in Fig. 25.5.

Following the model transformation scheme that we have presented in Fig. 25.4b, we obtain the simulation model specifically for AnyLogicTM shown in Fig. 25.6.

The simulation of DES model with the initial parameters presented above, would give the results as shown in Fig. 25.7. In view of these results, the business process modeler can make decisions, to do that, just run different simulations, with the only effort to make the business process and change, through the interface, the initial parameter values.

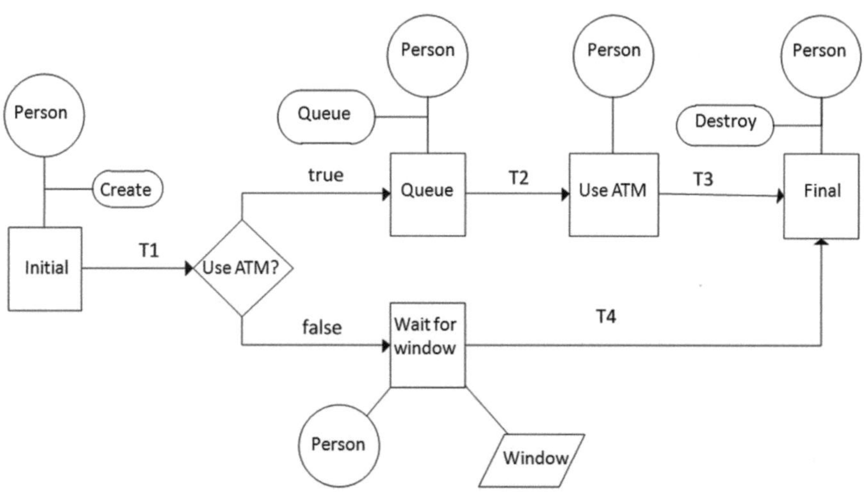


**Fig. 25.5** DES simulation model

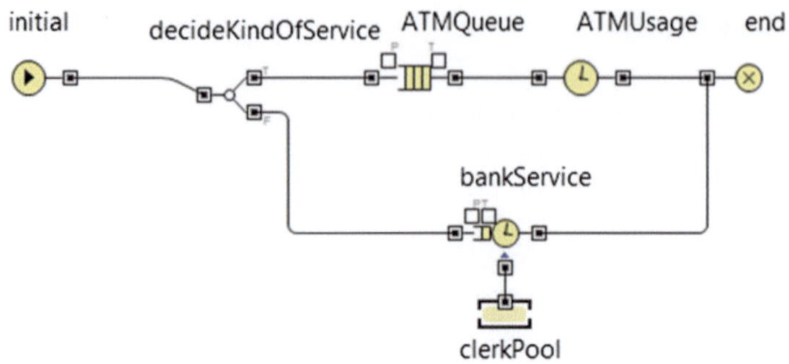


**Fig. 25.6** AnyLogicTM simulation model

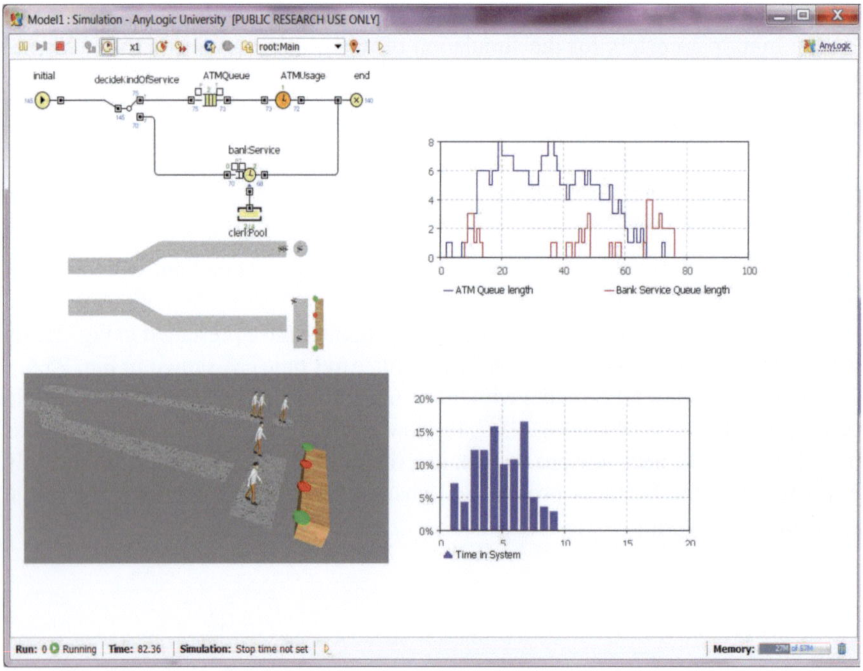


**Fig. 25.7** Simulation results

## 25.7 Conclusions and Future Work

Simulation models offer the opportunity to experience different decisions and analyze their results in systems where the cost or risk of actual experimentation are prohibitive. They also favor the study and decreased risks and advise management on the strategic, tactical and operational, therefore, they can be considered as a very important tool in business to facilitate decision making at the time of define and model business processes.

In this paper we proposed a metamodel in order to, from a BPMN 2.0 based description of business processes, automatically generate the simulation environment that can be executed using any of the existing simulation platforms. Our proposal focuses on reducing the gap between the team that defines the processes and technical staff that supports the simulation infrastructure.

The case study allowed us to conceptually validate the proposal, so that our future research will focus on: (1) generating a user interface that facilitates the designer configuring the simulation parameters; (2) making use of MDE by transformation rules for model-to-text (M2T) automatically generating the specific language model for various simulation platforms (3) developing the diverse metamodels and transformations model-to-model (M2M) which must provide an environment that integrates BPMN with DES following a MDE approach (4) approving the proposal with empirical studies and case studies in the industry.

**Acknowledgements** This research has been supported by the Tempros project (TIN2010-20057-C03-02), and Sim4Gest project (TIN2010-20057-C03-03) of the Ministerio de Economía y Competitividad, Spain, the NDTQ-Framework project and the Software Process Improvement and Formal Methods, both of the Junta de Andalucia, Spain (TIC-5789, TIC-195).

## References

1. van der Aalst W, Nakatumba J, Rozinat A, Russell N (2008) Business process simulation: how to get it right. BPM Center Report BPM-08-07, BPMcenter.org
2. Banks J, Carson JS, Nelson BL, Nicol DM et al (2005) Discrete-event system simulation
3. Cetinkaya D, Verbraeck A, Seck MD (2011) Mdd4ms: a model driven development framework for modeling and simulation. In: Proceedings of the 2011 summer computer simulation conference, Society for Modeling & Simulation International, pp 113–121
4. Garcia MT, Ruiz MVCC (2012) Como usar mde para obtener un modelo de simulacion a partir de un modelo de proceso de negocio. Actas de las Jornadas de Ingenieria del Software y Bases de Datos 2012 (JISBD 2012)
5. Guizzardi, G, Wagner G (2010) Towards an ontological foundation of discrete event simulation. In: Simulation conference (WSC), Proceedings of the 2010 Winter, IEEE, pp 652–664
6. Guizzardi G, Wagner G (2011) Can BPMN be used for making simulation models? In: Barjis J, Eldabi T, Gupta A (eds) Enterprise and organizational modeling and simulation. Springer, New York, pp 100–115
7. Kellner MI, Madachy RJ, Raffo DM (1999) Software process simulation modeling: why? what? how? J Syst Software 46(2):91–105
8. Law A, Kelton W (2000) Simulation modelling and analysis. 3a edicion. McGraw Hill. Boston 2000
9. Model BP (2004) Notation (BPMN) version 1.0. BPMI Notation Working Group. http://www.bpmn.org. Accessed April 2013
10. Model BP (2011) Notation (BPMN) version 2.0. OMG specification, Object Management Group. http://www.omg.org/spec/BPMN/2.0/PDF. Accessed April 2013
11. Onggo BS, Karpat O (2011) Agent-based conceptual model representation using BPMN. In: Proceedings of the Winter simulation conference, pp. 671–682. Winter Simulation Conference
12. Raffo DM, Wakeland W (2008) Moving up the CMMI capability and maturity levels using simulation
13. Weske M (2012) Business process management. Springer, Berlin
14. Zhang H, Kitchenham B, Pfahl D (2008) Software process simulation modeling: facts, trends and directions. In: Software engineering conference, 2008. APSEC'08. 15th Asia-Pacific, IEEE, pp 59–66

# Chapter 26
# A Pattern-Based and Model-Driven Approach for Deriving IT System Functional Models from Annotated Business Models

**Javier Berrocal, José García-Alonso, Cristina Vicente-Chicote, and Juan Manuel Murillo**

**Abstract** Companies increasingly demand IT systems to support (part of) their business. In order to maximize return on investment, these systems need to be perfectly aligned both with the business and with the legacy applications with which they will be integrated. In this vein, this paper presents a pattern-based and model-driven approach for deriving IT system functional models from annotated business models. First, we explain how to annotate the BPMN 2 business models with information useful for deriving the functionalities. Then, we introduce a set of patterns aimed to help identifying the functionalities and the relationships existing among them from the annotated business models. Finally, we present an ATL model transformation that, based on the previous patterns, allows us to obtain UML2 Use Case Diagrams.

**Keywords** Business-IT alignment • Business Process Models • Requirements Models • Requirements Relationships • Pattern-Based Model Transformations

## 26.1 Introduction

In the development of a software system, the business has to be deeply analysed in order to understand the organization's goals, its tasks and strategies, and the context with which the system has to interact [7]. Thus, the software engineer can clearly identify the system requirements aligned with the business [14].

Currently, there are proposals, such as BPCM [12] or BMM [5], offering some guidelines for documenting this knowledge, so that it can be later analysed to define

---

Additional material related to this paper can be found in the following website: http://bpmn2uc.zentipede.org.

J. Berrocal (✉) • J. García-Alonso • C. Vicente-Chicote • J.M. Murillo
University of Extremadura, Avda. de la Universidad s/n, 10003 Cáceres, Spain
e-mail: jberolm@unex.es; jgaralo@unex.es; cristinav@unex.es; juanmamu@unex.es

M. José Escalona et al. (eds.), *Information System Development: Improving Enterprise Communication*, DOI 10.1007/978-3-319-07215-9_26,

the system requirements aligned with it. In order to facilitate this analysis, some approaches, such as those proposed by Siqueira and Silva [16] and de la Vara and Sánchez [21], offer different mechanisms for deriving the functional requirements from the business specification. These requirements capture the intended behaviour of the system [22].

However, these approaches do not cover the derivation of the relationships between functionalities or between them and other business elements (e.g., legacy systems). Requirements relationships concern actions performed on one requirement that may affect or interact with other elements [9]. Identifying these relationships in the early stages of the development is crucial to achieve the business-IT alignment. Otherwise, the system might need to be re-implemented or even re-designed [9], usually meaning higher costs and project delays.

There are some approaches dealing with the identification of relationships between functional requirements. In particular [10, 18], like the present work, are focused on the analysis of business processes. While the aforementioned approaches adequately meet a range of possible relationships, they do not address some situations (e.g., Use Cases covering exception flows) and relationships (e.g., *include* relationships) that need to be covered to achieve a complete alignment.

The work presented in this paper tries to overcome some of these limitations by (1) providing an algorithm for annotating BPMN 2.0 [6] business models with information relevant for extracting IT system functionalities from them; (2) defining a comprehensive set of patterns for deriving the relationships existing among them from the annotated business models; and (3) a set of model transformations implemented in ATL [2] that, relying on the power of Model-Driven Engineering (MDE) techniques [17], and based on the previous patterns, allow us to obtain UML2 Use Case Diagrams. The proposed approach aims to improve the business-IT alignment reducing the risk of system integration problems from the earliest stages. This work is part of a more extensive work that advocates for making the relationships between business elements and requirements explicit, so that they can be reused during the overall system design process [4].

The rest of the paper is organized as follows. Section 26.2 motivates the research and presents the case study used to illustrate the proposal. Section 26.3 outlines the overall research line in which this work is framed. Section 26.4 proposes a method for annotating business processes. Section 26.5 presents the patterns and the ATL transformations defined for identifying the requirements relationships. Section 26.6 reviews some related works and Sect. 26.7 draws the conclusions and future works.

## 26.2 Motivation

Business Driven Development is a discipline aimed at developing IT systems aligned with the business [3]. Approaches, such as [5, 12], support this discipline by specifying how to document the business information. Business processes specified at a business analysis level with these approaches can be used for specifying the functional requirements required to support them [19]. Figure 26.1 shows a process,

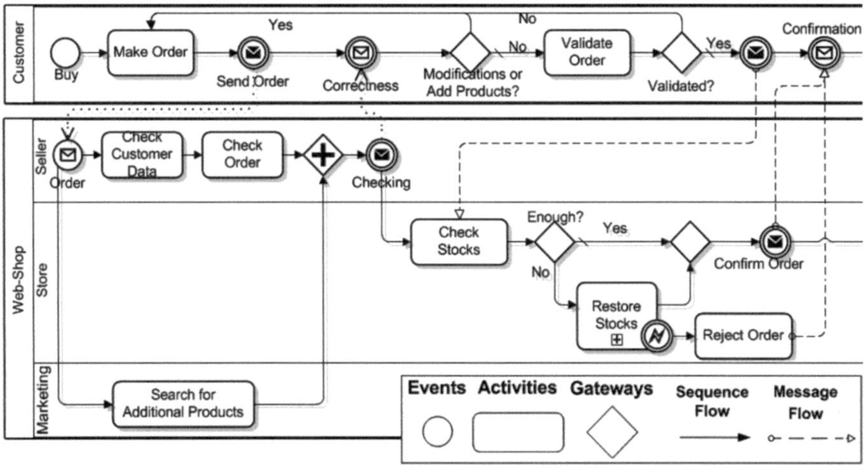


**Fig. 26.1** Business process for handling the orders of an online-shop

modeled with BPMN, for handling the orders of an online-shop. A description of this notation can be found in [1]. This figure only shows the part of the process relevant to the present study. This model gathers information related to the task sequencing, the roles responsibilities, and the input and output data.

Some approaches define methods that facilitate the derivation of Use Cases [10] or Tasks Descriptions [21] from business processes. For example, in [21], the authors define a new BPMN flow connector to specify which tasks are executed consecutively. All consecutive tasks are then mapped to a Task Description. For the online-shop example, the engineer could use this connector to map all the consecutive flow nodes concerning the placing of an order to a Task Description, and all the consecutive flow nodes concerning its validation to another one.

For improving business-IT alignment, it is not enough to extract functional requirements from business models; it is also essential to explicitly identify the relationships among them. For instance, in the online-shop example, let us associate a Use Case to each of the two Task Descriptions previously described (order placement and order validation, respectively). As shown in Fig. 26.1, if any order is modified before being validated, it is made again. In order to document this interaction, a relationship between the two Use Cases needs to be defined. If it is missed, a source of misalignment will be introduced that will need to be corrected in a later stage.

Currently, some works define patterns for deriving the relationships between Use Cases from process models. In [10] and [18] two patterns are defined. Although these patterns are really useful, some relations are beyond their control (e.g., the relations between Use Cases identified in exception or in parallel flows). For example, in the online-shop example, two flows are merged into one by a parallel gateway. If these flows are supported by different Use Cases, there should be an *include* relationship between them. This relation cannot be identified by the previous patterns.

In order to overcome this limitation, we have identified additional patterns addressing different situations for which relationships can be derived from the business models. Furthermore, some of these new patterns also take into account the relations between requirements and other business elements, such as legacy systems. These relations can help avoiding system integration problems in later development stages.

## 26.3 Annotating the Business Processes for Deriving the System Functionalities and Design

The research presented in this paper is part of a more extensive work aimed to make the relationships between business elements and between requirements explicit, so that they can be reused during the system design. A complete specification of the work can be found in [4]. This work proposes new activities focused on the specification and analysis of these relations. These activities are the following:

- *Model the context information*. The business processes are annotated with information, firstly, on the relationships between process tasks and business goals or legacy systems and, secondly, on the system functionalities that are going to support the business tasks. A set of annotations has been defined to model this information. These annotations are added to the model by the engineer.
- *Define the requirements relationships*. The relationships between the system requirements are specified. To identify these relationships from the previously annotated business processes, a set of patterns and ATL model transformations have been defined. In addition, these transformations automatically derive the system functional models.
- *Analyse the requirements relationships*. The requirements and their relationships are analysed with the aim of delimiting the architectural patterns that facilitate their fulfilment. We have defined model transformations to facilitate this task.

This paper focuses on part of the first and second activities. Specifically, for the first activity, it indicates how the business tasks are annotated with the functional requirements and the legacy systems that support them. For the second one, it details how to automatically derive the system functionalities and their relationships.

## 26.4 Annotating the Business Processes

This section describes the methodology guiding the business expert and the requirements engineer in the annotation of the business processes (activity 1 in Sect. 26.3). These annotations document the legacy systems and the Use Cases that are going to support each business task.

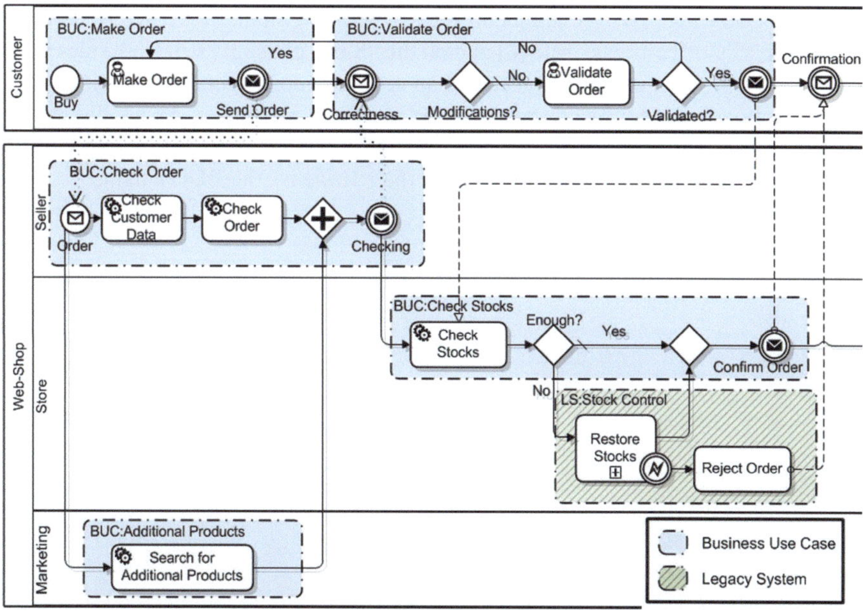


**Fig. 26.2** The online-shop example process annotated with the BUCs

Once the processes have been modeled, using some of the techniques previously indicated, first, the engineers should indicate what business tasks are already supported by legacy systems. In order to model this information the *Group* artifact is used. This artifact groups flow elements by a specific criterion. This criterion is specified by the *Category Value* attribute. Each Legacy System (LS) is modeled in the business process using a group with *Category* "LS" and *Value* "the LS's name".

Figure 26.2 shows the online-shop example process annotated with the legacy system "LS: Stock Control". This legacy system supports the "Restore Stocks" subprocess and the "Reject Order" task. Thus, the points in which the new system has to interact with this legacy system are also apparent in Fig. 26.2.

Subsequently, the support the new system is going to provide to each business task has to be defined. To do so, the engineers and the business experts have to discuss what tasks are going to be *automated*, which are going to be performed with the *support* of the system, and which are going to be done *manually*. To model this information, a task is annotated as *Service Task* if it is going to be automated, *User Task* if it is going to be supported, or *Manual Task* if it is going to be done manually.

Figure 26.2 shows the online-shop example process annotated with the kind of support that the system has to provide to each task. For example, the tasks "Make Order" and "Validate Order" have been defined as supported tasks.

The next step is the identification of the system functionalities required to support the process. For that purpose an algorithm is proposed. The algorithm ensures that

Business Use Cases (BUC)[1] are identified at a proper granularity. It is based on the Cockburn's "Coffee-break" rule [8] and on the "Step" concept [10]. Both rules indicate that a Use Case is a set of tasks that an actor can do without interruption.

The following pseudo-code presents the proposed algorithm. As we did before with the legacy systems, we use the *Group* artifact to model each identified BUC. The *Category* of all these groups is "BUC" and their *Value* is "the BUC's name".

### 26.4.1 Algorithm for the Identification of Business Use Cases

```
1. Begin with the first element of the business process.
2. All consecutive flow elements are part of the same BUC.
3. A BUC ends when:
   A. The control flow passes from one Lane to another.
   B. There is a time transition between tasks. This is:
      I.    There is a task marked as manual.
      II.   There is an intermediary event or a
            sub-process.
      III.  There is a communication with an external
            agent.This indicates the start (input of
            data) or end (output of data) of the BUC.
      IV.   There is a task performed by a legacy
            system.
   C. The end of the process is reached.
```

Figure 26.2 also shows the BUCs identified for the online-shop process. For instance, the "BUC: Check Order" groups the elements for checking the order until the result is sent to the customer. Modeling these elements in the BPMN diagrams provides the engineer with a first approximation to the system functionalities, and facilitates their discussion with the business experts.

The annotation of the business process is a guided set of activities in which the business expert and the requirements engineer have to analise them in order to annotate the business tasks. Nevertheless, the identification of the business tasks already supported by legacy systems can be semi-automated [15]. Also, the algorithm for the identification of BUCs can be implemented in order to provide a possible set of functionalities supporting the process tasks.

---

[1]The term "Business Use Case" is used here to group business tasks, even though it is similar to the term System Use Case [8].

## 26.5 Deriving the Functional Requirements and their Relationships

This section details how the new IT system functional requirements (defined using a Use Case Diagram) and the relationships among them are automatically derived from the annotated business processes by means of an ATL model transformation.

### *26.5.1 Deriving the Use Cases*

Once the BUCs have been modeled, it is possible to derive a Use Case Diagram from the annotated processes. First, the actors are extracted. These are the roles responsible for each business task (modeled in BPMN with Pools or Lanes), and the legacy systems with which the new IT system has to interact (modeled with the Group artifact and the *LS* category). Then, each BUC is mapped to a Use Case.

```
rule lane2Actor{
  from lane : inMM!Lane
  to actor: outMM!Actor ( name <- lane.name,
                  generalization <- generalization),
  generalization : outMM! Generalization(
                           general <- lane.getParent())
}
```

An excerpt of the ATL transformation implemented to derive Use Case Diagrams from the annotated business models is shown above. This excerpt presents a transformation mapping each lane to an actor. In addition, it creates a generalization relationship between the created actor and the actor extracted from the pool or the lane containing the processed lane. For the online-shop, an actor depicting the Store lane and a generalization between this actor and the Web-Shop actor are created.

Figure 26.3 shows the Use Case Diagram generated after applying the ATL transformation to the process illustrated in Fig. 26.2. The resulting diagram has been arranged with UML2Tools [11]. As shown in Fig. 26.3, all the pools, lanes and LSs have been transformed into actors, and all the BUCs have been transformed into Use Cases. Besides, the transformation also creates the relationships between actors and between Use Cases and actors.

### *26.5.2 Deriving the Relationships Between Use Cases*

Once the Use Cases have been extracted from the business processes, the relationships between them need to be identified. For this purpose, a set of process patterns has been defined identifying the most common situations for which Use Case

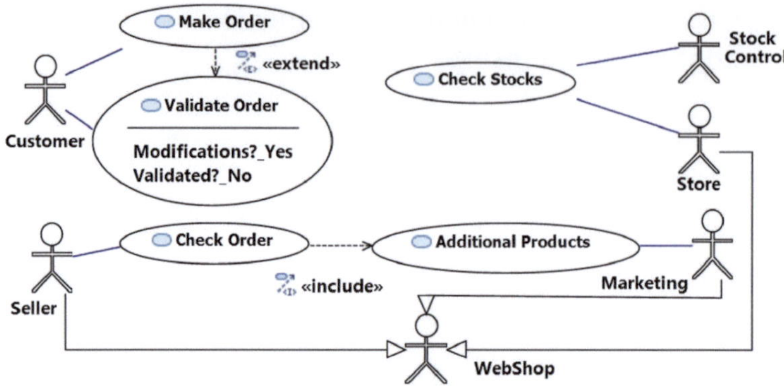


**Fig. 26.3** Use Case Diagram extracted from the annotated online-shop example process

**Table 26.1** Patterns for identifying requirements relationships

| Description | Source Pattern | Use Case Fragment |
|---|---|---|
| If there is a BUC grouping some tasks, an exclusive or inclusive gateway, and the gateway's default flow, and there is another business use case grouping the tasks defined in an alternative branch, then an *extend* relationship exists between the two use case, since the BUC in the alternative branch extends the normal flow to follow when the gateway is reached. | 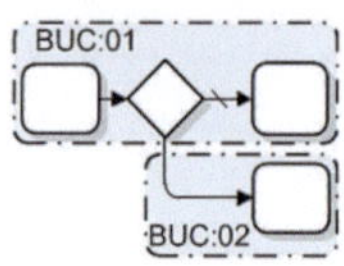  | 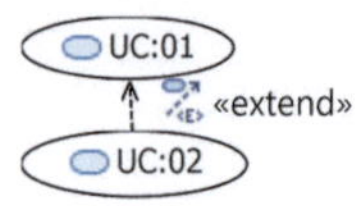  |
| When two or more control flows are merged into one by a parallel gateway and there is a BUC covering a branch and, at least, the parallel gateway, then there is an *include* relationship between the BUCs covering each branch and the BUC containing the gateway; since in order to continue the execution of the BUC containing the gateway, the other branches have to finalize. | 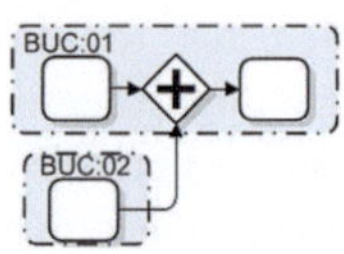  | 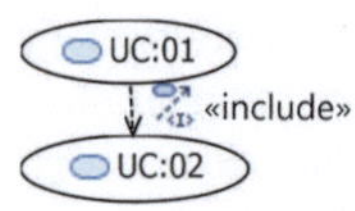  |

relationships need to be modeled. Table 26.1 contains the patterns relevant to the online- shop case study, and the Appendix gathers the most important ones identified as part of this work. These patterns can identify relationships between Use Cases and actors, between Use Cases and legacy systems, and *include* or *extend* relationships. To facilitate their comprehension, a description, the source BPMN pattern, and the derived use case fragment are detailed for each pattern.

Figure 26.3 shows the two relationships identified by applying the defined patterns to the annotated online-shop example process, depicted in Fig. 26.2.

The first one is an extend relationship between the "Make Order" and the "Validate Order" Use Cases. This relationship is identified by the first pattern defined in Table 26.1, since the "Validate Order" BUC contains an exclusive gateway from which an alternative branch (grouped by the "Make Order" BUC) starts. The "AlternativeExclusiveGateway" ATL rule, defined below, implements this pattern. This rule, first, identifies the source pattern in the business process. For each detected pattern, the extending and extended uses cases are identified. Then, an "extension point" is created in the extended Use Case. The *extend* relationship is created only if there is not another *extend* relationship already defined between the two use cases. Otherwise, the "extension point" is directly linked to the existing *extend* relationship. In the online-shop example process, this pattern is identified twice, because two different exclusive gateways are covered by the "BUC: Validate Order". The first time it is identified, an "extension point" and an *extend* relationship are created to reflect that, during the validation of an order, if a modification is required, a new order has to be placed. The second time, only the "extension point" is created (the *extend* is not duplicated) to reflect that if the order is not validated a new one can be placed.

```
  rule AlternativeExclusiveGateway {
     from seq : inMM!SequenceFlow ( seq.connectBUCs() and
         if  seq.sourceRef.oclIsTypeOf(inMM!ExclusiveGate
way) then
           if seq. sourceRef.default<>OclUndefined then
            seq.sourceRef.default <> seq
        else false endif
       else false endif)
     to ePoint  : outMM!ExtensionPoint  (
    name <- seq. sourceRef.name+' -' +seq. name, useCase<-
thisModule.resolveTemp(seq.sourceRef.getBUC () , 'uc'))
do {
       if (not seq.alreadyExtended (seq.targetRef, seq.
sourceRef) ){ thisModule .createExtendRelationship(seq);
    } else { seq.getExtend(seq.targetRef, seq.sourceRef).
extensionLocation  <-  seq.getExtend(seq.targetRef, seq.
sourceRef) .extensionLocation.append(ePoint) ; }
   }
```

The second relationship, identified by the second pattern defined in Table 26.1, is an *include* between the "Check Order" and "Additional Products" Use Cases. This pattern is applied because the "Check Order" BUC contains a merger parallel gateway and the "Additional Products" BUC covers the branch that has to be waited. The ATL rule "MergerParallelBranch", implementing this pattern, is presented below.

```
  rule MergerParallelBranch {
     from seq : inMM!SequenceFlow (seq.connectBUCs() and
       if    seq.targetRef.oclIsTypeOf(inMM!ParallelGate
       way) then
```

```
      seq. targetRef.gatewayDirection.toString () ='
      Converging' else false endif)
    do{
    if (not seq.alreadyIncluded(seq.targetRef, seq.
    sourceRef)){ thisModule .createIncludeRelationship
    (seq);}
  }
}
```

The correct identification of these relationships is crucial. Their underspecification, first, hinders the system design process and, second, increases the risk of poor integration. In the example presented above, if the relationships between the "Make Order" and "Validate Order" Use Cases were not explicitly defined then the architect would need to make an in-depth analysis of the business processes and the Use Cases in order to identify them. If even so the relationships remain undefined, then a business-IT misalignment will occur.

## 26.6 Related Work

Different works, such as [12] or [20], have proposed methodologies for modeling the knowledge of an organization and the requirements of the systems aligned with it. The specification of these requirements demands full attention during the in-depth analysis of the business models. Other works, e.g. [10, 16, 18, 21], focus on how to re-use these models to facilitate the requirements specification.

In [10], the authors start from processes modeled with activity diagrams. From these models they define mappings to identify the actors and the Use Cases, and a situation that leads to define *extend* relationships. Our proposal increases this one by defining a greater set of patterns, enabling the identification of other relationships (e.g., *include*) and situations (e.g., exception flows).

In [18], the authors extract certain parts of a Use Case, identified in a process, to another Use Case. Then, they present two patterns to identify the relationships between the extracted and the initial Use Cases. Our proposal increases this one by allowing the identification of relationships between consecutive Use Cases.

In [21], the authors define how to identify Task Descriptions in BPMN models. In order to specify them with a proper granularity, they group the flow elements that are performed consecutively. The identification of such elements is manually done by the engineers. This may lead to the identification of different granularity functionalities, depending on their experience. In this paper we have presented an algorithm aimed to help the engineers identify functional requirements of a uniform granularity.

Finally, [16] defines QVT transformations for deriving Use Case scenarios from BPMN diagrams. In this approach, each process is mapped to a Use Case. Conversely, our methodology enables the creation of one or more Use Cases (and the relationships between them) from each process, depending on their complexity.

## 26.7 Conclusions and Future Works

This paper presents a pattern-based and model-driven approach enabling software engineers to obtain the system functional models (defined as UML2 Use Case Diagrams) from annotated business models (defined in BPMN 2.0). The resulting models contain not only the functionalities but also the relationships existing among them. Making these relationships explicit improves the business-IT alignment and avoids potential integration problems.

Currently, we are working on a semi-automated (supervised) implementation of the algorithm for identifying BUCs (see Sect. 26.4). This implementation will offer the engineer a possible configuration of BUCs supporting the process tasks. Besides, we are working on the validation of the presented approach. To this end, it is being used in the development of an IT system which will support the Indra's business processes for quality assurance [13].

**Acknowledgements** This research was supported by the Spanish Ministry of Science and Innovation under Project TIN2012-34945, and from the Department of Employment, Enterprise, and Innovation of the Government of Extremadura as part of Project GR10129.

## 26.8 Appendix: Addition Patterns for Deriving Relationships from Annotated Business Models

| No. | Description | Pattern | Use Case Fragment |
|---|---|---|---|
| 01 | Each BUC must be related to the actor representing the Lane/Pool that contains it. In this way, the actor responsible for the Use Case is indicated | 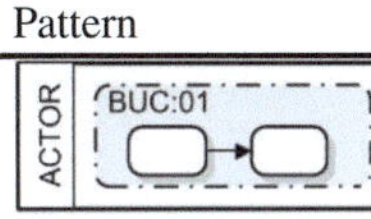  | 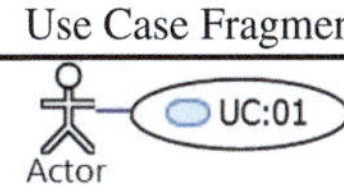  |
| 02 | When a BUC is preceded or followed in the control flow by a LS, there is a relation between them. This relation indicates that the BUC must receive or provide information to the LS | 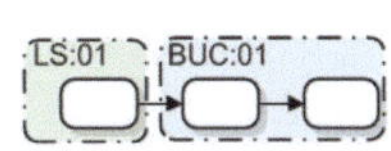  | 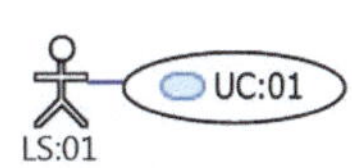  |
| 03 | If there is a BUC grouping some tasks, an exclusive or inclusive gateway, and a gateway's alternative flow, and there is another BUC grouping the default branch, then an include relationship exists between the two Use Cases, since the BUC in the default branch represents the normal functionality to follow when the gateway is reached | 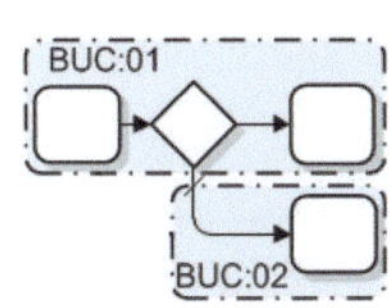  | 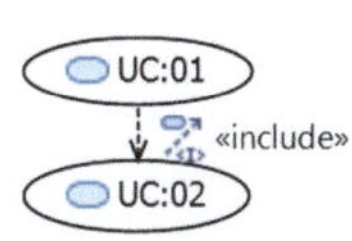  |

(continued)

(continued)

| No. | Description | Pattern | Use Case Fragment |
|---|---|---|---|
| 04 | If there are several BUCs connected by an exclusive or inclusive gateway, the BUC preceding the gateway is related to the BUC covering the default branch by an *include* relationship (since it represents the normal functionality to follow). Also, the BUC preceding the gateway is related to the BUC in the alternative flow by an *extend* (since it extends the normal flow) | 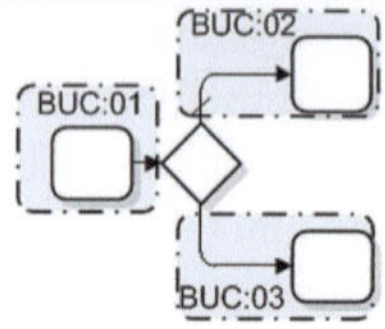 | 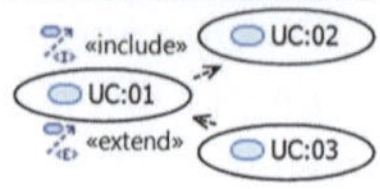 |
| 05 | When two or more control flows are merged into one by an exclusive gateway, the BUC identified after the merge has to be related to the BUCs of each branch by means of an include relationship. Thus, it reflects that whenever a BUC of a branch is completed the BUC following the merge has to be executed | 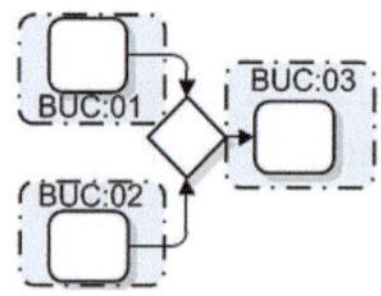 | 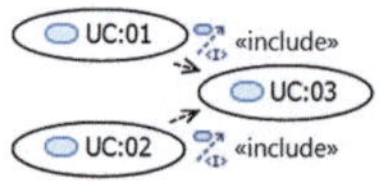 |
| 06 | If several business use cases are connected by a parallel gateway, then they have to be handled as isolated business use cases; since the execution of each branch is independent of the rest | 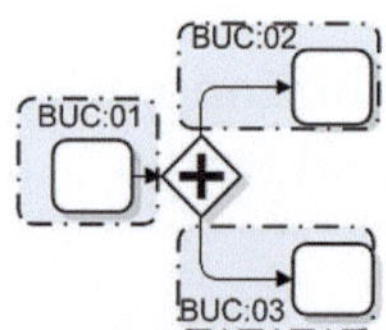 | 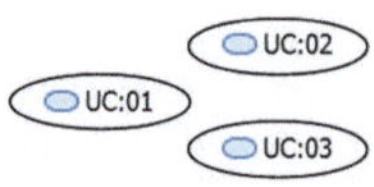 |
| 07 | When two or more control flows are merged into one by a non-exclusive gateway, and there is a BUC covering each branch and the flow nodes following the gateway, then there is no relationship between the business use cases; since the condition to start the BUC following the gateway is the completion of the BUCs in the branches | 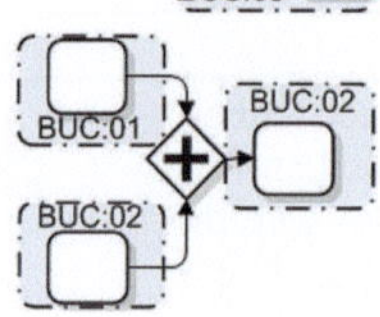 | 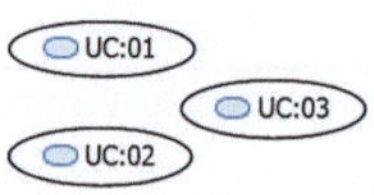 |
| 08 | Processing sub-processes. If in the control flow there are sub-processes, they will be treated independently. The business tasks detailed in the sub-processes will not be encapsulated in the Use Cases identified in the main process | 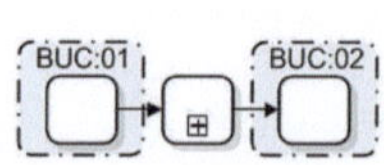 | 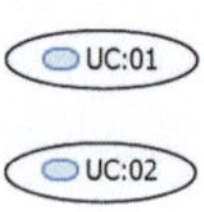 |

(continued)

(continued)

| No. | Description | Pattern | Use Case Fragment |
|---|---|---|---|
| 09 | Processing exceptions. If there is a sub-process in which the exceptions are handled, and all the sub-process's flows are covered by a single business use case, then the exception is added to the business use case as an exception scenario. The exception will thus be handled together with the business use case | 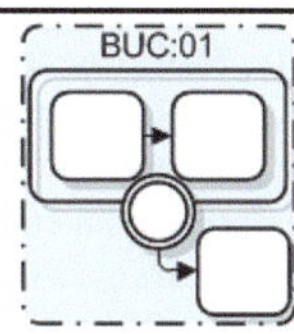  |   |
| 10 | Processing exceptions. If there is a sub-process in which the exceptions are handled, and the sub process's flows are covered by several business use case, then a new BUC must be created to control the exception flow. This has to be related by extend relationships to the BUCs covering subprocess's flows. The exception is thus encapsulated in a BUC that extends to the others | 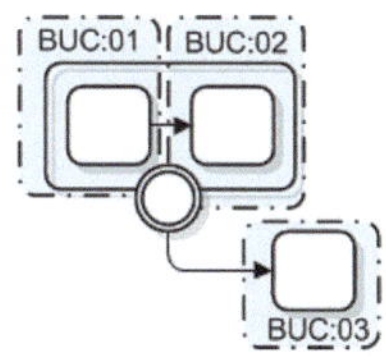  | 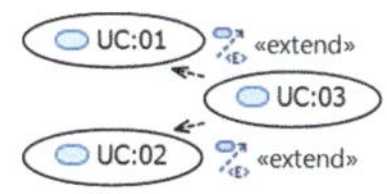  |

# References


1. Allweyer T (2010) BPMN 20: introduction to the standard for business process modeling. Books on Demand GmbH, Norderstedt
2. ATLAS team: ATLAS Transformation Language (ATL). http://www.eclipse.org/atl/
3. Baltzan P, Phillips A (2008) Business driven information systems, 2nd edn. McGraw-Hill, New York
4. Berrocal J, García-Alonso J, Murillo JM (2013) Modeling business and requirements relationships for architectural pattern selection. In: Kacprzyk J (ed) Software engineering research, management and applications 2013, Springer SCI. Springer, New York
5. BMM: Business Motivation Model. http://www.omg.org/spec/BMM/
6. BPMN: Business Process Modeling Notation 2.0. http://www.bpmn.org/
7. Cardoso E, Almeida J, Guizzardi G (2009) Requirements engineering based on business process models: a case study. In: Enterprise distributed object computing conference workshops, 2009. EDOCW 2009. 13th, pp 320–327
8. Cockburn A (2000) Writing effective use cases, 1st edn. Addison-Wesley Longman, Boston
9. Dahlstedt AG, Persson A (2003) Requirements interdependencies—moulding the state of research into a research agenda. In: Ninth international workshop on requirements engineering: foundation for software quality (REFSQ 2003), pp 71–80
10. Dijkman RM, Joosten SMM (2002) Deriving use case diagrams from business process models. Tech. Rep. TR-CTIT-02-08, University of Twente
11. Eclipse Plugin: UML2Tools. http://www.eclipse.org/modeling/mdt/?project=uml2tools
12. Gao S, Krogstie J (2009) A combined framework for development of business process support systems. In: Persson A, Stirna J, Aalst W, Mylopoulos J, Rosemann M, Shaw MJ, Szyperski C (eds) The practice of enterprise modeling, vol 39. Springer, Heidelberg, pp 115–129

13. Indra Company: http://www.indracompany.com/
14. Kotonya G, Sommerville I (1998) Requirements engineering: processes and techniques, Worldwide series in computer science. Wiley, Chichester
15. Perez-Castillo R (2012) Marble: Modernization approach for recovering business processes from legacy information systems. In: Software Maintenance (ICSM), 2012 28th IEEE International Conference on, pp 671–676
16. Siqueira F, Silva P (2011) Transforming an enterprise model into a use case model using existing heuristics. In: Model-driven requirements engineering workshop, pp 21–30
17. Stahl T, Voelter M, Czarnecki K (2006) Model-driven software development: technology, engineering management. Wiley, Chichester
18. Stolfa S, Vondrak I (2006) Mapping from business processes to requirements specification. Tech. rep., CUniversitat Trier
19. Traetteberg H, Krogstie J (2009) Enhancing the usability of BPM-solutions by combining process and user-interface modelling. In: Stirna J, Persson A, Aalst W, Mylopoulos J, Rosemann M, Shaw MJ, Szyperski C (eds) The practice of enterprise modeling, vol 15. Springer, Berlin, pp 86–97
20. Ullah A, Lai R (2011) Modeling business goal for business/it alignment using requirements engineering. Journal of Computer Information Systems 51(3):21–28
21. de la Vara J, Sánchez J (2009) BPMN-based specification of task descriptions: Approach and lessons learnt. In: Glinz M, Heymans P (eds) Requirements Engineering: Foundation for Software Quality, vol 5512. Springer, Berlin, pp 124–138
22. Wiegers KE (2003) Software requirements, 2nd edn. Microsoft, Redmond

# Chapter 27
# Applying Testing Techniques to Software Process Assessment: A Model-Based Perspective

**L. García-Borgoñón, R. Blanco, J.A. García-García, and M.A. Barcelona**

**Abstract** Software processes constitute a major asset for an organization. However, in many occasions there are differences between defined processes and executed processes. For this reason, organizations spend time and effort of their resources to find these non-conformances. The use of software testing techniques could be a useful way to reduce these costs. This paper proposes a model-based approach and shows how software testing techniques can be applied to evaluate the execution conformity in a software processes context, and also to evaluate the model designed. A real execution of a NDT methodology process by means of the process model included in NDTQ-Framework (a solution based on this approach that is currently being used in software development organizations) illustrates the final results. Finally, conclusions and future work are stated.

**Keywords** Model-Driven • Testing • Software Processes

## 27.1 Introduction

Software processes are recognized as fundamental assets in software-intensive organizations, since they support their capability to produce better products. Process definition, documentation, management and improvement have become in

L. García-Borgoñón (✉) • M.A. Barcelona
Aragón Institute of Technology, c/ María de Luna 7, 50018 Zaragoza, Spain
e-mail: laurag@ita.es; mabarcelona@ita.es

R. Blanco
University of Oviedo, Campus de Viesques, S/N 33204 Gijón, Spain
e-mail: rblanco@uniovi.es

J.A. García-García
University of Seville, Av. Reina Mercedes, S/N 41012 Sevilla, Spain
e-mail: julian.garcia@iwt2.org

M. José Escalona et al. (eds.), *Information System Development: Improving Enterprise Communication*, DOI 10.1007/978-3-319-07215-9_27,

organizations' key practices what is known as Business Process Management (BPM). Since business in a software-intensive organization is software development, software processes constitute the BPM focus within such organization [1].

A software process is a set of activities, methods and practices used in the production and evolution of software and the associated products [2, 3]. These processes and methodologies have always been described in appropriate terms to be used by a developer, but they are often described in manuals or books which project team follow as closely as possible [4].

However, differences can usually be noticed between organizations' defined processes and really executed processes in a specific project context. It occurs due to several causes, such as process described in unsuitable way, like could be natural language and misunderstandings which involves, or the existing gap between processes definition and execution. Thus, organizations draw on process and product assessment activities to solve this problem. Process assessment is a disciplined evaluation of an organizational unit processes against a process assessment model [5], which provides a set of indicators used for evaluating the effective process performance and management [6]. An assessment model can either represent the defined process or be based on one or more Process Reference Models [7].

These activities are manually executed, normally by quality offices, since processes orchestration are not widely used in software-intensive organizations. They verify and control, through a checklists set, that the process is followed properly, establishing a non-conformance record in cases of deviations between defined and executed processes. Non-conformances should be solved in a concrete time with a specific commitment. The cost of this kind of activity is usually called Quality Cost, and organizations assume it as necessary.

Besides, and particularly in periods of economic crisis like the one we are living, where optimization effectiveness and resource efficiency are essential, one of main objectives deals with reducing nonproductive costs.

In the last years, the Model Driven Engineering (MDE) [8] has been established as a common approach for software development [9], what has shaped the software industry to be model-centred. In addition, software testing is a very important phase in software development and maintenance, as it aims to find out faults in software products, thus helping developers to improve the quality of these products when the discovered faults are solved and reducing the cost produced by these faults.

As software processes are software too [10], this paper evaluates how a model-based approach working in liaison with testing software techniques can make easier activities related to performed processes evaluation, reducing their time and effort cost.

This paper is structured as follows: After this introduction, Sect. 27.2 shows the main work related to software processes evaluation, also known as Software Processes Assessment. Section 27.3 introduces some concepts related to testing techniques and describes how they are used in our approach. Then, Sect. 27.4 presents the proposed metamodel and Sect. 27.5 illustrates result in the NDTQ-Framework, a solution based on this approach that is currently being used in software-development organizations. Finally, Sect. 27.6 outlines conclusions and proposes future work in these lines of research.

## 27.2 Related Work

An organization's software process assessment is an activity especially related to software process improvement and, therefore, there are several proposals concerning it, particularly focused on integrating software process assessment and software process modeling. Later, more remarkable proposals on this topic will be pointed out.

OOSPICE is a project associated with the capability assessment space [11], although it also delivers methodology components in ISO 12207 [12], ISO 15504 and method engineering context. It copes with the concepts posed for assessing an organization's process enactment quality, through a metamodel with metaclasses such as Outcome as well as attributes relating to capability level on Process and Task. It is oriented towards capability appraisals against ISO 15504.

In [13] Hamann proposes an information model which integrates software process assessment and process modeling. It has basic elements based on process modeling such as Process, Product, Role and Tool, with the appropriate attributes and classes related to assessment information, such as Rating and Purpose. Makinen et al. [5, 14, 15] slightly modify Hamann's model, to make it more general and illustrative by classifying the elements into three categories: Assessment Model, Assessment Result and Modeling Result.

Henderson-Sellers and Gonzalez-Perez [4] propose a new standard metamodel to define and assess software processes with the same elements, but including two new concepts, powertypes and clabjects, as a way to solve problems derived from modelling both the methodology and project layers at the same time.

Despite evaluation concepts are similar, all these proposals are focused only in assessment based on a reference model, but they do not address the issue of assessing software process executed against software process modeled. That is the goal of our approach, which is presented in the following sections.

## 27.3 Applying Software Testing Techniques to Test Software Process

This section describes our approach with the aim of both, evaluating the executions conformity of the software process that has been modeled and measuring the level of acceptance of this model. Section 27.3.1 presents an overview of software testing and Sect. 27.3.2 describes the applications of software testing techniques to test a software process.

### *27.3.1 Software Testing Overview*

Software testing is part of the Verification and Validation process (V&V) [16]. It determines whether the developed products fulfill the requirements and satisfy the user's needs. Software testing deals with verifying the behavior of a system that is

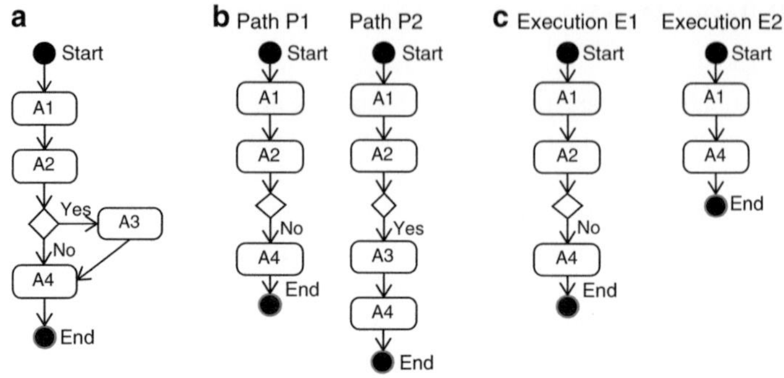


**Fig. 27.1** Introductory example

executed under specific conditions, against the expected behavior. This evaluation is carried out according to some aspects of the system [17].

Reactive and proactive approaches have been developed to achieve this goal. Proactive approaches try to detect faults in the Software Under Test (SUT) before they produce failures in an operational environment. For example, the SUT can be executed with a test case and, after that, the observed behavior and the expected behavior can be compared to detect any deviation (a test case is a set of test input data, execution conditions and expected results [17]). Reactive approaches identify faults after they produce failures. For example, a monitoring-based technique can detect any deviation from the expected behavior of the SUT by observing its real time execution.

On the other hand, several adequacy criteria have been defined [18] to specify the situations of interest to be tested (to be covered), which constitute the test requirements, and determine whether sufficient testing has been done. For instance, the path-testing criterion requires executing all or selected paths of the SUT (each path is a test requirement). Measuring the test coverage achieved, that is the degree to which the tests execute (cover) the paths of the SUT, it is possible to determine whether the testing process can be stopped.

### *27.3.2 Problem Approach*

As previously mentioned, software processes can be also considered software [10], therefore, it is possible to apply software testing techniques to test their conformity.

We will consider the introductory example depicted in Fig. 27.1, to illustrate the approach and objectives. We will use UML [19] for these examples because they are very intuitive, but other graphic representation language, such as BPMN [20], could be used. Part (a) shows a UML activity diagram that models a software process. The process model contains two correct activities paths (P1 and P2), which are shown in part (b). These paths represent the different scenarios of the software process modeled. Part (c) offers two UML activity diagrams that represent the real sequences of activities executed by the stakeholder (E1 and E2), which have been monitored. Analyzing the process model and the executions poses the following questions:

1. Do the executions E1 and E2 satisfy the process model?
2. If the path P2 is not followed (*covered*) by any execution, is it really necessary?

We have developed an assessment approach that combines reactive testing with path-testing criterion, to answer these questions. The goals of our approach are (1) to evaluate whether the executions of a software process satisfy the process model defined, and (2) to determine the paths of the process model that have not been covered by these executions and measures the degree of path coverage achieved. The following sections describe both goals.

#### 27.3.2.1 Evaluating Executions Conformity

A testing process is carried out to evaluate executions conformity with the process model. First, each transition of a monitored execution is classified as *valid* or *invalid*. It is considered valid when it also appears in the process model (for instance, the transition from A1 to A2 in the execution E1 in Fig. 27.1), otherwise it is considered invalid (for instance, the transition from A1 to A4 in the execution E2 in Fig. 27.1).

After that, our approach analyzes the classification of all transitions of the execution to determine its conformity with the process model. An execution satisfies the process model only if all transitions have been classified as valid (for instance, execution E1 in Fig. 27.1). Otherwise, the execution does not satisfy the process model (for instance, execution E2 in Fig. 27.1).

On the one hand, an execution that satisfies the process model can be considered a positive test, as it tries to check the behavior of a specific scenario of the process model. Thus, it is called *Positive Execution* in our approach. On the other hand, an execution that does not satisfy the process model can be considered a negative test, as it tries to cover a scenario for which the process model has not been designed. Consequently, it is called *Negative Execution* in our approach.

#### 27.3.2.2 Evaluating Path Coverage

Finding the paths of the process model that have not been followed by any execution can be useful to check the correctness of this model, and also to evaluate whether the software process is correctly executed. If a path is never covered, several reasons can be considered: (1) the path does not represent a scenario of the software process and therefore the model must be improved; (2) the path represents a scenario that is not very common but necessary, so the model is correct; and (3) the path represents a very common scenario that is not being considered by the stakeholder, so maybe the software process is not being correctly executed.

Our approach applies the path-testing criterion to the process model to address this issue. This way each path represented in the model is considered a test requirement, which is covered when a positive execution satisfies the path. Measuring the path coverage achieved allows observing the level of acceptance of the process model, due to the coverage is increased when a stakeholder executes a set of activities of the software process that is modeled by a path in the process model.

Considering the example in Fig. 27.1, the path testing criterion derives two test requirements: paths P1 and P2. Path P1 is covered by the positive execution E1 and path P2 is not covered by any positive execution then, the path coverage achieved is 50 %.

## 27.4 A Metamodel for Software Process Assessment

As mentioned above, many approaches have been developed in order to assess software processes with a reference model, but they have not been used to assess software processes executed against software processes modeled. This issue and the use of MDE to manage the conceptual complexity have been the basis of our proposal, that is, designing a metamodel for testing software process models.

This approach is presented in Fig. 27.2 and it is an extension of the software process metamodel presented in [21]. Besides, we present metaclasses from the extension metamodel.

The *Action* metaclass is the main class in the testing metamodel. We define an *Action* for each *Activity* aimed to test. It is possible not to test all activities in a process, so this relationship guarantees it. The attributes in this metaclass are: the start and end date of the action, the status of the action in a specific moment and the test result, where the result of the testing process performed on the action is recorded.

An *Action* has a set of preconditions and postconditions. The *Precondition* metaclass represents the previous action we need to check to be able to execute it. The *Postcondition* metaclass shows the following action and whether it is properly executed.

The *WorkProduct* metaclass represents a piece of the *Product* that is developed in each action, so that a *Product* could be considered as a collection of several *WorkProducts*. Finally, the *Stakeholder* represents someone or a tool that has actually executed the *Action*.

The main feature of our approach is that the metamodel extension allows us to test a process model which has been defined with the initial metamodel. This will enable us to establish testing points in the definition moment.

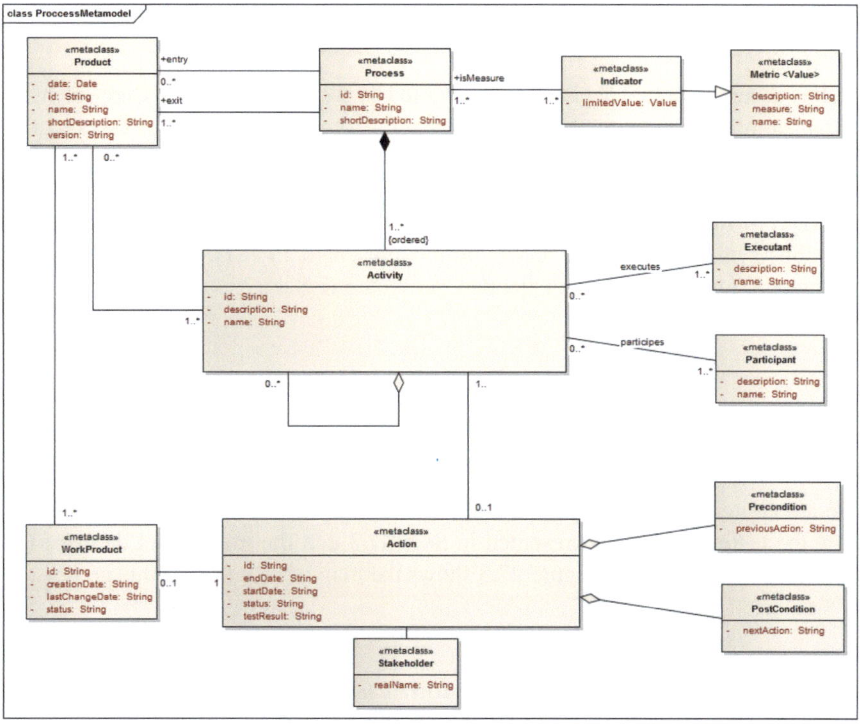


**Fig. 27.2** Testing process models metamodel

## 27.5 NDTQ-Framework: A Practical Example

This section includes a practical example to illustrate our approach. We use processes currently supported by NDT [22]. It is a methodology that defines metamodels for every phase of software development life cycle by providing a framework that make easier the use of new methods, standards and paradigms and as a result, it helps us improve software development quality. NDT uses different software processes, each one supported by main models, standards and rules related to the field where it is defined. These processes are classified in six groups:

- Software Development Processes. They are defined in terms of NDT life cycle.
- Software Maintenance Processes. They are based on ITIL [23] and CMMI [2].
- Testing Processes. They are based on ISO/IEC 29119 [24] standard.[1]
- Software Quality Processes. They are based on ISO 9001:2008 [25] and CMMI.

[1] ISO 29119 has not been yet approved completely, but they are based on the group of processes already defined.

- Project Management Processes. They are based on PMBOK [26] and CMMI.
- Security Processes. They are based on ISO 27001 [27] standard.

It is necessary a real deployment supported by tools to allow a software-intensive organization using this methodology to benefit from all its potential. This enhances its use and accomplishment. NDTQ-Framework was created with this goal in order to support all processes defined by NDT.

NDTQ-Framework is a framework implemented on Enterprise Architect tool, developed by means of the UML profile presented in [21]. It is based on the metamodel our approach has extended, presented in Sect. 27.4, to achieve the goal we are looking for.

### 27.5.1 A Process Example: Requirements Engineering Process

We are going to use the requirements engineering process as an example, to show how the testing approach presented in Sect. 27.3 and the metamodel described in Sect. 27.4 work together. Figure 27.3 shows the map of activities of this process.

### 27.5.2 Applying Testing Techniques to Requirements Engineering Process Assessment

We have considered that all activities have an action associated, and preconditions and postconditions are previous and next actions respectively. In this case, we are not going to use the WorkProduct concept to explain the testing techniques used.

We are going to consider the execution shown in part (a) of Fig. 27.4, which has been obtained from a working report tool, to illustrate how to evaluate a requirement engineering process execution conformity with the process model represented in Fig. 27.3. This tool registers the real data demanded by the process model, such as the start and end dates for each action performed as well as the status. Actions *RS01*, *RS02*, *RS03* and *RS10* represent the execution of the corresponding activities of the process model. The action *Condition1* represents the execution of the conditional activity after *RS02*, and the status indicates the answer obtained.

The result of the testing process performed is shown in part (b) of Fig. 27.4. Along this process, each action is evaluated as valid or invalid. We analyze the data registered by the working report tool and the knowledge obtained from the model about which are its previous and next actions in order to determine the test result.

For instance, the action *RS02* is valid because it starts after its previous action *RS01* has finished. That means that the execution of actions *RS01* y *RS02* represents a valid transition, as it is present in the process model, whereas the action *RS10* is evaluated as invalid. This action starts after the other actions of the working report have finished, nevertheless none of them constitutes its previous action.

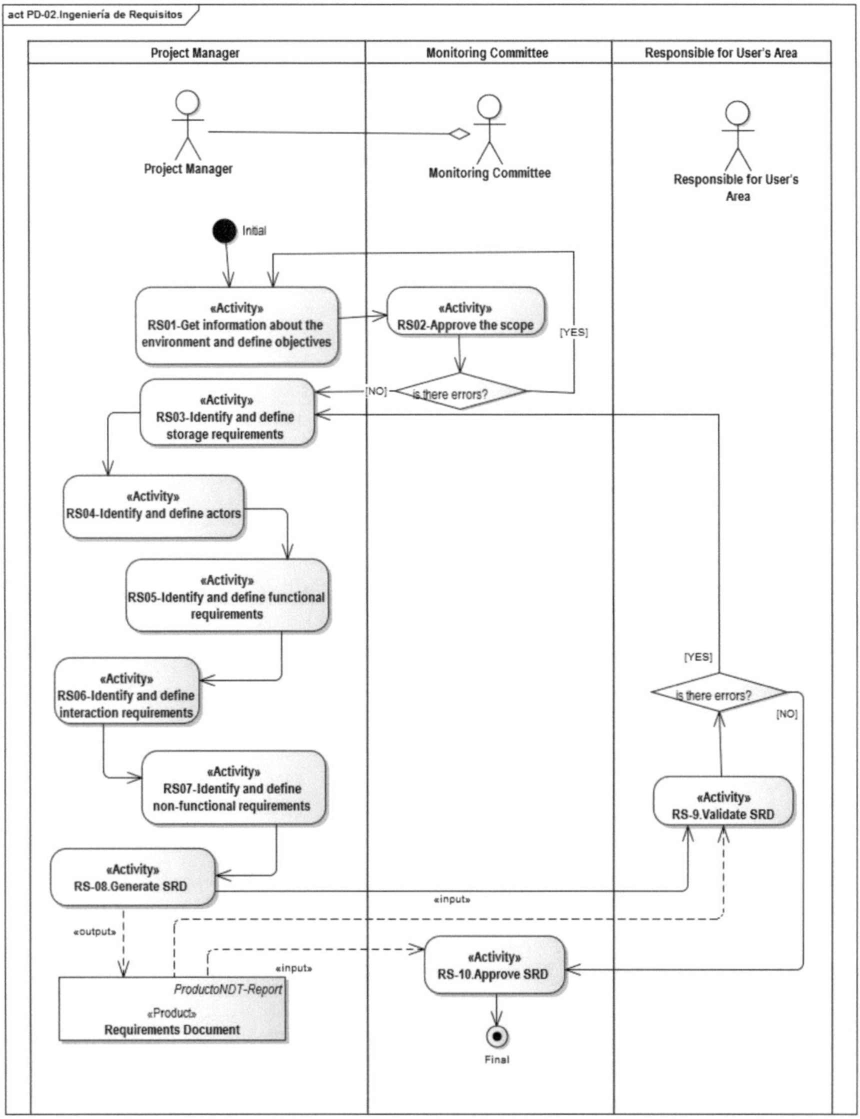


**Fig. 27.3** Map of activities of the requirements engineering process

The test result is invalid, because it is not possible to find a valid transition for action *RS10*.

As a conclusion, it should be mentioned that the execution does not satisfy the process model, since the test result of at least one action is invalid. Therefore, an inconsistency has been identified during the software process execution.

| *Action: RS01* | *Action: RS02* | *Action: Condition1* |
|---|---|---|
| **Start date:** 2013/04/01–9:00<br>**End date:** 2013/04/01–18:00<br>**Status:** Finished | **Start date:** 2013/04/02–9:00<br>**End date:** 2013/04/02–10:00<br>**Status:** Finished | **Start date:** 2013/04/02–10:00<br>**End date:** 2013/04/02–10:00<br>**Status:** No |
| *Action: RS03* | *Action: RS10* | |
| **Start date:** 2013/04/02–10:00<br>**End date:** 2013/04/02–15:00<br>**Status:** Finished | **Start date:** 2013/04/03–9:00<br>**End date:** 2013/04/03–10:00<br>**Status:** Finished | |

(a) Execution of the Requirement Engineering Process

| Test Results | | | | |
|---|---|---|---|---|
| **Action**: RS01<br>**Result:** Valid | **Action**: RS02<br>**Result:** Valid | **Action:** Conditional1<br>**Result:** Valid | **Action:** RS03<br>**Result:** Valid | **Action:** RS10<br>**Result:** Invalid |

(b) Result of the testing process

**Fig. 27.4** Example of an execution of a software process and the evaluation of its conformity

This example represents a negative test that is useful to identify inconsistencies in the process model. However, positive tests, which cover the different paths of the process model, must be considered to measure the path coverage that allows evaluating the level of acceptance of the process model.

## 27.6 Conclusions and Future Work

Nowadays, business processes constitute a very important asset for organizations in general and software-intensive organizations are not an exception. Defining, documenting and managing these processes require techniques and tools to support their application and maintenance. Nevertheless, these are not enough. Once the process is defined and deployed in an organization, it is mandatory to verify that it must be executed as it was defined. Usually, verification actions are mainly manual activities demanding an important effort and time cost. That is something that organizations assume as quality cost, although they need to reduce it as much as possible.

This paper presents a solution for automating this activities founded on a model-based approach and on the application of software testing techniques. This solution is offered by a metamodel and illustrated by a concrete solution named NDTQ-Framework.

This approach will improve in different ways as a future work. Firstly, we are working on obtaining more empirical data about its use in software-intensive organizations. This would allow establishing our approach as a continuous improvement mechanism according to the comments discussed in Sect. 27.3.2.2, which is widely recommended in many standards and good practices manuals. Getting data from these testing techniques may allow organizations to identify problems and make easier decision-making processes. Secondly, this approach can support an orchestration mechanism that conceives NDTQ-Framework as a whole solution for process definition and execution.

Finally, the approach can be used as assessment model in formal appraisals using a reference model like CMMI or ISO 15504, by helping obtain evidences as requested by these standards and models.

**Acknowledgements** This work has been supported by the projects TEMPROS (TIN2010-20057-C03-02) and Test4DBS (TIN2010-20057-C03-01) of Ministerio de Educación y Ciencia, Spain, and the NDTQ-Framework project (TIC-5789) of the Junta de Anda-lucía, Spain.

## References

1. Bendraou R, Gervais MP (2007) A framework for classifying and comparing process technology domains. In: International conference on software engineering advances, ICSEA 2007. Cap Esterel, French Riviera, France. IEEE, pp 5–5
2. Chrissis MB, Konrad M, Shrum S (2003) CMMi. Addison-Wesley, Boston
3. Humphrey WS (1989) Managing the software process (Hardcover). Addison-Wesley Professional, Reading
4. Henderson-Sellers B, Gonzalez-Perez C (2005) A comparison of four process metamodels and the creation of a new generic standard. Inf Softw Technol 47(1):49–65
5. Makinen T, Varkoi T, Soini J (2007) Integration of software process assessment and modeling. In: Portland International Center for Management of Engineering and Technology, Portland, PICMET 2007. IEEE, pp 2476–2481
6. Coletta A (1997) Process assessment using spice: the assessment activities. SPICE: The theory and practice of software process improvement and capability determination, Computer Society Press. IEEE, pp 99–122
7. ISO/IEC (2005) ISO 15504-1 information technology - process assessment - part 1 concepts and vocabulary
8. Ardagna D, Ghezzi C, Mirandola R (2008) Rethinking the use of models in software architecture. In: Quality of software architectures. Models and architectures. Springer, Berlin, pp 1–27
9. Schmidt DC (2006) Model-driven engineering. Computer (IEEE Computer Society) 39(2):25
10. Osterweil L (1987) Software processes are software too. In: Proceedings of the 9th international conference on software engineering. IEEE Computer Society Press, Los Alamitos, pp 2–13
11. Henderson-Sellers B, Stallinger F, Lefever B (2002) Bridging the gap from process modelling to process assessment: the oospice process specification for component-based software engineering. In: Euromicro conference, Dortmund, Germany, 2002. Proceedings. 28th. IEEE, pp 324–331
12. ISO/IEC (1995) ISO 12207 Information technology - software lifecycle processes
13. Hamann D (2006) Towards an integrated approach for software process improvement: combining software process assessment and software process modeling. Fraunhofer-IRB-Verlag
14. Lepasaar M, Makinen T (2002) Integrating software process assessment models using a process meta model. In: IEEE international engineering management conference, 2002. IEMC'02, Cambridge, UK, vol 1. IEEE, pp 224–229
15. Makinen T, Varkoi T (2008) Assessment driven process modeling for software process improvement. In: Portland International conference on management of engineering & technology, Portland, PICMET 2008. IEEE, pp 1570–1575
16. IEEE Standards Software Engineering (2004) Vol. 2. Process Standards. IEEE Std. 1012 2004. Standard for Software Verification and Validation Plans

17. ISO/IEC (2006) ISO/IEC 24765 software and systems eng. vocabulary
18. Zhu H, Hall PA, May JH (1997) Software unit test coverage and adequacy. ACM Comput Surv 29(4):366–427
19. OMG (2012). UML (unified modeling language). Last accessed 01-2013. http://www.omg.org/spec/UML/
20. OMG. BPMN, Business process modeling notation, version 2.0. Last accessed 01-2013. http://www.omg.org/spec/BPMN/2.0/
21. García-Borgoñon L, García-García JA, Ortega MA, Escalona MJ (2012) Software process management: a model-based approach. In: Proceedings of the 21st international conference on information systems development (ISD)
22. Escalona MJ, Aragon G (2008) Ndt. A model-driven approach for web requirements. IEEE Trans Softw Eng 34(3):377–390. http://doi.ieeecomputersociety.org/10.1109/TSE.2008.27
23. ITIL (Information Technology Infrastructure Library). Last Accessed 01-2013. http://www.itil-officialsite.com
24. ISO/IEC (2013) ISO/IEC 29119 software engineering – software testing standard. International Organization for Standardization
25. ISO/IEC (2008) ISO 9001:2008 quality management systems - requirements. International Organization for Standardization
26. Project Management Institute (2008) A guide to the project management body of knowledge (PMBOK guide)
27. ISO/IEC (2005) ISO 27001 information technology - security techniques - information security management systems - requirements

# Chapter 28
# A Model-Based Approach to Develop Self-Adaptive Data Visualizations

**Juan F. Inglés-Romero, Rober Morales-Chaparro, Cristina Vicente-Chicote, and Fernando Sánchez-Figueroa**

**Abstract** Nowadays, a growing number of applications make use of data visualization techniques to effectively present information to the users. The complexity of these techniques increases as the number of visualization devices, the amount and variety of the input data sources, and the user- and application-specific requirements grow and change more and more rapidly every day. This work illustrates how the combined use of model-based and adaptive software development techniques eases the design and implementation of the data visualization systems, providing them with the ability to dynamically adapt themselves to changing situations and requirements. As a core part of the paper we present a case study to help us introduce the main elements of the proposal and to discuss its benefits and limitations.

**Keywords** Data visualization • Adaptive systems • Visualligence • VML

## 28.1 Introduction

Data visualizations provide a valuable means to communicate information to users in a wide variety of application domains. A good graphical representation of the data allows the users to easily perceive magnitudes, tendencies, relationships, etc., in a fast and effective way. Conversely, a bad graphical representation can not only deceive the users expectations, but also lead them to misunderstand the data and, as a consequence, to make bad decisions on them.

---

J.F. Inglés-Romero (✉)
Departamento de Tecnologías de la Información y las Comunicaciones,
Universidad Politécnica de Cartagena, Edificio Antigones, 30202 Cartagena, Spain
e-mail: juanfran.ingles@upct.es

R. Morales-Chaparro • C. Vicente-Chicote • F. Sánchez-Figueroa
Quercus Software Engineering Group, Universidad de Extremadura,
Avda. de la Universidad S/N, 10003 Cáceres, Spain
e-mail: robermorales@unex.es; cristinav@unex.es; fernando@unex.es

M. José Escalona et al. (eds.), *Information System Development: Improving Enterprise Communication*, DOI 10.1007/978-3-319-07215-9_28,

Nowadays, the information society we live in provides us with a growing number of all kinds of data. These data are frequently updated, even in real-time. Moreover, the marketplace offers a wide variety of visualization devices (e.g., smartphones, PDAs, smartTVs, etc.), most of which include high quality screens that allow us to easily and intuitively interact with all kinds of applications, even via Web. In this diverse and changing environment, in which visualization applications gain an increasingly important role, providing them with the ability to adapt themselves to changing data, new devices, and different user profiles, may be essential not only to improve their performance, but also to improve the users' experience.

This work illustrates how the combined use of model-based and adaptive software development techniques eases the design and implementation of the data visualization systems, providing them with the ability to dynamically adapt themselves to changing situations and requirements. This work continues, integrates and extends some of our previous works related to: (1) the modeling of data visualization systems using *visualligence* [1]; and (2) the modeling of (self-) adaptive systems using *VML* (*Variability Modeling Language*) [2, 3].

The rest of the paper is organized as follows. Section 28.2 introduces the case study that will be used throughout the paper to discuss the proposal. Sections 28.3 and 28.4 describe the technologies used to model data visualization systems (*visualligence*) and their adaptation (*VML*), respectively. Section 28.5 details the implementation of the case study, and Sect. 28.6 reviews the related works. Finally, Sect. 28.7 draws some conclusions and lessons learned, and outlines our future research plans.

## 28.2 A Motivating Case Study

This section introduces the case study that will be used throughout the paper to illustrate our proposal. We will consider data visualizations based on bar charts. These visualizations will be adapted at run-time according to: (1) the features of the device in which the graphic is going to be displayed and, in particular, depending on the size of its screen; (2) the visual competency of the user, taking into account his visual acuity and if he is color-blind or not; and (3) the features of the data to be displayed, some of which can change at run-time (e.g., the number of categories or the values, i.e., the number of bars in the chart or their height). It is worth noting that, although this work is focused on bar charts, these three elements (device, user profile, and data) are sources of variability common to all data visualizations.

Regarding the features that can be adapted at run-time in the graphic, we consider the following ones: (1) the color palette; (2) the dimensions and spatial relationships among the elements of the graphic, e.g., the width of the bars and the padding between them; (3) the orientation of the graphic (horizontal or vertical bars); and (4) the selected visualization scheme that will determine the details to be displayed and how they will be arranged.

In order to determine which configuration (color palette, dimensions, orientation and visualization scheme) is the most appropriate for a given situation (depending on the device, the user profile and the data to be displayed), the goodness of each configuration will be evaluated in terms of two properties: the expressivity and the tangling of the graphic. A graphic will be considered more expressive the more information and details it contains. For instance, a visualization scheme that displays the numeric value associated to each bar is considered more expressive than one not containing this information. Regarding the second property, a graphic is considered tangled in the extent its elements come together or overlap, offering a less clear information to the user. In this vein, for instance, in order to display a set of data in a particular screen, the selection of certain graphic orientation could enable the inclusion of more details in the graphic than the other, thus improving its expressiveness. However, for users with a low visual acuity, it might be better to use a simpler (i.e., less detailed) visual scheme in order to reduce the tangling. As it can be seen, in general, improving one of the properties implies worsening the other. Thus, finding the optimal configuration for each situation implies balancing these two properties rather than trying to optimize them individually.

In certain situations, some properties might be more relevant than others. For instance, the lower the visual acuity of the user, the clearer the graphic must be (i.e., the less tangled), even if it means displaying less information (it makes no sense to display a lot of details that cannot be perceived by the user). On the other hand, certain contextual conditions may impose restrictions on the system configuration. For example, for color-blind users, there must be a rule that forces the selection of an appropriate color palette for them. Our proposal revolves around the implementation of adaptive data visualizations using the following approach. First, at design-time, we need to model: (1) the type of graphic we want to use to display the data (in our case, bar charts), using platform-independent primitives (see Sect. 28.3); and (2) the adaptation policy, taking into account the system variability, its execution context, and the rules and properties that will guide its reconfiguration at run-time to cope with changing situations and requirements (see Sect. 28.4). Next, from these two models, we will generate an implementation for the specific visualization platform planned to be used for displaying the data. At run-time, the context (selected device, user profile, and features of the data being displayed) will be monitored and evaluated. When a relevant change in the context is detected, the adaptation process will be launched. This process will select the optimal system configuration for the current situation by applying the rules and optimizing the properties described in the adaptation model (see Sect. 28.5).

## 28.3 Data Visualization

### *28.3.1 Background*

The main goal of data visualizations is to communicate information clearly and effectively through graphical means [4, 5]. The variety of data sources and visualization devices, the multidimensional nature of the data, and the continuous evolution of the user requirements make data visualization a complicated task which, nowadays, attracts an increasingly growing number of research groups [6].

```
module "barchart"
input data       : [{String, Integer}]
input height     : Integer
input width      : Integer
input margin     : Integer
input spacelabel : Integer
input barsize    : Integer
input padding    : Integer
input colors     : [Color]
loop BARS_AND_LABEL for data { |d,i|
   current_x <- ( padding / 2 ) <- current_x + barsize + padding
   output <- Rectangle {
       x <- current_x
       y <- height + margin
       height <- map( d[1], 0, height )
       width <- barsize
       color <- colors[ i \% colors.size ]
   }
   output <- Text { |d|
       x <- current_x + ( barsize / 2 )
       y <- height + margin + ( spacelabel / 2 )
       fixpoint <- ( center, center )
       text <- d[0]
   }
}
...
```

**Listing 28.1** Excerpt of the barchart pattern in visualligence

However data visualizations have been often considered second-class citizens from the Software Engineering point of view. This fact has led to bad praxis when developing data visualization applications [7]. Apart from some shy efforts aiming at automating the development of visualization systems [8], and some brilliant but technology dependent implementations [9], there are just a few proposals that apply the principles of Software Engineering to improve these applications.

With the aim of applying Software Engineering principles to data visualization we introduced ***visualligence*** in [1], a method that proposes a declarative and data-driven flow that allows experts in different domains to develop visualization patterns at a high level of abstraction. These patterns can be reused among different domains and code for different graphic implementation platforms can be generated from them. ***Visualligence*** incorporates well-known practices coming from Model-Driven Software Development (MDSD), proposing a Domain Specific Language (DSL) with not only a textual syntax but also a graphical one.

### 28.3.2 Graphic Modeling with Visualligence

Listing 28.1 shows an excerpt of the pattern used to model bar graphics (*barchart*) in ***visualligence***. The model includes the description of bars and labels, while other elements, such as axis or legends, have been omitted for lack of space.

The model in Listing 28.1 shows two of the most interesting features of ***visualligence***. On the one hand, one can observe that the language makes it possible to iterate over data, generating an output that it is not necessary to serialize. The principal iterator (@10–25) computes the bars (@12–18) and the labels (@19–24). It is assumed that these elements will be correctly rendered (independently of the visualization platform used) and that, in case of changes in data, all the attributes depending on those data will change accordingly. On the other hand, ***visualligence*** allows the use of linked hooks, as it is the case of line 11 in Listing 28.1. This makes it possible to initialize variables with a default value that can be updated in each iteration. In our example, the position of each bar in the X axis is being calculated as an increment of the position of the previous bar in the same axis.

### 28.3.3 Benefits and Limitations of Visualligence

***Visualligence*** offers a high level modeling language that gathers the common parts of different visualization platforms (calculation of spatial dimensions, generation of graphic objects based on data, etc.). This makes ***visualligence*** platform-independent. Thus, from a given visualization pattern one can obtain executable code for different platforms: Java for Android, Obj-C for iPhone, JavaScript for a Web browser, etc. However, ***visualligence*** does not provide, in a native way, a mechanism for modeling the adaptation of visualizations at run-time based on the context. As a consequence, the information related to the user profile, the rendering device or the data itself (often generated dynamically) is not taken into account.

Instead of letting the designer (that does not know the context a priori) or the user (that is not interested in improving our software) the decision on tasks such as the best color palette to use, the best width for the bar or the best graphic orientation, we propose modeling the adaptation process so that, at run-time and based on the context, the values for these parameters will be automatically computed in order to obtain an optimum visualization for the user.

Next section details how the use of ***VML*** (*Variability Modeling Language*) makes it possible to model not only the variability of a system but also the rules and properties that must be optimized to make itself adapt according to the context.

## 28.4 Modeling Variability with VML

***VML*** (*Variability Modeling Language*) allows modeling the adaptations of a system to improve its performance in a changing context. The current version of ***VML*** has been developed using a MDSD approach and it is supported by a textual editor created with *Xtext* [10]. This editor includes some advanced features such as syntax coloring and checking, and a completion assistant. Using the example presented in Sect. 28.2, the following subsections describe the basic elements of the language and the execution semantics of the ***VML*** models.

### 28.4.1 Modeling Primitives Available in VML

***VML*** offers, among others, primitives for modeling: (1) the variation points of the system; (2) the context variables; and (3) a set of rules and properties that allows calculate (1) based on (2). When creating a ***VML*** model, it is necessary to define adequately the variation points of the system (`varpoint`). In the same way that it occurs in Dynamic Software Product Lines [11], the variation points in ***VML*** represent software elements that can have different realizations (variants). Choosing one variant or another at run-time will lead to different configurations of the system. This way, the variation points will determine the decision space of ***VML***, that is, what parts of the system are susceptible of adaptation. Variation points, as well as the rest of the ***VML*** variables, are associated to a certain data type. ***VML*** gives support to different data types: enumerated, number ranges and boolean.

The model shown in Listing 28.2 includes seven variation points (@1—7), divided into two groups. The first group includes those variation points related to the 2D geometry of the graphic (Fig. 28.1):

1. *barsize* (integer): width of the bar,
2. *padding* (integer): space between bars,
3. *margin* (integer): margin between the graphic and the window border, and
4. *labelSpace* (integer): space reserved for showing the labels.

The second group includes the next three variation points related to the user and the execution environment:

5. *colorPalette* (enumerated): color palette used in the graphic. Two different variants are defined: FULL_COLOR and COLORBLIND_SAFE. The second one is useful for color-blind users although the fact of being monochromatic makes it less expressive than the first one.
6. *orientation* (enumerated): graphic orientation. Two different variants have been considered: VERTICAL and HORIZONTAL.
7. *visibilityScheme* (enumerated): five variants have been considered (SCHEME1,..., SCHEME5) associated to the five visual schemes shown in Fig. 28.2.

Once the variation points have been defined, we need to specify the context variables (`context`). These variables allow us to identify in which situations the system must be adapted, i.e., when its variation points need to be re-calculated. The five context variables considered in the case study (see Listing 28.2, @8—12), are:

1. *colorBlindness* (boolean): indicates whether the user is color-blind or not.
2. *visualAcuity* (integer from 0 to 100): visual acuity of the user
3. *availableWidth* and *availableHeight* (integer): size in pixels of the window in which the graphic will be displayed, and
4. *dataToDisplay* (integer): number of data categories to be displayed in the graphic.

These variables refer to three kinds of related contexts: user profile (1 and 2), visualization device (3), and features of the data to be displayed (4).

```
varpoint barsize : number [1:1:1000]
varpoint padding : number [1:1:1000]
varpoint margin : number [1:1:1000]
varpoint labelSpace : number [1:1:1000]
varpoint colorPalette : enum {FULL_COLOR, COLORBLIND_SAFE}
varpoint orientation : enum {VERTICAL (0), HORIZONTAL (1)}
varpoint visibilityScheme : enum {SCHEME1, ..., SCHEME5}

context colorBlindness : boolean
context visualAcuity : number [0:1:100]
context availableWidth : number
context availableHeight : number
context dataToDisplay : number

rule paletteRule: colorBlindness = true
  implies colorPalette = @colorPalette.COLORBLIND_SAFE
...
var remainingSpaceInAxis : number :=
  if (orientation = @HORIZONTAL)
     availableWidth + 2 * margin - labelSpace
  else availableHeight + 2 * margin - labelSpace

property expressivity : number maximizes {
  weight w := 0.5 * visualAcuity
  objective {
     case visibilityScheme = @visibilityScheme.SCHEME1 :
          obj := remainingSpaceInAxis
     case visibilityScheme = @visibilityScheme.SCHEME2 :
          obj:= remainingSpaceInAxis + 10
     case visibilityScheme = @visibilityScheme.SCHEME3 :
          obj:= remainingSpaceInAxis + 15
     case visibilityScheme = @visibilityScheme.SCHEME4 :
          obj:= remainingSpaceInAxis + 25
     default :
          obj:= remainingSpaceInAxis + 50
  }
}

property tangling : number minimizes {
  weight w:= 100 - 0.5 * visualAcuity
  objective { ... }
}
```

**Listing 28.2** Excerpt of the ***VML*** model designed for the case study

At this point it is necessary to establish how variation points are set in terms of the context variables. This is done in ***VML*** by means of properties (`property`) and ECA (*Event-Condition-Action*) rules (`rule`).

On the one hand, the ECA rules define direct relations between context variables and variation points. As shown in Listing 28.2, the left part of the rules establishes the triggering condition (values that must contain one or more context variables for the rule to be executed). The right part of the rule establishes the value that will get a concrete variation point under that situation. For example, the rule *paletteRule*

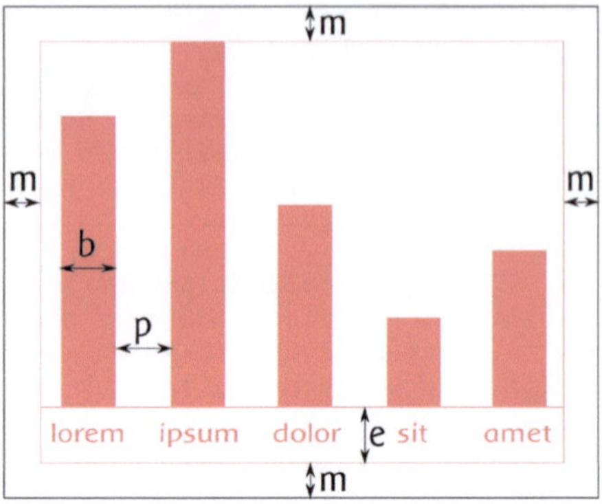


**Fig. 28.1** Geometric variation points: *b* (barsize), *p* (padding), *e* (labelSpace), *m* (margin)

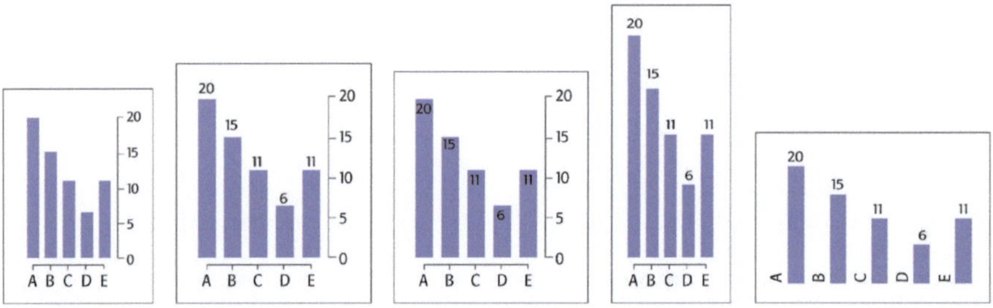

**Fig. 28.2** Visual diagrams considered

(@13–14) indicates that if the user is color-blind (*colorBlindness* = *true*), then the variation point *colorPalette* will be set to *COLORBLIND_SAFE*.

On the other hand, the properties specify system features that must be optimized (minimized or maximized). Each property is characterized by two functions. The first one defines the property in terms of variation points (objective function to be optimized, `objective`), while the second one defines the importance of the property in a given context (factor that weights the objective function, `weight`). It is worth noting that property definitions can be characterized using technical specifications, simulations or empirical data, while property weights are characterized in a more subjective way, depending on the designer experience.

In the ***VML*** model shown in Listing 28.2, two properties are defined: *expressivity* and *tangling*. The first property aims at maximizing the visualization expressiveness (i.e., the amount of information and the level of detail provided by the graphic), while the second one aims at minimizing the visual tangling generally produced by overlapping elements. This way, those visualization diagrams with more information have a higher score in the objective function of the *expressivity* property. Regarding the importance of the *tangling* property, it increases for users with a lower visual acuity. For these users it is more important to minimize the tangling of the visualization (i.e., to display a clear graphic) than to visualize all its details. Properties defined in ***VML*** models usually collide. In the case study, it is obvious that minimizing the tangling and maximizing the expressiveness are conflicting requirements. Thus, solving such optimization problems in ***VML*** must balance these requirements, looking for an overall optimum rather than optimizing each property individually. Next section, focused on the execution semantics of the ***VML*** models, details this optimization process.

### *28.4.2 Execution Semantics of the VML Models*

The execution semantics of the ***VML*** models is related to solving an optimization problem with constraints. ***VML*** uses syntactic sugar for describing the global objective function that optimizes the variation points and that, as a result, allows adapting the system. This global function is obtained adding the definitions of each property, once they have been weighted with their corresponding weights. Besides, the ECA rules impose constraints that must be satisfied.

With the aim of solving the optimization problem, ***VML*** models are transformed into *MiniZinc* [12] code. *MiniZinc* is a constraint programming language that can be executed by several constraint solvers. In our case G12 Constraint Programming Platform [13] is used. The adopted MDSD approach allows us to generate different artifacts from the ***VML*** models (e.g., by applying different model-to-model or model-to-text transformations). Thus, it would be possible to obtain implementations for constraint optimization languages other than *MiniZinc*. It would also be possible to obtain specifications in a formal language with the possibility of simulating or verifying the ***VML*** models [14], e.g., using model checking.

## 28.5 Details of the Implemented Solution

In the implemented solution, the visualization process is performed at the client side, while the adaptation process is executed at an external server (see Fig. 28.3). This architecture has been selected as, currently, it is not feasible yet to execute *MiniZinc* at the client side. This decision introduces certain limitations regarding the agility of the adaptation process (latency in the communications with the server)

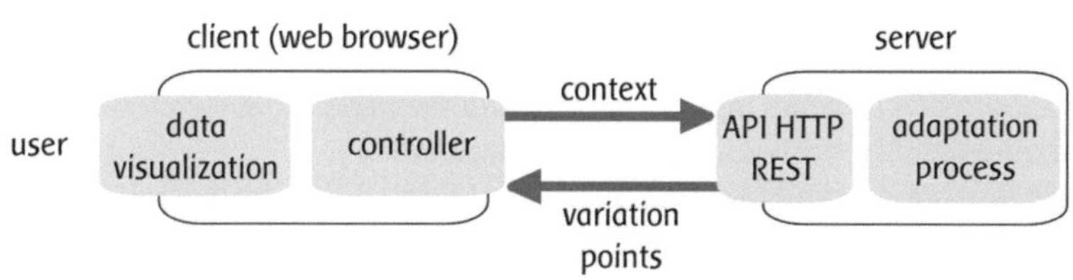


**Fig. 28.3** Architecture of the implemented solution

and scalability of the solution. However, it is important to note that the adaptation process is not continuous, but only executed when a relevant change in the context is detected. Thus, the performance of the overall system execution is not compromised.

The API built upon *MiniZinc* is a simple HTTP REST interface. With a single method, any client can remotely invoke the adaptation process. Usually, the client requesting the adaptation is the visualization process that asks for the best possible configuration of its parameters based on the current context. The API receives the current context as input (list of key-value pairs), and returns the configuration of the variation points (JSON object, serialized in the *MiniZinc* program).

Data visualization is built based on patterns. These patterns do not know anything about data sources or about the adaptation process. All the visualizations displayed in the same page are managed by a single controller. This controller is in charge of: (1) initializing the visualizations; (2) executing the adaptation process whenever a change in the context is registered; and (3) re-configuring the visualizations according to the result of the adaptation process.

It is important to highlight that, as a benefit of adopting the principle of *Separation of Concerns*, the visualization system is fully independent from the adaptation process. Thus, in case of failure of the server performing the adaptation, the system will continue displaying the data correctly, although in a sub-optimal way.

## 28.6 Related Works

There exist some works that evaluate the performance of different visualization techniques with the purpose of building adaptive visualizations according to the user profile. For example, Toker et al. [15] study the impact of four factors (perception speed, verbal memory, visual memory and user experience) in the effectiveness of two different visualization techniques: bar and radar diagrams. Goldberg et al. [16] propose the evaluation of different kinds of visualizations by following the user eyes. The empirical results provided by these works may improve the adaptation modeling in VML.

Voigt et al. [17] propose VISO (*VISualization Ontology*), an ontology that provides a vocabulary to annotate visualization components. These annotations allow recommending the most adequate visualization technique when a new data source appears. Moreover, VISO gives support to the configuration of user preferences and formalizes, by means of facts and ontology rules, the knowledge of visualization experts. This work allows somehow to model the variability of data visualizations. However, the fact of being highly coupled with the underlying technology makes it less flexible than the proposal presented in this paper.

Rodriguez-Pardo et al. [18] present a technique to adapt the color palette used in visualizations with the purpose of making it available for color-blind. Although the techniques used in [18] for selecting the appropriate palette are more complex than those used in our proposal, they could be easily integrated in our VML models.

One of the main constrains for adapting data visualizations comes from the plethora of rendering devices with different screens and sizes that are available nowadays in the market. Fuchs et al. [19] propose an adaptation strategy to automatically adjust the visualizations according to the features of the device. This work is focused on dispersion diagrams. In order to determine if, at a given moment, it is necessary to adapt the graphic, the visual effectiveness is estimated based on certain characteristics of the data and the size of the screen. When the visual efficacy of the graphic is poor for a given device and data, then an adaptation process is executed. In contrast to our proposal, this work does not take into account the user in the adaptation process. Moreover, unlike VML, it does not allow modeling the variability according to the domain and the application.

Finally, Rosenbaum et al. [20] present a generic proposal that allow adapting visualizations according to device requirements, user profiles and data. Considering that the visualization is pre-processed in a server, the technical foundations of this proposal are based on truncating the data flows in a progressive way. This allows offering the users the same graphic with more or less details, according to their requirements. In the same way that [19], this work offers less flexibility than our proposal in the sense of modeling and adjusting the adaptation according to the particularities of each application. Moreover, the adoption of a MDSD approach eases the simulation, validation and verification of the adaptation, making it possible to detect and correct existing problems in an early stage.

## 28.7 Conclusions and Future Work

In this work we have described our experience in modeling and implementing adaptive data visualization systems. It is worth noting that it has been possible to apply the advantages of software adaptation to the field of data visualization by ***combining two different approaches based on models: visualligence and VML***. The former is used for defining the visualization patterns, while the latter enables the specification of the system adaptation. The adoption of a MDSD approach enables platform-independent models from which we can obtain implementations for different platforms using model transformations. However, ***modeling some aspects such as expressiveness or tangling in VML has not been an easy task***. Both properties are quite subjective and it is not easy to find a precise definition. To alleviate this problem, and aligned with [15, 16], we consider to direct an empirical experiment with groups of users. This experiment will allow us to characterize and measure the goodness of different adaptations applied to visualizations.

Regarding the implementation performed for the case study, it is worth noting that the application of the ***Separation of Concerns*** principle allowed us to uncouple the visualization logic and the adaptation logic. Among other benefits, this separation improves the maintainability and extensibility of the solution. Although the prototype here presented only has the aim of demonstrating its viability, ***the implemented solution could be improved, allowing the adaptation process to execute***

***also at client side***. For this purpose different alternatives could be considered: using a *Constraint Solver* implemented in JavaScript or compiling to JavaScript some of the available Constraint Solvers using, for example, asm.js [21]. Asm.js is a strict subset of JavaScript, ready to be executed in browsers and that has been proved with success compiling from C and C++.

In the future we plan to continue exploring the benefits of applying adaptation mechanisms to data visualizations. Particularly, we plan to conduct an evaluation process with groups of users with two purposes. On the one hand, improving the adaptation modeling and, on the other hand, obtaining a measure of improvement in user experience and visualization efficiency.

**Acknowledgements** This work has been funded by MICINN under the grant MIGRARIA (TIN201 1-27430). Juan F. Inglés-Romero thanks Fundación Séneca-CARM for a grant (Exp. 15561/FPI/10). Rober Morales-Chaparro thanks FEDER + FSE for a grant.

## References

1. Morales-Chaparro R et al (2012) Desarrollo dirigido por modelos de visualización de datos para la Web. In: Jornadas de Ingeniería del Software y Bases de Datos
2. Inglés-Romero JF et al (2012) Dealing with run-time variability in service robotics: towards a DSL for non-functional properties. In: 3rd Int'l workshop on domain-specific languages and models for robotic systems
3. Lotz A et al (2013) Managing run-time variability in robotics software by modeling functional and non-functional behavior. In: 9th Int'l conf. on exploring modelling methods for systems analysis and design (EMMSAD), pp 441–455
4. Bertin J (1999) Semiologie Graphique: Les Diagrammes—Les réseaux—Les Cartes. Editions de l'Ecole des Hautes Etudes en Sciences, Paris
5. Tufte ER, Howard G (1983) The visual display of quantitative information. Graphics Press, Cheshire
6. Stanford Visualization Group. http://vis.stanford.edu
7. Few S (2004) Show me the numbers: designing tables and graphs to enlighten. Analitics Press, Berkeley
8. Mackinlay J (1986) Automating the design of graphical presentations of relational information. ACM Trans Graph 5(2):110–114
9. Bostock M, Ogievetsky V, Heer J (2011) D3: data-driven documents. IEEE Trans Vis Comput Graph 17(12):2301–2309
10. Xtext. http://www.eclipse.org/Xtext/
11. Hallsteinsen S et al (2008) Dynamic software product lines. Computer 41(4):93–95
12. Nethercote N et al (2007) MiniZinc: towards a standard CP modelling language. In: Proceedings of the 13th Int'l conference on principles and practice of constraint programming (CP), pp 529–543
13. G12. http://www.g12.csse.unimelb.edu.au
14. Inglés-Romero JF, Vicente-Chicote C (2013) Towards a formal approach for prototyping and verifying self-adaptive systems. In: 1st Int'l workshop on variability support in information systems (VarIS), pp 432–446
15. Toker D et al (2012) Towards adaptive information visualization: on the influence of user characteristics. In: Proceedings of UMAP' 12, pp 274–285

16. Goldberg JH, Helfman JI (2010) Comparing information graphics: a critical look at eye tracking. In: 3rd BELIV workshop: beyond time and errors: novel evaluation methods for information visualization, pp 71–78
17. Voigt M et al (2012) Context-aware recommendation of visualization components. In: Proceedings of the 4th int'l conference on information, process, and knowledge management, pp 101–109
18. Rodríguez-Pardo CE, Sharma G (2011) Adaptive color visualization for dichromats using a customized hierarchical palette. In: Proceedings of the SPIE-IS&T electronic imaging
19. Fuchs GA et al (2009) Device-based adaptation of visualizations in smart environments. In: Workshop collaborative visualization on interactive surfaces
20. Rosenbaum R et al (2011) A flexible, low-complexity device adaptation approach for data presentation. In: Proceedings of visualization and data analysis
21. asm.js. http://asmjs.org/spec/latest/

# Chapter 29
# A Language Oriented Extension to Toulmin's Argumentation Model for Conceptual Modelling

**Sebastian Bittmann, Balbir Barn, and Tony Clark**

**Abstract** The constructive step of modelling must account for the specific requirements of various stakeholders. Further, the quality of a model in terms of goodness-of-fit, completeness or other aspects may vary, because of future, as yet unknown, requirements. Natural language underpins the process of modelling as it is the predominant form through which models are acquired, negotiated and agreed. Whilst a model defines system requirements, it does not capture the assumptions, discussions and negotiations that led to the requirements. The ability to access this information, which is lost by most development processes, may significantly improve the utility of models with respect to changing requirements and system maintenance. This paper proposes a form of 'literate modelling' that can be used to capture the steps in model development and is based on Toulmin's Argumentation model. The paper contributes the design of an argumentation modelling language and a set of rules for integrating multiple languages (or domains) with the Toulmin approach. The language and approach is applied to the domain of business process modelling.

**Keywords** Conceptual modelling • Meta-model • Toulmin's argumentation model • Argumentation

S. Bittmann (✉)
University Osnabrueck, Osnabrück, Germany
e-mail: sebastian.bittmann@uni-osnabrueck.de

B. Barn • T. Clark
Middlesex University, London, UK
e-mail: b.barn@mdx.ac.uk; t.n.clark@mdx.ac.uk

M. José Escalona et al. (eds.), *Information System Development: Improving Enterprise Communication*, DOI 10.1007/978-3-319-07215-9_29,

## 29.1 Introduction

Natural language is typically used for expressing system requirements of associated conceptual models [1]. The complexity of systems and the ambiguity of natural language lead to a high risk of misinterpretation. Additionally stakeholders, such as development or maintenance engineers, encounter a conceptual model that is the result of requirements analysis by third parties [2]. This introduces a knowledge management problem whereby understanding the dependencies between requirements and their relationships to other system artifacts is difficult or hard to establish because the original multiparty reasoning process, or argument, has been lost. Further, the requirements stated by stakeholders undergo a translation from natural language into the modelling language being used. Ultimately, stakeholders are restricted to the end product of a complex and highly structured reasoning process in order to ensure that the model has incorporated their requirements in a satisfactory manner [3].

The conceptual model and associated requirements alone are insufficient to address this problem. Further explanations are necessary: the design decisions and the underlying reasoning cannot be inferred from the conceptual model alone. A conceptual model does not spring into life fully formed. It is the result of a series of arguments whereby individuals use their personal knowledge and beliefs to propose facts about a system and to challenge propositions made by others. Therefore, the model is constructed as a result of incremental steps, each of which fully explain the reasoning behind individual extensions and modifications applied to an initial collection of assumptions. The ability to trace model elements back to the parry and thrust of their original designers becomes clear when we consider subsequent maintenance of a system, perhaps arising due to a change in its deployment context. The context would be part of the assumptions used in the reasoning leading to parts of the model (that might otherwise seem to be unrelated to the change).

If the original model designers are available then they can retrace their chains of reasoning and identify those parts of the model that require change. However without the designers this is likely to be very difficult or more likely impossible, since initial design decisions are missing and therefore the identification of the parts of a model that have to be reconsidered becomes difficult [4, pp. 3–4]. The theories underlying a system model are normally neither explicated within the model nor are they accessible without natural language. The knowledge, insight or theory that the modeller uses is a theory in the sense of Ryle [5]. That is, a person who has a theory knows how to do certain things and can support the actual doing with explanations, justifications and responses to queries. That insight or theory is primarily one of building up a certain kind of knowledge that is intrinsic to the modeller whilst any auxiliary documentation such as the model itself remains a secondary product. Therefore discussions constituted by arguments that are formulated with a well-defined language, are a possible and more mature way of disclosing these theories. Such a disclosure promises three valuable features for universes of discourse. First, the evolution of the supplemented knowledge that is created in a specific universe of discourse becomes traceable through explicit arguments. Second, the underlying design decisions can be explained with reference to a given argument. Third, the knowledge, which is used to support an argument and therefore a design decision, is explicitly depicted. Hence the actual theories that have led to a certain conceptual model are made explicit and explainable.

Our contribution is to propose argumentation as a basis for conceptual modelling and to show how the foundational work by Toulmin on argumentation can be formulated as a meta-model that can be integrated with system development methods and tools. This leads to the choice of the foundational work of Toulmin that is reviewed and related to requirements and conceptual modelling in Sect. 29.3. Our main contribution is the definition of the language ArgML given as a meta-model in Sect. 29.5 and the proposition is briefly validated in terms of a system design process in Sect. 29.6. Finally in Sect. 29.7 we outline future research directions arising from the work presented in this paper.

## 29.2 Related Work

The construction of conceptual models as part of the requirements analysis phase of system development is described in [1, 6] where the authors address the dialogue between stakeholders and modellers. One approach is to use different representations to foster an understanding of the various stakeholders and their domains [7, 8]. A notable omission in these proposals is a justification of design decisions and how to account for them in the specification process. In particular, clarification of relationships between design decisions is not addressed. Most approaches represent requirements in natural language that does not take into account the different stakeholder domains. Together with the lack of dependencies and justifications, this can lead to ambiguity, misinterpretation, incompleteness and conflict.

Elsewhere research has tried to include the actual requirements given by stakeholders in a conceptual model, or in the modelling process, so that it becomes possible to capture design decisions leading from a requirement to a solution [9–12]. Unfortunately these approaches concentrate primarily on the management and change of requirements; the need to translate from natural language to the modelling language remains.

Proposals from the domain of artificial intelligence try to introduce a certain logical language for formulating arguments [13–21]. These approaches enable derivations of contradictions in arguments within a discussion. However every proposition given in an argument has to be specified with the general purpose language defined by the approach. So it is necessary to translate the domain-specific arguments given by stakeholders to a general purpose representation. In order to generate any implications and finally the conceptual model, it is further necessary to translate the result of the discussion back to the actual problem domain. Any domain-specific knowledge is therefore lost.

Finally, recently developed approaches try to introduce the reflections of a model, by means of extended concepts of the respective modelling language [22, 23]. These approaches are aware of the participation by multiple stakeholders and the need for discussions and reasoning. Unfortunately the concepts, which they try to introduce, are unaware of the problems with natural language. Furthermore actual discussions are not fully supported, since the concepts only offer to include certain reasons for a given design decisions, but the ability for reasoning about a design decision is missing. Moreover as soon as the reasons are stated, there is no facility to challenge them.

## 29.3 Toulmin's Argumentation Model

Toulmin's Argumentation Model (TAM) [24, 25] was originally derived from jurisprudence and will be used for the construction and structured justification of the design decisions underlying a conceptual model. Toulmin's argumentation approach uses the following concepts argument, claim, ground, warrant, backing, rebuttal and qualifier, as shown in Fig. 29.1. In overview, a claim is justified by an argument that uses agreed grounds as a basis. A claim may be supported by facts expressed as a warrant and limited in scope using a qualifier. Repeated use of these structures and their relationships leads to a structured argument. A claim asserts a proposition based on some grounds that are used as support. If the listener is not convinced by the grounds for the claim, then further grounds may be provided. Finally, a warrant may be required to show that grounds are indeed a basis for the claim. Warrants correspond to practical standards or canons of arguments and are likely to incorporate elements of domain-specific knowledge. Contrary to a claim and a ground, which address one or more particular entities, a warrant abstracts from a particular instance and embodies a more general statement [24, p. 100].

Circumstances might mean that grounds and warrants are insufficient. In such cases, a backing may form part of further arguments to provide additional support. Backings are evidence that the warrant is reliable. Grounds, warrants and backings may be subject to qualifiers which are used to assert the relationship between

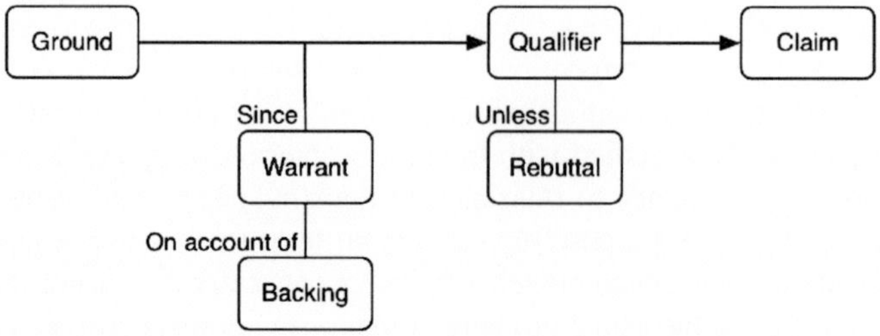


**Fig. 29.1** Toulmin Argumentation Model [24, p. 97]

grounds and claims. The concept of a qualifier serves to make a reflective statement about the validity of the corresponding claim. For example, a qualifier could define a claim as always valid, presumably valid or conditionally valid. It defines the conditions of the claim being valid when it was derived from its supporting warrants and the grounds related to the qualifier [24, p. 93]. There will be situations where a claim may not follow the grounds despite the production of warrants and backings; these situations are termed rebuttals. A rebuttal is a further claim that proposes statements that hold when the initial claim is invalid [24, p. 94]. Of note, is the importance of arguments that are field-invariant (holds for any domain) versus field-dependent (hold for a specific domain).

## 29.4 Toulmin Applied to Conceptual Modelling

Toulmin allows to classify and structure the various components of a discussion. However there are aspects impeding its application, especially in the domain of conceptual modelling. Firstly, TAM presented above relies on natural language to formulate arguments. Therefore there is a possibility for misinterpretation and misunderstanding of arguments by the different stakeholders. Secondly, it is difficult to derive the impact of a discussion. It may be that after a discussion certain convictions of certain theories are spread, but the actual impact on the universe of discourse and in particular on the actual knowledge cannot be specified [25]. There is the necessity to implicitly and explicitly reuse certain arguments in further discussions. Created knowledge should be used explicitly in further discussions to ground future claims. Furthermore the language should be affected by the knowledge and insights proposed by the arguments, thereby the implicit use of that knowledge through language would avoid the proposal of invalid propositions and reduce flaws in the process of modelling. These limitations require a number of TAM enhancements before it can be used as a vehicle for articulating theories relating to requirements and conceptual models. These include the use of languages which follows strict rules to allow for the unambiguous interpretation of propositions within a model instead of natural language.

## 29.5 The Argumentation Modelling Language

Before we can be in position to achieve the benefits of argument verification that are afforded by language formalization outlined in the rules of the previous section it is first necessary to better define the concepts underpinning TAM. One way to arrive at a better computationally useful understanding is to treat TAM as a language design activity, more specifically a model driven language design activity. Such an approach will utilise a suitable meta-language that can represent the various features of language such as the abstract syntax (defining the information structures

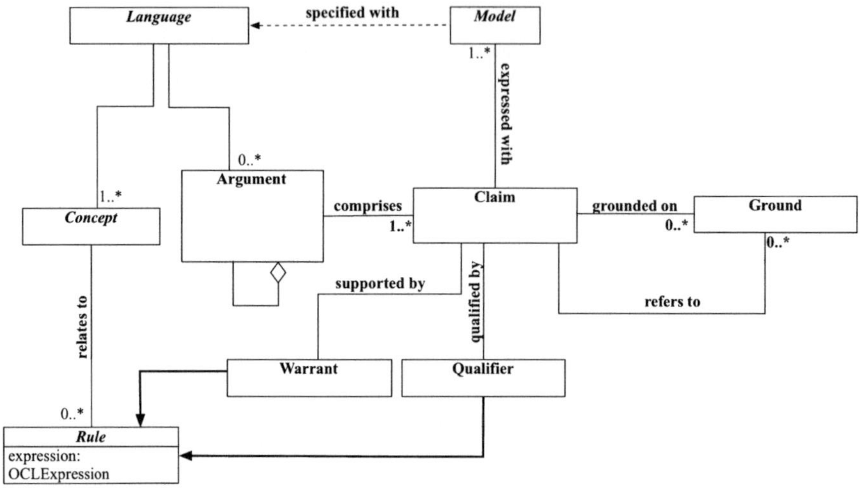


**Fig. 29.2** The Argumentation Modelling Language

to represent essential aspects of the language in a form suitable for machine processing); the concrete syntax (the appearance of the language on the screen or page) the various mappings necessary to relate the abstract syntax to the concrete syntax and the semantic domain [26].

Every argument is to be associated with a specific language that clearly defines the syntax and semantics to be used for representing claims, warrants, grounds and so on. Figure 29.2 shows the abstract syntax for the Argumentation Modelling Language (ArgML) drawn as a UML class diagram based on the language design approach used in [26, pp. 53–54]. Such a meta model of the abstract syntax provides a ready route for tooling using platforms such as MetaEdit+ [27]. Further, the use of meta-modelling techniques supports Domain Specific Languages (DSLs) that support the multiple expert stakeholders [28, pp. 26–27].

With the exception of a transformation and one removal, ArgML inherits every concept from the TAM except for the concept of a backing that is left for future work. Note that the concept of a rebuttal is not explicitly defined since any claim can rebuttal any other claim of the same argument. Direct linkage between claims by means of the concept of a rebuttal is not necessary, since all refutations of a claim are derivable based on the syntax of the underlying language.

The concept of a claim has two relations to the concept of a ground. First, grounds support a claim in order to justify them. Second, a ground refers to a claim of a previously settled argument, since any claim of a settled argument can be used to support further claims through a ground. However initially there is no knowledge embodied by the universe of discourse. The initial modelling language embodies the only available knowledge. This strictly results in the necessity for stating ungrounded claims in order to initiate the discussion. So the initial contributions to the knowledge embodied by the specific domain of discourse is based on tacit knowledge, which is formed by experience and is difficult to justify [29, p. 282]. ArgML is supported by additional mechanisms to express constraints embedded within claims, warrants and qualifiers by the use of the object constraint language (OCL) [30].

Based on this specification, the contribution of a discussion can be derived. The first derivation is a model that is constituted by the various claims. Every claim that is given in a universe of discourse must first be checked if it qualifies and then included in the model. Beginning with the very first assumptions given in the universe of discourse, incrementally every claim will be included to the derived model that qualifies based on its qualifier. The second derivation is a set of rules that will be embodied by a specific language. To derive these rules, respectively invariants, the various warrants and qualifiers that are interrelated through a contribution towards the same claim have to be selected. For each qualifier and warrant, which is used by the identical claim, an invariant can be derived. The invariant takes the form of an implication and states that if the underlying qualifier is satisfied, the warrant has to be satisfied as well. During an invariance check of a model, for every qualifier, whose expression was satisfied by the model, the corresponding warrant will be selected and checked against the model. The model satisfies the invariants if it satisfies all qualified warrants.

## 29.6 Application of the ArgML

This section provides a simple example based on a modelling language known as Event-driven Process Chain (EPC) [31]. Given a process model produced by a conventional design process, it is not possible to determine the underlying design decisions and convictions (that we refer to as theories about the model) that were used in its construction. Considering the example, the disclosure of design decisions in an 'order processing' business process becomes apparent through an argumentation based discussion. Figure 29.3 captures the underlying universe of discourse of such a process model. The respective model is depicted by means of the form already proposed by TOULMIN and which can be inferred from Fig. 29.1. Thereby, the various statements usually proposed in natural language are replaced by semi-formal model concepts.

The business process is initiated by an event called 'Order Received' that signals the receipt of an order. A claim is made as follows: After an order is received it will be processed by the function 'Process Order' that uses the billing system of the particular company. This claim is grounded on the initiating event and the fact that every event shall have at least a predecessor or a successor. After this claim, the argument proceeds in three different ways. First, it is claimed that: the function 'Process Order' needs to use the customer management system, since every function that uses the billing system needs to use the customer management system in order to access the required customer data (cf. top argument of Fig. 29.3). In the second branch, respectively the middle argument of Fig. 29.3, it is stated that: in

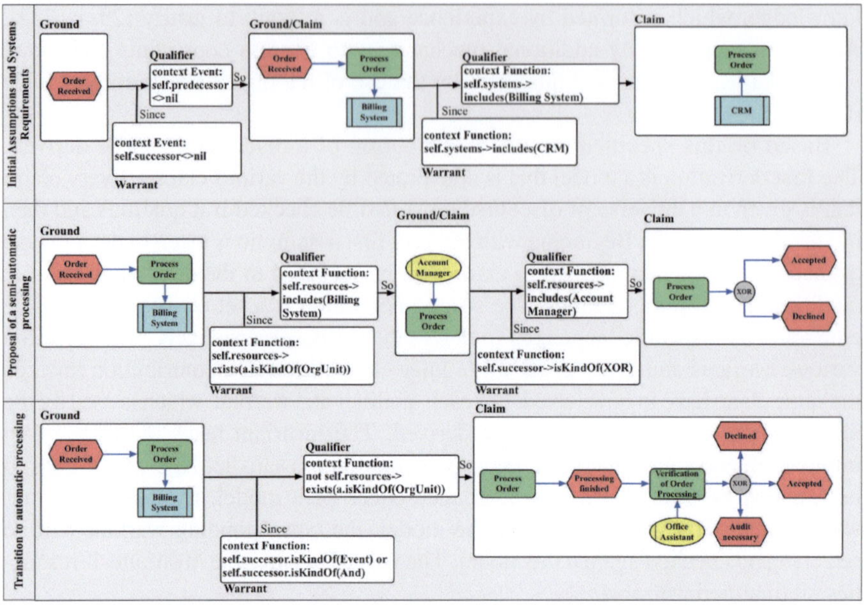


**Fig. 29.3** An exemplary universe of discourse regarding a business process

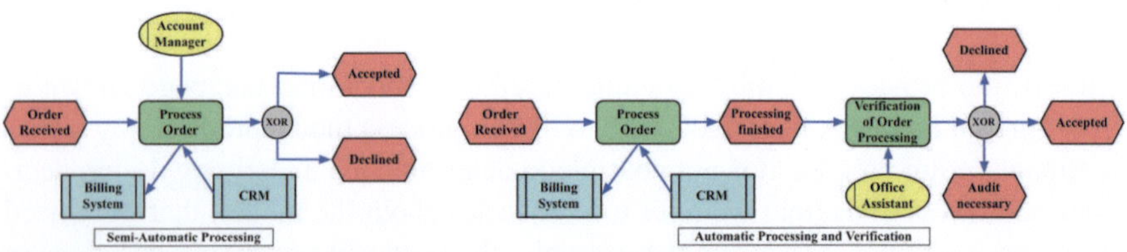


**Fig. 29.4** Derivation and evolution of conceptual models from discussions

order to use the billing system an organisational unit is required, namely the account manager, so the function 'Process Order' includes the account manager as a resource for its execution. Additionally with the acceptance of this argument, it becomes necessary that every function, which utilises the billing system, has to be operated by an organisational unit. Continuing the argument process, it was further stated that: if an account manager operates a function, then a decision has to be made following that particular function. Therefore during the processing of an order, the account manager has to decide about its acceptance. Disregarded the top branch of the discussion, a first business process model can be derived from these claims, illustrated on the left side of Fig. 29.4. During the evolution of this particular business process model, the account manager might become a superfluous component of order processing, because of the involvement of customer relationship management tools and IT infrastructure. As a result of the identification of redundancy within the process, it becomes necessary to check the automatically processed orders by an office assistant. Therefore the lower argument of Fig. 29.3 rebuttals the middle argument accordingly, which requires an adaptation, respectively an alignment, of the arguments and a continuation of the discussion. This alignment has to consider, e.g., where the validation has to be executed within the process. Furthermore there may exist design decisions that were decided with respect to the inclusion of an account manager. In the given example the decision succeeding the function 'Process Order' was justified with reference to the execution through an account manager. So with the exclusion of the account manager this design decision has to be reconsidered as well. Since the arguments of this path mainly rely on the consideration of an account manager, the middle argument of Fig. 29.3 needs to excluded or reconsidered. However, it is possible to infer which parts of the conceptual model can still be used based on the remaining arguments. So it is possible to distinguish between the valuable parts and the parts of a discussion that have to be excluded from the conceptual model due to the evolution of the business process. Furthermore, the language excludes the incorporated rules of the discussions as well, which were derived from the qualifier and warrants. With the elimination of the invalid arguments from the discussion it is possible again to specify a function that relies on the billing system, but does not rely on an organisational unit. So the enterprise wide rules for the application of the particular language have evolved as well.

## 29.7 Conclusion

This paper has presented a conceptual discussion for an extension to Toulmin's Argumentation Model to support the use of argumentation for addressing the need to reveal the theories embedded in conceptual models. Given the capstone nature of conceptual models to software development process, extracting such richness of design decisions has considerable implications for improvements to software practice. The paper has proposed formalising TAM into a language, ArgML using language engineering techniques as well as an initial abstract syntax and an approach to its use has been presented. Insights such as the recursive nature of arguments, the dependencies between arguments and how uses of the language end up extending the language itself have been described and illustrated with an appropriate example. However, the ArgML provides a frame of modelling and is applicable to any modelling language, which should be used for proposing claims and referring to grounds. Different applications scenarios evolve from the domain of enterprise modelling or enterprise architecture management as well as the conceptual modelling of software system, from which existing modelling languages can be used to formalise the different arguments and hence, the design decisions.

Future research will concentrate on a tighter integration of the ArgML with a language environment. The contribution of the ArgML towards domains, like enterprise modelling and enterprise architecture management, depends on the expressive power of the respective DSL. For now the expressiveness of the claims relies on the language, which is used for an argument. Moreover the expressivity of qualifiers and warrants relies briefly on the OCL. It is anticipated that the ArgML language will change as new requirements come to light as a result of further research.

## References


1. Hoppenbrouwers S, Proper HA, Van Der Weide TP (2005) A fundamental view on the process of conceptual modeling. In: Delcambre L, Kop C, Mayr HC, Mylopoulos J, Pastor O (eds) Proceedings of the 24th international conference on conceptual modeling, Lecture Notes in Computer Science, vol 3716. Springer, Berlin, pp 128–143. doi: 10.1007/11568322 9. http://dx.doi.org/10.1007/11568322 9
2. Lindland O, Sindre G, Solvberg A (1994) Understanding quality in conceptual modeling. IEEE Softw 11(2):42–49. doi: 10.1109/52.268955. http://ieeexplore.ieee.org/xpls/abs all.jsp?arnumber=268955
3. Shanks G, Tansley E, Weber R (2003) Using ontology to validate conceptual models. Commun ACM 46(10):85–89. doi: 10.1145/944217.944244. http://dl.acm.org/ft gateway.cfm?id=944244&type=html
4. Moran TP, Carroll JM (1996) Overview of design rationale. In: Moran TP, Carroll JM (eds) Design rationale: concepts, techniques, and use. Lawrence Erlbaum Associates, Mahwah, NJ, pp 1–20
5. Ryle G (1984) The concept of mind (1949). Hutchinson, London
6. van Bommel P, Hoppenbrouwers SJBA, Proper HA, van der Weide TP (2006) Exploring modelling strategies in a meta-modelling context. In: Meersman R, Tari Z, Herrero P (eds) On the move to meaningful internet systems 2006: OTM 2006 workshops, Lecture Notes in Computer Science, vol 4278. Springer, Berlin, pp 1128–1137. doi: 10.1007/11915072. http://dl.acm.org/citation.cfm?id=2148751.2148773 http://www.springerlink.com/index/10.1007/11915072
7. Mylopoulos J, Borgida A, Jarke M, Koubarakis M (1990) Telos: representing knowledge about information systems. ACM Trans Inf Syst 8(4):325–362. doi: 10.1145/102675.102676. http://dl.acm.org/citation.cfm?id=102675.102676
8. Rolland C, Proix C (1992) A natural language approach for requirements engineering. In: Loucopoulos P (ed) Advanced information systems engineering, Lecture Notes in Computer Science, vol 593. Springer, Berlin, pp 257–277. doi: 10.1007/BFb0035121. http://www.springerlink.com/content/d70229421277224t/
9. Ramesh B, Jarke M (2001) Toward reference models for requirements traceability. IEEE Trans Softw Eng 27(1):58–93. doi: 10.1109/32.895989. http://ieeexplore.ieee.org/xpl/articleDetails.jsp?arnumber=895989 http://ieeexplore.ieee.org/xpls/abs all.jsp?arnumber=895989
10. Rolland C, Prakash N (2000) From conceptual modelling to requirements engineering. Ann Softw Eng 10(1–4):151–176. doi: 10.1023/A: 1018939700514. http://hal.archives-ouvertes.fr/hal-00707077 http://dl.acm.org/citation.cfm?id=590679.590689
11. Egyed A, Grunbacher P (2002) Automating requirements traceability: beyond the record & replay paradigm. In: Proceedings 17th IEEE international conference on automated software engineering, pp 163–171. IEEE Comput Soc. doi: 10.1109/ASE.2002.1115010. http://ieeexplore.ieee.org/xpl/articleDetails.jsp?arnumber=1115010
12. Thomas O (2006) Management von Referenzmodellen: Entwurf und Realisierung eines Infomationssystems zur Entwicklung und Anwendung von Referenzmodellen. Logos, Berlin
13. Dalianis H, Johannesson P (1998) Explaining conceptual models—using Toulmin's argumentation model and RST. In: Third international workshop on the language action perspective on communication modelling, pp 131–140. http://citeseerx.ist.psu.edu/viewdoc/summary?doi=10.1.1.46.4918

14. Carbogim DV, Robertson D, Lee J (2000) Argument-based applications to knowledge engineering. Knowl Eng Rev 15(2):119–149. doi: 10.1017/S0269888900002058. http://dl.acm.org/citation.cfm?id=975740.975741
15. Prakken H (1995) From logic to dialectics in legal argument. In: Proceedings of the fifth international conference on artificial intelligence and law—ICAIL'95. ACM Press, New York, NY, pp 165–174. doi: 10.1145/222092.222230. http://dl.acm.org/citation.cfm?id=222092.222230
16. Kowalski RA, Toni F (1996) Abstract argumentation. Artif Intell Law 4(3–4):275–296. doi: 10.1007/BF00118494. http://www.springerlink.com/content/p782778515305p07/
17. Freeman K, Farley AM (1996) A model of argumentation and its application to legal reasoning. Artif Intell Law 4(3–4):163–197. doi: 10.1007/BF00118492. http://www.springerlink.com/content/g82373414263h146/
18. Wang M, Wang H, Vogel D, Kumar K, Chiu DK (2009) Agent-based negotiation and decision making for dynamic supply chain formation. Eng Appl Artif Intell 22(7):1046–1055. doi: 10.1016/j.engappai.2008.09.001. http://dx.doi.org/10.1016/j.engappai.2008.09.001
19. Gordon TF, Karacapilidis N (1997) The Zeno argumentation framework. In: Proceedings of the sixth international conference on artificial intelligence and law—ICAIL'97. ACM Press, New York, NY, pp 10–18. doi: 10.1145/261618.261622. http://dl.acm.org/citation.cfm?id=261618.261622
20. Bondarenko A (1997) An abstract, argumentation-theoretic approach to default reasoning. Artif Intell 93(1–2):63–101. doi: 10.1016/S0004-3702(97)00015-5. http://dl.acm.org/citation.cfm?id=257547.257553 http://linkinghub.elsevier.com/retrieve/pii/S0004370297000155
21. Dunne PE, Hunter A, McBurney P, Parsons S, Wooldridge M (2009) Inconsistency tolerance in weighted argument systems. In: Proceedings of the 8th international conference on autonomous agents and multiagent systems, vol 2, pp 851–858. International Foundation for Autonomous Agents and Multiagent Systems. http://dl.acm.org/citation.cfm?id=1558109.1558130
22. Koudri A, Champeau J (2010) MODAL: a SPEM extension to improve co-design process models. In: Münch J, Yang Y, Schäfer W (eds) New modeling concepts for todays software processes, Lecture Notes in Computer Science, vol 6195. Springer, Berlin, pp 248–259
23. Gallardo J, Bravo C, Redondo MA (2012) A model-driven development method for collaborative modeling tools. J Netw Comput Appl 35(3):1086–1105. doi: 10. 1016/j .jnca.201 1.12.009. http://dx.doi.org/10.1016/j.jnca.2011.12.009
24. Toulmin SE (2003) The uses of argument, updated edn. Cambridge University Press, Cambridge
25. Barn BS, Clark T (2011) Revisiting Naur's programming as theory building for enterprise architecture modelling. In: CAiSE'11 proceedings of the 23rd international conference on advanced information systems engineering. Springer, Berlin, pp 229–236. http://dl.acm.org/citation.cfm?id=2026716.2026740
26. Barn BS, Clark T (2010) A domain specific language for contextual design. In: Bernhaupt R, Forbrig P, Gulliksen J, Lárusdóttir M (eds) Human-centred software engineering, vol 6409. Springer, Berlin, pp 46–61. http://dl.acm.org/citation.cfm?id=1939212.1939219
27. MetaCase A. Domain-specific modeling with MetaEdit+. http://www.metacase.com/de/
28. van Deursen A, Klint P, Visser J (2000) Domain-specific languages : an annotated bibliography. ACM SIGPLAN Notices 35(6):26–36. http://cat.inist.fr/?aModele=afficheN&cpsidt=1448888
29. Koskinen KU, Pihlanto P, Vanharanta H (2003) Tacit knowledge acquisition and sharing in a project work context. Int J Proj Manage 21(4), 281–290. doi: 0.1016/S0263-7863(02)00030-3. http://dx.doi.org/10.1016/S0263-7863(02)00030-3
30. Cabot J, Gogolla M (2012) Object constraint language (OCL): a definitive guide. In: Bernardo M, Cortellessa V, Pierantonio A (eds) Formal methods for model-driven engineering, Lecture Notes in Computer Science, vol 7320. Springer, Berlin, pp 58–90. doi: 10.1007/978-3-642-30982-3. http://dl.acm.org/citation.cfm?id=2367348.2367351
31. Scheer AW, Thomas O, Adam O (2005) Process modelling using event-driven process chains. In: Dumas M, van der Aalst WMP, ter Hofstede AHM (eds) Process-aware information systems. Wiley, Hoboken, NJ, pp 119–146. doi: 10.1002/0471741442. http://doi.wiley.com/10.1002/0471741442

# Chapter 30
# Architecture Derivation in Product Line Development Through Model Transformations

**Javier González-Huerta, Emilio Insfran, Silvia Abrahão, and John D. McGregor**

**Abstract** Product architecture derivation is a crucial activity in Software Product Line (SPL) development since any inadequate decisions made during the architecture design directly impact on the non-functional properties of the product under development. Although some methods for architecture derivation have been proposed in the last few years, there is still a need for approaches that model the impact among architectural design decisions and quality attributes and use this information to drive the derivation of high-quality product architectures. This paper, presents a set of guidelines for the definition of pattern-based quality-driven architectural transformations in a Model-Driven SPL development environment. These guidelines rely both on a multimodel that represents the product line from multiple viewpoints as well as on a derivation process that makes use of this multimodel to derive a product architecture that meets the quality requirements. The feasibility of the approach is illustrated using a case study on the automotive domain.

**Keywords** Software product lines • Architectural patterns • Quality attributes • Model transformations

J. González-Huerta (✉) • E. Insfran • S. Abrahão
ISSI Research Group, Universitat Politècnica de València,
Camino de Vera s/n, Valencia, Spain
e-mail: jagonzalez@dsic.upv.es; einsfran@dsic.upv.es; sabrahao@dsic.upv.es

J.D. McGregor
Department of Computer Science, Clemson University,
312 McAdams Hall, Clemson, SC, USA
e-mail: JohnMc@cs.clemson.edu

M. José Escalona et al. (eds.), *Information System Development: Improving Enterprise Communication*, DOI 10.1007/978-3-319-07215-9_30,

## 30.1 Introduction

In recent years, Software Product Line (SPL) development has been widely adopted as a means of improving productivity and product quality. SPL is defined as a set of software intensive systems that share a common managed set of features, developed from a common set of core assets in a prescribed manner [1].

Software architectures support the achievement of business goals by satisfying functional and quality requirements. In SPL development, there are two architectures; one for the product line as a whole and another for each individual product. The *product line architecture* contains a set of variation mechanisms that support the quality and functional requirements of the entire set of products that constitute the product line. The *product architecture* is *derived* from the product line architecture by exercising its built-in architectural variation mechanisms [1]. However, when particular levels of quality attributes[1] are required, and cannot be attained via the architectural variation mechanisms, the product architecture can be transformed by applying architectural transformations [2].

In previous works [3–5] we have presented an approach with which to attain the desired quality attribute levels by deriving product architectures from the product line architecture using a multimodel that represents the different viewpoints of the system and the relationships among the elements in these viewpoints.

In this paper, we present a set of guidelines for the definition of pattern-based quality-driven architectural transformations in a Model-Driven Development (MDD) context. Pattern-based architectural transformations allow product architectures, derived from the product line architecture to be improved, when the required quality attribute levels cannot be achieved by exercising the product line architecture's built-in architectural variation mechanisms. This work generalizes and extends the work presented in [3] by providing guidance on how to define the model transformations by making use of this multimodel. The multimodel has also been extended in this paper to introduce a new viewpoint (the transformation view), which permits the explicit representation of architectural transformations and shows how its application impacts on the quality attributes.

The remainder of the paper is structured as follows. Section 30.2 discusses related works in the area. Section 30.3 describes the multimodel for the representation of SPLs. Section 30.4 introduces the guidelines for the definition of pattern-based architectural transformation. Section 30.5 illustrates the feasibility of the approach through a case study in the automotive domain. Finally, conclusions and future works are presented in Sect. 30.6.

## 30.2 Related Work

Several approaches that employ model transformations to deal with the improvement of software architectures have been reported (e.g., [6–9]).

[1] A quality attribute level indicates a threshold at which a system must perform to be useful to the stakeholders [20].

Merilinna proposed a tool that helps architects when selecting architectural transformations based on quality attributes [9]. The software architect determines and applies the best architectural transformation by considering the information gathered in the StyleBase database. Drago et al. proposed a framework with which to explore the solution space and select the architectural transformation that best fits the required quality attributes [6]. Maswar et al. presented an approach for improving the quality of architectures by applying architectural patterns based on the quality requirements [8]. The architecture is evaluated to check its conformance with certain quality requirements and the architect decides the refactoring pattern that will be applied via model transformations. Finally, Li et al. presented AQoSA, an approach for applying genetic algorithms to multi-objective architectural optimization [7] which relies on the exploration of the search space to optimize the allocation of components in computation nodes. However, neither [7] nor [8] consider relationships among alternative transformations and quality attributes. Although [6] and [9] define relationships among transformations and quality attributes, they do not establish the impact on quality that the application of a specific transformation has. This tradeoff information may be useful when deciding which transformation apply or for pruning the search space. Finally, none of the abovementioned approaches were defined to be used in SPL development.

There are, meanwhile, several approaches for the derivation of product architectures from product line architectures in SPL development (e.g., [10–12]). Botterweck et al. presented an approach for deriving the product architecture by means of a model transformation process taking a domain architecture model and a feature model as input and generating a product architecture model, by simply copying elements [10]. Similarly, Cabello et al. produced product architecture models by means of a QVT transformation that takes a feature model and the modular view of the architecture as input to generate the PRISMA component and connector view of the architecture [11]. Perovich et al. automate the derivation of product architectures by taking a feature configuration model as input [12] in order to populate the component and connector view of the architecture. However, none of them consider the quality attribute requirements when deriving the product architecture nor they consider the application of architectural patterns or architectural transformations.

In summary, most of the works in this area focus on improving software architecture quality by applying optimization techniques, rather than using the relationships among the architectural patterns and quality attributes to prune the solution space. With regard to the use of MDD techniques, there are some approaches that automate the derivation of product architectures by means of model transformations, but do not adequately integrate the quality attributes into the process. In our approach, we not only consider quality attributes in the derivation of the product architecture but we also conduct a quality-driven transformation approach. A priori knowledge is therefore used to apply only those architectural patterns that have a positive impact on the required quality attributes.

## 30.3 A Multimodel for Describing Software Product Lines

A *multimodel* is a set of interrelated models that represents different viewpoints of a particular system and permits the definition of relationships among elements in different viewpoints [13]. A *viewpoint* is an abstraction that yields a specification of the whole system restricted to a particular set of concerns. In any given viewpoint it is possible to define a model of the system containing only the objects that are visible from that viewpoint. Such a model is known as a *viewpoint model*, or a *view* of the system from that viewpoint [14]. A multimodel conforms to a meta-model that defines the viewpoint models, the relationships among elements within different viewpoints and the metaclasses required to define those relationships.

The multimodel for representing SPLs is composed of four interrelated viewpoints: *functional*, *variability*, *quality*, and *transformation.* In the last view we specify, among other aspects, the architectural patterns that model the design decisions which exist in the MDD-SPL transformation processes.

The multimodel plays a key role in the SPL development process at two separate levels: (i) in the *Domain Engineering* phase it expresses the impacts and constraints among variations, functional components and quality attributes, providing a more comprehensible and integrated model of the product line, and (ii) in the *Application Engineering* phase in which the final product is derived the relationships drive the different model transformation processes to produce the final product by reusing the information introduced during the domain engineering phase. Figure 30.1 shows an excerpt of the metamodel of the multimodel, which is composed of the individual metamodels of the feature, quality, functional, and transformation models. In addition to these four views, the metaclass *Impacts* represents the relationships that elements in the different views have on the quality attributes and has an attribute (*Weight*) for quantifying the relative importance of that relationship.

In order to represent the inter-viewpoint relationships, and to minimize the coupling among the metamodels involved, we use the proxy pattern [15] to extend the metaclasses that participate in these relationships. The proxy pattern is applied to the *Attribute, Feature, ComponentType* and *Alternative* metaclasses (defined in their corresponding metamodels) resulting in the *EAttribute,*[2] *EFeature, EComponentType* and

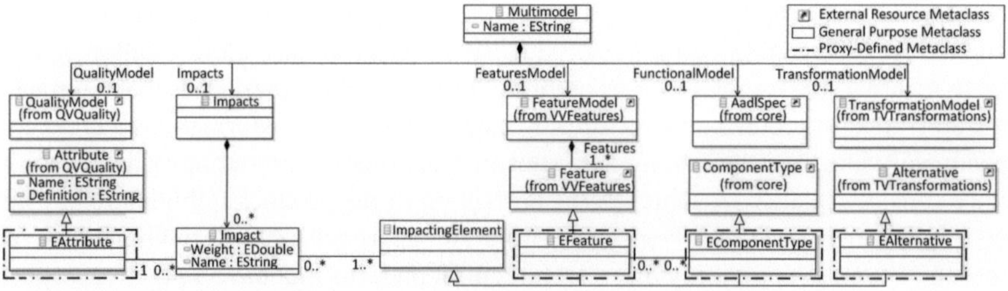


**Fig. 30.1** Excerpt of the multimodel metamodel

[2]The proxy metaclasses follow the naming convention E+ClassName, where ClassName is the base class that is extended.

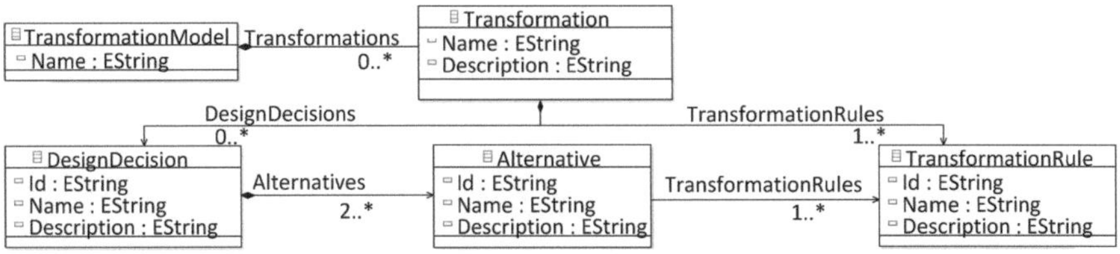


**Fig. 30.2** Transformation view metamodel

*EAlternative* metaclasses (defined in the multimodel's metamodel). The *EFeature, EComponentType and EAlternative* metaclasses also extend the *ImpactingElement* abstract metaclass to define the relationship with the *Impact* metaclass which allows us to define the impacts that *EAlternatives*, *EFeatures* and *EComponentTypes* have on the levels of the quality attributes.

Each of the views of the multimodel are introduced as follows: **Variability View:** The variability view has been defined using a variant [16] of the cardinality-based feature model [17], specifically defined for application in a model-driven SPL development context. The main element of this model is the *Feature*, which is a user-visible aspect or characteristic of a system [1].

**Functional View:** The functional view contains the structure of a system represented by the SPL architecture and the core assets (software components) that satisfy the requirements of the different features. The functional view has been defined using the Architectural Analysis and Design Language (AADL) [18]. AADL defines a textual and graphical representation of the runtime architecture of software systems as a component-based model.

**Quality View:** The quality view has been defined using a quality model for expressing non-functional requirements for SPL defined in [3]. This quality model is an extension of the ISO/IEC 2500 (SQuaRE) standard [13] which supports quality assurance and evaluation activities in SPL development. The quality view contains the hierarchical structure of quality in characteristics, sub-characteristics and quality attributes and also allows non-functional requirements to be expressed as constraints in this quality view.

**Transformation View:** The transformation view contains the explicit representation of design decisions (architectural alternatives) in the various model transformation processes that integrate the production plan for a MDD-SPL. Alternatives appear in a model transformation when a set of constructs in the source model admits different representations in the target model. The application of each alternative transformation could generate alternative target models that may have the same functionality but might differ in their quality attributes.

Figure 30.2 shows an *excerpt* of the transformation view metamodel, describing the hierarchical structure of *Transformations* and *DesignDecisions* that each transformation contains, *Alternatives* in a design decision and the *TransformationRules* constituting the transformation definition of each alternative.

The multimodel permits the definition of relationships among elements in different viewpoints such as composition, impact or constraint relationships [4]. In this work we focus on the impact relationship defined between elements in the transformation and quality views to represent how the selection of one alternative in a design decision can affect one or more quality attributes (e.g., applying the triple redundant pattern improves safety and reliability). The proxy metaclass *EAlternative* in Fig. 30.1 has been defined as a subclass of *ImpactingElement*, signifying that we can define *Impacts* between instances of *EAlternative* and instances of *EAttribute*. This relationship allows the quality attributes to become an additional decision factor when selecting from alternative architectural transformations.

## 30.4 Defining Pattern Based Quality-Driven Transformations

This section presents the guidelines for the definition of pattern-based quality driven architectural transformations.

### *30.4.1 Quality-Driven Architectural Transformations*

The architect designs the product line architecture to cover the entire set of products in the product line, including the variations permitted within the product line scope, and the architectural variation mechanisms needed to achieve them. Product line architecture is used to create the product architecture in the application engineering phase [1]. However, in some cases the product architecture has to be modified by applying architectural transformations to ensure the attainment of certain product-specific quality attribute levels that cannot be attained by exercising the architectural variation mechanisms [2].

In this work, we focus on *architectural patterns* [19], represented as *architectural transformations*, which specify how the system will deal with one aspect of its functionality, impacting directly on the quality attributes [20].

In our approach, three model transformation processes are used iteratively to obtain the product architecture that satisfies the required quality attributes: (i) the *product architecture derivation* in which the product architecture is generated as an instance of the product line architecture; (ii) the *product architecture evaluation* in which the measures defined in the multimodel's quality view are evaluated on the product architecture to determine its quality levels; and (iii) *product architecture transformation* in which the pattern-based architectural transformations are applied to the product architecture in order to meet the required quality attributes.

### *30.4.2 Guidelines for the Definition of Quality-Driven Architectural Transformations*

The application of the following guidelines facilitates the definition of architectural transformations that apply pattern-based quality-driven architectural transformations. A model transformation process relies on a transformation definition containing the transformation rules responsible for transforming a particular selection of the source model into constructs of the target model [21]. A transformation rule consists of two parts: a left-hand side (LHS), which accesses the source model, and a right-hand side (RHS), which expands in the target model [21].

In this work, we use transformation rules as specifications of graph relations between elements in the left-hand and right-hand models. The transformation process is controlled by matches. A *match* occurs when elements from the left-hand and/or right-hand model are identified as satisfying the constraints specified by the declaration of a transformation rule. A match triggers the creation (or updating) of model elements in the target model, and is driven by the declarative and/or implementation parts of the matched rule.

The guidelines are illustrated by using the QVT-Relations syntax but can also be applied by using any other transformation language (e.g., ATL, QVTOperational). The guidelines comprise the following steps:

(I) **Identification of Pattern-Based Architectural Transformations**
The quality attribute requirements that cannot be achieved using the built-in variation mechanisms are identified during the product architecture derivation process. Architectural transformations are selected to enhance these specific attributes. Each architectural transformation is previously associated with a set of transformation rules that affect the specific aspects of the architecture related to improving certain quality attributes.

Depending on the domain of interest, the set of applicable transformations can be reduced owing to the nature or the SPL architecture's topology.

(II) **Architectural Transformation Reorganization**
The next step is to reorganize and group the selected transformations into separate transformation processes, which are in turn organized into design decisions. Architectural transformations are grouped into the same design decision when they represent alternative transformations (i.e., they are applied to the same structures in the source model, sharing their left-hand side), or when they represent mutually exclusive design decisions (e.g., to apply dynamic or static allocation). The application of pattern-based architectural transformations can also be scheduled, to establish the sequence in which the architectural patterns should be applied.

The transformations are then represented in the transformation view of the multimodel as instances of the metaclasses supporting the hierarchical structure of *Transformations*, *DesignDecisions*, and *Alternatives* in each design decision (see Fig. 30.2). Finally, if the rules are reused in that particular transformation process, they are created as instances of the *Rules* metaclass in the transformation view.

(III) **Transformation Rules Definition**
After grouping the architectural transformations, the next step is to define the transformation rules in order to select and apply them. Two types of rules are distinguish in this approach: *pattern-rules* and *discriminant-rules*. The transformation rules responsible for applying each pattern are adapted to become a *pattern-rule* and a new *discriminant-rule* is added for each design decision.

*Pattern-Rules:* The *pattern-rules* are responsible for creating or updating elements in the target model. Pattern-rules are not executed automatically; they need to be invoked from a *discriminant-rule.* They consist of two domains: (i) the source model domain which defines the structures integrating the left-hand side of the rules and (ii) the target model domain which specifies the structures created in the target model (see Fig. 30.4b, in Sect. 30.5).

*Discriminant-Rules:* A new discriminant-rule is defined for each design decision, grouping all its alternatives. Discriminant-rules are executed automatically each time the left-hand side of the pattern is found in the source model. Discriminant-rules select one alternative in a design decision depending on how it supports the required quality attributes and invoke the pattern-rule which applies the selected architectural transformation. Discriminant-rules have three domains: (i) the source model domain, specifying the common structures on the left-hand side; (ii) the transformations model domain, specifying the design decision that the discriminant rule belong to; and (iii) the target model domain, specifying the common structures that are updated or created in the target model. The precondition of the rule specifies the query call which selects the alternative that best supports the required quality attributes levels $Q_a$,... $Q_n$, which are introduced in the transformation as external parameters. The query has access: (i) to the tradeoff results that express how each alternative supports the different quality attributes and (ii) to the parameters that express the quality attributes to be satisfied. The postcondition of the rule initiates the invocation of the corresponding *pattern-rules*, based on the value returned by the query on the precondition (see Fig. 30.4a, in Sect. 30.5).

(IV) **Alternative transformations Tradeoff**

During the tradeoffs, domain experts determine how the alternative supports each quality attributes, based on empirical evidence or on their experience. The tradeoffs analysis is performed by applying the Analytic Hierarchy Process (AHP) [22]. AHP is a decision-marking technique used to resolve conflict when it is necessary to address multi criteria comparisons. The results of the AHP is a weigh that shows the relative support for an alternative with regard to a quality attribute. These results are stored in the relationship defined between the transformation view and the quality view through the *Impact* metaclass in the multimodel. The transformation process uses these impacts to select and apply the alternative transformation that best fits the required quality attributes.

## 30.5 Case Study

This section illustrates the feasibility of the approach with a case study in the context of the safety-critical embedded systems of the automotive industry. This case study describes a Vehicle Control System containing several features such as the Antilock Braking System, the Traction Control, or the Cruise Control System.

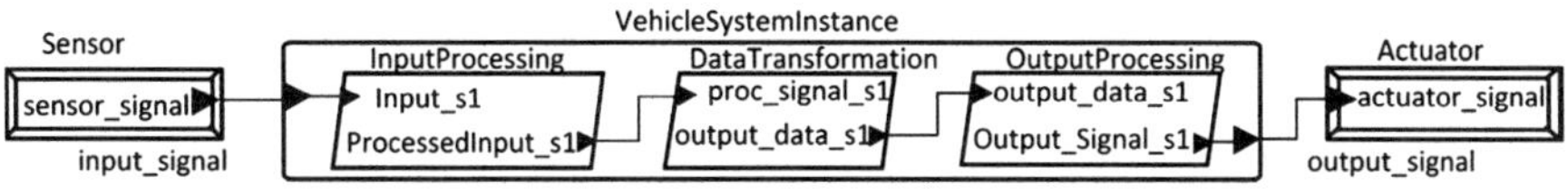


**Fig. 30.3** Architecture of the Vehicle Control Systems

**Table 30.1** Pattern-based architectural transformations catalog

| Name | Rationale | Left-hand side | Right-hand side |
|---|---|---|---|
| T1: Triple Modular Redundancy (TMR) | Only detects random faults. Since the channels are homogeneous, any systematic fault in one channel must be present in both of the others |  | 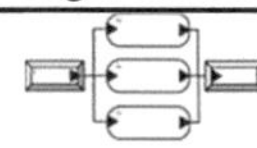 |
| T2: Sanity Check Pattern (SC) | Detects gross deviations from the controlled value to the actuator value. Provides minimal coverage against faults |  | 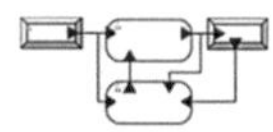 |
| T3: Watchdog Pattern (W) | Lightweight pattern rarely used in safety-critical systems. Best at identifying deadlocks |  | 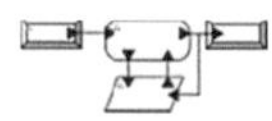 |

These features comprise a set of embedded systems present in modern automobiles. Figure 30.3 shows the architecture of each one of these systems, which is used to illustrate the guidelines defined in Sect. 30.4.2. It consists of capturing input signals from sensors, processing and transforming those inputs and sending the processed output to an actuator that potentially affects the state of other systems or mechanical parts of the car (e.g., engine, throttle position, brakes).

## *30.5.1 Identifying Pattern-Based Architectural Transformations*

The first step when defining quality-driven architectural transformations is to select the transformations to be applied in a specific domain. In this case study, we focus on architectural transformations for embedded systems [19], represented as pattern-based architectural transformations. Table 30.1 shows the catalog of architectural transformations considered. The catalog specifies the left and right-hand sides of the patterns, described using the AADL graphical syntax, and the rational associated with each pattern.

## *30.5.2 Architectural Transformation Reorganization*

The transformations catalog shown in Table 30.1 is reorganized in a unique transformation. A design decision *TransfGroup1* is defined grouping the three alternatives {T1, T2, T3}, since they are concerned with the safety and reliability aspects as is

a
```
top relation TransfGroup1
{  best_quality_support: String;
   checkonly domain Source: ... {...};
   enforce domain Target: ... {...};
   checkonly domain MultiModel Decision1:
               Transformations::DesignDecision{
                     Id='T1',
                     Name=' TransfGroup1',
                     Alternatives= best_alternative:
                        MultiModel::Alternative
                        {Id= best_quality_support} }
   when {  best_quality_support =
                       BestRankedAlternative(Decision1
                         ,QFaultTolerance, QLatency);}
   where { if (best_quality_support ='T1') then
                     TripleModularRedundancy(S,T)
              else if (best_quality_support ='T2') then
                     SanityCheck(S,T)
              else WatchDog(S,T) endif;endif;}
```

b
```
relation TripleModularRedundant
{ checkonly domain Source: ...
    {
       ...
    };
  enforce domain Target: ...
    {
       ...
    };
}
```

**Fig. 30.4** Discriminant (**a**) and pattern rules (**b**) excerpt

shown in Table 30.1, they share the same left-h and side. In this case study we are focusing in one design decision with three alternatives. However, the same schema can be followed to manage N design decisions with M alternatives.

After reorganizing the architect transformations, the next step is to create the transformation view of the multimodel with the structure of transformation, design decisions and alternatives. The safety and reliability transformation is defined as an insurance of the *Transformation* metaclass, which contains a *DesignDecision* instance *TransfGroup1* with three *EAlternative* instance {T1, T2, T3}.

## 30.5.3 Defining Transformation Rules

The definition of the transformation rules for the alternative in the *TransfGroup1* design decision includes the definition of a *discriminant-rule* in order to group all the alternatives. There are three pattern-rules associated with the three alternatives of this design decision: *TripleModularRedundancy*, *SanityCheck* and *WatchDog*.

Figure 30.4a shows an excerpt of the *discriminant-rule* implementation that makes the selection from the alternatives. Note that the rule group the alternative architectural transformations and contains the invocation to *BestRankedAlternative* query in the *when clause*. This query access the tradeoff results and the *QFaultTolerance* and *QLatency* required quality attribute levels, which are introduced as external parameters by the application engineer before executing the transformation. The query selects the alternative that best supports the required quality attribute levels on the basis of these two inputs. Finally, the *when clause* performs the invocation of the corresponding rule based on the value returned by the query.

### 30.5.4 Alternative Transformations Tradeoff

In the tradeoff analysis each alternative is compared as regards the $Q$ quality attributes by using the AHP technique, which is applied once for the entire product line in the domain engineering phase. For each quality attribute $Q_a$, the $N$ potential alternatives are compared in a pairwise comparison. In order to express how an alternative $Ax$ supports the quality attribute $Q_a$, in comparison to the alternative $Ay$, an AHP weight is assigned ranging from 1 for *equally important* to 9 for *extremely more important* [22].

For example, the domain expert defines that as regards *fault tolerance*, TMR is strongly more important (weight of 5) than SC, TMR is very strongly more important (weight of 7) than W, finally, W is strongly more important (weight of 5) than SC. As regards *latency*, W is moderately more important (weight of 3) than SC, and W is strongly more important (weight of 5) than TMR, finally, TMR is moderately more important (weight of 3) than SC.

The result of this comparison is an N x Q matrix that shows the relative support of the different architectural patterns to the quality attributes, as shown in Table 30.2a. Then, these values are normalized by applying the formula (30.1) to Table 30.2a in order to produce Table 30.2b, and finally, the *impact* that an architectural pattern has on a quality attribute $Q_a$ is calculated by applying the formula (30.2) to produce Table 30.2c [22]. The normalized impact values $I$ shown in Table 30.2c ranging from 0 to 1 are stored as *Weights* on the *Impact* instances in the multimodel.

$$NormQ_a[i,j] = \frac{Q_a[i,j]}{\sum_{k=1}^{n} Q[k,j]} \tag{30.1}$$

$$I[i] = \frac{\sum_{k=1}^{n} NormQ_a[i,k]}{n} \tag{30.2}$$

$$R_j = \sum_{i=0}^{k-1} Qi * Iij \tag{30.3}$$

### 30.5.5 Transformation Execution

In the application engineering phase, the application engineer introduces the quality attribute levels $Q$ that the specific product must fulfill as normalized weights ranging from 0 to 1 as external parameters when executing the transformation. For $k$ quality attributes, the transformation process calculates the ranking $R$ for each pattern $j$ by applying formula (30.3) to the impact values shown in Table 30.2c.

**Table 30.2** Architectural transformations and quality attributes tradeoff analysis

| | Fault tolerance | | | Latency | | | | Fault tolerance | | | Latency | | | | Impacts | |
|---|---|---|---|---|---|---|---|---|---|---|---|---|---|---|---|---|
| (a) | TMR | SC | W | TMR | SC | W | (b) | TMR | SC | W | TMR | SC | W | (c) | Fault tolerance | Latency |
| **TMR** | 1 | 5 | 7 | 1 | 3 | 1/5 | **TMR** | 1/1.3 | 5/11 | 7/8.2 | 1/6.3 | 3/7 | 1/5/1.5 | **TMR** | 0.68 | 0.24 |
| **SC** | 1/5 | 1 | 1/5 | 1/3 | 1 | 1/3 | **SC** | 1/5/1.3 | 1/11 | 1/5/8.2 | 1/3/6.3 | 1/7 | 1/3/1.5 | **SC** | 0.09 | 0.13 |
| **W** | 1/7 | 5 | 1 | 5 | 3 | 1 | **W** | 1/7/1.3 | 5/11 | 1/8.2 | 5/6.3 | 3/7 | 1/1.5 | **W** | 0.23 | 0.63 |
| **Sum** | **1.3** | **11** | **8.2** | **6.3** | **7** | **1.5** | | | | | | | | | | |

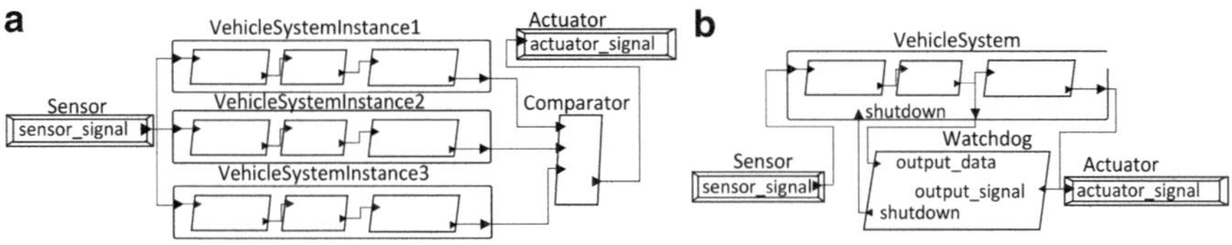


**Fig. 30.5** System architecture structure after applying the transformations

The transformation applied to the architecture shown in Fig. 30.3 results in the product architecture shown in Fig. 30.5a or Fig. 30.5b depending on the quality attributes selected by the application engineer.

If the application engineer selects the *fault tolerance* by introducing a weight of 1 for *fault tolerance* and 0 for *latency* the transformation process will select the TMR pattern (TMR: 1*0.68 + 0*0.24 > SC: 1*0.09 + 0*0.13 > W: 1*0.23 + 0*0.63). The resulting architecture is shown in Fig. 30.5a. However, if the application engineer selects the *fault tolerance* and *latency* by introducing a weight of 0.4 for *fault tolerance* and 0.6 for *latency* the transformation process will select the W (W: 0.4*0.23 + 0.6*0.63 > TMR: 0.4*0.68+ 0.6*0.24 > SC: 0.4*0.09 + 0.6*0.13). The resulting architecture is shown Fig. 30.5b.

## 30.6 Conclusions and Future Work

In this paper we present a set of guidelines for the development of pattern-based quality-driven architectural transformations. These guidelines rely on a multimodel that represents different viewpoints of a system, permitting the explicit representation of relationships between architectural patterns and quality attributes. The guidelines provide guidance on how to define pattern-based architectural transformations that use quality attributes as a decision factor when alternatives exist.

The feasibility of the approach has been illustrated with a case study in the automotive domain. The guidelines have been applied to define pattern-based transformations that select from architectural alternatives depending on the required quality attributes.

As future work we plan to develop a tool with which to support the definition of pattern-based quality-driven transformations. We also plan to apply this approach to other domains in which different patterns and quality attributes have been identified.

**Acknowledgements** This research is supported by the MULTIPLE project (MICINN TIN2009-13838) and the ValI+D program (ACIF/201 1/235).

## References


1. Clements P, Northrop L (2007) Software product lines: practices and patterns. Addison-Wesley, Boston, MA
2. Bosch J (2000) Design and use of software architectures. Adopting and evolving product-line approach. Addison-Wesley, Harlow
3. Gonzalez-Huerta J, Insfran E, Abrahao S, McGregor JD (2012) Non-functional requirements in model-driven software product line engineering. In: 4th International workshop on non-functional system properties in domain specific modeling languages, Insbruck
4. González-Huerta J, Insfran E, Abrahão S (2012) A multimodel for integrating quality assessment in model-driven engineering. In: 8th International conference on the quality of information and communications technology (Quatic2012), Lisbon
5. Insfrán E, Abrahão S, González-Huerta J, McGregor JD, Ramos I (2012) A multimodeling approach for quality-driven architecture derivation. In: 21st International conference on information systems development (ISD2012), Prato
6. Drago ML, Ghezzi C, Mirandola R (2011) Towards quality driven exploration of transformation spaces. In: 14th Conference on model driven engineering languages and systems, Wellington, LNCS, vol 6981, pp 2–16
7. Li R, Etemaadi R, Emmerich MTM, Chaudron M (2011) An evolutionary multiobjective optimization approach to component-based software architecture design. IEEE Congress on Evolutionary Computation, New Orleans, LA, pp 432–439
8. Maswar F, Chaudron M, Radovanovic I, Bondarev E (2007) Improving architectural quality properties through model transformations. In: 2007 International conference on software engineering research and practice, Las Vegas, NV, pp 687-693
9. Merilinna J (2005) A tool for quality-driven architecture model transformation. Ph.D. thesis, VVT Technical Research Centre of Finland, Vuorimiehentie
10. Botterweck G, O'Brien L, Thiel S (2007) Model-driven derivation of product architectures. In: 22th International conference on automated software engineering, New York, NY
11. Cabello ME, Ramos I, Gómez A, Limón R (2009) Baseline-oriented modeling: an MDA approach based on software product lines for the expert systems development. In: 1st Asia conference on intelligent information and database systems, Vietnam
12. Perovich D, Rossel PO, Bastarrica MC (2009) Feature model to product architectures: applying MDE to software product lines. In: 8th IEEE/IFIP and 3rd European conference on software architecture, Helsinki, pp 201–210
13. ISO/IEC 25000:2005 Software Engineering (2005) Software product Quality Requirements and Evaluation (SQuaRE). Guide to SQuaRE. http://www.iso.org/iso/catalogue_detail?csnumber=35683
14. Barkmeyer EJ, Feeney AB, Denno P, Flater DW, Libes DE, Steves MP, Wallace EK (2003) Concepts for automating systems integration. In: NISTIR 6928, National Institute of Standards and Technology, U.S. Department of Commerce, USA
15. Gamma E, Helm R, Johnson R, Vlissides JO (1994) Design patterns, elements of reusable object-oriented software. Addison-Wesley, Boston, MA
16. Gómez A, Ramos I (2010) Cardinality-based feature modeling and model-driven engineering: fitting them together. In: 4th International workshop on variability modeling of software intensive systems, Linz
17. Czarnecki K, Kim CH (2005) Cardinality-based feature modeling and constraints: a progress report. In: International workshop on software factories at OOPSLA, San Diego, CA
18. Feiler PH, Gluch DP, Hudak J (2006) The architecture analysis & design language (AADL): an introduction. Technical report CMU/SEI-2006-TN-01 1. SEI, Carnegie Mellon University
19. Douglass BP (2002) Real-time design patterns: robust scalable architecture for real-time systems. Addison-Wesley, Boston, MA
20. Bass L, Clements P, Kazman R (2003) Software architecture in practice, 2nd edn. Addison-Wesley Professional, Boston, MA
21. Meta Object Facility (MOF) (2003) 2.0 Query/View/Transformation Specification. http://www.omg.org/spec/QVT/1.1/
22. Saaty TL (1990) The analytical hierarchical process. McGraw-Hill, New York, NY

# Chapter 31
# Collaborative Modeling Through the Integration of Heterogeneous Modeling Languages

**Francisca Pérez, Pedro Valderas, and Joan Fons**

**Abstract** Models are becoming the new programming code to specify software products. However, most tools that support model descriptions are oriented to a single-user with specific skills, which avoids the involvement of different roles to perform collaborative modeling (such as domain experts, stakeholders or end-users). In this paper, we present an approach to achieve collaborative modeling by bridging heterogeneous modeling languages. Specifically, our approach enables an existing modeling language to involve a different role who uses a different modeling language. On the one hand, we apply interoperability mechanisms to bridge heterogeneous modeling languages. On the other hand, we apply variability mechanisms in a novel way to set gaps in partially instantiated models that are completed using models of the different modeling language. We show the feasibility of the approach through a sample scenario that bridges two existing modeling languages to develop web information systems.

**Keywords** Collaborative system development • Model-driven development • Model transformation • Variability management • Domain-specific languages • Collaborative modeling

F. Pérez (✉)
Centro de Investigación en Métodos de Producción de Software, Universitat Politècnica de València, Camino de Vera, s/n, 46022 Valencia, Spain

Escuela Politécnica Superior, Universidad San Jorge,
Autovía A-23 Zaragoza-Huesca Km. 299, 50830 Villanueva de Gállego, Zaragoza, Spain
e-mail: mfperez@usj.es

P. Valderas • J. Fons
Centro de Investigación en Métodos de Producción de Software, Universitat Politècnica de València, Camino de Vera, s/n, 46022 Valencia, Spain
e-mail: pvalderas@pros.upv.es; jjfons@pros.upv.es

M. José Escalona et al. (eds.), *Information System Development: Improving Enterprise Communication*, DOI 10.1007/978-3-319-07215-9_31,

## 31.1 Introduction

Models [1] have replaced source code as the main ingredient of the development process. This model-centric schema has been increasingly used in the last few years by emerging modeling languages in different domains. Therefore, there is a broad agreement in the fact that it is important and useful to involve different roles in the construction and modification of models [2] (i.e., domain experts, stakeholders or end-users). Thus, the nature of collaborative modeling is widely accepted as a process where a number of people actively contribute to the creation of a model [3]. However, most modeling languages are oriented to create a model by a single-user with specific skills [3], so different roles cannot be involved in collaborative modeling. Moreover, most modeling languages are not designed to interoperate with other modeling languages, which makes them isolated alternatives.

Interoperability mechanisms is a growing trend [4] and it can provide several benefits, i.e., existing modeling languages can be used as complementary alternatives to perform the modeling tasks from different domains (such as system design, business processes, etc.), from different roles (such as project managers, system designers, domain experts, etc.), and from different software representations that could have a different abstraction level (such as a visual language that helps the description of models). There are approaches such as [5–7] that achieve interoperability to perform modeling tasks from two different modeling languages. However, these approaches are focused on transformations from one model to another and they do not tackle collaborative modeling to allow a new user to describe subsets of a model.

We propose a method named Medem that enables collaborative modeling by bridging two different modeling languages. Specifically, Medem enables the user of an existing modeling approach to define a set of gaps. Another user can fulfill these gaps using models of a different modeling approach. To do this, we combine interoperability and variability mechanisms in a non-intrusive way for the existing meta-models. On the one hand, we apply interoperability mechanisms by means of both a weaving model that links model concepts of each approach, and model transformations that obtain model descriptions from one model to another. On the other hand, we apply variability mechanisms in a novel way to determine gaps that may be fulfilled by the new user. We show the feasibility of Medem through a sample scenario to develop web information systems.

The remainder of the paper is structured as follows: Sects. 31.2 and 31.3 describe the sample scenario and how Medem is initialized and used. Section 31.4 presents the lessons learned. Section 31.5 presents the related work and Sect. 31.6 concludes.

## 31.2 A Sample Scenario

In the Valencian Regional Ministry of Infrastructure, there are different modeling languages that are being developed to represent organization' needs. These languages are supported by a FREE CASE that is built on Eclipse. Among these languages,

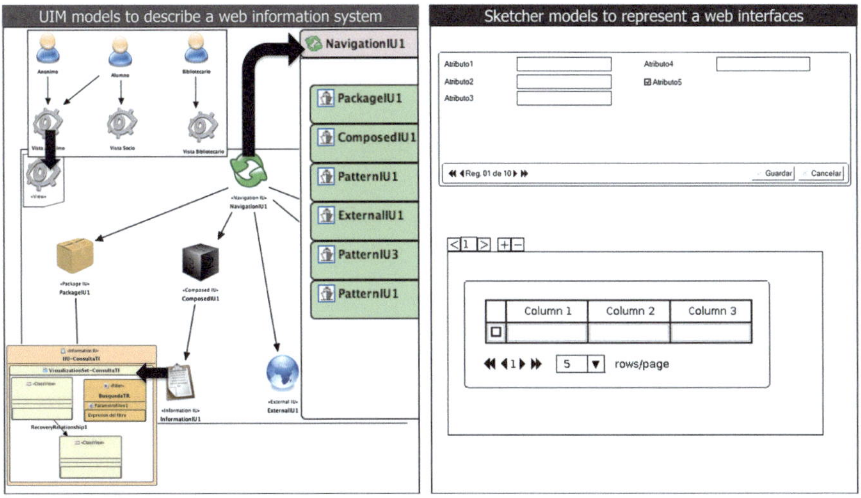


**Fig. 31.1** Snapshots of UIM and Sketcher models

there are two that are focused on the development of web information systems User Interface Modeler (UIM) [8] and Sketcher [9]:

- **UIM** supports organization' needs on web information systems (e.g., online procedures that citizens may complete to request subventions, grants, etc.). Specifically, UIM allows software development experts to specify interfaces and how they are related in an abstract way with the information system. UIM uses concepts in its models such as ClassView, Visualization Set, Filters, Patterns and Package Unit Interaction, so the use of UIM may be complex and it requires that software development experts spend time in learn the modeling language. The left side of Fig. 31.1 shows several snapshots of different UIM models from the UIM editor. Although UIM use visual elements, its complexity is related to the amount of concepts, properties, and the navigability among models.
- **Sketcher** expresses an initial representation of user interfaces in a closer way to users. Sketcher is based on the standard notation of sketching tools, which favors the users' validation in early stages of development. The right side of Fig. 31.1 shows a snapshot of simple Sketcher models, which specify the elements of a web interface.

In particular, software development experts use UIM to develop web information systems but they also need to collaborate with domain experts in documentation. Domain experts in documentation are in charge of designing user interfaces because they know the information that citizens have to complete to carry out the organization' procedures. However, domain experts do not use UIM as software development experts do, domain experts use Sketcher to express the initial representation of user interfaces. Afterwards, software development experts manage all Sketcher interface descriptions to integrate them in UIM models. Thus, the web information system is developed including descriptions of both roles. Therefore, we have detected two main problems: (1) software development experts are forced to translate and integrate descriptions from Sketcher models to existing UIM models; and (2) domain experts are not guided during descriptions, so isolated features can arise and software development experts are forced to integrate them.

## 31.3 The Medem Method

For the sample scenario presented above, interoperability mechanisms can be applied to transform model descriptions using Sketcher (domain experts in documentation) to UIM (software development experts). However, domain experts do not have the knowledge to describe features of the web information system as software development experts have using UIM, so collaborative modeling mechanisms are required.

To tackle this, we propose the Medem method that is made up of four main steps: (1) $User_1$ creates from scratch or reuses a model conforms to the model structure (meta-$model_1$) of a modeling approach using the *Medem interface for* $model_1$. This model is identified as *partially instantiated* $model_1$ (see Fig. 31.2 (1)); (2) $User_1$ also defines a list of gaps in the *partially instantiated* $model_1$. Each gap should be described by $User_2$ (see Fig. 31.2 (2)); (3) $User_2$ describes each gap using the *Medem interface for model2*. Each gap description is stored in a model conforms to the model structure (meta-$model_2$) of another modeling approach (see Fig. 31.2 (3)); and (4) Gap descriptions are transformed in models1 to complete the previously defined gaps. Thus, a*fully instantiated* $model_1$ is obtained (see Fig. 31.2 (4)) by including description of both $User_1$ and $User_2$ in a collaborative way.

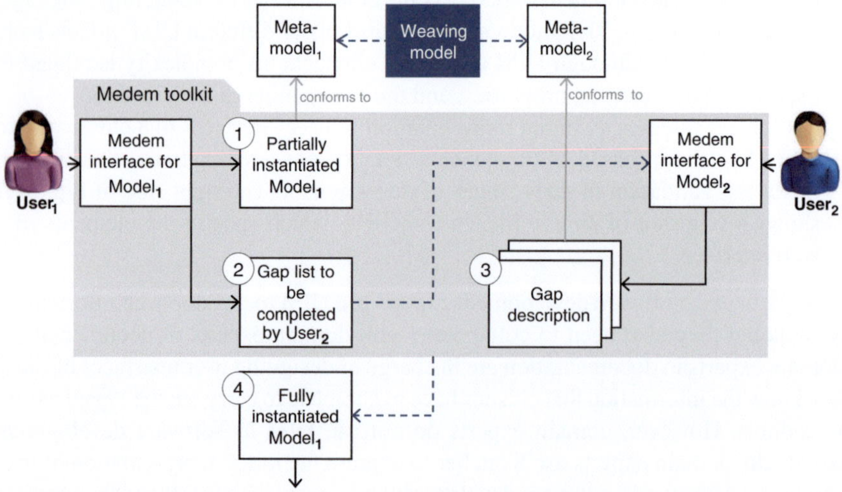


**Fig. 31.2** The Medem method

To support these steps and to validate the feasibility of the Medem method, we have developed a toolkit prototype as the underlying tool framework. Next, we describe both how the Medem toolkit is initialized and how the toolkit is used in each step by applying the sample scenario.

## *31.3.1 Initializing the Medem Toolkit*

The initialization of the Medem toolkit is carried out by a user who has knowledge in modeling languages (from now onward meta-model Architect). It is worth noting that the Medem toolkit initialization is only carried out once at the beginning, soit is reused even though the steps of Medem are reapplied (i.e., if $User_1$ or $User_2$ change model descriptions). To initialize the Medem toolkit, the Architect carries out three phases as the left side of Fig. 31.3 shows: *Existing*, *Feasibility* and *Initialization*.

### 31.3.1.1 The *Existing* Phase

This phase represents the required input elements to initialize the Medem toolkit. The top of Fig. 31.3 shows that this phase takes as input two ecore meta-models[1] ($metamodel_1$ and $meta\text{-}model_2$). Each meta-model is a model that describes the structure of models from a different modeling approach. Moreover, this phase takes as input two model editors (Model $editor_1$ and Model $editor_2$) that enable the creation or modification of models conform to a meta-model ($meta\text{-}model_1$ and $meta\text{-}model_2$ respectively).

In the sample scenario, we take as input both the UIM and Sketcher ecore meta-models and the UIM and Sketcher model editors.

### 31.3.1.2 The *Feasibility* Phase

The Architect determines if meta-models are able to exchange model descriptions (interoperable) (see the center of Fig. 31.3). To do this, the Architect looks for strong differences among meta-model concepts that avoid $model_1$ descriptions are transformed to $model_2$ descriptions. For example, strong differences can be found if $meta\text{-}model_1$ concepts cannot be transformed in $meta\text{-}model_2$ concepts, so they are not able to exchange model descriptions. If strong differences are found, the meta-models are not interoperable and Medem is not feasible. Also, the Architect looks for structural heterogeneities among concepts. According to [10] structural heterogeneities can be found in internal properties (such type, cardinality, etc.) and

[1] An ecore meta-model is a model of a model using the Eclipse Modeling Framework (EMF). EMF implements the Object Management Group's (OMG) Meta Object Facility (MOF) specification, which standardizes a meta-model for object oriented analysis and design.

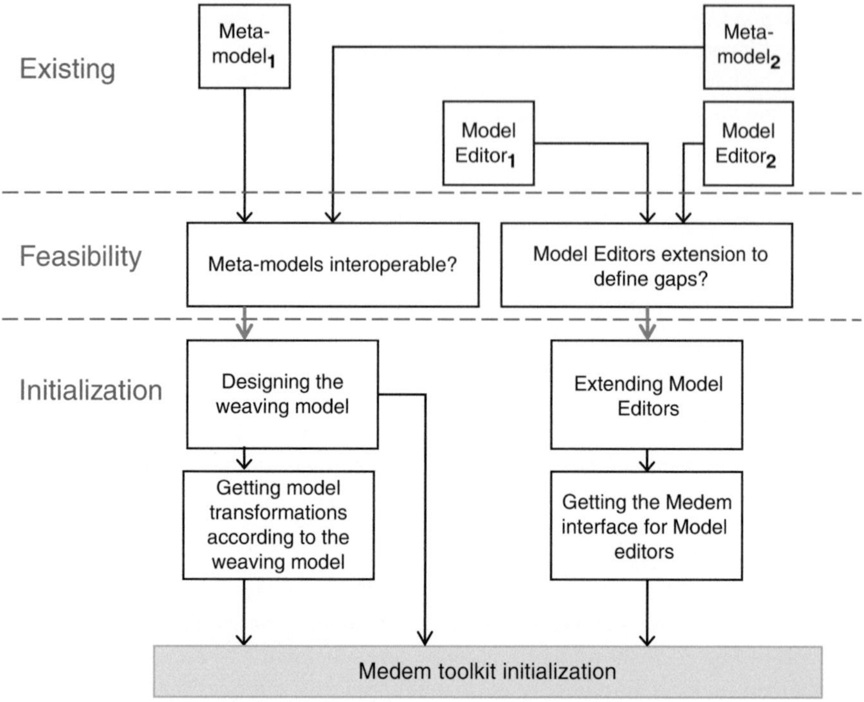


**Fig. 31.3** Initializing the Medem toolkit by the meta-model architect

element relationships (such as inheritance) and they can be resolved with model transformations. For example, structural heterogeneities may emerge if a concept of the meta-model$_1$ can be described using two concepts of the meta-model$_2$. If structural differences are found or not, the meta-models are interoperable and Medem is feasible. Next, the Architect determines the feasibility of Medem if model editors can be extended to support gap descriptions.

In the sample scenario, we have not found strong differences among meta-model concepts. By contrast, we found some structural heterogeneities in internal properties and elements relationships. For example, a *MenuItemAction* UIM concept is linked with two Sketcher concepts (*Navigation* and *TargetIU*). In addition, the UIM and Sketcher editors are online available and they can be extended to support gap descriptions. Then, Medem is feasible in UIM and Sketcher.

#### 31.3.1.3 The *Initialization* Phase

The Architect initiates the Medem toolkit by (1) designing a weaving model to link concepts between meta-models, (2) obtaining model transformations according to the weaving model, and (3) extending the model editors to support Medem (see the bottom of Fig. 31.3).

**Table 31.1** Relation between Medem and CVL concepts

| Medem | CVL | Description |
| --- | --- | --- |
| Gap list | CVLmodel | The gap list is stored in a CVL model using an input model$_1$ (base model). The CVL variability model indicates which gaps must be described and how these gaps must be involved and related to other elements of the model$_1$ |
| Gap | Placement fragment | Each gap described by User$_1$ is stored as placement fragment in the CVL model |
| Gap description | Replacement fragment | It contains model descriptions that fit in a gap |

1. **Designing the weaving model.** The weaving model is a model that contains different kinds of relationships to link meta-model elements. To design the weaving model, we use the existing weaving approach Atlas Model Weaver (AMW) [11], which consists in defining a specific mapping model (called weaving model) between the meta-model$_1$ and meta-model$_2$. The weaving model provides Medem with a bi-directional way to link elements of the meta-models involved. In addition, the weaving model is non-intrusive with the meta-models, so we do not need to modify the meta-models which sometimes can be unfeasible. Instead of using AMW to create the weaving model, the modeler can use different weaving approaches such as the proposed by Didonet et al. [10] that provides matching transformations, which automatically create a weaving model. For the sample scenario, we define a weaving model to link UIM and Sketcher meta-model concepts. As the weaving model covers many concepts of both UIM and Sketcher, a full description of each link included in the weaving model is out of scope of this paper.
2. **Obtaining model transformations.** The weaving model contains abstract and declarative links that are used to produce model integration transformations. Model transformations are used to enable interoperability between the model$_1$ and model$_2$ descriptions. To support the sample scenario, we use the weaving model to obtain model transformations using ATL[2] as the transformation language. As we described in the design of the weaving model, the modeler can also use a different approach to obtain model transformations according to the weaving model. For example, the approach proposed by Didonet et al. [10] also provides mechanisms to semi-automate the development of transformations. Moreover, there are approaches such as the proposed by Cuadrado et al. [12] that brings generic model transformations to promote reusability. This can accelerate the development time of transformations [10], or it can favor the reuse of model transformations if the modeling languages change.
3. **Extending the model editors for supporting the definition and description of gaps.** To do this, we use a proposal sent by IBM, Thales, Fraunhofer FOKUS and TCS for the OMG Common Variability Language (CVL) [13] Request For Proposal (RFP). CVL proposes two main concepts to express variability: placement fragment and replacement fragment. Table 31.1 shows the main Medem concepts, the CVL concept that supports them, and a brief description.

[2] http://www.eclipse.org/atl/

Therefore, CVL allows Medem to: (1) manage the placements of a base system model; (2) manage the components that can fit into the placements; (3) define a set of boundary points that give the boundary between a placement and the rest of the model; and (4) express gaps independently of the $m_1$ and the $m_2$ models. Moreover, CVL provides tool support to display CVL concepts (such as placements) in a model editor. We identify each extended editor as *the Medem interface for model$_1$ or model$_2$*. On the one hand, the Architect extends the model editor$_1$ to enable the CVL graphical editor by implementing a set of CVL APIs. Thus, the editor supports the creation of placement fragments by selecting model elements. On the other hand, the Architect extends the model editor$_2$ to include the following areas: Components (shows a guide with all the placement fragments of the variation model that should be completed), Components personalization (provides an empty template where a placement fragment can be completed using $m_2$ concepts), and Information (helps with the description of components).

We omit implementation details for space limitations, but it is worth noting that the toolkit supports gap description using both the model transformations previously obtained and the EMF Model Query framework,[3] which provides an API to build and execute query statements. These query statements are used to discover and modify model elements. For example, we have implemented a *select* statement by using EMF Model Query to obtain all the placement fragments of a given resource (the CVL variability model):

```
SELECT statement =
 new SELECT(
  new FROM(resource.getContents()),
  new WHERE(new EObjectTypeRelationCondition(
   CVLPackage.Literals.PLACEMENT_FRAGMENT,
   TypeRelation.SAMETYPE_OR_SUBTYPE_LITERAL)));
```

### 31.3.2 Using the Toolkit

The previously described Medem steps are automatically supported using the initialized toolkit (represented by a grey square in Fig. 31.2) as follows:

**(Step 1 and Step 2) User$_1$ uses the Medem interface for Model$_1$.** Following the sample scenario, the software development expert defines gaps in a reused UIM model that should be described by User$_2$. The toolkit automatically supports the creation of both the partially instantiated UIM model (Step 1) and the gap list (Step 2) as we previously described. For example, left side of Fig. 31.4 shows a subset of a UIM model. This model includes different UIM concepts that represent a new editable information view to a new procedure. These elements are represented as a class with two attributes (name and id), and a filter to select those elements that have

[3] http://www.eclipse.org/modeling/emf/?project=query

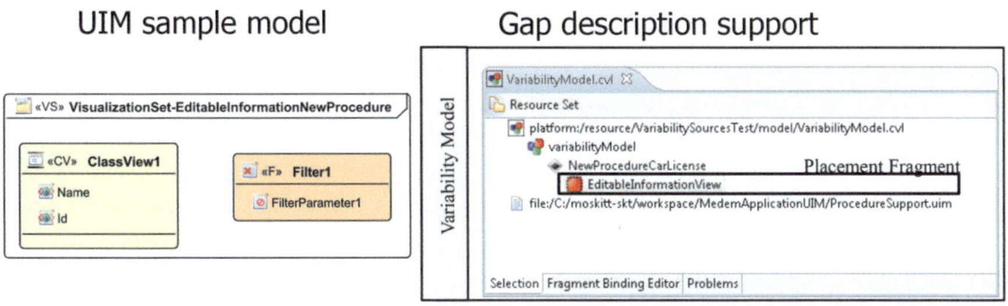


**Fig. 31.4** An example of an UIM model and gap description support

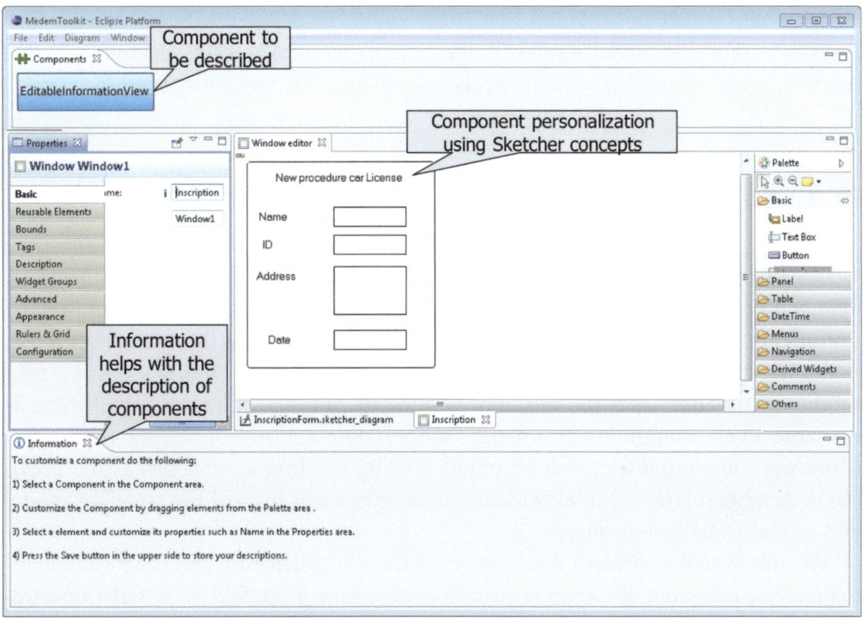


**Fig. 31.5** The Medem interface for $Model_2$

to be shown to the user. The right side of Fig. 31.4 shows the CVL variability model that has been obtained by the toolkit to store a gap (placement fragment). Thus, $User_2$ will be able to describe the *EditableInformationView*.

**(Step 3) $User_2$ uses the Medem interface for $Model_2$.** Figure 31.5 shows a snapshot of the Medem interface for model2 that instantiates the toolkit to support domain expert's descriptions. The top of the figure shows the *Components* area with the placement fragment that $user_1$ defined above. The center of Fig. 31.5 presents the *Components personalization* area to describe each gap (placement) using Sketcher concepts. The figure shows a description of the *EditableInformationView* placement fragment. According to the description presented in the figure, the *EditableInformation View* description will have several visible attributes that allow citizens to provide the required data for the new procedure.

**(Step 4) Transformation of $m_2$ model descriptions into $m_1$ model descriptions to complete gaps.** The toolkit also automatically supports this step once gap descriptions are completed. Following the new procedure example description, the toolkit performs the following: (1) it uses the model transformations to obtain a UIM model; (2) it creates a new replacement titled *EIViewReplacement* in the CVL model; (3) it updates the *EIViewReplacement* to include the concepts obtained from the UIM model. To do this, the toolkit uses EMF Model Query update queries; (4) it creates a substitution element and a resolution element in the CVL model to relate the *EditableInformationView* placement to the new replacement fragment. The substitution element and the resolution element are CVL concepts that determine the set of decisions of the model; and (5) it uses the CVL model to execute a transformation that is provided by CVL, which obtains a resolution model. The resolution model is an m1 model (in the running example a fully instantiated UIM model) that is completed with the decisions of the CVL variability model.

## 31.4 Lessons Learned

On the one hand, we consider that we provide solutions to the problems that were detected in UIM and Sketcher in Sect. 31.2. On the other hand, we detect that the initialization phase of Medem using the toolkit (designing the weaving model, obtaining the model transformations and extending the model editors) may require more time than the remaining steps of Medem. However, this phase is reusable even though users change model descriptions or create new ones. Moreover, this initial time can be reduced using existing approaches as the previously described [10, 12] to accelerate the development time of the weaving model and model transformations.

We also detect a problem that appears in the transformation that the toolkit automatically carries out if Sketcher descriptions cause a conflict with UIM descriptions. For example, the same component is already used in UIM descriptions, or the same component is used to different purposes. To solve this, we have extended the Medem toolkit to show an information message in the Medem user interface for model2 if this conflict occurs after gap descriptions are transformed. To do this, we define a set of rules and we also use EMF Model queries between gap descriptions and the m1 model.

## 31.5 Related Work

There are related works [5–7, 14–17] that provide interoperability mechanisms in different modeling languages in order to achieve different goals. Some modeling approaches were carefully designed to interoperate from a different modeling approach in order to provide a different language to edit models. For example, Berti et al. [14] describe CTTE, which is a system that helps domain users convert natural language descriptions of tasks and scenarios into a hierarchy of subtasks. Lin and Landay [15] describe Damask, which is a system that prototypes ubiquitous computing applications and test them with domain users. Giachetti et al. [5] propose a process, which integrates Unified Modeling Language and Domain-Specific Language models in a unique Model-Driven Development solution.

Other solutions use model transformations and weaving models to interoperate in order to simplify the implementation of the integration components needed. For example, Opdahl [6] presents a modeling approach that facilitates language interoperability in the context of business processes. Klar et al. [7] show how interoperability can be used to support a complete development process. However, these approaches are focused on providing mechanisms to interoperate among fully instantiated models and they do not tackle collaborative modeling to allow a new user to describe subsets of a model.

## 31.6 Conclusions

In this paper, we have presented a method called Medem that enables collaborative modeling by bridging two different modeling languages. Specifically, Medem enables the user of an existing modeling approach to define a set of gaps that another user can fulfill using models of a different modeling approach. On the one hand, we apply interoperability mechanisms by means of both a weaving model that links model concepts of each approach, and model transformations that obtain model descriptions from one model to another. On the other hand, we apply variability mechanisms in a novel way to determine gaps that may be fulfilled by the new user. To do this, Medem uses the CVL variability language to define a model that expresses placements and completes them independently of both $m_1$ and $m_2$ models.

Thus, Medem offers a balance between keeping the properties of m1 models and having the flexibility to describe properties using $m_2$ models. Medem also uses an API to build and execute query statements for discovering and modifying elements of both the weaving model and the CVL variability model. We would emphasize that both the weaving model and the CVL variability model are non-intrusive with the meta-models of the modeling languages.

We showed the feasibility of Medem by applying a sample scenario to enable collaborative modeling between two different roles (software and domain experts) who use different existing modeling languages (Moskitt UIM and Sketcher respectively). The practical application is supported by a toolkit prototype that we have developed to make the application of Medem feasible. Moreover, the lessons learned triggered an extension of the Medem toolkit to show information messages if a conflict occurs in the transformation of $model_2$ descriptions in the model1 descriptions.

The main innovation of Medem is the combination of both variability and modeling techniques to (1) delimit the concerns that may be described by each role by describing gaps and completing them, and (2) reuse the Medem initialization if a new model descriptions are carried out. We believe that Medem is especially useful for both taking advantage of the experience of different roles in modeling tasks and preventing those roles are forced to use different modeling primitives in order to participate in the same project.

**Acknowledgments** This work has been developed with the support of Ministerio de Ciencia e Innovación (MICINN) under the project EVERYWARE TIN2010-18011 and co-financed with European Regional Development Fund (ERDF).

## References


1. Mellor SJ, Clark AN, Futagami T (2003) Guest editors' introduction: model-driven development. IEEE Softw 20(5):14–18
2. Renger M, Kolfschoten GL, de Vreede G-J (2008) Challenges in collaborative modeling: a literature review. In: Advances in enterprise engineering I, held at CAiSE 2008, vol 10, Montpellier, pp 61–77
3. Rittgen P (2008) Coma: a tool for collaborative modeling. In: CAiSE Forum, vol 344 of CEUR workshop proceedings, CEUR-WS.org, pp 61–64
4. Brunelière H, Cabot J, Clasen C, Jouault F, Bézivin J (2010) Towards model driven tool interoperability: bridging eclipse and microsoft modeling tools, chap 5. In: Modelling foundations and applications, vol 6138, Springer-Verlag, pp 32–47
5. Giachetti G, Marín B, Pastor O (2009) Using uml as a domain-specific modeling language: a proposal for automatic generation of uml profiles. In: Proceedings of the 21st international conference on advanced information systems engineering, CAiSE'09. Springer, Berlin, pp 110–124
6. Opdahl AL (2010) Incorporating uml class and activity constructs into ueml. In: ER workshops, vol 6413, Springer, pp 244–254
7. Klar F, Rose S, Schürr A (2008) A meta-model-driven tool integration development process. Vol 5 of Lecture Notes in Business Information Processing, Springer, pp 201–212
8. Valencian Regional Ministry of Infrastructure. UIM basic concepts. www.moskitt.org/fileadmin/conselleria/documentacion/Manual_Usuario/1. 1.3/ManualMetodologicoUIMIngles.pdf
9. Valencian Regional Ministry of Infrastructure. Sketcher plugin. www.moskitt.org/eng/moskitt-1320/
10. Del Fabro MD, Valduriez P (2009) Towards the efficient development of model transformations using model weaving and matching transformations. Softw Syst Model 8(3):305–324
11. Del Fabro MD, Bézivin J, Valduriez P (2006) Weaving models with the eclipse amw plugin. In: Eclipse Modeling Symposium, Eclipse Summit Europe
12. Cuadrado JS, Guerra E, Lara J (2011) Generic model transformations: write once, reuse everywhere. In: Cabot J, Visser E (eds) Theory and practice of model transformations, vol 6707 of Lecture Notes in Computer Science. Springer Berlin Heidelberg, pp 62–77
13. Common Variability Language Revised Submission (2012) http://www.omgwiki.org/variability/lib/ exe/fetch.php?id=startcache=cachemedia=cvl-revised-submission.pdf
14. Berti S, Paternó F, Santoro C (2006) Natural development of nomadic interfaces based on conceptual descriptions. In: Lieberman H, Paternò F, Wulf V (eds) End user development, vol 9, chap 7. Springer, Dordrecht, pp 143–159
15. Lin J, Landay JA (2008) Employing patterns and layers for early-stage design and prototyping of cross-device user interfaces. In: CHI'08, New York, pp 1313–1322
16. Schnieders A, Puhlmann F (2007) Variability modeling and product derivation in E-business process families, Springer Netherlands, pp 63–74
17. Raedts I, Petkovic M, Usenko YS, van der Werf JMEM, Groote JF, Somers LJ (2007) Transformation of bpmn models for behaviour analysis. In: MSVVEIS-2007. INSTICC Press, Milan, pp 126–137

# Chapter 32
# Positing a Factorial Model for Consumer Trust in Mobile Payments

**Ahmed Shuhaiber, Hans Lehmann, and Tony Hooper**

**Abstract** Payments via mobile and wireless terminal devices (i.e. cell phone/smartphone handsets) have been introduced by businesses for more than a decade now—alas mostly with less than desired success. There are numerous barriers to widespread adoption and the required wireless architectures are often prone to competitive as well as technological incompatibilities. However, one important acceptance/appropriation determinant is users' trust. A review of current literature indicates insufficient levels of consumer trust in mobile payments—despite its widely acknowledged potential. Moreover, this seems to be accompanied by a dearth of research aimed at establishing, isolating and operationally defining the factors that influence consumer trust in these payments. This study explored such trust factors in an explorative, qualitative study using focus group sessions in a significant Middle Eastern country. The results are formulated in a factorial framework based on five main conceptualisations: user/consumer characteristics; environmental (social, business and cultural) influences; provider profiles; mobile-device specifications; and the nature/level of perceived risks.

**Keywords** Consumer trust • Mobile commerce • Mobile payment • Trust in mobile-payment services and applications

## 32.1 Introduction

We have entered a new 'all mobile' era, in which mobile phones are used as phones, internet connections, organizers, jukeboxes, games consoles, messaging devices, shopping tools and others. This diverse range of options over mobile phones has

A. Shuhaiber (✉) • H. Lehmann • T. Hooper
School of Information Management, Victoria University of Wellington, Wellington, New Zealand
e-mail: ahmed.shuhaiber@vuw.ac.nz; hans.lehmann@vuw.ac.nz; tony.hooper@vuw.ac.nz

M. José Escalona et al. (eds.), *Information System Development: Improving Enterprise Communication*, DOI 10.1007/978-3-319-07215-9_32,

made it possible for mobile users to expect their devices to function as an all-in-one wallet. Thus, the popularity of mobile devices is increasing day by day. Utilising mobile devices in the business and commerce fields leads to the concepts of mobile business and mobile commerce. Mobile commerce (m-commerce) refers to exchanging products and services via mobile telecommunications networks [1, 2]. M-commerce has many applications, such as mobile shopping, mobile marketing, mobile banking, mobile ticketing, mobile entertainment and others.

In order to complete an m-commerce transaction, a customer needs to exchange values, goods and services with a wireless mobile device. This monetary transaction that is associated with m-commerce is called a mobile payment. A mobile payment (m-payment) is defined as "a payment where a mobile device is used to initiate, authorize and confirm an exchange of financial value in return for goods and services" [3, p. 141].

## 32.2 Trust in Mobile Payments

For more than a decade, telecommunications companies have offered mobile payment services. However, and despite the potential of m-commerce and m-payments, trust is a major obstacle in its adoption and development [4–6]. Trust is a multi-disciplinary term, and has many meanings, dimensions and characteristics. Moreover, trust has been studied in psychology, management, communication, sociology, economics and political sciences. Trust in m-payments has many facets and dimensions: psychological, social, cultural, technological and technical aspects. Thus, trust in m-payments is complex and is not easy to understand.

Many scholars in the field of electronic and mobile commerce have argued that there is a lack of trust in m-payments worldwide [5, 7, 8]. Other scholars argue that customer trust in m-payments at least needs to be developed [9–11]. These problems facing trust in m-payment can be explained by a lack of understanding of the factors that influence customer trust in m-payments.

Trust and culture are closely related constructs [12, 13] and probably one cannot be properly understood without the other [14, 15]. Previous research, which examined culture and online trust across cultures, suggests the need to include culture in the framework because trust, and its antecedents, changes across cultures [16–19]. With respect to the Arab and Gulf countries, and the United Arab Emirates (UAE; "Emirates") specifically, no scholarly research has yet been found that discusses the factors that influence trust in m-payment. Instead, there are very few relevant studies with regard to m-payments adoption and development in few Arab countries, which discuss trust as an independent construct. The literature points out the importance of trust when adopting m-payments, without clarifying how this trust could be achieved, and what factors influence it. As a result, we do not understand the factors that influence customer trust in m-payments in Arab countries. This study aims to understand the factors that influence customer trust in Consumer–to-Business (C2B) m-payments. The Emirates is a Middle Eastern developing country, and is one of the Arab and Gulf states. Trust has been indicated as an important factor that impacts the adoption of m-payment services in the Emirates and other Arab countries [20–23], but its specific relevance to m-payments has not been studied, and its factors have not yet been assembled. A resultant framework could shed valuable insight into this important field, and through a distinct cultural lens.

The next section presents the research methodology, followed by the section of the findings demonstration. Finally, the last section rounds off with discussion and conclusion.

## 32.3 Methodology and Data Collection

As mobile payments are a relatively new research area in the UAE with little previous empirical work on the subject, a qualitative approach using focus group interviews was chosen to explore consumer trust in m-payments. The focus group technique has been used in social research, but possibly it's most obvious use has been applied to investigate consumer habits and preferences [24], and to test customer reactions to products or services [25].

The focus group discussions followed a semi-structured guide, which was tested with a pilot group of five participants. The group was consistent with the research design, and thus there was no need for major modifications to the guide. Four focus group sessions took place in the Emirates, in the four main emirates (Dubai, Abu Dhabi, Sharjah and Al-Ain). Three of these sessions were conducted in Arabic, while one was in English.

To ensure proper discussion and interaction during the sessions, the four naturally forming groups were selected for the current study. In order to maximise disclosure among focus group participant, occupation-based homogeneity between each focus group members was followed as a key consideration in establishing selection criteria for individual groups. In other words, members knew each other as friends, classmates, co-workers, or through a common hobby. The following groups were interviewed: professional adults, academic staff and students, professional young adults, and adults of parents and young workers. The group sizes varied between six and seven participants, consistent with academic suggestions [24, 26]. The interviews lasted between 70 and 90 min.

In total, 27 multi-national volunteers (Emirati, Jordanian, Syrian, Palestinian, Egyptian, Sudanese, Indian, Pakistani, Bangladesh, Spanish, Australian and American) aged between 19 and 52 participated in the sessions. They shared an awareness of m-payments concepts, technologies and experiences in m-payments services in the Emirates. With participants' permission, all sessions were recorded and some notes were taken to commentate on the participants' perceptions. The sessions were then transcribed and sent back to the participants for checking and validation. Afterwards, Arabic transcripts were translated into English and prepared for analysis.

The participants were reached through three main entities: telecommunication companies, banks and online social groups and forums. These entities have been particularly selected because the term m-payment is closely related to telecom companies and banks, and their customers are seen to be the most familiar with the m-payment concept and characteristics, have some knowledge about the concept, or already have used it.

## 32.4 Findings and Analysis

The key themes are identified from the focus group discussions process. The information provided is considered to be appropriate due to its relevance to the stated research objective, and its value in revealing patterns, themes and concepts relating to the factors influencing customer trust in mobile payments.

To analyse the discussions' data, we used the Constant Comparative Analysis (CCA) strategy, which is widely applied in qualitative research [24]. The CCA strategy consists of three processes: (1) open codes, (2) axial codes, and (3) selective codes. The first step begun with is open coding, which is the process of identifying concepts, and discovering their dimensions by breaking down, comparing, conceptualizing, and categorizing data [27, 28]. Afterwards, axial coding is used to group the codes developed during open coding into categories. In axial coding, the coding occurs around the axis of a category, linking categories at the level of properties and dimensions [27]. The transcripts are inspected for similarities or differences and grouped into groups of conceptual units. In the last step, selective coding is used to integrate and refine the concepts emerged during axial coding.

By applying the CCA approach, analysis revealed many important aspects that play a role in affecting customer trust in mobile payments, such as: customer past experience in m-payments, social influences, security and technical risks, reputation of the m-payments provider, and other issues associated with the mobile device. In more details, the factors emerged during the focus group discussions were categorized into five main groups: customer characteristics, environmental (social and cultural) influences, provider characteristics, mobile-device characteristics, and perceived risks. Next sequenced sections describe these groups and the factors included in each group.

### *32.4.1 Customer Characteristics*

The focus group discussions exposed some personal characteristics and intrinsic values of customers that could influence their trust in m-payments. For instance, the majority of all participants argued that customer past experience in using m-payments services and other online payment methods can strongly influence their trust in conducting an m-payment. For instance, one participant argued: "Previous experience in m-payments is a factor of trusting it… My personal experience in conducting an m-payment determines my trust in it". Another participant argued that trust in m-payments could be gained through a cumulative practices and past experiences of this payment method, by saying: "Trust increases or decreases by usage. To me, at the beginning of launching m-payment services in the Emirates, I had some concerns in using it, but these concerns were away when I started using this service, and trust has become higher and higher". Discussions further noted that the first experience in and usage of an m-payment service or application significantly influence trust in m-payments. Moreover, the relationship between customer past experience in m-payments and trusting it was illustrated, depending on positive and negative practices and past experiences he/she had with m-payment services and applications.

Away from past experiences and practices, the majority participants argued that customer' knowledge in and awareness of m-payments services and applications would influence their trust in m-payments. In a participant's words, "the significant issue which is related to trust in m-payments is that to what extent customers are aware of this service and how much they know about it". According to one participant, awareness of m-payments means "to be aware of m-payment services, applications, its characteristics and details". The interviews specified that awareness could be associated with several sources of knowledge, such as education, profession and culture.

Two demographic variables were identified as contributing factors of trust in mobile payments through the discussions. Some participants indicated that the age of customers could influence their trust in m-payments. For instance, one participant argued that age is associated with trust in technologies in general, and with m-payments in specific, in a direct relationship. "I would say age plays a significant role in trust in m-payments… I would say the older the age the less the trust". Other participants preferred to use the term generation instead of age. For instance, a participant argued: "I agree that age is a factor of trust. I cannot say the older the people the less the trust, perhaps it is a matter of generations. Our generation has grown with this technology… The older generation are more conservative and take a while to pick up and use such a new technology". Arguably, few participants indicated a relationship between gender and trust in m-payments. For example, a participant argued: "I feel males are more likely to trust in m-payments than females". Another person disagreed with him, arguing that "females tend to trust in m-payments without deeply thinking… They may not think of security and technical troubles when conducting the payment, so they trust in it more than males".

## *32.4.2 Environmental (Social and Cultural) Influences*

The majority of all participants argued that social influences, especially the word of mouth, are factors of trust in m-payments. For example, a participant argued: "I trust my friends. If I trust a friend, I trust his/her perceptions and experience in m-payments". Similarly, another participant argued: "I think here [in the Emirates] word-of-mouth is very effective… In the past, I did not trust in buying air tickets online by my mobile in the past, but I was encouraged by a friend.

I tried it and started trusting it since then. His advice resulted in my trust…". Participants illustrated the effect of word of mouth and the social interactions among customers by the significant others, such as family members, relatives, friends, colleagues and others.

Discussions identified mass media as another source of social influences. Participants argued that media can make customers aware of m-payments characteristics, the thing that can lead to trusting in it. "My trust could be affected by Media means… Media provides me with the awareness and knowledge required to understand this technology and start trusting it". Similarly, another participant agreed and continued: "Media is very affective on my trust… Media can offer customers the consciousness of m-payment services and affect my mind without I know…when getting such information about the service; I will naturally feel that I have a positive attitude towards this service and makes me trust it". Interviewees further noted that media could play a major role in trusting m-payments, especially if the service is newly adopted and not many customers had tried it before. Other interviewees considered this influence as conditional, subject to the advertiser (service provider) and how trustworthy it is. However, the influence of the provider is mentioned later in the paper.

Many participants suggested that external parties could influence customer trust, such as governmental legislations for customer protection, and supporting the service by an external party. For instance, a participant argued that "if m-payment services are supported by governmental legislations and policies then customer trust in these services would increase dramatically… banks could also play that role by monitoring payments and protect their customers". Another participant agreed and said: "Such policies protect customers and provide compensations in case of troubles". In addition to the regulations legislated by governmental agencies, participants supported the acting of financial institutes as a third mediated party between the provider and the customer, which indicate the importance of a third party certificate when conducting an m-payment.

Participants also considered the culture factor and its impact on trust in m-payment. Some of the participants viewed culture as the background of a person, or as a literacy, while others defined culture as the atmosphere around a person. The majority of the participants considered culture as an important factor that influences customer trust in m-payments. A participant argued: "culture shapes a person's behaviour. If you come from a background or a culture that tend to trust new technology easily, you will be initiative to trust in it". Similarly, another participant said: "there are people who are open minded to new technologies where others may not… it could be also related to the environment surrounding them… I think the multi-cultural environment here makes the people somehow willing to trust new technologies".

Many participants argued that the availability of m-payment services and the abundance of its application in a country could impact customer trust in m-payments as a cultural influence. For instance, a participant argued: "to what extent m-payment services and applications are available for usage and how they are spread impact a person's trust… The more abundant the service the more trustworthy it will be". Another participant agreed and continued: "I agree. Involving some public institutions, such as power and water companies, in the m-payment services would increase trust in m-payments… It becomes a norm and a prevailing trend, and this automatically increases trust in these services".

### 32.4.3 Provider Characteristics

The discussions indicated that trusting the provider of m-payment services is significant for trusting in m-payments. The provider in this context, as determined by the participants, is a telecom company (two main telecom companies were specified). The majority of the participants pointed out that the reputation of the provider is strongly related to their trust in m-payments. For instance, a participant argued: "to me, trusting in m-payments is a result of how I see the service provider and what I heard about it. For example, I do not trust in the m-payment service provided by the company because I do not trust in the company itself, and this is because I hear about its problems frequently". Individual participants associated the provider's reputation with several issues, such as the efficiency of the employees' staff, its system processes, its services and products, or the number of years in business.

Some participants agreed that the number of years in business could influence their trust in m-payments. For instance, one participant argued that the number of years the provider in business, or provider's age as he described it, results in more knowledgeable and trustworthy its staff members. He said: "provider's reputation, which I think is an important factor of trust, is associated with the provider's age. The older the company, the more experienced its staff members and the higher ability to deal with problems". Likewise, another participant argued: "the older the company the more experience in the industry and the more professional and solid in business, and accordingly the more trustworthy will be".

Another characteristic of the provider that was discussed during the focus group sessions was the size the company. Some participants demonstrated the size the provider (the telecom company) as the number of branches it has, the extent of availability and the degree of spanning domestically or abroad, while others related the size of the company to the number of the company's employees and customers. Many of the participants argued that the size of the provider affects their trust in m-payments. For instance, one participant said: "…the more the customers and staff members, the more the branch numbers, the higher availability and spanning of its services, the higher trust".

### 32.4.4 Mobile Device Characteristics

A minority of the participants argued that the brand of the mobile device has an influence on their trust, and that there are some mobile devices that are more secured than others. For instance, a participant argued: "some mobile devices are more secure than others, by its operating systems, and some other devices are more vulnerable to security breaches. For example, Galaxy mobile devices are less secured than iPhone or LG. Therefore, I would trust in iPhone or LG for my m-payments more than Galaxy mobiles". Similarly, another participant argued: "In the Emirates, some mobile devices are provided with some settings to connect them with banks for shopping. For instance, BlackBerry has special built-in software that connects the user with the telecom and the Abu Dhabi National Bank to ease the process of purchasing online. This service distinguished BlackBerry from other mobile devices such as Nokia or Samsung, and made m-payments more trustworthy". Thus, some participants consider some sort of mobile device brands to be more trustworthy than other devices.

The design of the mobile device is related to trust of few participants in m-payments. For example, one participant argued: "the device could play a role in trust, especially when using a touch screen mobile". Another participant agreed and gave some justifications, by arguing: "my concern in the device is the touch screen option. This can cause in entering wrong numbers and amounts of money, or can transfer the amount to someone else's account". The discussions further indicated that lack of trust in m-payments could be related to the mobile software than its hardware, and other individuals related the latency of the device and its battery life to trust in conducting an m-payment.

### 32.4.5 Perceived Risks

The participants perceived financial risks as the amount of money to be paid by the mobile device, while they associated technical risks with mobile networks and telecommunications. Security risks were perceived as hacking possibilities, fraud cases, and stealing credit card numbers, whereas privacy issues were discussed as exposing personal information and details when conducting an m-payment transaction. Participants' main concerns were about financial and security issues. Technical issues came second, whereas few of them showed privacy concerns.

The majority of all participants argued that their trust in conducting an m-payment is associated with the amount of money they will pay by their mobile devices. They preferred conducting micro m-payments more than conducting macro m-payments. Around half of the participants argued that they would not trust in paying more than 500Drhs (≈USD137) by their mobiles. One participant argued: "the amount of the m-payment I am a going to conduct is important to me and impact my trust. I do trust in m-payments but I have concerns regarding paying big amounts of money by my mobile. It could be the same case for my other online payments…". Another participant agreed, and continued to describe the relationship between the amounts of money to be paid by the mobile and his trust in m-payments. He said: "The bigger amount of money paid through the mobile, the more risky transaction will be, the lower trust people will have in m-payments".

Similar to financial risks, the majority of the participants agreed that security risks influence their trust in m-payments, such as hacking, fraud, and steeling credit card numbers. For instance, one participant argued: "I have some hacking concerns. A hacker can steel my credit card info and other personal details". Likewise, another participant argued: "I think there are lot of security breaches while conducting an m-payment, hacking and so on. Hacking is spread all around the Emirates, and people here are aware of it… I have other concerns, such as dealing with Visa Cards and some information about the card owner". The participants indicated that the mentioned security issues are also related to other online payment methods.

Many of the participants indicated that there are some technical issues associated with m-payments, and these issues can influence their trust. For instance, one participant argued: "…I cannot trust in m-payments blindly. Although the technology of m-payments is well developed here in the Emirates, most times I am scared of technical problems, such as getting disconnected or network malfunctioning". Participants identified some forms of technical risks that are associated with mobile payments, such as system is down, fault in the service, uncompleted processes, or an instantly loosing coverage when moving from one coverage area to another.

Few participants concerned about privacy concerns and argued that it can influence their trust in m-payments. For instance, one participant said: "I do have some privacy concerns. I understand their need for my credit card information, but I think that further details such as the balance of my bank account, or unneeded data such as my age are part of my privacy that I do not like to share with others publicly". Similarly, another participant argued: "…I know some people who fear from privacy breaching when using their mobiles for payments. They feel like others tracking their activities and behaviour online and accordingly they may not trust in paying by their mobiles".

## 32.5 Discussion and Conclusion

The purpose of this paper was to explore factors that influence consumer trust in mobile payments. The findings, as summarized in Table 32.1, lists these contributing factors categorized into five main groups (trust determinants) along with their positive or negative influences, and the degree of consensus on them.

The findings suggest that the trust in mobile payments are related to specific group of factors; customer characteristics, environmental (social and cultural) influences, provider characteristics, mobile-device characteristics and perceived risks. In addition, it is found that customer past experiences, customer awareness, word-of-mouth, uncertainty avoidance, provider's reputation, financial and technical risks are the most agreed on factors that influence customer trust in m-payments. Furthermore, the findings indicate that the mostly used applications for mobile payments include paying bills (power and water), Car parking, traffic registrations and fines, and mobile-banking payments. The most trustworthy amount of payment to

**Table 32.1** Factors affecting consumer adoption of mobile payments

| Trust determinant | Contributing factor | Proposed effect on trust | Degree of consensus |
|---|---|---|---|
| Customer characteristics | Past experiences | +, − | Vast majority |
| | Customer awareness | + | Majority |
| | Customer demographics | | |
| | Age | + | Some |
| | Gender | + | Few |
| Environmental (social and cultural) influences | Word-of-mouth | + | Majority |
| | Mass media | + | Some |
| | Third-party certificate | + | Many |
| | Uncertainty avoidance | + | Majority |
| | Prevalence of m-payments | + | Many |
| Provider characteristics | Reputation | + | Majority |
| | Number of years in business | + | Some |
| | Size | + | Some |
| Mobile-device characteristics | Brand | + | Minority |
| | Security | − | Minority |
| | Design | − | Few |
| Perceived risks | Financial risks | − | Majority |
| | Security risks | − | Majority |
| | Technical risks | − | Many |
| | Privacy risks | − | Few |

be conducted using a mobile device varied from micro-payments (AED100 ≈ USD28) to low end macro-payments (AED500 ≈ USD137). Participants were also asked to give score of their general trust in m-payments in the Emirates in a 0–10 scale. The average score of their trust was 7.13 out of 10.

These findings suggest that in order to deal with the lack of trust, customers need to be better familiarized with current m-payment services and applications. A wide adoption of m-payment services and application could result in increasing the customer awareness in this relatively new type of payment, and consequently becomes part of their culture. Providers can associate governmental agencies or financial institutes in the payment provide, which in turn could make the customers feel more confident and more trusting in m-payments, regardless the perceived potential risks.

In addition, these findings are found consistent with trust in electronic and mobile payments literature. For instance, customer past experience and awareness of m-payments were found significant to customer trust in m-payments in several studies [5, 7, 8, 29]. Another example is the word-of-mouth, third party recognition, provider's reputation, security and privacy risks were also evident to have significant influence on trust in m-payments [4–10, 30]. However, what is found original to our knowledge and to date are: mass media, provider's size and age, technical risks, and all factors associated with the mobile device (brand, security, and design). The last bunch of factors are seen to be exclusively associated with m-payments because of their direct relation to the device of payment, whereas the remaining others could be related to the Emirati technological culture and environment.

This study provides important theoretical contributions to the existing trust research, by providing a comprehensive overall picture of factors influencing trust in m-payments from a customer perspective. The existing trust models, however, focused on certain aspects of these factors. Testing and validating a framework of the factors would be considered for future quantitative work.

## References

1. Abu Bakar F, Osman S (2005) Towards the future of mobile commerce (m-commerce) in Malaysia. In: Proceedings of IADIS: IADIS international conference, web based communities 2005, Algarve, Portugal
2. Tarasewich P, Nickerson RC, Warkentin M (2001) Wireless/mobile e-commerce: technologies, applications, and issues. In: Americas conference on information systems, no. 7
3. Au YA, Kauffman RJ (2008) The economics of mobile payments: understanding stakeholder issues for an emerging financial technology application. Electron Commer Res Appl 7(2): 141–164
4. Meng D, Min Q, Li V (2008) Study on trust in mobile commerce adoption—a conceptual model. Paper presented at the 2008 international symposium on electronic commerce and security
5. Siau K, Sheng H, Nah F (2003) Development of a framework for trust in mobile commerce. Paper presented at the proceedings of the second annual workshop on HCI research in MIS, Seattle, WA
6. Sreenivasan J, Noor MNM (2010) A conceptual framework on mobile commerce acceptance and usage among Malaysian consumers: the influence of location, privacy, trust and purchasing. WSEAS Trans Inf Sci Appl 7:661–670
7. Corbitt BJ, Thanasankit T, Yi H (2003) Trust and e-commerce: a study of consumer perceptions. Electron Commer Res Appl 2(3):203–215
8. Yeh YS, Li Y-M (2009) Building trust in m-commerce: contributions from quality and satisfaction. Online Inf Rev 33(6):1066–1086
9. Lu J, Yu CS, Liu C (2005) Facilitating conditions, wireless trust and adoption intention. J Comput Inf Syst 46(1):17–24
10. Qingfei M, Decai M, Qiuyan Z (2008) An empirical study on trust in mobile commerce adoption. Paper presented at the service operations and logistics, and informatics, IEEE/SOLI 2008. IEEE international conference
11. Siau K, Shen Z (2003) Building customer trust in mobile commerce. Commun ACM 46(4):91–94
12. Doney PM, Cannon JP, Mullen MR (1998) Understanding the influence of national culture on the development of trust. Acad Manage Rev 23(3):601–620
13. Fukuyama F (1995) Trust: the social virtues and the creation of prosperity. The free, New York
14. Benbasat I, Gefen D, Pavlou PA (2008) Special issue: trust in online environments. J Manage Inf Syst 24(4):5–11
15. Hofstede G (2001) Cultural consequences: comparing values, behaviours, institutions and organizations across nations, 2nd edn. Sage, Thousand Oaks
16. Gefen D, Rose GM, Warkentin M, Pavlou PA (2004) Cultural diversity and trust in IT adoption: a comparison of USA and South African e-voters. J Global Inf Syst 13(1):54–78
17. Gefen D, Heart T (2006) On the need to include national culture as a central issue in e-commerce trust beliefs. J Global Inf Syst 14(4):1–30

18. Pavlou PA, Chai L (2002) What drives electronic commerce across cultures? A cross-cultural empirical investigation of the theory of planned behaviour. J Electron Commer Res 3(4):240–253
19. Strong K, Weber J (1998) The myth of the trusting culture. Bus Soc 37(2):157–183
20. Al-Mamari M (2007) Mobile commerce development in Oman. Unpublished master's thesis, The University of Sheffield. Available from http://dagda.shef.ac.uk/dissertations/2006-07/External/Al-MamariMohammed MScI S.pdf
21. Bhatti T (2007) Exploring factors influencing the adoption of mobile commerce. J Internet Bank Commer 12(3):1–13
22. Manochehri N-N, AlHinai YS (2008) Mobile-phone users' attitudes towards mobile commerce & services in the Gulf Cooperation Council countries: case study. In: International conference on service systems and service management
23. Rouibah K (2007) Does mobile payment technology Mnet attract potential customers? The case of Kuwait. In: Proceeding of the 18th Australian conference on information systems, pp 199–211, 5–7, Toowoomba, Australia
24. Pickard AJ (2007) Focus groups. Research methods in information. Facet Publishing, London, pp 1–329
25. Bryman A, Bell E (2007) Business research methods, 2nd edn. Oxford University, Oxford
26. Gibbs A (1997) Focus groups. Social Research Update 19:1–5. Available from http://www.soc.surrey.ac.uk/sru/SRU19.html
27. Denzin NK, Lincoln YS (1994) Introduction: entering the field of qualitative research. In: Denzin NK, Lincoln YS (eds) Handbook of qualitative research. Sage, Thousand Oaks, pp 1–17
28. Strauss AL, Corbin JM (1998) Basics of qualitative research: grounded theory procedures and techniques, 2nd edn. Sage, Thousand Oaks, CA
29. Lee MKO, Turban E (2001) A trust model for consumer internet shopping. Int J Electron Commer 6(1):75–91
30. Park J, SuJin Y (2006) The moderating role of consumer trust and experience: value driven usage of mobile technology. Int J Mobile Mark 1(2):24–32

# Chapter 33
# Multipoint Web Real-Time Communication

**Ruben Picek and Samuel Picek**

**Abstract** One of the most significant current challenges for the Web is enabling human real time communication via voice and video known as a Real Time Communication (RTC). Many web applications with this functionality need some kind of plugins, downloads or installs. This paper introduces a novel approach on how to integrate real-time multimedia communication into *web browser* as the most natural way. After analyzing problems and addressing open issues of browser RTC, paper focuses on one identify issue—multiple peer connections, presenting potential network topologies with their pros and cons in terms of implementation, cost and quality. Our research is then focus on implementing one multipoint network topologies which enables videoconferencing. Proposed solution is based on using a WebRTC standard as work in progress with the biggest positive echo in research and developers in that field [12].

**Keywords** Web Real-Time Communication • WebRTC • Real-time audio/video in browser • P2P and multipoint communication

## 33.1 Introduction

Researchers and developers have devoted many efforts, often in a nonstandard way and without documentation, for finding appropriate solutions for RTC [9]. RTC has been corporate and complex, requiring expensive audio and video technologies to be licensed or developed in house. Integrating RTC technology with existing content, data and services has been difficult and time consuming, particularly on the web. Many web services already use RTC, but need downloads, native apps or

R. Picek (✉) • S. Picek
Faculty of Organization and Informatics, University of Zagreb, Zagreb, Croatia
e-mail: ruben.picek@foi.hr; samuel.picek@foi.hr

M. José Escalona et al. (eds.), *Information System Development: Improving Enterprise Communication*, DOI 10.1007/978-3-319-07215-9_33,

proprietary plug-in such as Adobe Flash or Microsoft ActiveX. This includes Skype, Facebook (which uses Skype) and Google Hangouts (which use the Google Talk plugin) [6, 8]. Downloading, installing and updating plug-in can be complex, error prone and annoying. Plugin can be difficult to deploy, debug, troubleshoot, test and maintain—and may require licensing and integration with complex, expensive technology. It's often difficult to persuade people to install plugin in the first place [6]. These are just a few reasons which stresses the importance of standardize solution. *Can we solve these open issues implementing RTC in web browser?* WebRTC is an upcoming standard which enables that idea but in peer to peer communication.

This article examines the biggest issue of WebRTC—problem of multiple peer connections. Next sections of article will first, explore WebRTC in more details with his architecture, components and open issues. Then, we present several network topologies for multipoint communication and consider their operating costs and quality of service across different domains in terms of scale. Based on one proposed network topologies, we provide a solution for WebRTC multipoint communication, implementing video conferencing system as one most important (multipoint) field of use. We consider our solution a valuable and significant contribution for the future research in this area.

## 33.2 Web Real Time Communications

This section presents a short review of Web Real-Time Communication (WebRTC) in elements of definition, evolution, component structure and architecture while some very important open issues are discuss in Sect. 33.3. But let's first explain what WebRTC is?

WebRTC is an upcoming standard that aims to enable real-time communications capabilities like voice calling, video chat and P2P file sharing, among new generation of Web browsers via simple Javascript APIs without any plug-ins, downloads or installs [8]. Its purpose is to help build a strong real-time communication platform that works across multiple web browsers and across multiple platforms [3]. WebRTC "system" includes web servers, browsers running various operating systems on various devices including desktop PCs, tablets, and mobile phones, and other servers. Additional elements include gateways to the Public Switched Telephone Network (PSTN) and other Internet communication endpoints such as Session Initiation Protocol (SIP) phones and clients or Jingle clients. WebRTC enables communication among all these devices [8].

Enabling RTC in browsers is under joint development of two main Internet standardization bodies IETF (Internet Engineering Task Force) and W3C (World Wide Web Consortium). Each of them establishes a working group with their aims. These two different but interrelated working groups are: RTCWeb within I ETF and WebRTC within the W3C [6, 8, 9].

RTCWeb group has focused on the *protocols* and *interactions* that the I ETF must address, including interoperability with legacy systems (such as existing

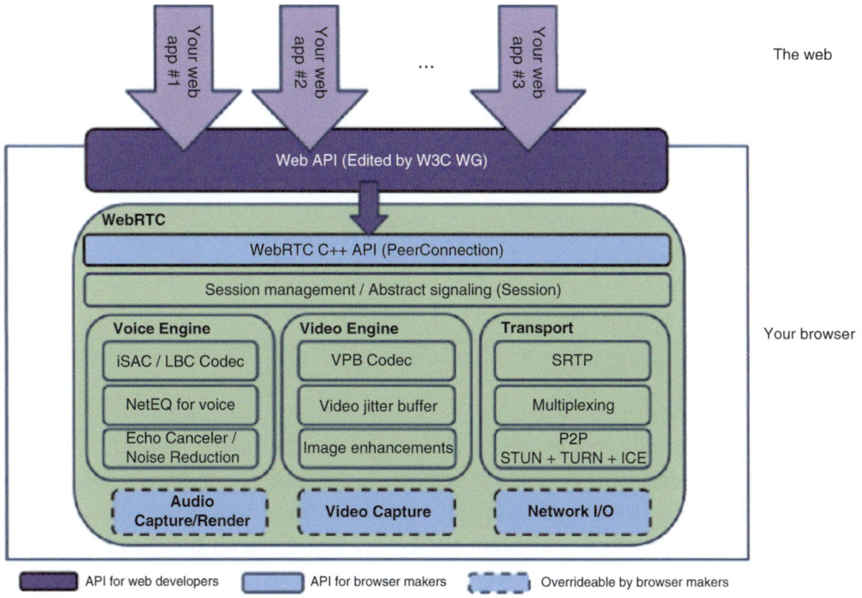


**Fig. 33.1** WebRTC architecture

telecommunications systems). WebRTC group is working to define an *APIs* allowing browsers and scripting languages to interact with media devices (microphones, webcams, and speakers), processing devices (encoders/decoders), and transmission functions, all called with term WebRTC [8, 9]. Currently, WebRTC is a work in a progress with advanced implementations in the Chrome and Firefox browsers [6].

### *33.2.1 Architecture*

According to [3] the WebRTC architecture is presented on Fig. 33.1.

Figure 33.1 represents the overall architecture of WebRTC where the first layer, *Web API* is the main interface designed towards the web application developers rand their applications. This layer is standardized by W3C and it defines the interfaces which are exposed to client side (browser) javascript engine. *WebRTC C++ API (PeerConnection)* layer defines an API which every browser maker must consume if they wish to provide a support for WebRTC standard. Behind this *PeerConnection API* are the classes and interfaces which *PeerConnection* handles and those are: *Session management*, *Voice Engine*, *Video Engine* and definitions for *Transport*. Each of these modules is targeted to solve specific set of problems for RTC, such as echo cancellation, automatic gain control, noise reduction, dynamic jitter buffers, error concealment, network traversal and session setup. The bottom layer defines three components which are overridable by browser makers and those are:

*Audio capture renderer*, *Video Capture* and *Network I/O*. These three components define the basis for *get UserMedia* inside browser and allow browser to access audio and video hardware devices.

### 33.2.2 WebRTC Components

According to specification [1, 2, 6], WebRTC currently consist of three major components:

(1) getUserMedia (aka. MediaStream)
(2) RTCPeerConnection
(3) RTCDataChannel

All of those components are available for end-user developers through Javascript API's. First component, *get UserMedia* is the main component which is responsible for access synchronized multimedia streams (video, audio or both) from local devices (IP cameras, web cameras, microphones). Each *MediaStream* object, when created, has input, which is a *LocalMediaStream* (representing an audio and video data received from local computer), and an output which might be a video element or an *RTCPeerConnection* object. To represent its data on the screen we use video element tag, and to send it over network an *RTCPeerConnection* has to be created beforehand. The main goal behind this API is to enable a *MediaStream* for any streaming data source, currently it only supports camera and microphone, but experimental code has already exists. It has plenty potential fields of use, such as surveillance, recording or real-time communication, better customer service support, education and so on.

To exchange media data which is received from *MediaStream* between browsers (communication channel between peers),WebRTC uses a second component, mechanism to establish communication known as *RTCPeerConnection*. This component is most complex of the big-three, because its task is to provide a stable and efficient communication of streaming data. It handles voice (capture and render), video (capture) and transport (network I/O) processing. Codecs and protocols used inside WebRTC do all the work which makes real-time communication possible and solves a lot of implementations problems for which VoIP developers does not have to worry about anymore, such as: echo cancellation, bandwidth adaptively, packet loss concealment, dynamic jitter buffering, automatic gain control, noise reduction and suppression, image 'cleaning', etc.

This mechanism uses a process known *Signaling* which is not part of *RTCPeerConnection*. WebRTC application developers have to specify which messaging protocol they want to use, such as SIP, XMPP, and Websockets (or any other duplex communication channel should fulfill the requirement). According to [6], the purpose of signaling is to exchange these types of data:

- session control messages: to initialize and/or close communication channel
- network configuration: get computers IP address and port
- media capabilities: which codecs and resolutions are handled by two browsers which uses a WebRTC communication.

Signaling must be done before RTCPeerConnection can be established. Third component, *RTCDataChannel* enables peer to peer transport of any arbitrary data, besides audio and video, with high throughput. The major potential fields of use of this API are gaming, file transferring, remote applications, decentralized networks, etc. The idea behind *RTCDataChannel* is to leverage existing *CPeerConnection* to enable powerful and flexible peer-to-peer communication.

It supports multiple simultaneous channels, different delivery semantics (reliable/unreliable), built-in security (datagram TLS), etc.

Because of lack of space, a detail implementation of each component is not presented here, but code snippet can be found in [6].

## 33.3 Open Issues of WebRTC

A cross-industry effort to create a new/common platform of RTC in browser is not without problems. Some open issues discussed in this section are based on standpoints of industry experts in this filed found on web, authors of scientific papers and books and us as a system architects.

### *33.3.1 Security, Privacy and Technical Issues*

One important open issue is *security*. There are a number of possible ways where WebRTC application can break the security model. For example, unencrypted data might be intercepted with some of the man in the middle attacks, application might record and distribute video or audio without user knowing, malware might be installed with WebRTC application which can cause big problems. To avoid these issues WebRTC implements some basic line of defence. According to [6, 11] current implementations of WebRTC use secure protocols such as Datagram Transport Layer Security (DTLS) and Secure Real-Time Transport Protocol (SRTP), for all WebRTC components (including signaling mechanism) usage of encryption is required. All of the components of WebRTC runs inside of browsers sandbox (not as a plugin—separate process), does not require installation and they are updated with the browser which makes the security model of the RTC much secure and reduces the issues to minimum.

Both working groups have to consider *privacy* issues that arise when exposing local capabilities and local streams, *technical* issues on implementing data channels in particular [2], defining set of protocols, enhancing security model, etc.

### *33.3.2 Network Topology Issue: Multipoint Communication*

According to [4, 5, 8, 10] there are network topologies for *point to point* (P2P) session between two browsers, or between a browser and another endpoint called *Triangle* and *Trapezoid* indicating the shape of Signalling.

But, one of the most common problems in architectural design of the WebRTC application is choosing the right *network topology* for data exchange to support multipoint or conferencing sessions involving multiple browsers. This is known as *multipoint videoconferencing problem*. The lack of documentation and research/design patterns in this area leave the architect empty handed which forces him to do such research on their own. WebRTC only defines connection between two endpoints in the network, and videoconference requires support for the multiple endpoints connected simultaneously. This is straightforward but complex problem: *how to connect multiple endpoints over different devices on the web?*

Enabling video conference is dependent on very robust network for handling the demands of the video, audio and other arbitrary data which takes place on it. Although, WebRTC addresses a lot of problems which are connected with robust network, network challenge becomes even well-marked as the number of endpoints in the video conference grows. There are two standpoints from which we need to consider this problem. First is *economic standpoint*: more endpoints to support, the bigger operating cost is, therefore we need to consider different network topologies which will reduce the operating cost of the video conferencing system thereat not endanger the second standpoint which is *quality of service* (QoS). These standpoints are argumented with Earon [7], where he with different scenarios describes different operating costs and quality of service factor for different topologies.

This issue—choosing the right network topology for multipoint communication is the focus of our research and in next sections we propose several multipoint communication network topologies and present one of the possible solutions for the videoconferencing problem using WebRTC standard.

## 33.4 Multipoint Network Topologies

This section describes the multipoint network topologies in detail. Proposed topologies for multipoint web real time communication are: Media Server, Mesh Topology, DAG Topologies.

### *33.4.1 Media Server Topology*

Media server topology is one of the topologies with high operating costs at scale but very efficient and easy to implement video conferencing application with WebRTC standard. When we talk about media server topology in WebRTC context we need to mention several server cluster topologies which we used in our research.

*Single-server topology* provides a fully functional application server which is capable of serving a small to medium community. In this architecture databases (persistent and caching), message queues, application (RESTful API) and web servers are defined on one machine which handles all the traffic.

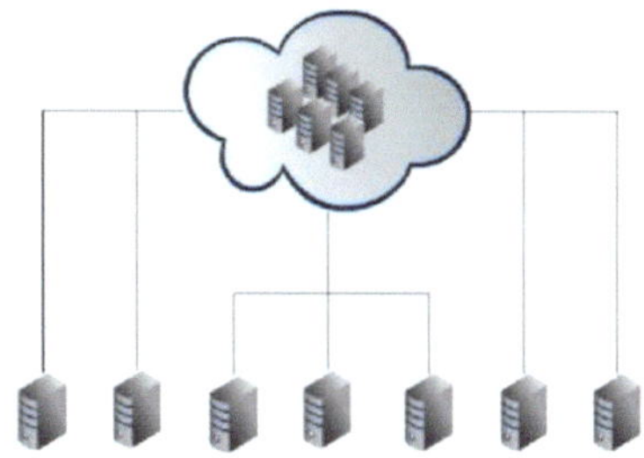

**Fig. 33.2** Media server topology

In *stand-alone topology* all the components which define the system for handling RTC requests are on different physical machines. The combinations of machines which form a complete distributed system are behind the load balancers in server cluster. This topology is useful for medium to large community.

The Fig. 33.2 describes the most general case, in which clients connect to our server cluster which exposes the API for handling *Rooms*, *Tokens* and *Users*.

The advantage of this topology is that at small scale is very efficient and provides the initial grounds for further testing with its easy to implement interface, but at scale operating costs goes up which is main deficiency of this topology. However higher operating costs means that we can handle more simultaneous sessions while minimizing the amount of data for processing by each browser. In terms of scaling this topology, we can choose vertical or horizontal scaling.

To *scale horizontally* we need to add more nodes to server cluster. As hardware prices drop and performance goes up, this is considering a low cost for a commodity which we are getting. This is also known as scaling out.

To *scale vertically* means to add more resource to single node in the server cluster. This typically involves adding more CPU or memory power to single machine. This enables existing system inside server cluster to use virtualization technology more effectively. This is also known as scaling up.

Both scaling up and scaling out rises our operating costs, but to conclude does it scale the quality of service or experience we have to consider the bottleneck. The main bottleneck in this architecture is the network traffic/bandwidth between our server cluster endpoints and our clients, because in this architecture every client in the current video conference session connects to the server cluster which is responsible for handling payload on each connected client. The payload from one client gets distributed to all other clients in the current session through this server cluster and therefore cluster must be equipped to handle high payload. So, all the operating costs for scaling must go to solving this main bottleneck. When solved efficiently, this architecture's provides high reliability from end user perspective and sustainability from providers' perspective. Quality of service and quality of experience are very high on small number of video conferencing participants (5–10 people) on which we have concluded the tests using the WebRTC standard. So, according to [8] this topology has the advantage of being able to scale to very large sessions while also minimizing the amount of processing

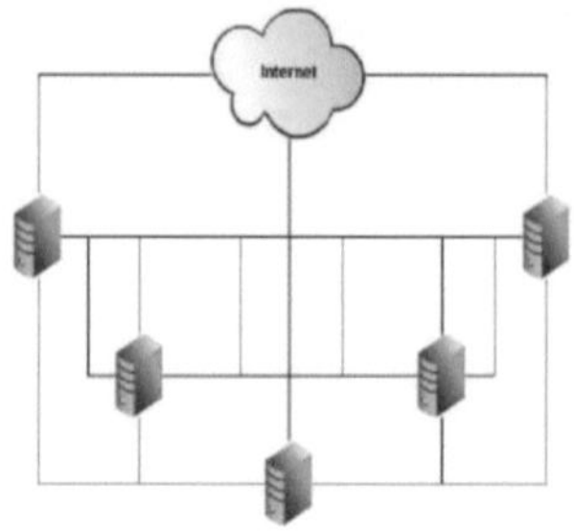


**Fig. 33.3** Mesh topology

needed by each browser when a new participant joins the session, although it is perhaps inefficient when only one or a small number of browsers are involved, such as in peer-to-peer gaming.

### *33.4.2 Mesh Topology*

Another way to implement multi-party sessions in WebRTC is mesh topology. To implement this, each browser in our video conference session has to establish a peer connection with every other browser in the session. Saving all of the peer connection objects inside client browser application is going to produce very fat clients, as the number of the participants inside video conference session grows. This can produce a huge bottleneck on the client side and the performance may drop significantly because different users will have different bandwidth speed.

The Fig. 33.3 shows the topology. As mentioned, every client is connected to every other client. This has advantage of no Media Server infrastructure therefore operating cost is low while quality of service (media latency and video/audio quality) for small number of participants remains intact. This topology may not be suitable for large number participants in video conference session because bandwidth required to sustain the overall session grows for each new participant. Because network connections are not symmetrical inside one session, users with low bandwidth cannot handle video conference session with big number of participants [8]. Compared to *Media Server Topology* this topology cannot scale at price of operating cost.

### *33.4.3 DAG Topologies*

Directed Acycling Graph topologies are one of the most interested ones because it gives ability to boot performance of the overall network, but they are the hardest to implement. This topology is used in very specialized cases such as games or event streaming. The idea behind this topology is that top level parent is the creator of the room and the initiates the video conference session. Every child has a peer connection directly to its parent. The Fig. 33.4 shows the DAG topology.

Implementation of this topology need to be investigate in future work.

Fig. 33.4 DAG topology

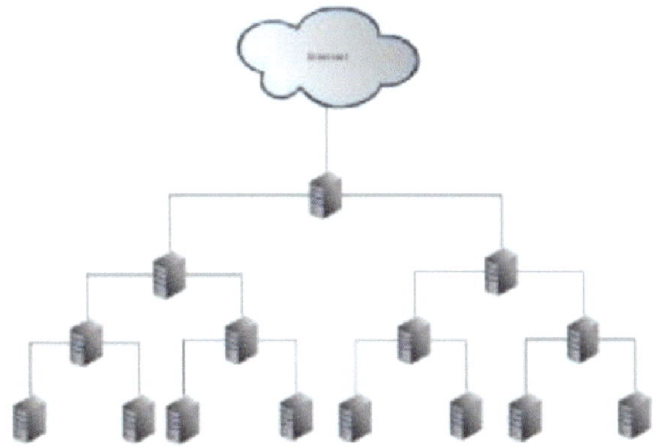

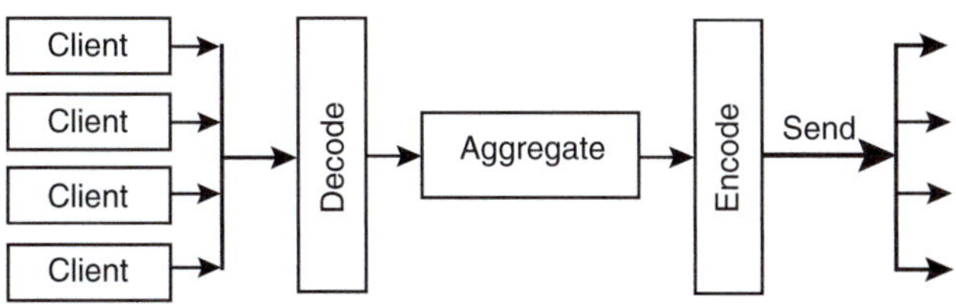


Fig. 33.5 Media server API architecture

## 33.5 Media Server Topology Implementation for Multipoint Videoconference

After providing a detail view of proposed topologies we describe design solutions which we used as *Media Server API*. The reason why we decided to implement *Media Server* architecture for our research is low operating cost and ability to carry out additional stress test's at scale. When we discuss server side RTC multipoint solution there is couple of ways to go about it. *First*, server will handle all video and audio processing, sending every endpoint exact video stream they should be viewing. S*econd*, server sends multiple streams to one endpoint, while some of the work will be delegated to them like decoding multiple incoming streams. In this case server will act as a relay of the media. We choose the first case, and the Fig. 33.5 and the following text describes it.

The server side will receive all media streams, and decode them. It will make the layout of the end video stream, encode it and send it to all other participants which are subscribed. This is classic solution for enterprise videoconferencing because it requires very small amount of effort from each endpoints.

After we have decided which case we will use, we have to elaborate the designed API. Already we have briefly mentioned three main components of which our API consists, those are: *Room*, *Token* and *User. Room class* (Fig. 33.6) represents a video

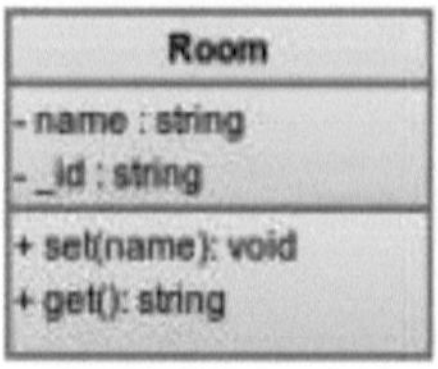


**Fig. 33.6** Room class

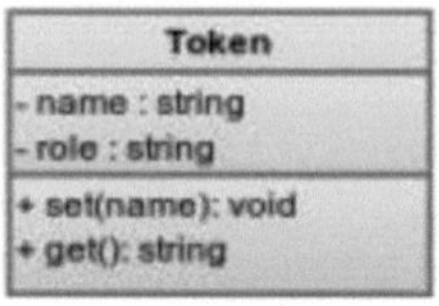


**Fig. 33.7** Token class

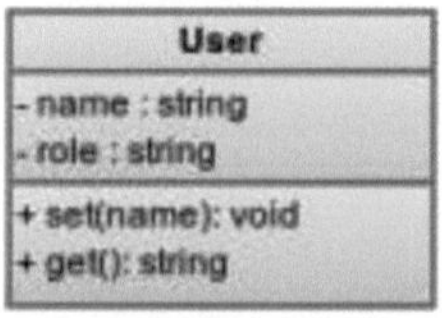


**Fig. 33.8** User class

conference session itself. When created this object represents a conference room. Then in this room users can interchange their streams. Each client in room can publish his stream and/or subscribe to other streams which are published in this room. Each room has attributes: unique identifier, name—and methods: get and set. A *Token* is string object that allows you to add more/new participants to specific room. When new client wants to connect to a specific room, a token need to be created which client will consume and use it to connect to a specific room. Token class (Fig. 33.7) consists of name and role attributes. The name attribute represents the user which is connecting and role attribute specifies user role inside the system. *User class* (Fig. 33.8) represents the participant which is connecting to room with specific token. Before creating the user object, a token has to be created because user class accepts the same attributes—name and role of the participant which requested connection to specific room.

After defining main objects, we have to explain the main interface for manipulation of those objects. With those three, there is also a main *Service Manager* interface which handles initialization and CRUD operations of those objects. The *Service Manager* interface (server side) implements methods from *RoomManager*, *TokenManager* and *UserManager* classes. This interface is represented by UML Class diagram shown at Fig. 33.9.

This interface will be behind developers RESTful service and atop of implementation which enables every client to make a *PeerConnection* towards the server. Then the server will act as an aggregation unit of all those *PeerConnections* inside one room. Inside room each user can subscribe to any or all streams inside room and it can publish its own media stream. Each of the manager classes will handle the

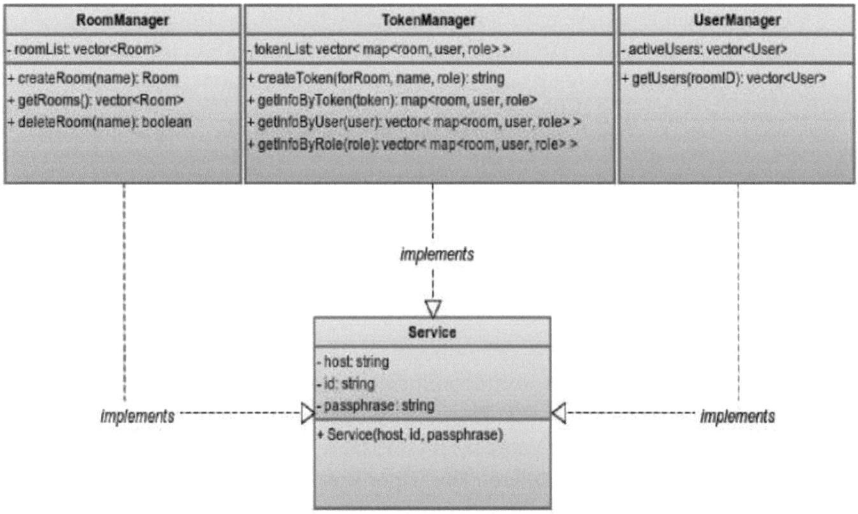


**Fig. 33.9** UML class diagram for Service Manager interface

persistency of data across the sessions with backend database. Any production ready database could work in this case, but we used noSQL database—mongoDB, because we wanted the entire web stack to be in JSON. Such webstack is really viable for web environment because it minimizes the need for serialization/deserialization of data types and it minimizes the traffic which is generated. Now that we have defined this API, we need to define access towards it—we have done this with RESTful API which consists of these created routes:

- */new _room* - POST method
- */new _token/<room_name>* - POST method
- */rooms* - GET method
- */users/<room>* - GET method

When we make a POST request towards*/new_room* route we have to send new rooms name. Then this procedure will get triggered and it will call our API for creating a new room—*createRoom*. Then the *Room Manager* will create a new Room class on the socket and it will save this room to database. When POST request is directed to*/new_token/<room_name>* route it will execute *create Token* method inside *Token Manager* and it will create SHA1 universally unique identifier which will represent key for connecting to specified room. Now on the client side when we wish to connect to specific room we have to obtain this token which we connect with *PeerConnection* initialization towards the server. GET request on our RESTful API are for retrieving some important data. Route*/rooms* will retrieve all of the rooms which are active inside that service [Fig. 33.9]. Route*/users/<room>* will retrieve all the users inside specific room. This feature is usually necessary for user interface design—to see who is online.

This implementation (API) is embedded to NODEHALA project which go alive (26.04.2013) at conference event at Faculty of organization and informatics, called *Skills exchange*, where were present 35 participants. We stream this event as video-conference for session of 4 h and have 41 people on stream. During this period of time, we have 25 concurrent connections. Everything was smooth and only problem was interference of white color, what is a big success of our research.

## 33.6 Conclusions and Future Work

In this paper authors present short review of current state in Real Time Communication with an emphasis on WEBRTC as upcoming standard which will enable Real Time Communication into web browser. The paper points out how problem of multipoint network topologies fits into the context of WEBRTC and present potential network topologies for solving this problem. The paper presents an implementation of one possible solution which enables videoconferencing. Our research—developed API, shows that at least one multipoint network topologies is possible and can be implement, enabling browser to browser communication via simple Javascript APIs and HTML what is valuable and significant contribution. Future work our research will be focused on DAG network topology and finding a way of their implementation.

## References

1. WebRTC (2012) WebRTC 1.0: real-time communication between browsers. http://dev.w3.org/2011/webrtc/editor/webrtc.html
2. RTCWeb Working Group (2012) WebRTC data channel protocol. http://tools.ietf.org/html/draft-jesup-rtcweb-data-protocol-00
3. WebRTC Code and API, specification from WEBRTC organisation, http://www.webrtc.org/reference/architecture
4. Ada Bo X, Yin L, Ruimin S, Wenyin L, Changjun J (2004) PipeSeeU: A scalable peer-to-peer multipoint video conference system. Springer, Berlin, pp 475–479
5. Davies M, Zeiss J, Gabner R (2012) Evaluating two approaches for browser-based real-time multimedia communication. In: MoMM '12 proceedings of the 10th international conference on advances in mobile computing & multimedia, ACM
6. Dutton S (2012) Getting started with WebRTC. http://www.html5rocks.com/en/tutorials/webrtc/basics/#toc-security
7. Ann Earon S. Video conferencing network issues: costs and options. http://www.slideshare.net/GlobalETM/video-conferencing-networkissues-options
8. Johnston AB, Burnett DC (2012) WebRTC: APIs and RTCWEB protocols of the HTML5 real-time web. Digital Codex LLC, St. Louis
9. Loreto S, Romano SP (2012) Real-time communications in the web issues, achievements, and ongoing standardization efforts. IEEE Internet Comput 16:68–73
10. Ponec M, Sengupta S, Chen M, Li J, Chou PA (2009) Multi-rate peer-to-peer video conferencing: A distributed approach using scalable coding. http://research.microsoft.com/pubs/81366/ponec.pdf
11. Rescorla E (2011) Proposed WebRTC security architecture, IETF 82
12. Smith A (2013) Get started with WebRTC. http://www.netmagazine.com/tutorials/get-startedwebrtc

# Chapter 34
# Affecting Decision-Makers' Attention through Mobile BI: Implications for Mobile BI Design Process

**Olgerta Tona and Sven A. Carlsson**

**Abstract** Mobile business intelligence (BI) enables the mobile workforce to attain knowledge by providing access to information assets anytime anywhere. Mobile BI's main aim is to support decision-makers during the decision making processes. In decision situations many issues may be relevant, but it is impossible for decision-makers to attend to all of them as decision-makers are constrained by their limited capacity of attention. This paper uses the attention-based view theory to explore mobile BI and discuss its implications in terms of decision-makers' attention and design process. The results of this study are based on analyses of published material, e.g. trade press and white papers, and semi-structured interviews with important stakeholders in the mobile BI field. This paper shows that the access to issues and answers anytime anywhere via mobile BI and its alerting capabilities affect the focus of attention of the decision-makers. The stakeholders should be actively involved in the design process so that the efficiency and effectiveness of mobile BI in the distribution of the decision-makers' attention will be enhanced.

**Keywords** Mobile business intelligence • Decision-making • Attention-based view

## 34.1 Introduction

Commentators on decision support and Decision Support Systems (DSS) have called for serious discussion of the discourses underpinning decision support and DSS [1, 2]. Carlsson [3] taking part in these discourses—which are critical to the advancement of the DSS field as well as to DSS practice—proposes the attention-based view

O. Tona (✉) • S.A. Carlsson
Department of Informatics, Lund University School of Economics and Management, Lund, Sweden
e-mail: olgerta.tona@ics.lu.se

M. José Escalona et al. (eds.), *Information System Development: Improving Enterprise Communication*, DOI 10.1007/978-3-319-07215-9_34,

as a basis for DSS design. It is important to design DSS which are congruent in how an organization channels and distributes the attention of its decision-makers. Along these lines, attention becomes an essential construct. The attention problems such as its limited time and capabilities are highlighted and brought into the centre of theories and models in some of the organizational decision-making literature. In many decision situations many issues are relevant to consider, therefore decision-makers receive too many signals but not every signal can be attended to at once or attended at all. The attention-based view theory of the firm explicitly links structure, activity and cognition and the view stresses that organizational decision-making is affected by both the limited attentional capacity of humans and the structural influences on a decisionmaker's attention.

Since 1970s, in an attempt to support the decision-making process, a wide area known as Decision Support System (DSS) has emerged. During the history of DSS, different sub-fields have emerged such as: Personal DSS, Group Support Systems, Negotiation Support Systems, Intelligent DSS, Knowledge Management-Based DSS, Data Warehousing and Enterprise Reporting and Analysis Systems [4]. Among the sub-fields most present in practice, business intelligence (BI)—categorized in Enterprise Reporting and Analysis Systems—has been heavily implemented in different industries [4]. BI combines data gathering, data storage, knowledge management with analyses and provide the right information to different users in an organization, enabling them to take decisions and act [5]. Traditional BI delivers the information via web-based portals or desktop applications. Therefore, the employees can access it via their PC or laptop devices usually connected to the network of the organization.

However, the advancements in technology are not limiting the users to their PCs and laptops in the office, but also are empowering them while 'on the road' [6]. Hosack et al. [7] did a study about the future of DSS and one of the trends pointed out is mobile computing. They argue that mobile devices offer a platform which challenges the traditional DSS. It allows the users to interact with the data anytime and anywhere. The mobile devices in this case create a new platform where the size, speed and reach of data lead to technological advances to decision-making. Due to the new technological developments, mobile BI as a new sub-field of BI has emerged. Mobile BI enables the mobile workforce to attain knowledge by providing access to information assets anytime anywhere. Along these lines mobile BI users are encouraged to take decisions 'on the move'.

O'Donnell et al. (2012) found that senior executives have already started to implement mobile BI in their companies for a variety of operational purposes and developing BI on mobile devices is one of the main topics of concern among practitioners. Nowadays, decision-makers are faced with an information overflow generated from different sources. Sometimes, many sources may contain conflicting information and decision makers have to decide which source to trust and above all where to focus their attention. On the other hand, based on the discussion of Carlsson [3], DSS shall be designed and managed to regulate and govern organizational attention in direction which are in line with the organization's strength, weakness, opportunities and threats. Additionally, he calls for more studies on how attention-based view can guide the design of DSS. Considering the limited academic research conducted in this new sub-field (mobile BI), the main contribution of this paper is to explore mobile BI using the attention-based view of the firm as an evaluation framework. Consequently, this paper will shed light on (1) the effects of mobile BI on decision-makers' attention and (2) the implications of the attention-based view in the mobile BI design process.

The remainder of the paper is organized as follows. The theory of attention-based view, which will be adopted throughout this study, is presented. Section 34.3 discusses the methods used followed by the discussion. Conclusions are presented in the final section.

## 34.2 Attention-Based View Theory

In order to explain firm behaviour one shall actually explain how firms and their structures channel and distribute the attention of the firm's decision-makers [8]. According to Ocasio [10], attention includes noticing, interpreting, and focusing of time and effort by decision-makers on different issues and answers. Issues are related to the problems, opportunities and threats faced by the company whereas the answers are the actions taken such as projects, proposals and routines. Attention plays also an important role in strategic renewal. "While it is difficult to specify the conditions under which the identification or creation of opportunities will occur, we can state that innovations and solutions cannot be created without organizational attention… Organizational attention refers to the allocation of information processing capacity within the organization to a defined issue or agenda" [11].

According to Ocasio [10] the attention-based view of the firm is based on three interrelated theoretical principles: (1) focus of attention, (2) situated attention, and (3) structural distribution of attention. *The focus of attention* principle says that what a decision-maker is doing depends on what issues and answers a decision-makers focus. The decision-makers will select from the repertoire of issues and answers, the ones they find appropriate to attend at a specific time. This means that they are selective and they can also ignore other issues, which they may find irrelevant. *The situated attention* principle says that what issues and answers decision-makers focus, and what they do, depends on the specific context, setting and the situation they find themselves in. Therefore, the situation with which decision makers are encountered will influence their focus of attention and consequently their actions. *The structural distribution of attention* principle says that what specific context, setting, and situation a decision-maker finds herself in, and how she attends to them depends on how the firm's rules, resources, and social relationships regulate and control the distribution and allocation of issues, answers, and decision-makers into specific activities, communications and procedurals. Hence, the attention of the decision-makers is influenced and shaped by the organization as well.

Simon has noted that information and information systems might have a significant negative impact upon decision makers' attention focus: "What information consumes is rather obvious: it consumes the attention of its recipients. Hence a wealth of information creates a poverty of attention" [12]. If information and IS can have a negative impact on decision-makers' attention it is most likely that they can also have a significant negative impact upon organizational actions and moves and ultimately firm performance.

According to Carlsson [3], the attention-based view is a promising view of organizational decision making and this view can be a bases for DSS design. He discussed the implications of attention-based view for the decision support portals (DS-P) to channel and distribute the attention of organizational decision-makers. A DS-P can be seen as a personalized front end through which a user (decision- maker) can access all the information, applications and services needed to perform decision and knowledge related work and activities. According to the attention based view of the firm, the design of DS-P shall be based on the three theoretical principles. A DS-P affects a decision-maker's attention and the decision-maker's actions and moves by providing access to issues and answers (*focus of attention* principle). A DS-P is a part of a decision-maker's context, setting and situation (*the situated attention* principle) and is also a part of the organization's rules, resources, and social relationships which regulate and control the distribution and allocation of issues, answers and decision-makers into specific activities, communications, and procedurals (*structural distribution of attention* principle). It can be argued that DS-P differences lead to changes in the procedural and communication channels and that these changes may have significant impact on organizational actions, moves and performance.

## 34.3 Research Method

In order to explore the impact of mobile BI on decision-maker's attention, two complimentary main sources are used: publications and interviews with main stakeholders in the mobile BI field.

In terms of published material, the searches have been concentrated to two main business literature databases: ABI/Inform's Global and the Business Source Complete. The keyword used during the search was "*mobile business intelligence*." ABI/Inform's Global yielded 43 articles whereas Business Source Complete 20 articles. Most of them formed the practitioner publications where trade press, magazines, product reviews were included. Additionally, a search on the key mobile BI vendors' websites produced five white papers. Of 68 articles from the databases and white papers only 45 articles were further analysed. The excluded articles were omitted because of: (1) duplicates in the databases, (2) referring only to BI, and (3) referring only to mobile applications in general.

In terms of publications, Fig. 34.1 shows their distribution over the years. An increase of mobile BI publications is observed during 2010, which corresponds with the release of the tablets in the same year.

In addition to the mobile BI publications, other empirical data were collected through seven interviews with representatives from mobile BI vendor companies and a consultancy (see Table 34.1). These companies were selected as they are the main players in the mobile BI field. The interviewees at the mobile BI vendor

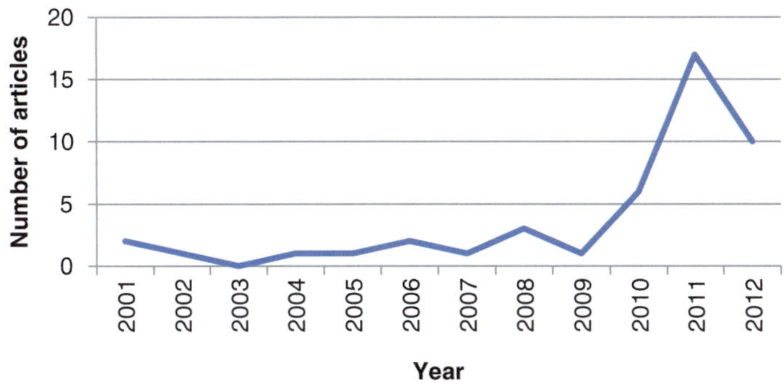


**Fig. 34.1** Published articles over the years

**Table 34.1** List of interviewees

| No. | Cases | Responsibility of the interviewee | Location |
|---|---|---|---|
| 1 | Japersoft | Director Product Marketing | San Francisco, US |
| 2 | Qliktech | Product Manager | Lund, Sweden |
| 3 | QNH | Mobile BI Consultant | Amsterdam, Netherlands |
| 4 | Smart eVision | Vice President | Naperville, IL, US |
| 5 | Tableau | Product Management Director | Seattle, WA, US |
| 6 | Transpara | Vice President/Founder | Pleasanton, CA, US |
| 7 | Yellowfin | Strategic Alliances and Technical Account Manager | Melbourne, Australia |

companies and consultancy are persons with adequate knowledge of the area. The interviews ended once the thematic saturation was reached, where according to Creswell [13] that point comes where no new information can be obtained to add understanding of the main categories and themes emerging.

The participants were contacted via e-mail. Due to their different locations most of the interviews have been conducted via Skype. The interviews lasted for approximately 1 h each. Interviewees' experiences, perceptions and ideas on mobile BI were the main drivers of the interview guide. Semi-structured interviews were chosen because it made it possible to add or ask the questions in different ordering as the interviews unfold. The interviews were recorded with the consent of the interviewees. They were transcribed and e-mailed back to the participants for comments or feedback. Once confirmation was obtained, the transcripts were ready to be analysed.

The analysis technique used is content analysis, which, based on Krippendorff [14], valid inferences from texts will enhance the understanding of mobile BI. All the articles and interview transcripts have been read carefully, being attentive to the metaphors, stories, and keywords used to get a rich picture of the impact of mobile BI on decision-makers' attention. During the reading, following Miles and Huberman [15], the content was organized into chunks of words, sentences or paragraphs with the same meaning. Based on the meaning, chunks of information representing the same theme were labeled and coded.

## 34.4 Discussion

In this section we will discuss mobile BI using the three main principles of the attention based view: focus of attention, situated attention, and structural distribution principle.

### *34.4.1 Focus of Attention*

The access of information anytime anywhere via mobile BI has been accentuated by nearly all the interviewees and the business articles. According to Yuan et al. [16] this is mainly due to the capabilities of mobile devices which are wireless, connect easily to the networks and thus access the information faster. Consequently, having information in the palm of the hand enables 'decisions on the go'.

Based on the analyses of the data and the attention-based view theory, we argue that mobile BI adds value in triggering the focus of attention. Having access to the necessary information anytime anywhere, a decision-maker's attention and actions are affected. He will focus his attention on a limited number of issues and answers—problems, opportunities or threats—that the organization is facing and need to be handled by the decision maker. Mobile BI provides him the necessary access to the repertoire of the issues and answers, and depending on the context he will choose to respond or ignore them.

Strongly related to the focus of attention, another feature accentuated in business articles [17–19] and among the interviews conducted is the alerting/notification feature of mobile BI. '*Armed with mobile BI apps, store managers can receive messages from the corporate- exception reporting tool, alerting them to fast- or slow-moving merchandise, allowing them to fine-tune assortments and displays on the fly.*' [19]. Based on the attention-based view of the firm [10], the store managers' attention is captured by the alerts and they are provided with the necessary information to take further actions. Alerting seems to have an important role in shifting the decision-makers' attention to the most important issues they need to be aware of or take further actions.

Another example provided by one of our interviewees support the argument of how mobile BI by means of its alerting function can be used to address an emergency.

> So, for example, in a bio tech company in San Francisco, somebody got an alert on their phone and was able to see the data that one of the batches of the drug was going bad; the temperature has dropped below or something went wrong and they were home as it was weekend or something; they were able to rush in and save the batch and it was around a 500,000 dollars batch. So, one instance… software pays for 10 times over at least and I will say most of our customers do that. (Interviewee, Transpara)

This case is consistent with the findings of Gebauer and Shaw [20] and Yuan et al. [16] where the use of notifications in real time is correlated to the need for handling emergency situations. For this bio-tech company time is critical and alerting on time empowers the users by giving them access to the essential information at the right time. The users had the necessary 'weapons' to take decision about fixing the batch, although they were not in the office; a fast decision which resulted in reduction of decision cost [21]. The importance of being alerted depends upon the type of the organization. The bio-tech company discussed above is an operational company where the data must be up to date 24/7, known as mission critical data. On the other hand there are some financial companies where in most cases you have to wait till the end of the month to make sense out of the data and alerting significance fades away: '*Financial thinks in months periods, book-keeping periods. Everything that happens in between is like in a time vacuum. Just… not happening until the end of the month. So, you can get real time financial information but it tells you nothing.*' (Interviewee, Consultant).

However, alerting can be twofold. Besides the benefits described above, sometimes, the alerts can be interruptive and have a negative impact on the decision-makers focus. For instance, if he is working in an important task and receives an alert, his focus will shift. He will deal with this alert and eventually that will consume some time before he continues with his previous task, although the former might be more important than the latter for which he has been notified.

We conclude that during the design of mobile BI special attention shall be directed on its alerting feature—based on its effects—in order to focus and distribute the attention of decision-makers in directions that are "congruent" with the organization's strengths, weaknesses, opportunities and threats. During the design of mobile BI it is important to consider the culture and social aspect of the organization where mobile BI is being implemented. In general decision makers will consider and attend the issues with greater legitimacy, value and relevance. Therefore it is crucial to include them in the design process where the main Key Performance Indicators (KPIs) and their corresponding thresholds, which the users need to monitor and be notified are identified. The rules of the games [10], which guide the decision makers in understanding, interpreting, and taking actions within a specific organization shall also be taken into consideration in order for mobile BI to be effective in the decision maker focus of attention.

### *34.4.2 Situated Attention*

According to this principle, the context triggers the focus of attention of the decision-maker and thus depicts its actions. By means of mobile BI, the decision-makers have access to the information anytime anywhere, but according to Ocasio [10] they will concentrate their efforts on specific issues based upon the situations they find themselves in, as well as the positions and responsibilities they have in the company.

For instance, one financial distributor, an early adopter of mobile BI, described how salespeople can check '*the latest activities with a customer they are about to visit, such as the amount of revenue historically generated by that customer*' [22]. In this case we see the salesperson accessing the large data in his mobile device on his way and takes the necessary actions based on the customer profile. The decision-maker attention focus is affected by the specific context: customer site. Based on the context, the attention of the sale person is focused in knowing some general information about that specific customer such as the amount of revenue generated. At this point mobile BI gives him the possibility to access the data at the right moment and consequently mobile BI becomes integrated in the context and situation. Its main role is to support the user to make decisions and take further actions: such as how to behave or what to offer to his customer based on the background check. Hence, mobile BI supported the user to concentrate his energy and efforts on that specific issue, without wasting time or channelling his attention to other areas such as: whom to call or ask for that specific customer. Therefore we find this innovation part of a decision-maker's context, setting and situation and as such it helps the decision-maker to take decisions in different contexts.

Another example is when the salespersons are "*…going to a client site, in a meeting, they use their mobile device to show some products or use some data to their clients or partners*" (Interviewee, Smart eVision). In this instance we notice a joint collaboration with the customer where they can both take a decision based on the information provided by mobile BI. Again, we find mobile BI becoming part of the situation where the decision-maker finds himself in. It supports the collaboration between the users by proving the necessary information at the right time and place, affecting their attention in that specific situation. Consequently it enhances communication among participants who are responsible for the decision making process [21]. This leads to decision time reduction, faster customer care service and as a result a higher satisfaction. These benefits deriving from its usage in different contexts and situations may be in line with organization' s objectives.

Therefore we conclude that mobile BI is an integrated part of the context and situation that decision-makers attention focus is affected by. According to Carlsson and Tona [23], we argue that mobile BI has great potential to emerge with its users and tasks leading to the creation of an emergent entity where it will be hard to distinguish it from its users, work, processes and contexts. Being part of that context, mobile BI supports its users to take decisions faster, collaborate, communicate and keep their focus on the specific issue they have decided to address. It is important that during the design of mobile BI the designers and other stakeholders work iteratively through the model of situated attention and make specific design decisions. So, it is essential to know mobile BI users and the responsibilities they have in the company. Trying to customize it based upon the role each user have will support them during the decision making process in the different situations and contexts which affect their focus of attention.

### *34.4.3 Structural Distribution of Attention*

According to Carlsson [3], in order to support structural distribution of attention mobile BI shall be part of the organization's rules, resources, and social relationships which regulate and control the distribution and allocation of issues, answers and decision-makers into specific activities, communications, and procedurals.

The main mobile BI users in an organization are the mobile users. By means of this innovation the organization wants to empower especially the users who are in the field and in need of information more or less continuously. In an effort to classify IS mobile users, Dahlbom and Ljungberg [24] classified them as wanderer, traveller and visitor. In addition to this classification, Andersson [25] suggested the ranger as a fourth category. Wandering is related to the local mobility within the working environment; a traveller usually travels from one place to another; the visitor spends some time in another location; a ranger is completely detached from the organization and very rarely visits the organization. In terms of mobile BI users three main categories are observed: executives, sales employees and field personnel. We classify executives as travellers since they are usually away on business trips, on the move, e.g. in cars, on trains and planes, or in meetings. The sales employees have a resemblance to the ranger, as they need to go to different locations and meetings with their customers and clients and being away from the office most of the time. Field personnel are rangers who very rarely go to the office and by having access to mobile BI they have all the information they need. Based on Andersson [25], mobile BI use seems to be complimentary to traditional BI use in the cases of executives (travellers). For sales employees (rangers) and field personnel (rangers) traditional BI is not an option.

Mobile BI becomes part of the organization's resources in supporting mobile users to be in contact with the issues and answers concerning the organization during the time they are out in the field. The attention of decision makers is influenced by the regulations and procedures settled by the organization such as: being supplied with mobile devices supporting mobile BI, having access anytime anywhere on the plethora of information, being alerted or notified in real time for the main key issues and answers. Additionally, mobile BI becomes part of the communication channel used by the company integrated in the spatial dimension [10] as technology available. It helps during different meetings—within the company and outside it—to allocate the firm's attention.

> And now with mobile BI what the CFO can do is to open his iPad in the morning when he is in the back office. He favorites or marks the information that he is going to show that afternoon to the report session, so in that way he is controlling, self-servicing, he feels confident with the system and the information, and when he is presenting it, he is also confident in front of people, the board, the financial people...so it is really important for those guys. (Interviewee, Consultant).

During a meeting session, mobile BI provide the users with key information which may drive the meeting and as well will help to define the issues and answers on which the decision-makers have to focus more. Also, they may handle the issues on time, without involving the other BI analysts to provide the necessary reports which will result in waste of time.

Additionally it supports the interaction among participants within the communication channel. One of the interviewees argues that 'k*ey to our mobile app is collaboration and you can effectively collaborate in a very effortlessly*.' (Interviewee, YellowFin). It relates to the finding of Mayer and Weitzel [26] where communication is more accentuated in smaller devices and as such this functionality adds value to the usage of mobile BI by enhancing the communication between the decision-makers.

We conclude that the designer shall work closely with the stakeholders to distinguish the mobile BI main users' profiles to be feed with information in order to channel their attention towards specific issues and answers, which need to be handled. As mobile BI will become part of the politics, procedures and communication channels of the organization, its impact on the distribution of attention among its users shall be evaluated and the best approach for a seamless integration shall be embraced.

## 34.5 Conclusions

Mobile BI has emerged recently as a sub-field of BI, in response to the new business needs of the market in terms of updated, real time information anywhere at any time. This paper discusses mobile BI by using an attention-based view framework based upon three main principles: focus of attention, situation attention and structural distribution of attention. Throughout this study we also argue that the design process of mobile BI shall be driven by the attention-based view theory.

This paper concludes that mobile BI influences the focus of attention of decision makers due to its capabilities in providing access anytime anywhere to the answers and issues on which decision-makers need to focus their attention. Additionally, mobile BI becomes part of the situations and contexts the decision-makers find themselves and support them during the decision-making process in terms of decision reduction time. It enhances the interaction and collaboration among the participants within the channels as a result of its integration on the organization protocols, procedures and communication channels to 'direct' the attention of decision-makers.

During the design of mobile BI, we argue that alerting—as an essential feature—should be considered carefully (because of its drawbacks) to increase the efficiency and effectiveness of mobile BI in the distribution of the decision-makers' attention. Therefore, the involvement of decision-makers during the design process is important in: (1) determining the most relevant and legitimate issues and answers which require their focus and attention, (2) identifying the main KPIs and their thresholds for which they should be notified on time, and (3) defining the frequency of the alerts to avoid certain drawbacks like interruption and shift of the attention to unimportant issues. Additionally, mobile BI user profiles should be identified so that customization can be applied based on each profile, need and most probable situations and contexts in which decision-makers will find themselves in. Along these lines, the attention of decision-makers will be influenced to remain focused in the most relevant and legitimate issues based on the role they play inside the company.

Moreover, the features of mobile BI shall be aligned with the procedures, protocols and communication channels of the company, as it will become an integrated part of them. Having a better comprehension of the company regulations, where mobile BI is going to be implemented, will improve the chances of an efficient and effective mobile BI usage.

## References


1. Humphreys PC (1998) Discourses underpinning decision support. In: Berkeley D, Widmeyer G, Brezillon P, Rajkovič V (eds) Context-sensitive decision support systems. Chapman & Hall, London
2. Stabell CB (1987) Decision support systems: alternative perspectives and schools. Decis Support Syst 3:243–251
3. Carlsson SA (2008) An attention-based view on DSS. Encyclopedia of decision making and decision support technologies. IGI Global, Hershey
4. Arnott D, Pervan G (2008) Eight key issues for the decision support systems discipline. Decis Support Syst 44(3):657–672
5. Negash S, Gray P (2008) Business intelligence. In: Burstein F, Holsapple C (eds) Handbook on decision support systems 2. Springer, Berlin Heidelberg
6. Cowie J, Burstein F (2007) Quality of data model for supporting mobile decision making. Decis Support Syst 43(4):1675–1683
7. Hosack B, Hall D, Paradice D, Courtney J (2012) A look toward the future: decision support systems research is alive and well. J Assoc Inform Syst 13(5):3
8. O'donnell P, Sipsma S, Carolyn W (2012) The "Hot" Issues in Business Intelligence: The View of Practitioners. In: Respício A, Burstein F. (eds.) Fusing Decision Support Systems into the Fabric of the Context. Amsterdam: IOS Press
9. Simon H (1947) Administrative behaviour. Macmillan, New York
10. Ocasio W (1997) Towards an attention-based view of the firm. Strat Manag J 18(Summer special issue):187–206
11. Simons R (1995) Levers of control: how managers use innovative control systems to drive strategic renewal. Harvard Business School Press, Boston
12. Simon H (1997) Designing organizations for an information-rich world. In: Lamberton DM (ed) The economics of communication and information. Edward Elgar, Cheltenham
13. Creswell JW (2007) Qualitative inquiry & research design: choosing among five approaches. SAGE Publications, Thousand Oaks
14. Krippendorff K (2012) Content analysis: an introduction to its methodology. SAGE Publications, Thousand Oaks
15. Miles MB, Huberman AM (1994) Qualitative data analysis: an expanded sourcebook. SAGE Publications, Thousand Oaks
16. Yuan Y, Archer N, Connelly CE, Zheng W (2010) Identifying the ideal fit between mobile work and mobile work support. Inform Manag 47(3):125–137
17. Pabbisetty M (2010) Mobile BI: raising the IQ of the enterprise, one device at a time. Saama white paper
18. Gentile B (2011) Taking intel to the pineapple fields [online]. http://www.channelweb.co.uk/crn-uk/opinion/2118696/taking-intel-pineapplefields. http://www.chainstoreage.com/article/focus-business-intelligence Accessed Feb 2013
19. Amato-Mccoy D (2012) Focus on: business intelligence. Store managers harness big data with mobile apps. Chain store age—news and analyses for retail executives, Accessed June, 2013 http://www.chainstoreage.com/article/focus-business-intelligence
20. Gebauer J, Shaw MJ (2004) Success factors and impacts of mobile business applications: results from a mobile e-procurement study. Int J Electron Commerce 8(3):19–41
21. Holsapple CW, Sena MP (2005) ERP plans and decision-support benefits. Decis Support Syst 38(4):575–590
22. Pickett M (2006) Lincoln financial distributor implements mobile, real- time business intelligence. DM review
23. Carlsson SA, Tona O (2012) Using business intelligence to prevent and solve crime in Wallander land. J Decis Syst 21(3):217–230

24. Dahlbom B, Ljungberg F (1998) Mobile informatics. Scand J Inform Syst 10(1):227–234
25. Andersson B (2011) Harnessing handheld computing—managing is support to the digital ranger with defensive design. In: Jain H, Sinha A, Vitharana P (eds) Service-oriented perspectives in design science research. Springer, Berlin
26. Mayer JH, Weitzel T (2012) Appropriate interface designs for mobile end-user devices—up close and personalized executive information systems as an example. 45th Hawaii International Conference on System Science (HICSS)

# Chapter 35
# Generating a REST Service Layer from a Legacy System

**Roberto Rodríguez-Echeverría, Fernando Maclas, Vlctor M. Pavón, José M. Conejero, and Fernando Sánchez-Figueroa**

**Abstract** Web 2.0 phenomenon, REST APIs and growing mobile service consumption are leading the development of web applications to a new paradigm, named cross-device web application. Most organizations often possess legacy systems which should face an ongoing evolution process to enhance its accessibility and interoperability. In such scenario, a REST API plays a key role, defining the interaction layer between the legacy system and all its heterogeneous frontends. This work presents a model-driven approach to derive a REST API from a legacy web application within the frame defined by a modernization process. In this work we detail the API generation process and provide a sample implementation instrumenting one of the studied web development frameworks to evaluate the suitability of the approach.

**Keywords** Web services • Software modernization • Model driven engineering

## 35.1 Introduction

Since the publication of the REST architectural style [1], the design and development of RESTful web services has been gaining momentum and now it may be considered the *de facto* standard to define the interaction between Web 2.0 frontends and their backends [2, 3]. Actually, REST services may be conceived as one of the key factors in the success of a web application, so a great effort is dedicated to its development and evolution. REST services define, on the one hand, a way of lean integration among a service provider and other applications (consumers), and

---

R. Rodríguez-Echeverría (✉) • F. Maclas • V.M. Pavón • J.M. Conejero
F. Sánchez-Figueroa
University of Extremadura, Caceres, Spain
e-mail: rre@unex.es; fernandomacias@unex.es;
victorpavon@unex.es; chemacm@unex.es; fernando@unex.es

M. José Escalona et al. (eds.), *Information System Development: Improving Enterprise Communication*, DOI 10.1007/978-3-319-07215-9_35,

on the other hand, a mean to get a cross-device web application, spreading the original scope of the web application.

Legacy systems have always considered web service technology as a proper mean to gain interoperability and to decrease their evolution related costs. Since RESTful services became mainstream fueled, first, by the Web 2.0 adoption and, then, by ongoing mobile service consumption raise, legacy applications have searched for alternatives to evolve their interaction layers (web or RPC services) to this new standard and become cross-device web applications.

The main goal of this work is to present an approach for the automatic generation of a REST service layer that provides an alternative access to a legacy web system. This process is framed within a whole modernization strategy: the MIGRARIA project [4]. This project defines a systematic and semi-automatic process to modernize legacy non-model-based data-driven web applications into RIAs. The process starts with the generation of a conceptual representation of the legacy system by using reverse engineering techniques. According to that project, this conceptual representation is based on the MIGRARIA MVC metamodel that allows the specification of legacy web applications developed by means of MVC-based web frameworks. Once the conceptual representation is obtained, the modernization engineer must decide the legacy system functionalities that should be part of the target system, i.e. the new RIA client. In other words, she must specify the set of components or subsystem to modernize. This decision is carried out in our approach by the selection of the different views included in the subsystem within the generated conceptual model of the legacy application. Note that the modernization process defined by MIGRARIA does not aim at sustituying the current system by a new one. Indeed, its main purpose is to complement the system by the generation of new ways of access and interaction such as a RIA client [5] and its corresponding server connection layer, a REST service layer in this work. In that sense, the selection of views to identify the subsystem to be modernized is the strategy most aligned with the modernization process defined.

In this work we present, on the one hand, the architecture of a REST support layer defined as an extension of the Struts v1.3 web development framework and, on the other hand, the model-driven process to adapt such layer to obtain a REST service layer conformed to the legacy subsystem to be modernized. It is worth mentioning that the presented approach defines a conceptual process to generate a REST API from a legacy web application. The selection of a concrete technological environment, mainly defined by Strus v1.3, is just for illustration purposes.

The rest of the paper is structured as follows. Section 35.2 introduces our approach to generate REST services from a legacy web application. The related work is discussed in Sect. 35.3. Finally, main conclusions and future work are outlined in Sect. 35.4.

## 35.2 The Approach

Based on the selected subsystem to modernize, the process must generate, on the one hand, a RIA client and, on the other hand, a REST service layer that allows the communication between the new RIA client and the legacy web application.

In this paper, we focus on the approach defined to generate the REST services in the context of the modernization scenario defined.

In the specification of the REST service layer we may identify two different parts clearly separated: (1) a part of the system depending on the web framework used to develop the legacy application; (2) a part specific for the legacy system being modernized. The former includes all the extensions and adaptations performed in the framework to add it the REST capabilities aforementioned. The latter mainly concerns to the specification of the behavior of the service layer for a particular web application, namely: (a) the mappings among REST requests and legacy controllers; (b) the input (request) and output (response) data conversion among the formats expected by, on the one hand, the RIA client and, on the other hand, the controller. Obviously, while the first part remains unchanged for different legacy web applications, the second one depends on the concrete legacy system being modernized. The process followed to specify the web application dependent part consists of the next activity sequence:

1. Identification of the resources related to the subsystem to be modernized.
2. REST URI[1] generation for the identified resources.
3. Identification of the available operations over the resources based on the selected views, their requests and the controllers that manage these requests.
4. Generation of request mappings for the operations available for each resource and for a particular data format.
5. Delegation of the REST request handling to the controllers available in the legacy application and providing the corresponding data conversions.

A detailed description of both parts of this REST service layer generation process is provided in the next sections.

### 35.2.1 Framework-Dependant REST Service Layer Specification

In order to generate a REST service layer from the legacy application, the framework used to develop the legacy system must be extended conveniently to enable the handling of REST requests. From a modernization point of view, we may rely on two different approaches to obtain this REST service layer:

1. To generate the REST service layer from its conceptual description by using model-to-code transformations.
2. To manually implement the REST service layer and, later, automatically generate its configuration based on the information extracted from the legacy application and the functionality to be included into the target RIA client.

The main disadvantage of the first alternative concerns to the need of adapting (or redefining) the transformations once and again for each web framework considered

[1] Uniform Resource Identifier.

in the approach even, in many cases, for its different versions. Moreover, based on the structural properties of this REST service layer which clearly contains a static part, it does not seem to make sense to completely generate it from scratch by means of transformations. Thus, the second alternative seems more suitable although it implies implementing and maintaining different versions of the REST service layer according to the different frameworks and versions. In other words, this second approach requires different implementations for the framework-dependant part for different frameworks and versions whilst the variable part is automatically generated by means of model-to-text transformations. This variable part represents, in our case, the REST service layer configuration and, thus, it will be automatically generated by the model transformations. The main inconvenience of this second approach (implementation and maintenance of the REST layer for different frameworks) is alleviated by the low frequency of changes (really low or even null) due to framework evolution.

In order to illustrate how a particular web framework is instrumented to provide REST support according to our approach, an implementation for the Struts v1.3 web framework has been developed. This implementation is graphically described in the process shown in Fig. 35.1. This figure is organized according to the lifecycle of a request in the Struts framework extended with the REST support. Note that each step in the lifecycle has been enumerated in the figure. The figure shows, on the one hand, the request-response lifecycle for a request generated by the legacy web application (dashed line) and, on the other hand, the request-response cycle for a REST request generated by the target RIA client (solid line). This comparative illustration allows appreciating the modifications performed to the web framework in order to incorporate the REST support. Notice that the white boxes indicate modified or generated elements. It may be easily observed that the extension has basically required the modification of the classes: *ActionServlet* (that implements the Front Controller[2] in Struts) and *RequestProcessor* (which handles the request lifecycle within the framework). On the one hand, the *CustomActionServlet* class has the next responsibilities: (1) adding support for PUT and DELETE HTTP requests; (2) loading the specific configuration of the REST service layer for a concrete web application; and (3) deciding which *RequestProcessor* must handle the request. On the other hand, *RESTRequestProcesssor* is responsible of: (1) generating the needed context for the legacy controller invocation (step 3-a); (2) generating the response object according to the specified data format (step 3-b); (3) delegating to the legacy controller (*Action*), according to the REST-controller mapping specified in the configuration, the handling of the request (step 3-c); and (4) delegating the response generation in the corresponding component according to the specified data format (step 3-d).

In addition to the commented extensions, the *JSONWriter* class has been added, as an alternative new response composer. In this case, it generates a JSON-valid response from the data generated by the invoked controller (*Action*). Obviously, the generation of responses in other data formats would require the implementation of different Writer classes.

[2] http://struts.apache.org/development/1.x/apidocs/org/apache/struts/action/ActionServlet.html

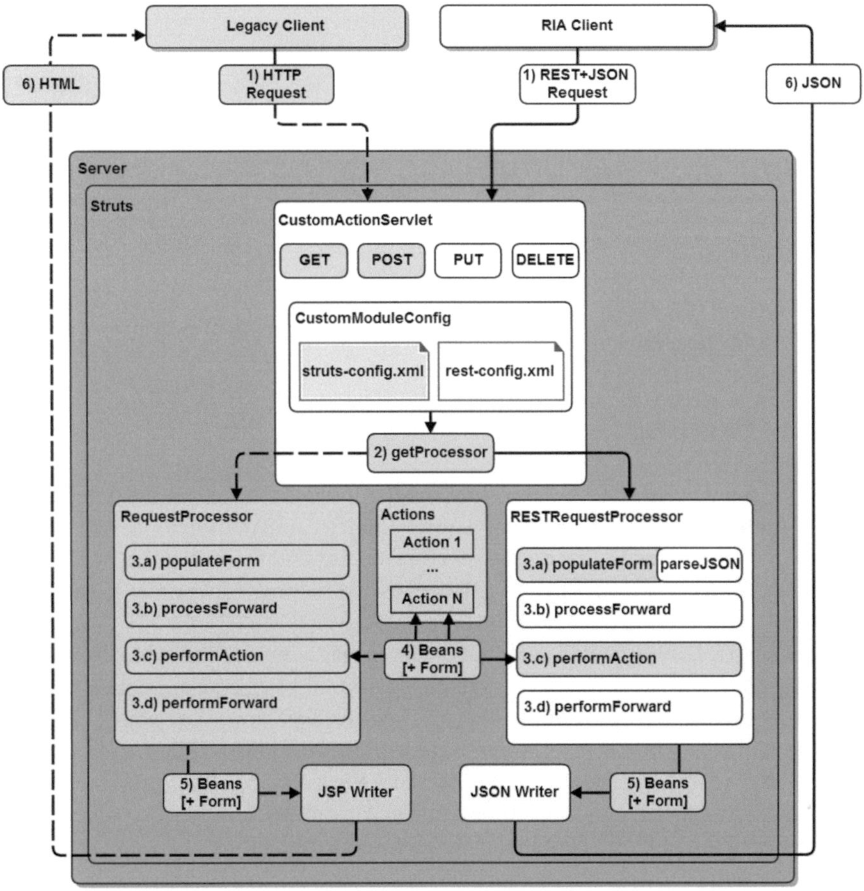


**Fig. 35.1** REST layer for Struts v1 0.3.X

## 35.2.2 Application-Dependant REST Service Layer Specification

Next, we describe the main steps of our approach to extract the information to set up the framework-dependant REST service layer implementation. Such information is obtained from the MIGRARIA MVC model of the legacy web application.

To illustrate the approach we have used an excerpt of our case study named *Conference Review System*.[3] In concrete the subsystem dealing with paper submission by authors.

[3] http://www.eweb.unex.es/eweb/migraria/?Case_Studies:Conference_Review_System

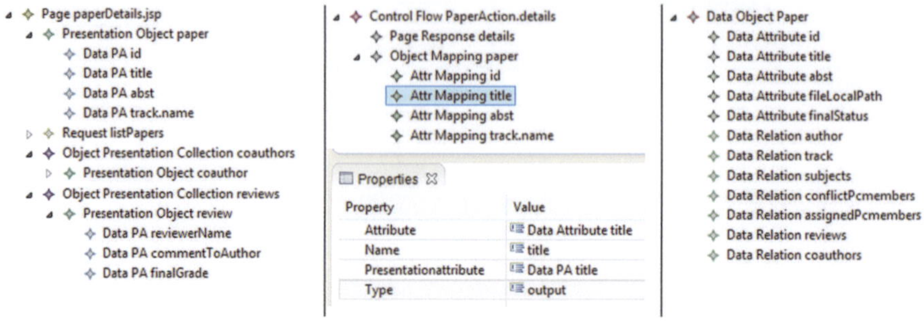


**Fig. 35.2** Resource identification on a MVC model

#### 35.2.2.1 Identifying Resources

The specification of a REST service layer implies to redesign the interaction between client and server from a resource-centric point of view. In that sense, the adequate identification of the involved resources and their relations becomes a fundamental step in the REST service layer generation process. In order to carry out this activity, the modernization engineer must analyze and query the MVC model to find out the resources related with those views that compose the subsystem to be modernized. Each view container (page) is composed by one or more presentation objects that define a concrete data view over the corresponding data objects defined at the model component, e.g. presenting some of their attributes. These presentation objects may represent input or output data and they may be either presented individually or organized into collections. In order to identify the model objects, an analysis of the controller component is required to identify the actual mappings defined among presentation and model objects within a concrete view.

Figure 35.2 presents the presentation object *paper*, included in the view *paperDetails*, and its mapping with the data object *Paper*, defined in the controller *paperAction.details*. As the final result of the analysis of the MVC model, the engineer must obtain: (a) the data objects involved in the subsystem to be modernized; (b) their relationships; and, moreover, (c) the different ways of being presented by views and consumed by controllers. All this information is gathered in the potential resources column of the Table 35.1 which shows the results for the paper submission subsystem in our running example. This table contains also the relations established among the collected resources.

#### 35.2.2.2 Generating REST URIs

Taking as input the information regarding the resources and their relations obtained in the previous step, all the feasible REST URI combinations for each obtained resource are generated by means of the next process. The URIs column of Table 35.2 shows a brief sample of the collection generated for our running example which illustrates the main construction blocks of our process.

**Table 35.1** REST resources and relations

| Views | Potential resources | Resources | I/O | Collection | Relations |
|---|---|---|---|---|---|
| paperCreate | paperForm | Paper | I | NO | 1 Track: N Paper |
| | track | Track | O | YES | 1 Paper : N Coauthor |
| paperEdit | paperForm | Paper | I | NO | 1 Paper: N Review |
| | track | Track | O | YES | 1 Author: N Paper |
| | coauthorForm | Coauthor | I | NO | 1 Author: N Review |
| | coauthor | Coauthor | O | YES | N Author : M Track |
| paperDetails | paper | Paper | O | NO | |
| | coauthor | Coauthor | O | YES | |
| | review | Review | O | YES | |
| paperList | paper | Paper | O | YES | |
| | author | Author | O | NO | |

**Table 35.2** REST configuration process result summary

| URIs | Operations | Actions | REST requests |
|---|---|---|---|
| /paper | Create | paperAction.new | POST |
| | Read | paperListAction | GET |
| | Update | | |
| | Delete | | |
| /paper/{id} | Create | | |
| | Read | paperAction.details | GET |
| | Update | paperAction.save | PUT |
| | Delete | paperAction.delete | DELETE |
| /paper/{id}/coauthor | Create | coauthorAction.new | POST |
| | Read | paperAction.details | GET |
| | Update | | |
| | Delete | | |
| /paper/{id}/coauthor/{id} | Create | | |
| | Read | | |
| | Update | coauthorAction.save | PUT |
| | Delete | coauthorAction.delete | DELETE |

The steps of the process are described as follows:

1. Generate base URI pair for each resource:
   a. /resources
   b. /resources/{id}
2. Generate base URI pair for each one-to-many relationship:
   a. /resources one/{id}/resources many
   b. /resources one/{id}/resources many/{id}
3. For each many-to-many relationship, locate the master resource and execute the previous step taking the master resource as the main one.
4. Complete the URIs by adding to the query string any control parameter that appear in the original requests.

#### 35.2.2.3 Identifying Available Operations

Considering that our modernization process takes as input a previously implemented legacy application, obviously the set of operations for each resource that the REST service layer must provide is limited by the operations available in the original system. Therefore, we need to identify the operations previously defined over these resources to be modernized in order to properly filter the set of possible REST requests to be provided. Again, this information is obtained by the modernization engineer by analyzing and querying the MVC model. This analysis allows, in this case, to obtain all the requests related with the set of views that compose the subsystem to be modernized. Although most of these requests are included in these views, there may exist other ones not included in the views but being responsible for generating a view, e.g. a request generated by a menu link. Note that the MVC model allows the identification of the controller that must handle a particular request but also the controller responsible for the generation of a particular view. The Controller component of a MVC model allows the specification of the operation call sequence that a controller executes to manage the data objects of the legacy application. Based on the analysis of this sequence, the operations being performed over a resource within a controller may be identified. A comprehensive analysis of the controllers that handle the set of requests related with the modernization provides, as result, the list of available operations for each resource.

To illustrate the operation identification process, we consider here a specific request contained in one of the views to be modernized of the running example, concretely the *paperDetails* request of the *paperList* view. Based on this request, the controller (*ControlFlow*) that handles it is identified by means of the existing relationships between the *View.Request* and *Controller.ControlFlow* elements. In this case, the *paperDetails* request is handled by the *paperAction.details ControlFlow* which contains the operation call sequence described in Fig. 35.3. In this case, for example, the operation *get* is invoked over the data object *Paper*, passing as parameter the *id* of the paper (request parameter) to fetch and receiving a *Paper* object as result. As shown, in this model excerpt, all the operation calls are related to concrete data operations defined in a concrete data object. In a MVC model, all the data operations are classified according to its defined mission among the following categories: *create*, *update*, *remove*, *get*, *getAll* and *getFilter*. Such classification provides the modernization engineer with a common vocabulary to identify unequivocally the list of data operations performed by a controller. Such identification is necessary to match properly those operations with the correct HTTP method to use in a REST request. Table 35.2 shows such mapping.

#### 35.2.2.4 Generating REST Requests Mappings

Following a similar approach to the generation of REST URIs, our process generates all the possible combinations of URIs and HTTP methods, producing four different request mappings for each REST URI. Next, that collection is filtered by the set of

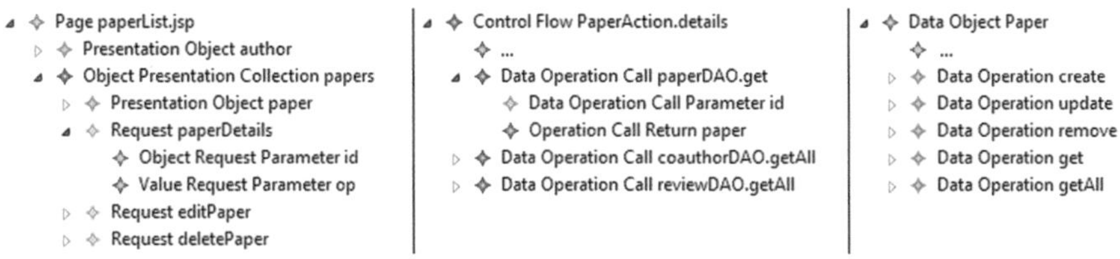


**Fig. 35.3** Operation identification on a MVC model

available operations, so only the REST requests with a proper matching with a legacy controller is maintained, while the rest is removed. Table 35.2 summarizes the results obtained for the REST URIs selected in our illustrative example.

#### 35.2.2.5 Delegating REST Request Processing to Legacy Struts Actions

It comprises the following steps:

1. The definition of the mappings between REST requests and existing controllers. This mapping is derived from the previously obtained information regarding the relation request-controller. Therefore, for instance and according to Table 35.2, the management of an HTTP GET request for the path*/paper/*{*id*} will be delegated to the *paperAction. details* controller.
2. The conversion of the input data (parameters or message body) of the request to the corresponding data format expected by the controller. Since legacy controllers are reused in the process, the context that they expect must be built in order to ensure a right execution. Thus, it could be needed, for instance, that a particular object that represents the information sent by an HTML form (expected by the controller) must be populated with the data obtained from the REST request (e.g. in JSON format). The generation of this context needed by the controller requires to know, on the one hand, the parameters of the source request and, on the other hand, the presentation objects related with the source request. All this information may be also easily gathered from the MVC model of the legacy application by analyzing the structure of the view that contains the source request. In our running example, the *new* request contained in the *paperCreated* view is related with the input presentation object named *paperForm*. Therefore, the object expected by the controller must be generated to handle that input presentation object. Figure 35.4 presents, on its left side, the structure of the JSON message for the creation of a new paper, and on its right side, the expected object by the Action *paperCreate* which will be created and initialized conveniently from the JSON message received.
3. The generation of the response in a particular data format (e.g. in JSON format) instead of generating an HTML page as response. REST service layer clients usually expect responses in a particular data format (JSON, XML) in order to

```
{
  "paperForm" : {
    "id" : ""
    "title" : "My New Paper"
    "abst" : "Abstract..."
    "fileLocalPath" : "mypaper.pdf"
    "author" : "5"
    "track" : "3"
  }
}
```

```
public class PaperForm extends ActionForm {

    private Long id;
    private String title;
    private String abst;
    private String fileLocalPath;
    private long author;
    private long track;
```

**Fig. 35.4** JSON input to form bean

properly manage them at client tier. Therefore, the response generated by the legacy application, as an HTML page, is not useful in this case and it must be replaced by a response generated in the previously accorded data format. In MVC based web development frameworks, controllers used to delegate to a different framework component the composition of the final response, i.e. a presentation template management system. Thus, controller main responsibilities are to generate the needed output objects to populate the template selected to compose the response. Based on this approach, we may know the presentation objects that a particular controller will generate (and their structure) just analyzing the MVC model.

#### 35.2.2.6 Generating REST Service Layer Configuration

In this work, the (semi)automatic configuration is generated by means of a model to text transformation described in ATL [6]. That transformation implements the process depicted above and generates an XML file as output. The schema of such file has been defined as an extension of the one used by the configuration file of Struts v1.3, so its deserialization can be implemented as an extension of the actual Struts configuration module. For the sake of brevity, we do not explain the concrete extensions performed.

## 35.3 Related Work

Concerning the reverse engineering process defined by the MIGRARIA project, although our intention is to follow the guidelines proposed by Architecture Driven Modernization (ADM) [7], we have declined to use Knowledge Discovery Meta- model (KDM) because of its complexity to specify MVC-derived semantics. We have also declined to use one of the ripe existing MDWE approaches [8] cause most of them are designed within forward engineering approaches. The abstract concepts they define are difficult to match with the information resulting of a reverse engineering process. In our opinion, the semantic gap may be wide enough to motivate the definition of a new language to bridge the technology platform of a web application and MDWE approaches.

Concerning the work presented herein, different approaches and case studies exist about generating a REST service layer from a legacy application. [9] describes a common process for re-engineering legacy systems to RESTful services which is focused on data-driven systems. The process description is performed at a conceptual level. Parts of its process have been used to derive the process presented in this work. Our approach has extended and adapted that process to a concrete scenario of the modernization of a web application. We also provide a realization of this process in a specific web development technology. Both approaches share a common focus on URI definition [10] describes the case study of an existing legacy application for Internet bidding generalized and replaced by a RESTful service layer. Instead of reusing existing implementations the authors designed a new protocol and implemented it in different languages. The main common points is the focus on data-centric CRUD operations. Finally, [11] develops a service layer oriented to resources and CRUD operations, by the transformation of an object-oriented model of a legacy application into a resource-oriented one, hereafter generating the URIs is a simple step (M2M transformation on the contrary to our M2T translation). Although those three model-based approaches have been proposed to define REST services from a legacy application, to our knowledge neither of them is contextualized within a complete model-driven modernization process and nor provides extensions of legacy technology as the approach presented in this work.

## 35.4 Conclusions and Future Work

This work tackles a part of the modernization process defined within the MIGRARIA project which faces the evolution of the interaction of a legacy web system towards a Web 2.0 new RIA client. This process requires not only the generation of the new RIA client but also the creation of a new connection layer for enabling the data interchange between the client and the original system functionalities. In this paper we have presented how this connection layer is generated. In particular, the generation of a REST service layer is described. It provides the client with a service layer to handle the user requests. By means of a running example (excerpt of a fully-functional case study) we have detailed the activities that comprise the process that are mainly divided into two main steps: a) framework-dependant REST service layer specification and b) application-dependant REST service layer specification. While the former is performed just once and reused for applications developed in the same MVC web framework, the latter depends on the particular application being modernized so that, in this case, model-driven techniques come to scene to systematically automate the process providing benefits regarding reusability and formalization. Note that the work presented here takes as input the results obtained in the previous steps of the process that generates a conceptual representation of the legacy system in terms of models (conformed to a new defined metamodel).

As main lines for future work on MIGRARIA we consider the following: (1) improve the tool support to simplify the developer's intervention in the process; (2) enrich the process with domain semantic information to improve the resource and relations identification and the REST request generation.

**Acknowledgements** Work funded by Spanish Contract MIGRARIA—TIN201 1-27340 at Ministerio de Ciencia e Innovación and Gobierno de Extremadura (GR-10129) and European Regional Development Fund (ERDF).

## References

1. Fielding RT (2000) Architectural styles and the design of network-based software architectures. PhD thesis, University of California, Irvine
2. Programmable Web. API dashboard. http://www.programmableweb.com/apis. Accessed 12 June 2013
3. Tilkov S. Is REST winning? http://www.infoq.com/news/2007/05/is-rest-winning. Accessed 12 June 2013
4. Rodríguez-Echeverría R, Conejero JM, Clemente PJ, Preciado JC, Sanchez-Figueroa F (2012) Modernization of legacy web applications into rich internet applications, vol 7059/2012, pp 236–250
5. Rodríguez-Echeverría R, Conejero JM, Clemente PJ, Pavón VM, Sanchez-Figueroa F (2012) Model driven extraction of the navigational concern of legacy web applications. In: Grossniklaus M, Wimmer M (eds) Current trends in web engineering. Lecture notes in computer science, vol 7703. Springer, Berlin, pp 56–70
6. Jouault F, Kurtev I (2005) Transforming models with ATL. In: MoDELS satellite events, pp 128–138
7. Ulrich W, Newcomb P (2010) Modernization standards roadmap in information systems transformation architecture-driven modernization case studies. Morgan Kaufmann, Burlington, Massachusetts, USA, pp 45–64
8. Rossi G, Pastor O, Schwabe D, Olsina L (2007) Web engineering: modelling and implementing web applications (Human-computer interaction series). Springer-Verlag London
9. Liu Y, Wang Q, Zhuang M, Zhu Y (2008) Reengineering legacy systems with RESTful web service. In: Proceedings of the 2008 32nd annual IEEE international computer software and applications conference (COMPSAC' 08). IEEE Computer Society, Washington, DC, USA, pp 785–790. doi: 10.1109/COMPSAC.2008.89. http://dx.doi.org/10.1109COMPSAC.2008.89
10. Engelke C, Fitzgerald C (2010) Replacing legacy web services with RESTful services. Proceedings of the first international workshop on RESTful design—WS-REST'10, vol 5, p 27
11. Szymanski C, Schreier S (2012) Case study: extracting a resource model from an object-oriented legacy application. In: Alarcon R, Pautasso C, Wilde E (eds) Proceedings of the third international workshop on RESTful design (WS-REST '12). ACM, New York, pp 19–24. doi: 10.1145/2307819.2307825; http://doi.acm.org/10.1145/2307819.2307825

# Chapter 36
# Effective and Considerate Change Notifications in Multidisciplinary Engineering Projects

**Estefanía Serral, Richard Mordinyi, and Stefan Biffl**

**Abstract** In multidisciplinary engineering projects the effective and considerate propagation of changes is essential to ensure a consistent view of the project and to minimize defects and risks. To achieve this, the changes coming from one discipline need to be communicated and coordinated with the participants of the disciplines where those changes are relevant. In this paper, we introduce a user-centric context-adaptive notification manager (UCAN), which makes use of modelling techniques at design and runtime to derive effective change notifications that are delivered in a considerate manner. We present a prototype implementation of UCAN based on industry use cases and discuss its benefits and limitations based on the feedback from industry partner domain experts.

**Keywords** Change management • Adaptive notifications • Multidisciplinary engineering projects

## 36.1 Introduction

Multidisciplinary engineering projects require communication and coordination among engineers coming from heterogeneous disciplines. In these projects, the expert knowledge is often embodied in several overlapping discipline-specific data models involved in individual processes, methods, and tools. The different local representations of common concepts in these data models hinders efficient

E. Serral (✉) • R. Mordinyi • S. Biffl
Christian Doppler Laboratory for Software Engineering Integration for Flexible Automation Systems, Vienna University of Technology, 1040 Vienna, Austria
e-mail: estefania.serral@tuwien.ac.at; richard.mordinyi@tuwien.ac.at; stefan.biffl@tuwien.ac.at

M. José Escalona et al. (eds.), *Information System Development: Improving Enterprise Communication*, DOI 10.1007/978-3-319-07215-9_36,

collaboration and interaction between disciplines, and makes it very difficult to support an essential and success-critical task in these projects: change management [1]. If the changes introduced in a discipline are not notified to the affected engineers of other disciplines, several discipline models may become inconsistent, increasing the risks of costly defects and development delays [2]. For instance, a change in the alarm indicator of a critical measure in a mechanical plan (e.g., temperature level alarm indicator) should be notified to software engineers (alarm indicator implementation), electrical engineers (sensor wiring), and process engineers (processes where the temperature sensor is involved).

To manage changes across disciplines, current approaches rely on the manual use of notification tools, such as emails. These tools do not consider either project-relevant aspects (e.g. stakeholders responsible of an artefact, change dependencies, urgency of propagating changes) or important engineer-specific information (e.g. engineer availability, notification preferences). This leads to risks of failing to correctly communicate changes or overwhelming engineers with notifications they do not need to receive. For instance, in a typical current setting, when an engineer makes a change, he or she creates the change notification and sends it to the rest of project engineers without considering whether they are affected by the changes. Also, those engineers that are affected may be notified in an unsuitable manner (e.g., sound notification during meetings or without the needed information for understanding and propagating the changes). Thus, the inappropriate demand of attention and interruption of the engineers may cause them to ignore notifications even when they are important; consequently, changes may be found not taken into account at critical points in the project.

Therefore, concurrent updates require more sophisticated means of project communication [1] that are aware of the relevant information for managing changes. In this paper, we introduce a user-centric context-adaptive change notification manager (UCAN). UCAN represents the relevant information for change management in a machine-understandable format, enabling automatic change notifications that are effective (i.e., containing the relevant information and being addressed to the appropriate recipients) and considerate (i.e., requering only attention when it is needed and using the appropriate communication resources) [3]. In these notifications, UCAN provides the engineers with all the needed information for effectively carrying out the consequent synchronization and consistency-checking tasks. Furthermore, UCAN dynamically adapt the way change notifications are delivered to properly inform the appropriate engineers using the most suitable communication resources for each situation.

The remainder of the paper is as follows: Sect. 36.2 summarizes related work. Section 36.3 presents research issues. Section 36.4 introduces a real-world use case where UCAN is applied. Section 36.5 describes the models for capturing the knowledge relevant to achieve adaptive change management notifications. Section 36.6 describes UCAN notification process. Finally, Sect. 36.7 discusses the results and Sect. 36.8 concludes the paper and proposes further work.

## 36.2 Related Work

This work presents UCAN - a software manager capable of managing change notifications in an effective and considerate way. Therefore, the related work comes from the following areas: (1) change management in multidisciplinary projects and (2) notification adaptivity to the user context.

Change management in multidisciplinary projects is currently a time consuming task since it is necessary to search what has been changed and determine the effects on other disciplines [1]. Although available tools are strong in supporting engineering disciplines in a multidisciplinary project individually, change management across the different disciplines involved in a project is still ineffective and inefficient [4].

A well-known example of multidisciplinary projects is mechatronic engineering projects [4], in which several approaches have been proposed to detect inconsistencies between models by using batch processes and consistency rules [5]. However, these approaches do not scale and consume significant execution time. Other approaches based on instant consistency checkers have been proposed [6]. Although these approaches observe user activities and evaluate model changes, they do not take into account project information or the use of consideration notifications to inform project members about relevant changes.

Regarding Notification Adaptivity, current approaches to adapt interactions/ notifications to the context of use are mainly focused on minimizing unnecessary interruptions to the user [7]. However, the amount of studies concentrating on the design of unobtrusive interactions is still limited. Approaches in the area of Considerate Computing [3] are mainly focused on predicting acceptability by analysing the needed user attention and the cost of interruption. The efforts of these approaches have been put on minimizing unnecessary interruptions, overlooking how the interactions must be provided.

Regarding adaptation of system interactions with users, the approaches have been focused on dealing with user interface adaptation in response to context changes. Calvary et al. [8] give an overview of different modelling approaches to deal with user interfaces supporting multiple targets in the field of context- aware computing. DynaMO-AID [9] and Blumendorf et al. [10], define runtime architectures to process context data that supports migration, distribution and multi-modality. These works define the adaptation space in terms of the environment and the platform. In contrary, the work presented in [11] deals with adapting interaction according to human limitations by reconfiguring mobile user interfaces. Our work uses its consideration framework for dealing with the management of change notifications in a considerate way.

## 36.3 Research Issues

In this paper, we deal with the challenge of providing effective and considerate change notifications in multidisciplinary engineering projects. From this challenge, we derive the following key research issues:

***RI-1: Modelling the required knowledge for effective and considerate user notifications***: What kind of information in a specific project context is relevant to achieve effective user notifications and deliver them in a considerate way? How to model this information in a human and machine understandable format in order to support automated user notifications?

***RI-2: Designing an effective and considerate change notifications process***: Who should a certain notification be addressed to? Which priority should it have? Which are the most appropriate notification resources and capabilities to deliver a certain notification to each involved engineer in each context?

For investigating these research issues we gathered requirements from interviews with industry partners, and developed a prototype of UCAN.

## 36.4 Industry Use Case

This section describes a multidisciplinary engineering use case from an industrial partner who develops, creates, and maintains hydro power plants. The life cycle of a power plant is divided into several phases that have to be managed and are concurrently administrated in different discipline-specific tool data models by globally distributed engineers. Specifically, after gathering the requirements and specifications for building the plant in the requirements engineering discipline, the following phases are developed: (1) specifying the physical devices (sensors and actuators) that are needed in the plant: this phase is performed in the mechanical engineering discipline; (2) drawing the topology of the system and designing the circuit diagrams that connect the physical devices: this phase is performed in the electrical engineering discipline; (3) creating and testing the Programmable Logic Controller (PLC) software to define the behaviour of the device controllers: this phase is performed in the software engineering discipline.

While in theory changes are applied only within one phase and in a sequential order, the reality shows that changes in any phase are applied at any time, mostly due to time restrictions. Thus, concurrent development requires frequent synchronizations between various tool models. The way of keeping the view on the plant among the various engineering tools consistent depends on the development culture in the company. With respect to our industry partners, changes need to be discussed and agreed by all engineers who are affected by a change before being applied to every affected tool [1]. Therefore, involved engineers have to be aware of the changes that affect their field of responsibility. Appropriate notification mechanisms automatically triggered by introduced changes need to be placed into that process in order to allow effective and considerate change notifications, and traceable decision making processes.

## 36.5 Modelling the Required Knowledge

To enable effective and considerate change notifications, based on discussions with industry partners, we propose a modelling language to capture: (1) the tool data models used in the different disciplines and how they are interconnected: to be able to analyse the effects of a change; (2) notifications and their context: to describe and compose all the information needed for effectively notifying changes; (3) notification resources and when they must be used: to describe the available resources that must be used to deliver a notification in a considerate way.

### *36.5.1 Tool Data Models and Their Interconnections*

To specify the tool data models and their interconnections in a formal way, we use the EKB framework introduced in [12], which is a data-modelling approach successfully applied in multidisciplinary projects for allowing the representation and querying of project information, and establishing how the different disciplines are interconnected. The EKB uses common engineering concepts identified beforehand as basis for mappings between discipline-specific tools and project-level engineering knowledge (i.e., the common model), and represents the tool models and the common model as well as their data instances (i.e., individuals) using ontologies. An ontology is "a specification of a conceptualization" [13]. It allows semantically specifying (1) the concepts and relationships that characterize a domain, and (2) specific situations in that domain. In this context, change propagation must be performed by first mapping the changes from a tool ontology into the common ontology, and from there into the other tool ontologies.

Figure 36.1 illustrates the application of the EKB framework to the use case. The figure shows very simplified versions of the data models of three of the disciplines

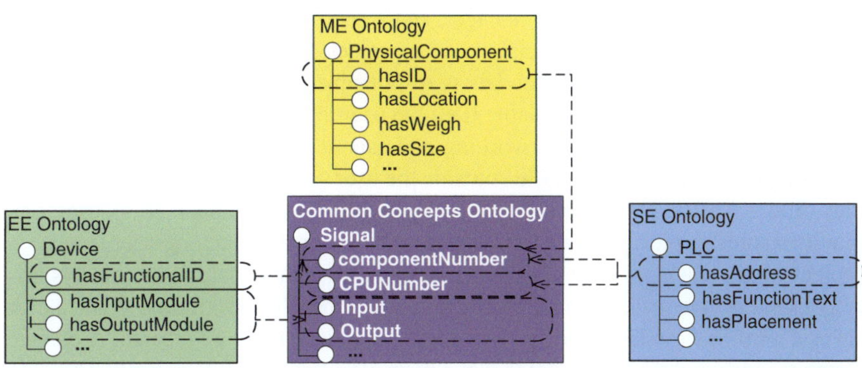


**Fig. 36.1** EKB in the hydro power plant project

involved in the development of hydro power plants: mechanical engineering (ME), electrical engineering (EE) and software engineering (SE). The upper side shows the concept Physical Component from the ME domain, which describes the components of the plant. The left side shows the concept Device from the EE domain, which describes how the devices are connected. The right side shows the concept PLC from the SE domain, which describes the controllers' behaviour of the devices. From discussions with the industry partner it is known that one of the common concepts between these disciplines is Signal, shown in the middle of the figure. Also, the mappings between each tool model and the common model are shown in an abstract way; e.g., the attributes has Functional ID of the Device concept, has ID of the Physical Component concept, and has Address of the PLC concept, are mapped to the component Number of the Signal concept.

Using the EKB, the tool ontologies, the common ontology, and the mappings between both are explicitly represented. The ontologies are described in OWL (Web Ontology Language)[1] which is a W3C standard markup language that is machine-understandable and facilitates data sharing and integration. The mappings are represented using SPARQL[2] Construct queries. The EKB also provides services to access concepts, data and mappings at design and run time.

### 36.5.2 *Notification Context Modelling*

We propose an OWL ontology-based model for formally representing the relevant information of the change and the notification itself, as well as the context information of the users involved in the project. We select OWL to represent this knowledge in a semantic way and to facilitate its integration with the EKB.

Specifically, the designed ontology defines the following concepts. The Project concept defines the information of a project and has: an identifier, a start and end date, the phase in which the project is, a goal, the identifiers of the data models created in the project, a project manager, and a set of engineers. The concept User describes each one of the participants of a project. Each user is described by its contact information, the EKB models is responsible of, its location, whether it is accompanied, and its state: available (i.e., the user is working but can receive notifications), busy (i.e., the user is working and cannot receive notifications, e.g., the user is in a meeting), or absent (i.e., outside the user working hours). A user can also play different roles. For instance, for the running example (see Sect. 36.4), users can play two roles: project manager and engineer. A user can be an engineer in several projects, but a manager only in one. A user can also change data of the models it is responsible of. These changes are represented using the Change concept. When a user saves a modified model in the system, the changes instances are created. A change instance represents an individual that has been modified, and it is described by who

[1] OWL Web Ontology Language (OWL)—www.w3.org/TR/owl-ref/

[2] SPARQL—http://www.w3.org/TR/rdf-sparql-query/

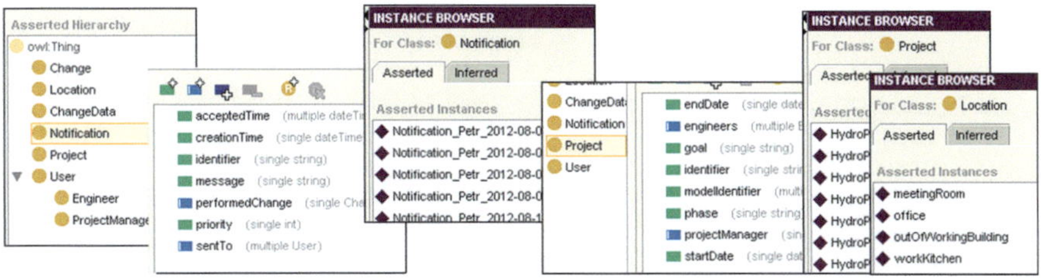


Fig. 36.2 A context model example in Protégé.

has performed the change and when, the reference to the change data, and the change type (update, delete, or create).

Finally, the Notification concept describes a change management notification. When a modified model is saved, a notification is created by each affected model and addressed to the engineer responsible of the model. The notification contains a set of dependencies that the engineer must check. The dependencies are the mappings that are affected by the changes and that are defined from the changed model to the corresponding affected model. Therefore, a dependency contains a SPARQL construct mapping, the set of changed instances that affect the mapping, and the set of concepts in the target model whose instances are affected by the mapping. In addition, the notification also stores its priority, when it is delivered and when the changes are accepted. The priority can be configured according to the project. For the prototype, it is preliminary set by the number of changed instances grouped in the notification, divided by the phase of the project (e.g., see phases in Sect. 36.4).

Figure 36.2 shows our current ontology for context modelling in multidisciplinary systems. The figure shows the main classes of the model, their main attributes and relationships, and some individuals in a tree form. The specific context of the project is represented as instances of the classes defined in the ontology. For instance, the engineer Petr is defined as an instance of the User class, office is defined as an instance of the Location class, etc. The information about projects, users and locations is manually created. The dynamic information about the users, like its location, can be automatically updated by a context monitor like the one presented in [14], while the changes and notifications are managed by UCAN, the notification manager.

### 36.5.3 Notification Resources and Their Activation

To represent the notification resources in a technology-independent way, we make use of feature modelling[15]. Feature modelling is widely used to describe a system configuration and its variants by means of features (coarse-grained interaction functionality). The feature model shown in Fig. 36.3 describes the notification resources provided by the running project (see Sect. 36.4). According to the model,

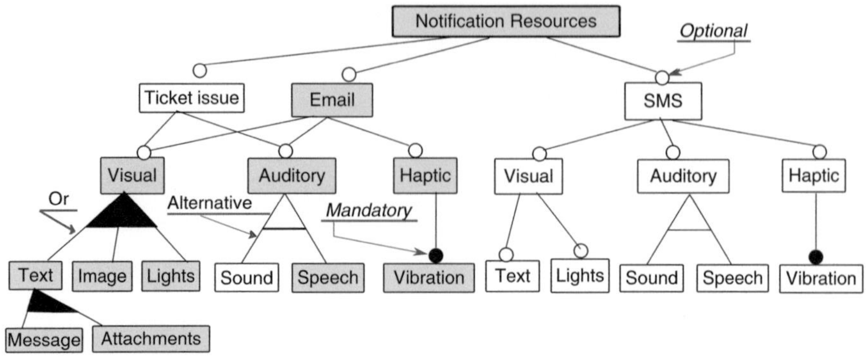


**Fig. 36.3** Notification resources' modelling

three notification services can be used: ticket issue (mechanism for sending traceable messages to a certain set of receivers), email and SMS. For instance, the notification using email service can be provided by one or several of the following resources: a visual element, which can be a combination of text (message and/or attachments), image and a flashing light; an auditory element, which can be sound or speech (i.e., reproducing the message); and a haptic element that is provided with vibration. Note that the model can be easily reused, adapted or extended to support the notification requirements of other projects.

Depending on the degree of attention required, different notification features will be activated to deliver the notifications in a considerate manner [3], i.e., using the most appropriate resources according to the user situation and preferences. To do this, we use the conceptual framework presented in [11] in which six consideration levels are defined according to the initiative and attention factors. In the initiative factor, interaction can be reactive (the user access the notifications) or proactive (the system notifies the user). With regard to the attention factor, an interaction can take place at the background (unadvised interaction), in a completely-aware way (the user is fully conscious of the interaction), or in a slightly-appreciable way (some effort is needed to perceive the interaction).

Using this framework, we define which features should be activated for each consideration level and for each user. We refer to this set of features as Feature Configuration (FC). For example, the FC of the feature model shown in Fig. 36.3 (represented by the grey boxes) is as follows: FC=Email, Visual, Text, Message; Email, Visual, Text, Attachments; Email, Visual, Image; Email, Visual, Light; Email, Audiotory, Speech; Email, Haptic, Vibration. This FC is assigned to the proactive completely-aware consideration level of the user Petr. The feature selection can be based on the studies of the effects of each modality and modality combinations on cognitive load. Note that it is not necessary to define all the considerations levels for all the users (see, for instance, Fig. 36.4), and that the defined considerations levels can be reused for users with the same preferences.

In addition, we also define the context conditions that make the system to choose the appropriate consideration level each moment for each user (see Fig. 36.4).

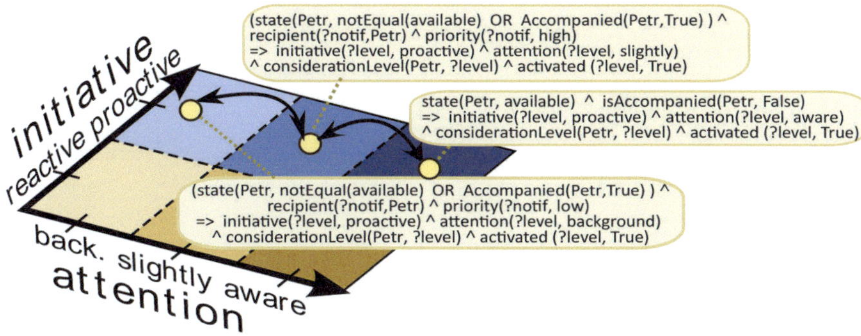


**Fig. 36.4** Example of consideration model for the engineer role

By means of these conditions, the consideration model is enriched with context semantics that define the criteria to deliver notifications in a context-adaptive way. For instance, the completely-aware consideration level is chosen when Petr is available and not accompanied.

We have implemented the feature model using the OWL ontology proposed by Zaid et al. in [16]. We extended this ontology with the Consideration Level concept, which is defined by: the initiative and attention factors, the user/s for which the level is defined, and the feature configuration that should be activated for the level. For instance, we specify the three proactive consideration levels shown in Fig. 36.4 for the user Petr. The context conditions are specified as SWRL[3] rules, which basing on the information of the context model, automatically set which consideration level is active for each user.

## 36.6 UCAN Notification Process at Runtime

In order to notify the changes in an effective and considerate way, UCAN analyses and manages the described models at runtime by using SPARQL queries, which is the W3C recommended language for querying ontologies, and can also be used for updating and creating individuals. To be aware of the changes, UCAN is developed on top of the Engineering Service Bus (EngSB) platform [17], which enables data exchange in multidisciplinary systems by using the EKB. Using this platform, UCAN follows the notification process shown in Fig. 36.5.

The process starts when an engineer changes the data stored in an engineering tool model and makes the changes public by checking them in into the EngSB. The EngSB then (1) stores who has performed the changes and when, (2) creates a new version of the tool ontology model where the data changes are applied, (3) executes the SPARQL Construct queries creating a new version of the common model,

[3] Semantic Web Rule Language (SWRL)—http://www.w3.org/Submission/SWRL/

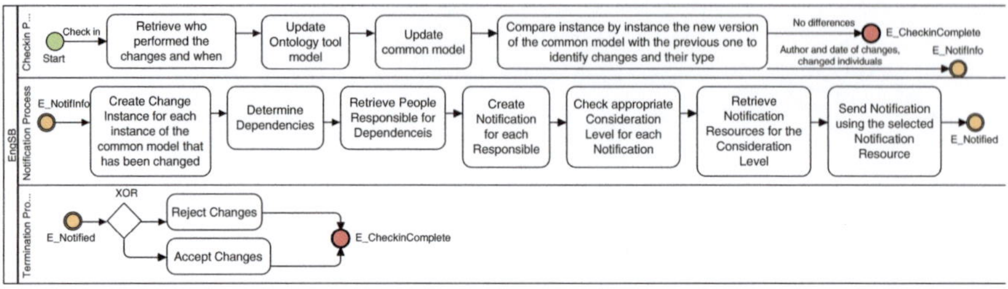


**Fig. 36.5** Process of creating considerate Notification.

and (4) compares the new model with the previous one to identify the changes and their type (i.e., create, update, delete). This information, i.e., author and date of the changes and the individuals that have been changed in the common model, is sent to UCAN -our notification manager-, which creates a Change individual (in the context model) for each individual that has been changed. Then, UCAN analyses the mappings from the common model (specified in the EKB) to determine the dependencies of the changed instances with other tool models (see Sect. 36.5.2). Based on this information, UCAN asks to the context model the engineers responsible for working with those tool models.

Once the context and the dependencies of the changes have been made clear, UCAN creates an instance of a notification for each affected model, addressing the notification to the responsible engineers. Next, UCAN sets the creation time of the notification, a default message with instructions for managing the changes [1], the priority of the notification, and the set of dependencies.

After creating the notification, UCAN queries the notification resource model to get the FC of the considerate level currently active for the recipient (or its role) of the notification. Finally, UCAN sends the notification using the resources specified in the retrieved FC. For instance, if a notification is sent to Petr when he is alone and available, UCAN will use the notification resources for the proactive completely-aware consideration level (resources highlighted in grey in Fig. 36.3) since it would be his active level according to the SWRL rule shown in Fig. 36.4. However, if Petr is available, is accompanied and the notification priority is high, UCAN will notify Petr using the resources assigned for his proactive slightly-appreciable consideration level.

Finally, when an engineer receives a notification, he/she can reject the changes or accept them.

## 36.7 Discussion

In this section we discuss the effectiveness and consideration of change management notifications using UCAN. We analysed empirical results on changes performed in the hydro power plant engineering project presented in Sect. 36.4.

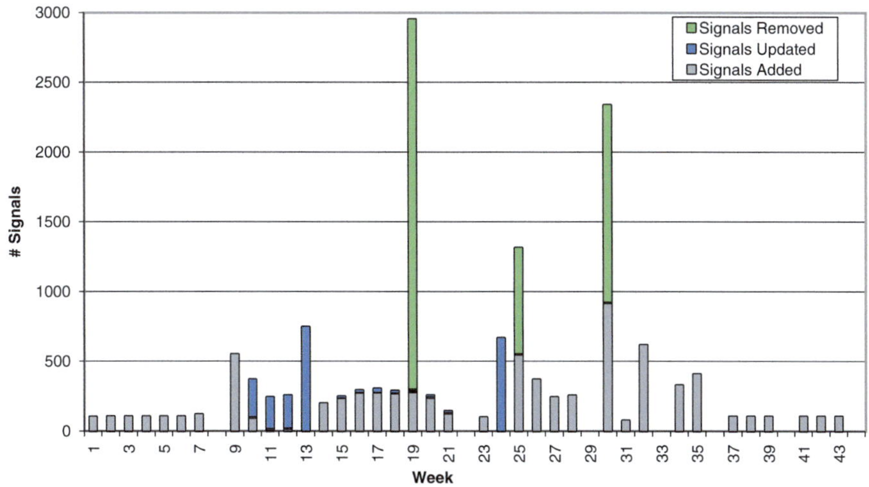


**Fig. 36.6** Weekly modifications on the signals of the project

Figure 36.6 shows the overall number of changes performed every week in a selected plant part. The shown changes have an impact on instances of the common concept Signal, and therefore, require change propagation to other disciplines. The figure shows the high number of changes and, therefore, the need for a reliable change management system.

To improve the change management process, we developed a prototype of UCAN and applied it to the project. We discussed the approach with industry partners. Even though they consider that UCAN introduces an added modelling complexity, they do believe that UCAN can provide more effective and considerate notifications than their traditional tools. The main reasons were that (1) all notifications are triggered and delivered automatically to and only to the affected engineers, (2) the notifications contain the needed information for understanding the changes and automatically propagate them, (3) enables traceability and analysis of notifications for further processing, and (4) notification are delivered using the appropriate resources for each situation, and consequently, more notifications are accepted.

Taking into account (1) the high number of change notifications that are usually managed in multidisciplinary projects and also that are sent to not affected engineers, and (2) that the overall effort of understanding a change and check its dependencies in current approaches is approximately 5 min; it is expected that UCAN brings significant benefits in change management. However, further evaluations are needed to analyse the specific levels of effectiveness and consideration that UCAN achieves in a real environment. To do this, it is necessary to evaluate: (1) the real increase of notifications that are properly attended, as well as the decrease of the required time to propagate the changes; and (2) the user acceptance of the received notifications.

## 36.8 Conclusion and Future Work

Multidisciplinary projects require sophisticated methods for managing changes that have effects across different disciplines. In this work, we have presented UCAN - a user-centric context-adaptive change notification manager that allows multidisciplinary projects to manage change notifications in an effective and considerate manner. UCAN is aware of the tool data models managed by the different disciplines, their interconnections, the changes performed and the context of these changes. Based on this information, the manager notifies the engineers affected by each change in a considerate manner, providing them with all the information they need for managing synchronization and consistence- checking issues in an effective way. Thus, UCAN can bring significant benefits for improving and accelerating change management.

As further work, we plan to extend the UCAN process so that, when the changes are rejected, a notification is automatically sent to the rest of affected engineers to inform them that the changes need to be discussed before being applied. On the other hand, if all the engineers accept the changes, it will be possible to automatically apply the changes by executing the SPARQL constructs from the common model to the rest of models. This will create new versions of the target models, which will be set as the current model versions of the project. In this context, it will be also necessary to extend UCAN to support the management of concurrent changes.

In addition, we plan to evaluate the UCAN prototype in real environments for checking the specific effectiveness and consideration improvements in the change notification process. Future work will be also focused on designing an end-user interface to allow engineers to easily personalize the consideration model. It is worth noting that, since UCAN interprets the models at runtime for delivering the notification using the opportune resources, this tool will not have to change any line of code to apply users' personalization, but only will have to change the consideration model of the corresponding user.

## References

1. Winkler D, Moser T, Mordinyi R, Sunindyo WD, Biffl S (2011) Engineering object change management process observation in distributed automation systems projects. In: EuroSPI 2011, IEEE, pp 1–10
2. Lüder A, Estévez E, Hundt L, Marcos M (2011) Automatic transformation of logic models within engineering of embedded mechatronical units. Int J Adv Manuf Tech 54:1077–1089
3. Gibbs WW (2005) Considerate computing. Sci Am 292(1):54–61
4. Schäfer W, Wehrheim H (2007) The challenges of building advanced mechatronic systems. In: 2007 Future of Software Engineering, IEEE, pp 72–84
5. Hehenberger P, Egyed A, Zeman, K (2010) Consistency checking of mechatronic design models. In: ASME and IDETC/CIE 2010 conferences, pp 1–8
6. Egyed A (2011) Automatically detecting and tracking inconsistencies in software design models. IEEE Trans Software Eng 37(2):188–204

7. Ramchurn SD, Deitch B, Thompson MK, Roure DCD, Jennings NR, Luck M (2004) Minimising intrusiveness in pervasive computing environments using multi-agent negotiation. In: MOBIQUITOUS 2004, pp 364–372
8. Calvary G, Coutaz J, Thevenin D, Limbourg Q, Bouillon L, Vanderdonckt J (June 2003) A unifying reference framework for multi-target user interfaces. Interact Comput 15(3):289–308
9. Clerckx T, Vandervelpen C, Coninx K (2008) Task-based design and runtime support for multimodal user interface distribution. In: Engineering interactive systems. Springer, Berlin, Heidelberg, pp 89–105
10. Blumendorf M, Lehmann G, Albayrak S (2010) Bridging models and systems at runtime to build adaptive user interfaces. In: EICS 2010, ACM, pp 9–18
11. Gil M, Giner P, Pelechano V (2012) Personalization for unobtrusive service interaction. Pers Ubiquit Comput 16(5):543–561
12. Moser T, Biffl S (2012) Semantic integration of software and systems engineering environments. IEEE Trans Syst Man Cybern C Appl Rev 42(1):38–50
13. Gruber TR (1993) A translation approach to portable ontology specifications. Knowl Acquis 5:199–220
14. Serral E, Valderas P, Pelechano V (2013) Context-adaptive coordination of pervasive services by interpreting models during runtime. Comput J 56(1):87–114
15. Batory D (2005) Feature models, grammars, and propositional formulas. In: International software product line conference. Lecture notes in computer science, vol 3714, p 720
16. Zaid LA, Kleinermann F, De Troyer O (2009) Applying semantic web technology to feature modeling. In: SAC 2009, ACM, pp 1252–1256
17. Biffl S, Schatten A (2009) A platform for service-oriented integration of software engineering environments. In: Proceedings of the 2009 conference on new trends in software methodologies, tools and techniques, pp 75–92

# Chapter 37
# Using Agile Methods for Infrastructure Projects: A Practical Experience

**C.J. Torrecilla-Salinas, J. Sedeño, M.J. Escalona, and M. Mejías**

**Abstract** Agile approaches have emerged in Software Development projects during the last decade becoming a real alternative for organizations developing software. These approaches have been used successfully in several and different projects: from small teams of in-site developers to teams comprising hundreds of developers all over the world. Nevertheless, Agile approaches are not only suitable for Software Development projects, but these methodologies can also be considered to be a framework to plan, estimate and manage any kind of project s focused on quick-response, adaptation to customers' needs and early delivery of value to costumers. This paper presents the result of an experience dealing with using an Agile framework, based on Scrum, in an infrastructure project applied to a Spanish Public Administration. According to the results of the project, the paper takes out the main lessons learned and proposes further lines of research.

**Keywords** Agile methodologies • Scrum • e-Government • Public Administration • Infrastructure

## 37.1 Introduction

Agile methodologies are becoming a solid alternative in the field of software development [1, 2], as they provide support to some special needs required by these kind of projects, like reduction of "time-to-market," adaptability and quick response to

C.J. Torrecilla-Salinas (✉) • M.J. Escalona • M. Mejías
Department of Computer Languages and Systems, University of Seville, Seville, Spain
e-mail: carlos.torrecilla@iwt2.org; mjescalona@us.es; risoto@us.es

J. Sedeño
Department of Computer Languages and Systems, University of Seville, Seville, Spain

Agencia Andaluza de Instituciones Culturales. Junta de Andalucía, Sevilla, Spain
e-mail: jorge.sedeno@juntadeandalucia.es

M. José Escalona et al. (eds.), *Information System Development: Improving Enterprise Communication*, DOI 10.1007/978-3-319-07215-9_37,

changes [3, 4]. However, some of the Agile approaches, like Scrum, can be considered to be a framework to support project management rather than a software development methodology [5]. This fact enables thinking that Agile approaches can be used to handle any kind of project s, whenever their needs can relate to quick adaptation to changing needs and early delivery of value. Normally, the work carried out by an IT department is not only associated with software development projects, but also with technology infrastructure projects, among others, therefore the feasibility of expanding the Agile approach beyond software development can allow standardizing practices and processes among these departments, by reducing effort and cost.

Based on the foregoing, this work presents the result of applying a Scrum-based Agile approach to support the processes of estimating, planning and managing an infrastructure project developed by a Spanish Public Administration, with the following objectives: assess the feasibility of using Agile approaches in IT projects unrelated to software development, with a special focus on the project management aspects and take out the main lessons learned with the project, in order to identify further lines of research.

This paper is organized into the following sections. After this introduction, Sect. 37.2 offers an overview of the related work and the proposed Agile approach and Sect. 37.3 presents the project and its environment. Then, Sect. 37.4 states the results of the project and lastly, Sect. 37.5 identifies the main lessons learned and proposes possible future lines of work.

## 37.2 An Overview of the Agile Framework

This section presents an overview of the related work and the Agile framework used during the project. The approach was based on Scrum and included elements of Agile estimating, planning and project management techniques. Figure 37.1 summarizes it as follows:

### *37.2.1 Related Work*

Although the majority of research in the field of Agile is focusing on software development projects, there are also approaches trying to use Agile approaches in other fields of System Development. For instance, there are approaches centered on

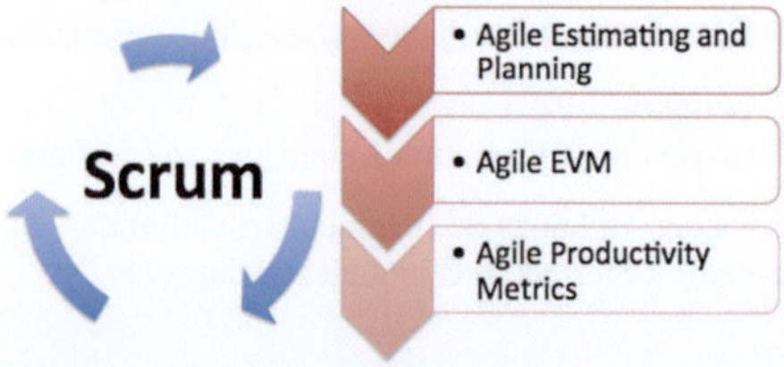


**Fig. 37.1** Agile framework

the use of Agile techniques in hardware development projects included in the so called Agile System Engineering [6], but they are more focused on assessing whether Agile approaches are suitable enough for developing hardware systems than on describing the details of the management process in that project. There are also proposals regarding the use of Agile approaches for rapid hardware modeling [7], but this kind of projects are more related to the use of software to model hardware than to the assessment of Agile in non-software projects. In addition, there are other approaches dealing with the use of Agile approaches in infrastructure projects, as the one proposed by Debois [8], which describes some case studies and takes out patterns concerning the introduction of Agile in this type of projects. However, this work is not considering aspects such as how the projects are estimated, planned and managed. In our work, we present an agnostic framework suitable for different types of projects covering elements such as project estimation, project plan, project management and productivity improvement, and assess it in a non-software development project with the aim of extracting the lessons learned.

### 37.2.2 Scrum

In 2001, a work known as the "Agile manifesto" [9] was published including the main principles and values characterizing Agile approaches. One of the most popular Agile methodologies is Scrum [4], proposed by Jeff Sutherland and Ken Schwaber [10]. It is a framework that supports product development [5] founded on an iterative and incremental approach for project management. Scrum proposes a development process divided into Sprints, time-boxed periods lasting 2–4 weeks that will be repeated during the project. The Scrum process uses an artifact called Product Backlog, containing all the features to develop during the project, ordered by priority. The team selects the features to develop and commits to it at the beginning of each Sprint, creating what is called the Sprint Backlog. The progress is checked daily once the Sprint is planned. The resultant work is reviewed in liaison with the main stakeholders after each Sprint in a Sprint demo and the development process is reviewed in a Sprint retrospective.

### 37.2.3 Agile Estimating and Planning Techniques

Scrum does not define any technique to establish and estimate the pieces of functionality that are part of the Product Backlog. Thus, some other Agile techniques are normally used to fill these gaps. In this particular case, the features included in the Product Backlog are "user stories" [11, 12]. Two techniques will be used to estimate user stories. On one hand, the development team will calculate the size of each story by means of an ideal unit called "story points" [13]. The "*Planning poker*" technique will be used to perform estimates [13, 14]. On the other hand, the value of each story will also be estimated by means of the "Value point analysis" technique [15]. For this value, the customers or their representatives should perform the estimation. With these two values, we can obtain a relation between the cost and benefit of every user story (it is called "Return of investment" or ROI) by dividing the value in value points by the size in story points. This magnitude can be used to order the Product Backlog and establish its priority. The process of estimating and planning will be constant during the project, reserving a slot of time to review the Product Backlog content during each Sprint. This process, known as Product Backlog Grooming [16], can take up to 10 % of the available time on each Sprint.

### 37.2.4 Agile Earned Value Metrics

An Agile-based approach to the Earned Value Management techniques [17] can be used in order to control the planned schedule and cost the project entails. These techniques deal with measuring the relationship among cost, scope and schedule along the project [18], based on calculating two main indexes, the *Cost Performance Index* (CPI), which measures the relation between estimated and real cost, and the *Schedule Performance Index* (SPI), which measures the relation between estimated and real schedule. At the end of each Sprint, the proposed Agile approach to EVM measures these indexes to calculate them, through both, the estimated and completed story points and the estimated and dedicated work hours.

### 37.2.5 Agile Productivity Metrics

A number of productivity metrics will be used to facilitate the team's performance process of continuous improvement. They will be obtained by means of planned and finished story points and dedicated work hours, with the aim of keeping a lightweight process. These data are also obtained through the EVM calculation. Metrics were proposed by Downey and Sutherland [19] and they are calculated per Sprint. The following ones will be used:

- **Team Velocity (in Hours)**: It calculates the product of the velocity in story points by the team's average amount of hours per story point.
- **Work Capacity**: It deals with the total amount of hours worked by the team during a Sprint.
- **Focus Factor**: It refers to the result of dividing the Team Velocity (in Hours) by the Work Capacity.
- **Percentage of Accepted Work**: It is the result of dividing the hours dedicated to accepted work by the Work Capacity of the team during a Sprint.
- **Target Value Increase (TVI+)**: It refers to the finished story points of certain Sprint divided by the average story points of all finished Sprints.

## 37.3 The Project and Its Environment

This section presents the environment in which the project was developed, the project itself and its main results, which will be discussed in the following sections.

### *37.3.1 The Environment: The Regional Government of Andalusia*

The project presented in this paper was developed by the regional government of Andalusia, a region located in the south of Spain. Junta de Andaluc'a is the name of Andalusia's regional government; counting with more than 200,000 employees, it is one of the main economic actors in the region [20]. This project was developed by one of the ministries of the Junta de Andaluc'a: the regional Ministry of Culture and Sport, which is in charge of developing and coordinating public policies both in the cultural and sports areas. Some of its duties are related to the management of public museums, archives, sports clubs, art galleries and theaters. It is also responsible for supporting regional cultural and sports industries, together with other public and private stakeholders. The ICT Department, one of the departments of the Ministry of Culture and Sport, is in charge of this project, being responsible for all Information Technology policies in the Ministry. It is mainly responsible for providing the entire ICT infrastructure needed to operate all Ministry internal systems.

It is important to highlight that, as in the last years citizens' demands on quick and easy access to public services have increased, governments have launched different initiatives to cover these needs, such as e-Government public services [21, 22]. Due to the commitment of the regional Ministry on the e-Government strategy, the number of systems under the responsibility of its ICT Department has increased dramatically.

### *37.3.2 The TOPOS Project*

As it was mentioned, the ICT Department of the regional Ministry of Culture and Sport is in charge of providing the entire ICT infrastructure that supports all the Ministry systems. On the one hand, in the last years, the number of systems under the responsibility of the Ministry highly increased due to the e-Government strategy of both the regional and the national government. On the other hand, the effects of today's economic crisis made the organization increase the efforts on costs rationalization and optimization. These two main reasons made the ICT Department of the Ministry start with a project to improve its internal infrastructure and provide a better service with fewer resources. The project was called TOPOS and it lasted more than a year, starting on January 2012 and finishing on February 2013. The main goals of the project were:

- Reorganize the environments where the systems were deployed, including a development, pre-production and production environment for each system, using virtualization solutions. It intended to reduce the number of physical servers, in order to save space and energy consumption on the datacenter.
- Uniform the version of the software products used to support the systems (operative systems, databases, application servers…), in order to reduce the maintenance costs required to run these systems.
- Update the versions of some homemade software products that support the general systems of the Ministry, in order to eliminate some errors and security risks.
- Clean and decommission some obsolete systems, in order to clean and free resources that could be used to support new services.
- Get high availability configuration of all e-Government services operating into the regional Ministry to improve the service offered to citizens.

As TOPOS is an infrastructure project, it is affected by the business projects of the regional Ministry and their different priorities, whereby to be able to respond to possible changes on requirements and priorities, an Agile approach was established, with Scrum as a base framework.

A team of five members, partially working on the project and coming from both the system management team and the e-Government bureau of the ICT Department, developed the project by means of internal resources. Then, the following roles were played among the members of the team:

- A *Scrum Master*, whose goal was to ensure the adherence of the team to Scrum practices and help it identify and remove detected errors.
- A *Product Owner*, who tried to identify the business needs and prioritize them using the Product Backlog.
- A *Development Team*, who aimed to develop each of the requested features during the selected Sprint.

It has to be mentioned that a particular aspect of TOPOS project, as it concerned the basic infrastructure of the regional Ministry, deals with its impact both on the internal and external users (mainly citizens), for they will be affected, for instance, by services stops when performing the identified tasks. This element demanded an extra effort to work in liaison with the different departments of the Ministry and also to manage changes.

## 37.4 Results of the Project

TOPOS project started in January 2012 with the development of a business case approved by the ICT Department board. After the business case approval, a number of workshops were conducted during this month to identify its initial scope and to

**Table 37.1** Initialforecast velocity

| Available hours | Estimated initial velocity | Hours per story point |
|---|---|---|
| 137 | 7 | 20 |

**Table 37.2** Summary of initial project plan

| Magnitude | Value | Uncertainty (±25 %) | Initial forecast |
|---|---|---|---|
| Total story points | 70 | 18 | 70 ± 18 |
| Velocity | 7 | 2 | 7 ± 2 |
| Number of iterations | 10 | 3 | 10 ± 3 |
| Sprint length | 30 days | N/A | 30 days |
| Project length | 300 days | 90 days | 300 ± 90 days |
| Hours per Sprint | 137 h | 34 h | 137 ± 34 h |
| Hours per project | 1,370 h | 340 h | 1,370 ± 340 h |
| Cost per iteration | €3,584.30 | €887.01 | €3,584.30 ± 887.01 |
| Total project cost | €35,843.00 | €8,870.10 | €35,843.00 ± 8,870.10 |

draft an initial project plan. An initial Product Backlog was created containing 21 user stories, with a total value of 70 story points, as a result of these workshops. The number and value of the stories varied during the project due to different aspects, such as new user's needs or stories re-estimation after the clarification of uncertainties. The Product Backlog included 23 user stories with a value of 64 story points at the end of the project. With the initial Product Backlog, the team established a Sprint length of 30 days, which would be stable during the project. Then, due to the lack of historical data, the team forecasted its velocity, estimating it in 7 story points and 137 initial available hours. Table 37.1 summarizes these estimations:

Based on these initial forecast s, a project plan, which included an original cost estimation founded on story points, was developed by means of the techniques proposed by Mike Cohn [13], obtaining €25.90 as the average cost per hour. It must be pointed out that, as it was developed with internal resources, all costs cited in this paper are estimations to manage the project and not real expenditures. An uncertainty percentage was added to these estimates according to McConnell's "cone of uncertainty" [23]. Table 37.2 summarizes the initial project plan:

The usual Scrum Sprint-based cycle started after the approval of the initial project plan. Each Sprint began with a Sprint planning meeting, where, depending on the available work capacity and the business priorities, the team selected the user stories that should run during the Sprint and committed to them. Moreover, each story was divided into tasks and each task was estimated. Finally, a Sprint Backlog was developed as a result of this meeting.

This artifact was the main management tool during the Sprint. Both Product and Sprint Backlog were created using Excel Spreadsheets and shared using a network folder. Before the start of the Sprint, all identified tasks were uploaded in a ticketing tool called Redmine. The team members should take responsibility for the different tasks during the Sprint upon their completion. Besides, the work performed and a

**Table 37.3** Project results against project estimations

| Magnitude | Initial forecast | Final value |
|---|---|---|
| Total story points | 70 ± 18 | 64 |
| Velocity | 7 ± 2 | 5.8 |
| Number of iterations | 10 ± 3 | 11 |
| Sprint length | 30 days | 30 days |
| Project length | 300 ± 90 days | 330 days |
| Hours per Sprint | 137 ± 34 h | 67.72 |
| Hours per project | 1,370 ± 340 h | 744.94 |
| Cost per iteration | €3,584.30 ± 887.01 | €1,777.74 |
| Total project cost | €35,843.00 ± 8,870.10 | €19,293.95 |

daily estimation on the remaining work was included in the tool on a daily basis. The Scrum Master was in charge of creating a Sprint burndown chart every day and distributing it among the team members. Then, the team worked on the identified tasks during the Sprint, using this Sprint burndown chart [13] as a tool to track the remaining work in order to accomplish the committed work. It is worth mentioning that, the team was not physically placed, thus, it was not possible to hold the Daily Scrum meeting to coordinate the work. However, some electronic tools were used to be in contact such as the mentioned ticketing tool, wikis, e- mails and instant messaging tools. During each Sprint, 10 % of the available time was spent to revisit the Product Backlog intending to update estimations or priorities, for instance, in what is known as "Product Backlog Grooming" [16].

The team met twice after each Sprint: the first time for a Sprint Review, to present the results of each Sprint and the other for a Sprint Retrospective, to review the process itself. It must be highlighted that retrospectives were based on the principles of Agile retrospectives [24], including techniques such as Ishikawa Diagrams, Five Whys and other innovation games. The Sprint retrospective started with a general assessment of the work performed during the Sprint, giving it a score between 1 and 5. After that, a radar diagram was developed, assessing four axis of the work: team, technology, process and environment. The team selected the most and less valuable axis and brainstormed things that should be improved or maintained. Afterwards, and using dot voting, the main problems of the Sprint and the best practices are pointed out. Lastly, Ishikawa Diagrams and Five Whys were used to find the root cause of the main problems and some improvements were proposed for each root cause. These improvements were followed up during the later retrospectives to ensure the resolution of the identified root causes.

The project lasted 11 iterations, being finished in February 2013. Table 37.3 shows the final project results and compares them with the initial estimations:

As it can be observed, the real data related to the duration of the project and the amount of work to carry it out remained within the margins of the initial forecasts.

That was not the case of the cost and dedicated hours forecasts, probably because of the absence of historical data. Therefore, this initial forecast was corrected at the end of each Sprint, as it will be shown later. Figure 37.2 shows the burnup chart of the project with the finished story points against the remaining story points of the

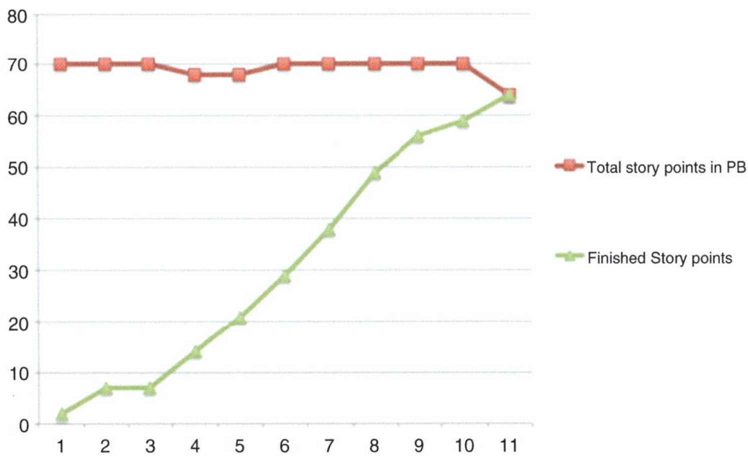


**Fig. 37.2** Project burn-up chart

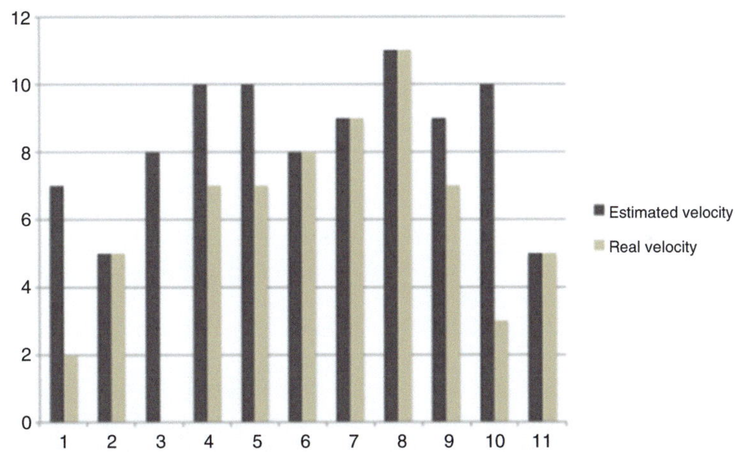


**Fig. 37.3** Velocity evolution through Sprints

Product Backlog, and Fig. 37.3 represents the evolution of the real velocity against the velocity planned through the different Sprints:

As it is noticed in both figures, after the initial Sprints, the team velocity tends to stabilize and to be almost aligned with the forecasts. Table 37.4 shows the results of each Sprint, including EVM calculations:

Additionally, the cost estimation per Sprint, and therefore, the cost estimation of the project, was updated at the end of each Sprint, in order to adjust them to the real data. This also entailed, as Table 37.4 represents, the updating of the estimated hours per Sprint. Due to this fact, the estimating ability of the team improved, as both CPI

**Table 37.4** Results of Sprints

| Sprint | Estimated story points | Finished story points | Total story points | Expected percent completed[a] | Actual percent completed[b] | Estimated hours | Real hours | Estimated cost per Sprint[c] | Planned value[d] | Earned value[e] | Actual cost[f] | Cost Perf. index[g] | Schedule Perf. index[h] |
|---|---|---|---|---|---|---|---|---|---|---|---|---|---|
| 1 | 7 | 2 | 70 | 10.00 % | 2.86 % | 137 | 74.4 | €3.548,30 | €3.903,13 | €1.115,18 | €1.926,96 | 0.58 | 0.29 |
| 2 | 5 | 5 | 70 | 10.00 % | 10.00 % | 136 | 78.6 | €2.724,68 | €2.997,15 | €2.997,15 | €3.962,70 | 0.76 | 1.00 |
| 3 | 8 | 0 | 70 | 21.43 % | 10.00 % | 105 | 48.65 | €2.227,40 | €5.250,30 | €2.450,14 | €5.222,74 | 0.47 | 0.47 |
| 4 | 10 | 7 | 68 | 25.00 % | 20.59 % | 101 | 102.7 | €1.959,66 | €5.389,06 | €4.438,05 | €7.882,67 | 0.56 | 0.82 |
| 5 | 10 | 7 | 68 | 35.29 % | 30.88 % | 120 | 80.82 | €2.198,13 | €8.533,93 | €7.467,19 | €9.975,90 | 0.75 | 0.88 |
| 6 | 8 | 8 | 70 | 41.43 % | 41.43 % | 75 | 45.55 | €1.986,40 | €9.052,31 | €9.052,31 | €11.155,65 | 0.81 | 1.00 |
| 7 | 9 | 9 | 70 | 54.29 % | 54.29 % | 106 | 102.92 | €1.985,86 | €11.858,45 | €11.858,45 | €13.821,28 | 0.86 | 1.00 |
| 8 | 11 | 11 | 70 | 70.00 % | 70.00 % | 89 | 51.55 | €2.015,80 | €15.521,64 | €15.521,64 | €15.156,42 | 1.02 | 1.00 |
| 9 | 9 | 7 | 70 | 82.86 % | 80.00 % | 67 | 67.25 | €1.876,86 | €17.106,22 | €16.516,35 | €16.898,20 | 0.98 | 0.97 |
| 10 | 10 | 3 | 70 | 94.29 % | 92.19 % | 77 | 56.5 | €1.889,25 | €19.594,22 | €19.158,17 | €18.361,55 | 1.04 | 0.98 |
| 11 | 5 | 5 | 64 | 100.00 % | 100.00 % | 46 | 36 | €1.777,54 | €19.552,95 | €19.552,95 | €19.293,95 | 1.01 | 1.00 |

[a] Estimated percentage of story points to be completed during certain Sprint
[b] Real percentage of story points completed during certain Sprint
[c] Calculated as the sum of the real dedicated hours of the previous Sprints plus the estimated dedicated hours of the actual Sprint divided by the number of Sprint
[d] Calculated as the number of Sprints multiplied by the Estimated Cost per Sprint and by the Expected Percent Completed
[e] Calculated as the number of Sprints multiplied by the Estimated Cost per Sprint and by the Actual Percent Completed
[f] Calculated as the Real Hours multiplied by the estimated cost per hour
[g] Calculated as the Actual Cost divided by the Earned Value
[h] Calculated as the Planned Value divided by the Earned Value

**Table 37.5** Productivity metrics

| Sprint | Velocity | Average hours per point story[a] | Velocity in hours [b] | Work capacity (hours)[c] | Focus factor[d] | Hours dedicated to accepted work[e] | % of accepted work[f] | TVI+[g] |
|---|---|---|---|---|---|---|---|---|
| 1 | 2 | 37.20 | 74.40 | 74.4 | 100.00 % | 58.9 | 79.17 % | 100.00 % |
| 2 | 5 | 21.86 | 109.29 | 78.6 | 139.04 % | 60.9 | 77.48 % | 142.86 % |
| 3 | 0 | 28.81 | 0.00 | 48.65 | 0.00 % | 0 | 0.00 % | 0.00 % |
| 4 | 7 | 21.74 | 152.18 | 102.7 | 148.17 % | 51.5 | 50.15 % | 200.00 % |
| 5 | 7 | 18.34 | 128.39 | 80.82 | 158.86 % | 40.85 | 50.54 % | 166.67 % |
| 6 | 8 | 14.85 | 118.82 | 45.55 | 260.85 % | 34.73 | 76.25 % | 165.52 % |
| 7 | 9 | 14.04 | 126.39 | 102.92 | 122.80 % | 78.16 | 75.94 % | 165.79 % |
| 8 | 11 | 11.94 | 131.37 | 51.55 | 254.84 % | 38.75 | 75.17 % | 179.59 % |
| 9 | 7 | 11.65 | 81.56 | 67.25 | 121.27 % | 41 | 60.97 % | 112.50 % |
| 10 | 3 | 12.02 | 36.05 | 56.5 | 63.80 % | 17.5 | 30.97 % | 50.85 % |
| 11 | 5 | 11.64 | 58.20 | 36 | 161.66 % | 22.25 | 61.81 % | 85.94 % |

[a] Calculated as the sum of the dedicated working hours on the finished Sprints divided by the number of finished Sprints
[b] Calculated as the sum of the completed story points in certain Sprint multiplied by the average number of hours per story point
[c] Total number of working hours on a certain Sprint
[d] Velocity in Hours divided by Work Capacity (Hours)
[e] Working hours dedicated to finished stories
[f] Hours Dedicated to Accepted Work divided by Work Capacity (Hours)
[g] Calculated as the finished story points of the current Sprint divided by the average story points of finished Sprints

and CSPI show, becoming stable as the project advanced. Table 37.5 shows the results of the productivity metrics through the different Sprints of the project:

It is observed that, on the one hand, the average percentage of Accepted Work was almost 60 %, meaning that a significant amount of work during the Sprints was dedicated to deliver value to customer and, on the other hand, the average TVI+ was around 120 %, meaning that, on average, the team was able to deliver more story points than in the previous Sprints. Lastly, the average Focus Factor was around 140 %, meaning that the team tended to overestimate the efforts needed to achieve the expected results. These metrics were used during the Sprint retrospectives to guide the discussion and improve the development process.

## 37.5 Lessons Learned and Future Work

During the previous sections, an example of the application of an Agile framework, based on Scrum and including several other Agile techniques, has been presented. This framework has been relevant to estimate, plan and manage an infrastructure project developed within the Public Administration through 11 Sprints as well as to help the team continuously deliver value to the customers. The selected framework included the techniques to support the estimating, planning and managing efforts, as continuous processes along the project and not as an initial phase. The example shows how a long-term project can be planned using the Agile techniques, despite uncertainties, and based on the plan, be managed in an Agile way.

The team has obtained results during the project founded on an Agile framework. One important element to highlight is that this approach allows adapting the forecasts at the end of each Sprint, according to the acquired knowledge of the team. Using Agile EVM was very useful, as it improved the plan at the end of each Sprint. The team was able to learn through the project, attending to the average cost of the past Sprints and the actual forecast.

Using productivity metrics was also very useful as a guide during the Sprint retrospectives to constantly enhance the development process, pointing to certain problems in certain Sprints (for instance, whether a high percentage of working hours was not dedicated to finish and deliver stories to the customer) and to effectively know the amount of the teamwork that it is needed to deliver a value. These metrics also allow quantifying how much (or how less) the team improved during the project.

The possibility of extending and systematizing the proposed approach in order to define a general Agile framework, suitable for IT projects, at project management level, should be useful to face projects with uncertainties on requirements and needs of frequent feedback. The possibility of assessing the framework against a maturity model, as CMMI [25], will provide customers with a relevant idea of the quality of the followed development process.

**Acknowledgements** This research has been supported by the project QSimTest (TIN2007-67843-C06_03), by the Tempros project (TIN2010-20057-C03-02), and by the project NDTQ-Framework (TIC-5789) of the Junta de Andaluc'a, Spain. We would also like to thank the Ministry of Culture and Sport of Junta de Andaluc'a for letting us issuing these data.

## References

1. Ambler SW (2002) Lessons in agility from internet-based development. IEEE Software 19:66–73
2. Begel A, Nagappan N (2007) Usage and perceptions of agile software development in an industrial context: an exploratory study. In: Proceedings of the first international symposium on empirical software engineering and measurement ESEM '07, IEEE, Madrid, Spain, 21–27 September 2007
3. Glazer H, Dalton J, Anderson D, Konrad M, Dhrum S (2008) CMMI or agile: why not embrace both!. CMU/SEI-2008-TN-003, Pittsburgh
4. Pikkarainen M et al (2008) The impact of agile practices on communication in software development. Empir Software Eng 13:303–337
5. Sutherland J, Schwaber K (2011) The scrum guide: the definitive guide to scrum: the rules of the game. http://www.scrum.org/Scrum-Guides. Accessed 1 Mar 2013
6. Huang PM, Darrin AG, Knuth AA (2012) Agile hardware and software system engineering for innovation. In: Proceeding of IEEE aerospace conference 2012, Big Sky, MT, USA, 3–10 March 2012, IEEE

7. Debois P (2008) Agile infrastructure and operations: how infra-gile are you? In: Proceeding of agile conference 2008, AGILE'08, Toronto, Canada, 4–8 November
8. Allen JN, Abdel-Aty-Zohdy HS, Ewing RL (2009) Agile hardware development with rapid hardware definition language. In: IEEE international conference on electro/information technology, 2009. EIT '09, 7–9 June 2009, pp 383–388
9. Beck K et al (2001) Manifesto for agile software development. http://www.agilemanifesto.org. Accessed 1 Mar 2013
10. Schwaber K (1995) Scrum development process. In: Proceedings of the 10th annual ACM Conference on object oriented programming systems, languages and applications, OOPSLA'95, Austin, TX, USA, 15–19 October 1995. ACM
11. Beck K, Andres C (2004) Extreme programming explained: embrace change, 2nd edn. Addison-Wesley, Boston
12. Cohn M (2004) User stories applied: for agile software development. Addison-Wesley, Boston
13. Cohn M (2005) Agile estimating and planning. Addison-Wesley, Upper Saddle River
14. Smith G, Sidky A (2009) Becoming agile in an imperfect world. Manning Publications, Greenwich
15. Highsmith J (2009) Agile project management: creating innovative products, 2nd edn. Addison-Wesley, Upper Saddle River
16. Cohn M (2009) Succeeding with agile using scrum. Addison-Wesley, Boston
17. Sulaiman T (2007) AgileEVM: measuring cost efficiency across the product lifecycle. http://www.infoq.com/articles/agile-evm. Accessed 1 Mar 2013
18. Project Management Institute (2012) Project management book of knowledge (PMBOK) guide, 5th edn. Project Management Institute, Newtown Square
19. Downey S, Sutherland J (2012) Scrum metrics for hyperproductive teams: how they fly like fighter aircraft. In: Proceedings of 45th Hawaii international conference on system science, HICSS 2012, Maui, Hawaii, USA, 4–7 January 2012
20. Pérez García F et al (2011) Las Diferencias Regionales del Sector Público Español. Fundación BBVA, Madrid
21. European Parliament (2010) Communication from the Commission to the European Parliament, the Council, the European Economic and Social Committee and the Committee of the Regions. A digital agenda for Europe. http://eur-lex.europa.eu/LexUriServ/LexUriServ.do?uri=COM:2010:0245:FIN:EN:HTML. Accessed 1 Mar 2013
22. Spanish Parliament (2007) Law 11/2007, 22th June, on citizens electronic access to public services. http://www.boe.es/boe/dias/2007/06/23/pdfs/A27150-27166.pdf. Accessed 1 Mar 2013
23. McConnell S (2009) Software projects survival guide. Microsoft, Redmond
24. Derby E, Larsen D (2006) Agile retrospectives. Making good teams great. The Pragmatic Bookshelf, Dallas
25. Torrecilla-Salinas CJ, Escalona MJ, Mejías M (2012) A scrum-based approach to CMMI maturity level 2 in web development environments. In: Proceeding of international conference on information integration and web-based applications & services 2012, iiWAS'12, Bali, Indonesia, 3–5 December 2012. ACM

# Chapter 38
# The Uncertainties Management Framework: Eliminating Uncertainties from the Process

**Deniss Kumlander**

**Abstract** Building software under current pressure of uncertainties coming from quickly changing global world, increased flexibility of requirements and changing customers' expectations is a central challenge even using agile methodologies. Sources of uncertainties include requirements, communication process, involved people and the software engineering process itself. This paper is dedicated to revising the software engineering process and presenting ways we could eliminate uncertainties leaking through the process. This is done within the scope of the uncertainty management framework, which is an addition to common project management practices designed to identify, manage and control uncertainties within all stages and aspects of software engineering projects. The process related uncertainties management uncertainties' elimination techniques include errors injection, stress testing, certainty testing of units, integration, features, usability and transportation components. It also concentrates on handling objects, which are subject for uncertainties like debt, recognised uncertainties, decisions and self-organised teams' abilities. The software engineering process verification on uncertainties is a multistage process including planning, decision to execute, acting and monitoring, analysis, modification, acceptance and re-evaluation.

**Keywords** Software engineering • Methodologies • Uncertainties • Process

## 38.1 Introduction

The main goal of software engineering is to implement tools for customers accordingly to their past and present wishes, vision and business environment requirements. These tools will let the customers to achieve their goals faster, easier with

D. Kumlander (✉)
Department of Informatics, Tallinn University of Technology,
Rja Str 16, Tallinn, Estonia
e-mail: kumlander@gmail.com

M. José Escalona et al. (eds.), *Information System Development: Improving Enterprise Communication*, DOI 10.1007/978-3-319-07215-9_38,

less cost. The modern software engineering business faces a lot of new challenges due permanently increasing competition on the market, high expectations of customers requiring constantly increasing quality, shortening developing cycle and increased flexibility defining requirements on the software to be built. The flexibility becomes more and more important as the global business environment is starting to change with an incredible speed. All this reflects into the software development not only in a form of agile methodologies, but also producing a lot of uncertainties ignoring which is impossible nowadays for companies hoping for success. The customer satisfactions' reviews show that despite of modern software engineering approaches the gap between delivered and expected software is sufficient—up to 27% of all projects fail because of that gap [1]. Besides, only 20% of functionality in average is reported to be used "often" or "always" [2].

It is important to explore reasons why the shortened development cycle with constants demos and enhanced collaboration are not able to bridge this gap. The most serious failures are related to different kinds of uncertainties. The more uncertainty in a problem, feature or task, the less precise or correct we can be in solving or understanding that problem and the more far away we are from the final implementation. Most analysts agree that we should understand and assess the uncertainty implementing functionality, making estimates, and balance that uncertainty against the precision in customer expectations we seek [3]. Therefore the uncertainties management becomes very important in order to ensure software engineering projects success [4].

The aim of this paper is to extend the uncertainties management framework introduced earlier and discuss the project management process related uncertainties. The paper is organised as following: in the second chapter the past works are discussed to show the paper starting position, ideas proposed earlier and lay the ground for the paper proposals. In the third chapter the process evaluation is presented aiming to eliminate the uncertainties introduced by the process itself. Here different tests are proposed serving to assure the process functionality, handling of objects and phases of the process verification. The question of individuals and tools reviewed next. The chapter ends with dependability and project management system attributes discussion.

## 38.2 Related and Past Works

### *38.2.1 Related Works on Uncertainties*

Uncertainties sometimes are equated with lack of knowledge or risks, but neither of those are correct assumptions. The best was to describe the topic is to present the classification. Uncertainties include not only the lack of knowledge, but also unpredictability also known as ontological uncertainty, variability and ambiguity (presence of multiple frames or view on the same topic) [5, 6]. Generally uncertainties can be classified by objects (for example, the project assumptions, team, perception by customers, environment at the completion phase, impact of the new solution on the existing systems, interfaces etc.) or nature. The last includes incompleteness of knowledge (lack of information, unreliable information, ignorance or lack of understanding), unpredictability (how the expectations will change during the project, competitors behaviour, technological innovations on the market) and multiple understanding (conflicting views, different values or judgements, lack of solid understanding on how the problem should be solved or what problem we are solving). Sometimes those are divided into variability, unforeseen uncertainties, foreseen uncertainties and chaos [7].

The strategies of dealing with uncertainties greatly depend on the project subject within those arise and are not general enough to be comprehensive and coherent. At the same time the subject got a sufficient attention during the past decade and the following strategies can be outlined [5, 8, 9]:

1. Producing/recreating the knowledge through uncertainty assessment, reduction and case study;
2. Ignoring or using the reactive approach;
3. Interacting—managing the expectations, communicating uncertainties, using dialogical learning, persuasive communication and oppositional modes of actions;
4. Preparing by finding the worst sequence of events and training to deal with it, limiting negative impacts within it etc;
5. Building flexible enough or resilient solutions.

### *38.2.2 Uncertainty Management Framework*

Uncertainties management is the crucial part of modern software engineering practices, which is mostly ignored by management and modern software development practices or dealt with reactively [10]. In our past works we have presented an uncertainty management framework which is composed of the following elements:

1. Transparency
2. Light Agility
3. Handling
4. Communication Ambassadors
5. Well-defined Work Procedures

*Transparency*—it is the most crucial element handling uncertainties since those occur or grow when transparency is missing. Agile development key elements are designed for increased transparency of the process [11, 12] and handling changes in customer expectations [13]. Transparency can be increased by implementing a central information repository:

- Containing enough information about uncertainties nature, time-frame when they are likely to be resolved and possible resolution alternatives;

- Providing access for different teams to this information on constant basis;
- Capturing evolution of the uncertainty including history of changes, discussions and feedbacks from customers and experts [14].

*Handling*—the transparency of uncertainties provides a basis for handling them. Uncertainties can be permanent or temporary and the kind of uncertainty defines the handling technique. Assessing, planning activities to resolve uncertainties, defining the schedule basing on a time moment when an uncertainty can be resolved and synchronisation with other tasks is required to handle uncertainties correctly.

*Light-agility*—The agility still plays an important role handling temporary uncertainties that can be resolved via visualisation of the end result. Simulation achieved by shortened work-cycle help us to deal with such uncertainties [12, 15, 16]. The light modifier in the term refers to the fact that not all organisations are ready to adopt agility practices fully, but are ready to modify their process implementing the most crucial set of agile techniques to handle uncertainties.

*Communication ambassador*—unfortunately the desired constant collaboration [10, 17] is impossible in many cases—consider for example distributed organisations. Here we are not only unable to resolve uncertainties efficiently using enhanced communication, but also introduce uncertainties by communication gaps or slow communication [18]. Besides, the communication is not only the message transferring act, but also a process that should be managed. This becomes very important in large or distributed organisations and includes chasing other people to provide communication, do reviews etc.; supervising communication flow and ensuring that context is also transferred; directing the flow to the right persons; ensuring none-zero communication and monitoring the communication from the time-perspective; synchronising efforts of different teams working with different issues in different projects and been in different time-zones.

*Well-defined work procedures*—undefined or unclear procedures introduce a lot of uncertainties and therefore should be resolved before executing the software engineering process or fixed during it in case any modifications are needed. Besides, well-defined procedures are required to cope with issues resolving uncertainties. It should be specified who is responsible for what and what are procedures deciding on alternatives.

### 38.2.3 Test Based Verification of the Process

Uncertainty can be introduced not only by external factors or people but also by the software development process. Even having the well-established process the organisation cannot be certain that uncertainties are minimised since the formal process, which is not followed, can be equally dangerous as a missing process since can hide problems and so increase the ratio of uncertainties we have to deal with at the project final stages. During our past works we proposed an idea that the software development process can be verified by stressing or injecting errors in order to discover uncertainties leaks.

The idea of the errors injection is to try unbalancing the system by injecting problems ourselves into the process and see how the process (the team) will react/recover from this and so either detect leaks of the process or prove its reliability [10]. It is possible to inject errors in different artefacts and so possible to test every stage of the process affecting documents of that stage.

The other technique we proposed is stressing the team with a special task to verify its resistance and correctness of the reaction [10].

Both elements can be seen as a unit testing of a process stage where we examine an individual phase on uncertainties proneness. Below we are going to extend this idea and model it as a subsystem of the uncertainty management framework.

## 38.3 Process Assurance

The earlier described methodology was applied in different organisations including the highly distributed organisations and branches of leading software vendors, which gave us a relatively broad feedback. Analysing it we have found that the used terms were both unusual, hard to understand and quite frightening to teams and managers. The only well accepted one was the *stress test* since it corresponds to managers' earlier experience with other test procedures. Most problems applying the proposed method were related to the error injection as nobody felt secure enough adding extra errors into the process. Therefore we propose both extend and change the terms to align them to the standard testing terminology. In the next chapter we first present an overview and then expand it in details in succeeding subchapters.

### *38.3.1 Certainty Assurance*

The processes of verifying the used methodology on introducing uncertainties can be divided into the following:

Certainty element testing:

- Certainty unit testing;
- Certainty integration testing;

The next stages can be called certainty system testing as next we concentrate on system as a whole, although following only one concept or property only sometimes.

- Certainty features testing;
- Certainty performance testing;
- Certainty usability testing;
- Certainty transportation testing;

Let's now describe in more details each one:

*Certainty unit testing*: the test of every stage as it was described earlier in our past works;

*Certainty integration testing*: investigating the border area between neighbour stages in order to find communication and other gaps. First of all notice that individuals bounded to different stages are most likely to have completely different backgrounds and experience, use different terminology and concepts. Therefore the information flow tends to corrupt floating from one stage to the other and our aim is to ensure that the organisation has implemented enough protection mechanisms against such gaps. It could include a mandatory feedback, team building work, mixing teams and many others. For this paper perspective it is not important—the important statement is—we should monitor such border areas and check what happens there.

*Certainty features testing*: the process dedicated to specific kind of activities like for example the risk management. Generally saying here we could isolate different aspects of the software development process and monitor how this aspect works through the entire process. Does it has a proper initiation, receives information and result in any actions clear defined and bounded to responsible persons and is finished?

*Certainty performance testing*: performance of the system and its ability to recover. This question has been raised also in our earlier works and can be measured using stress testing. It is important to ensure that any sudden increase of work load is not hidden from other project members risking to get severe problem at final stages, but is handled and managed;

*Certainty usability testing*: this test divides into two. The first one is ability of the process management system to store enough details to recover the knowledge later on demand. It is quite a wide problem in the modern agile software development practices—zero or zero-close documentation leads to losing a lot of details and decisions made initiating, correcting or finalising features development. Sometimes even during the same feature development the same logical or behavioural aspects of the feature are revised several times wondering about past decisions mainly because reasoning of those is not remembered any longer. If the team is forced into such cycle then we may have severe problems of our ability to use the software development and specification system correctly or it is missing absolutely.

The second one is the usability of the uncertainties management routines and procedures. The more complex and less transparent is such system, the less people use it to handle uncertainties. They may prefer to leave uncertainty to be unhandled believing it is easier to cope with its consequences than with the uncertainty management routines and practices. Such luck of user friendly routines can be called "uncertainty in the square" as produces uncertainties about uncertainties and so can add power to existing lack of certainty.

*Certainty transportation testing*: ability of channels to carry on the information back and forward without corrupting it. Sometimes channels in use are not ideal and pushed to us by the organisational IT or easiness to install and maintain. At the same time information corruption can be high in such “simple to use” channels producing a lot of uncertainties and so in the end of ends more complex to handle in total. For example communicating over the phone is known to lose up to 45% information in compare to face to face dialog. This channel can be enhanced by using modern video streaming devices (more complicated than just a webcam) allowing having a view on the entire meeting room.

### 38.3.2 Uncertainties Handling

- Handling Debts
- Handling Known Uncertainties
- Handling Decisions
- Handling Self-Organised-Teams

Talking about uncertainties sources and verifying the project process we should examine not only processes or activities, but also objects of our processes.

One common uncertainty source is the accumulated uncertainty. A small problem, which has a minimum effect, can be postponed even beyond the release border to the next versions, but start to accumulate/grow and hit us seriously once. Therefore it is important to examine how do we handle technical or functional debt? The agile software development let us release business software without handling all open questions. Moreover the business environment forces us to release none error/debts free software to meet key events on the market and hit important customers. All this is acceptable if we not only postpone solving certain problems to the future, but also record, handle and resolve them later. If we have no such system and tend to rediscover our debts then we never are able to plan our load correctly as those threads can hit us any moment in the future increasing the load dramatically. Here we should ask product owners and software architects—whether they are able to list the debts of the previous releases. If they are unable to answer then the software vendor has a serious problem. If their answer doesn’t correspond to the chief tester or chief developer problem the organisation again has a serious problem.

The other important question to revisit examining the process is how we handle known uncertainties. Do we have a list of them, with an action plan to resolve it, responsible person and potential date when it can be cleared out? It can happen than even such simple request is not fulfilled and we again know uncertainties, but do not handle them and so have no uncertainty management framework in place. The standard uncertainties management system should have the following subparts and functionalities:

- *Uncertainties inventory*: at the earliest possible stage it should be possi-ble to record an uncertainty to proceed with assessment and quantification. The inventory

is organised to store all relevant information regarding different kinds of uncertainties including choices, available data/information/knowledge to characterize each items, discussion logs, possible resolution paths and responsible people.
- *Uncertainties classification and clarification schemes collection*—the knowledge and the know-how database helping to clarify immediately uncertainties or providing the guidelines on dealing with them;
- *Uncertainties propagation system*: a nonlinear graph based system to visualize uncertainties interconnections and how do they bound to different features requests from the product backlog.

Thirdly we need to examine how we record our decisions, both functional and process related. Again the most common source of uncertainties is the lack of organisational memory both short and long term. Theoretically we should never rediscover our short term decisions, but the complex software with many details doesn't let product owners and other team members remember every detail on their past decisions and the reasoning of such decisions. Therefore, if the decision is communicated and accomplished by the business scenario explaining its need, it is important to register it either directly or using secretary help (manual or automated) in form of Wiki or any other system, to be able to revisit it. Obviously the same will become crucial if any key person is leaving the company as one month education of the person replacing him is not enough to transfer all details and knowledge acquired during many years or work experience.

Finally it is important to visit our self-organised teams to ensure that any lack of autonomy doesn't produce uncertainties. Are they capable to decide ourselves or are waiting for manager input and managers assume they are self-organised? Are they able to communicate any problems or discoveries openly or they take autonomies as an absolute responsibility to deal alone with raising problems without informing product owners that some features may become much more complex than initially estimated and so not actually worse to be built up?

### 38.3.3 Phases

Each process verification test should pass the following phases:

- Plan
- Decide to execute
- Execute and Monitor
- Analyse and Modify
- Accept
- Re-evaluate

Even if we have a clear understanding of problems in our organisation, it is not advisable to execute the verification process without planning. We need to understand the size of required certainty tests, agree on responsible people and how we will measure and record results, so everyone involved into the process is aware and ready to follow the plan.

After we have collected possible verification tests scenarios, we should revise them, align to current activities in executed software projects and decide whether it is a feasible time to execute them or not. The high load in the current projects can be advantage (we can efficiently measure) or disadvantage (if our activities will interfere too much and so affect important deadlines).

The next stage is to execute our actions' plans accordingly to the decided earlier. It is important to fulfil the plan during this stage in all details as re-execution can be a costly process. It is also important to document the stages and actions to be able not only to monitor and measure the results, but also to avoid faulty reaction of the system later after test completion in case such reaction can be produced by our verification actions. For example, it is important, stressing any aspect of the process, to eliminate later any environmental changes producing such stress, i.e. not to leave it to be permanent.

After we completed our execution plan we transfer to the analysis stage where the collected information and logs are reviewed and measured. Quantification of discovered uncertainties and gaps on this stage is a key element. Most likely, during this stage we will come up with the proposal on how the current software project planning and execution process can be improved. During this stage we discuss these changes first with responsible managers and the team. We need not only to correct, but also to give to the team explanation why changes are done and what those are designed for. Otherwise the formal implementation of new rules could not lead to the desired improvements. It is also advisable to record changes we do for later tracking.

The modification phase is followed by the accept stage where we monitor how the implemented proposed modifications are working—we monitor here how the new process handles uncertainties and accept the modification in case the behaviour is correct and no new pitfalls are discovered.

The final stage is the re-evaluation. Unfortunately all modifications tend to change into formal routines which are followed by team members without understanding why those are needed. Therefore we can observe a negative effect, that our improvements stop working after some time. Therefore it is important to reevaluate closed uncertainties gaps periodically to ensure that those were not opened again, i.e. keep the closed gaps also under control.

### 38.3.4 Individuals and Responsibilities

Individuals and interactions—the testing process is dependent on the individuals who decide the format of the test, who lead the test, and who actually perform the tests [19]. Therefore it is very important to form a team of people to measure our performance handling uncertainties from those who understand the importance of the work and uncertainties nature. Ideally these people should have had an experience dealing with uncertainties drawbacks, which will appear at the final stage of a project, to understand full importance of the task.

The question of responsibilities has two sides: external and internal. Under internal responsibility we mean the assignments of people to carry on uncertainties identifying test on the current project. Such an assignment is usually permanent, i.e. he or she should redo tests periodically to ensure no new uncertainties sources or gaps are introduced after the previously found are fixed.

Under the external responsibility we mean the assignments of people to deal with different kinds of uncertainties. It is important information for planning and carrying the process verification tests. Who is responsible for the risk management? Who is dealing with feature related uncertainties? Who is responsible for technological threads?

Finally, it is important, assessing individuals involved into software engineering process, to leverage their:

- Domain Knowledge
- Experience in the responsibility area
- Technical knowledge in the responsibility area
- Communication skills
- Awareness on internal and external standards—Team integration level
- Learning ability

This way we can rank people by ability to produce certainty or vice versa—by the probability they will add uncertainties, for example producing unreliable code, unreliable reports on fully tested areas and so forth. Basing on this we can decide to include such person into:

- Education program
- Pair programming/testing initiative
- Supervise by senior position (code review initiative)
- Demand producing documentation on activities.

### 38.3.5 *Tools*

The usage of tools is a standalone and interesting questions. We need tools to:

- Record our tests and actions to be done next in order to not forget, at the end of the verification process, to roll back verification specific modifications;
- Measure the verification process and monitor the uncertainties handling improvements efficiency in time;
- Register and handle uncertainties discovered during the verification process;
- Register and handle features, descriptions, decision etc. to avoid uncertainties;

Every company will find its own set of tools to be used. It is important to have them in place and make everybody aware of the tools we use including the training. One of the common reasons for adding uncertainties to the projects, which is related to the process, is using several tools (each sub-team/member insists on own). In this case nobody knows which tool information is the most correct and full. People may start doing decisions on faulty data. Therefore it is important to avoid using different tools for the same purpose and agree on something particular, as described next.

### *38.3.6 Dependability*

It cannot be proved that the software or process is absolutely correct or uncertainties free. It is possible to detect problems sometimes by falsifications, but if the behaviour turned out as predicted, the model is only confirmed and cannot be proved [20] to be uncertainties free. Therefore we definitely should use continuous testing i.e. a continuous attempt to "falsify" the company believe into the well- established project management system they have.

The validation by means of testing increases our dependability on the current model although this dependability can never be established with absolute certainty [20].

Nevertheless we can measure the current system attributes using qualitative or quantitative measures:

- Availability
- Reliability
- Integrity

The *availability* is important attribute of all system components and is described as system's readiness to correctly serve the organisational needs. All missed, unregistered or unavailable information and statuses produce sufficient uncertainty of different kinds. A similar question can be stated about the *reliability* of the system work or produced output. All the incorrect or false information obtained from the system leads the organisation in the wrong direction and force to implement incorrect features showing a clear lack of certainty of what we do. The absence of *integrity* in the project management system could mean a different view point on the current status, requests or expectations. The more subsystem we have, the more duplication of information and absence of integrity the more gaps we produce. The different views, opinions are reports is produced from different systems creating a chaos in the project.

Other system attributes are important for evaluating the system, but less important for uncertainties management framework: *Safety*, *Maintainability* and *Confidentiality*.

## 38.4 Conclusions

The most reliable, well-defined process doesn't always guarantee that no uncertainties are introduced by its components. The process is implemented and followed by people, who do not always remember to do it correctly, do not understand the reasons of the way it is designed or have enough time to follow it correctly since are stressed by other responsibilities and tasks. Therefore it is very important to verify the software engineering process we use on uncertainties leakage and deal with discovered problems immediately.

The process verification is not a complex process. It just needs to be planned and organised correctly and accordingly to the presented guidelines to avoid missing any components or producing a chaos on the live project. It contains verification of routines and sub-processes, objects and individuals involved into the engineering process. The question of used and required tools is also vital both handling uncertainties and validating the process.

## References

1. Bennatan EN, Emam KE (2005) Software project success and failure. Cutter consortium. http://www.cutter.com/press/050824.html. Accessed 29 June 2013
2. Khan AA (2004) Tale of two methodologies for web development: heavyweight vs agile. Postgraduate Minor Research Project, pp 619–690
3. Booker JM, Ross TJ (2011) An evolution of uncertainty assessment and quantification. Sci Iran 18:669–676
4. Kumlander D (2006) Software design by uncertain requirements. In: Proceedings of the IASTED international conference on software engineering, pp 224–2296
5. Brugnach M, Dewulf A, Pahl-Wostl C, Taillieu T (2008) Toward a relational concept of uncertainty: about knowing too little, knowing too differently, and accepting not to know. Ecol Soc 13:30
6. Schrader S, Riggs WM, Smith RP (1993) Choice over uncertainty and ambiguity in technical problem solving. J Eng Tech Manag 10:73–99
7. De Meyer A, Loch CH, Pich MT (2010) Managing project uncertainty: from variation to Chaos. MIT Sloan Management Review
8. Raadgevera GT, Dieperinkb C, Driessenb PPJ, Smitb AAH, van Rijswickc HFMW (2010) Uncertainty management strategies: Lessons from the regional implementation of the Water Framework Directive in the Netherlands. Environ Sci Pol 14(1):64–75
9. Walker WE, Harremoes P, Rotmans J, van der Sluijs JP, van Asselt MBA, Janssen P, Krayer von Kraus MP (2003) Defining uncertainty: a conceptual basis for uncertainty management in model-based decision support. Integrated Assess 4:5–17
10. Kumlander D (2009) Uncertainties management framework—foundational principles. In: Proceedings of the 11th international conference on enterprise information systems. INSTICC Press, Porto, pp 103–108
11. Beedle M, Schwaber K (2001) Agile software development with SCRUM. Prentice Hall, Englewood Cliffs
12. Cockburn A (2002) Agile software development. Addison Wesley, Reading
13. Kumlander D (2012) Uncertainty management framework elements—test based verification of the process. In: Recent advances in computer engineering series, WSEAS, vol. 5, pp 517–521

14. Rauterberg M, Strohm O (1992) Work organisation and software development. Annu Rev Automat Program 16:121–128
15. Braithwaite K, Joyce T (2005) XP expanded: distributed extreme programming. In: 6th International conference on eXtreme programming and agile processes in software engineering. Springer, Berlin, pp 180–188
16. Stapleton J (1997) DSDM dynamics system development method. Addison Wesley, Reading
17. Forsgren O (2006) Churchmanian co-design—basic ideas and application examples. In: Nilsson AG, Gustas R, Wojtkowski W, Wojtkowski WG, Wrycza S, Zupančič J (eds) Advances in information systems development: bridging the gap between academia and industry. Springer, New York, pp 35–46
18. Kumlander D (2006) Bridging gaps between requirements, expectations and delivered software in information systems development. WSEAS Trans Comput 5(12):2933–2939
19. Winter J, Rönkkö K, Ahlberg M, Hotchkiss J (2011) Meeting organisational needs and quality assurance through balancing agile and formal usability testing results. In: Proceedings of the third IFIP TC 2 Central and East European conference on software engineering techniques. Springer, Berlin, pp 275–289
20. Gagliardi F (2007) Epistemological justification of test driven development in agile processes. In: Proceedings of the 8th international conference on Agile processes in software engineering and extreme programming. Springer, Berlin, pp 253–256

# Chapter 39
# Towards the Development of a Complex Adaptive Project Environment Assessment Tool

**Corina Raduescu and Asif Qumer Gill**

**Abstract** Managing complex software and information systems (IS) development projects in a continuously changing business environment is an arduous task. We argue that the success of a complex project depends on the underlying capability of the specific project environment to be adaptive to changes. Organizations have therefore a need to assess the capability of the underlying project environment to identify how well it supports the management of a complex project. In this paper we present the design of an initial "complex adaptive project environment assessment" (CAPEA) tool, which has been developed based on the eight key characteristics of the complex adaptive systems (CAS): Autonomy, Interdependence, EcoSystem Structure, Context Awareness, Non-Linearity, Emergence, Self-Organization, and Adaptation. The evaluation of the CAPEA tool is conducted by assessing two project methods in order to provide an initial indicative proof of concept. The evaluation of the CAPEA tool indicates that it can assist organizations to assess and improve the management of complex software or IS project methods, and reduce costly consequences associated with failed projects.

**Keywords** Complex project • Complex adaptive systems • Assessment tool • Adaptive environment

C. Raduescu (✉)
Discipline of BIS, University of Sydney, Sydney, NSW, Australia
e-mail: corina.raduescu@sydney.edu.au

A.Q. Gill
School of Software, University of Technology Sydney, Sydney, NSW, Australia
e-mail: asif.gill@uts.edu.au

M. José Escalona et al. (eds.), *Information System Development: Improving Enterprise Communication*, DOI 10.1007/978-3-319-07215-9_39,

## 39.1 Introduction

In today's society organizations develop, use, and rely upon software and technology that are integral part of their functioning. Specifically, they help organizations in various fields such as: design, production, and delivery of products and services to their customers. Despite recent developments in both technical and research areas, we hear many stories of failed projects [1–3]. The success rate of software development and Information Technology (IT) or IS projects is still low.

The following success and failure rates have been reported by IT Project Success Survey [2]: ad-hoc projects: 49 % are successful and 14 % are failures; iterative projects: 61 % are successful, and 11 % are failures; *agile projects*: 60 % are successful, and 12 % are failures, and *traditional projects*: 47 % are successful, and 17 % are failures. According to Feldmann [1] we witness a decrease in the use of formal project management methods from 70 % in 2010 to 58 % in 2012.

Given the changing nature of the environments, the technology and IS need to adapt and respond quickly to organizations' changing or new requirements. Specifically, we argue that software and IS are developed and maintained in continuously changing environments, and are therefore characterized by a high degree of uncertainty and complexity. Based on the above reports, more adaptive and flexible project methods start to be the preference nowadays, an argument supported by their higher rate of success.

To help improve the overall rate of projects success, in this paper we turn our attention to complex adaptive systems (CAS) theory and view software and IS projects as complex projects that display characteristics of CAS [4]. We argue that the success of a complex project depends on the underlying capability of the specific project environment to be adaptive during the management of such project. Specifically we suggest that organizations (1) need to assess the underlying capability of the project environment to identify how well it supports the management of a complex project, and (2) tailor suitable project management method/s to successfully manage complex projects.

Our aim in this paper is to presents a "complex adaptive project environment assessment" (CAPEA) tool that can be used by organizations to assess the underlying capability of the project environment to support the management of a complex project. The research work presented in this paper is part of an on-going project in which we investigate the factors that contribute to the complexity of a project. As a result, the current version of the tool captures a limited number of initial factors as key criteria for assessing the project environment capability. We use a design science approach to develop and test an initial version of the CAPEA tool. In this paper the evaluation of the tool is conducted by assessing the capability of two project management methods in order to provide an initial proof of concept. The tool will evolve as our research progresses, and is expected to capture and a larger set of components of a project environment in order to fully assess the entire project environment. We believe that CAPEA tool can assist organizations to assess and improve the management of complex projects, and reduce costly consequences associated with failures of such projects.

This paper is organized as follows: first, we present the key characteristics of CAS in the context of software development and IS projects. Second, we present the research method. Third, we discuss the development of the CAPEA tool. Fourth, we present the assessment of the tool as an initial proof of concept. Fifth, we discuss the limitations of the current work. Last, we conclude with the contributions of our current work to the body of knowledge and future research directions.

## 39.2 CAS Characteristics of Software Development Projects

It has been observed that software and IS development projects are increasingly facing unplanned and unregulated interconnectedness and they began to undergo growth in problems of design, management, maintenance and decommissioning [5, 6]. As the scale and interconnectedness of required software and technologies continues to grow, such projects become more complex and their ability to deliver solutions tailored to particular requirements appears to be hindered and more difficult to predict.

To address the complexity inherent in such projects, we refer to complex adaptive systems (CAS) theory. Based on prior work, software development projects are structured and behave like CAS. Prior literature [7–11] notes that CAS have specific characteristics that make them more difficult to understand and manage. CAS theory can therefore be the appropriate lens to help understand and improve to the management of complex software and IS projects [4, 8].

Based on a review of CAS literature and referring to the diversity and complexity encountered in such projects, we have identified *eight* key CAS characteristics relevant to our context: Autonomy, Interdependence, Eco-System Structure, Context Awareness, Non-Linearity, Emergence, Self-Organization, and Adaptation. Below we introduce the specific CAS characteristics found in software and IS development project environments. These characteristics and the factors contributing to complexity form the basis for developing the assessment criteria used in CAPEA tool. Note that we refer to a small number of factors contributing to complexity for each CAS characteristic as they are part of our on-going research project.

### *39.2.1 Autonomy*

Components or agents in CAS have a degree of freedom and are autonomous to some extent [4, 7–9]. They are subject to simple, localized rules, as opposed to centralized rules belonging to the whole system. There is a limited control in the overall system, however control is considered a mechanism that maintains the system together.

In software and IS projects we observe a number of centralized norms and rules that balance the freedom, and control the agents' behavior. These rules and norms are part of the processes that help the project moving forward, e.g., having an overall project plan, deadlines, milestones, performing resources allocation, communication, coordination, etc. Project norms and rules are captured in project management methods and are examples of control that typically maintain the stability of the relationships among the project components, as well as manage their behavior [6]. Factors assessing and contributing to the degree of autonomy might be: identifying whether the project environment supports the structure of an organization or project to be hierarchical or flat, and whether it supports fixed or flexible rules driven behavior.

### 39.2.2 Connectivity and Interdependence

Connectivity refers to the inter-relatedness of individual components within a system, as well as to the relatedness among systems [6, 8, 9]. In software and IS projects, deliverables of one team are connected and impact on the outcomes of the other teams, and consequently on the whole project overall [6]. For that reason, software project teams must bring together resources from different parts of the organization/s and ensure they are engaging in communication and exchange the vital project information to deliver the project.

Factors assessing or contributing to the degree of connectivity and interdependencies might be: identifying whether there is a significant number of business units or teams involved in a project, and whether there is a high level of interdependence among these units/teams. Specifically are the teams involved in a project relying upon the best resources they need, are the teams co-located or distributed, and are there clear processes in place such as communication and coordination to support the exchange of vital project information.

### 39.2.3 Eco-System Structure

Each system has sub-systems and is part of a larger system/s. Each system operates in a nested structure of sub-systems that are interconnected and interact with each other; thus a CAS has an eco-system structure [8, 9, 11].

In software and IS projects, usually there are a number of teams (sub-systems) working in parallel to deliver different parts of the project. The interactions among the project members and/or various teams can be viewed as belonging to the whole project, that is, they belong to an eco-system. The benefit and main aim of having an eco-system structure is to break down major project activities into manageable blocks; e.g., project milestones [6]. Factors assessing or contributing to the degree of the eco-system structure may include: the number of project teams and sub-teams involved in the whole project and how these teams interact with each other; that is, whether and what processes are in place to support their collaboration, interactions, and coordination.

### 39.2.4 Context Awareness

Small differences in initial conditions or situation of a system can produce different or unanticipated effects [10, 11]. A situation is a position or environment characterized by a set of factors called contingencies or situational factors [12].

In software and IS development environments, the same team delivering the same project in different contexts can achieve different performance levels and results [6]. It is therefore critical to acknowledge the characteristics of the environment in which the project is conducted because they will impact on its trajectory, progress, and ultimately on its delivery. Factors assessing or contributing to the context may include project or domain factors (e.g., size, complexity, users, and resources), technical factors (e.g., development platform, infrastructure), expertize, and external factors (e.g., law, organization culture) [12]. We suggest that understanding these situational factors and their dynamics is critical to managing the projects.

### 39.2.5 Non-linearity

Non-linearity is associated with the concept of change. The assumption is that the components of the system interact in less ordered or non-linear way, and can lead to unexpected or unknown behavior [7–9]. One consequence of this characteristic is that the overall system behavior is very difficult to be anticipated and planned for.

In software and IS projects, due to the dynamic nature of organizations environments, often we see ambiguous, unclearly determined, and changing requirements [4]. This leads to unplanned events and actions to be taken during the project in order to address and deliver solutions that meet the business requirements. Factors assessing or contributing to the degree of non-linearity can be: identifying whether the environment allows or supports a high or low level of uncertainty, variability, ambiguity, rate of change, ambiguity, novelty, and diversity in requirements and overall behavior.

### 39.2.6 Emergence

Emergence is related to the concept of the "whole." A CAS needs to be understood and studied as a complete and interacting whole, rather than as an assembly of distinct and separate components. In other words it is important to acknowledge that the system displays an aggregate behavior. Emergence is a process that is dependent on the internal behavior of the system.

Prior research on success and failure of projects has highlighted that most software and IS projects do not necessarily follow the initial project plan. In many instances ad-hoc or unexpected behavior arises as a result of non-linear interactions among the components of the project [4]. Factors assessing or contributing to the degree of emergence can be: identifying whether the project environment welcomes variability, whether there are strong interactions among components resulting in new system's behavior and structure, as well as how easily or not change can happen.

### *39.2.7 Self-Organization*

A CAS has the ability to suddenly take on a new form in response to changing conditions [7–9]. The process leading to the spontaneous order reached by the system in response to changing conditions is called self-organization.

In a software or IS project, self-organization can be described as the spontaneous coming together of a group to perform a new (often unplanned) task (or for some other purpose) [6]. The group decides what to do, how and when to do it; and no one outside the group might direct these activities. The group is autonomous to a certain extent; that is, the group will still have to obey certain project rules and norms, given the group aim is the overall project delivery. Another example of self-organization in software projects is re-assessing and planning in response to variation in available resources [4]. Factors assessing or contributing to the degree of self-organization can be: identifying whether the governance of the project is tight or not, whether it allows for spontaneous organizing, and the degree of freedom to act in such situations.

### *39.2.8 Adaptation*

As a response to changes in environment, a CAS responds and adapts to accommodate and use favorably these changes in order to maintain or improve itself [7–9]. In today's turbulent and changing environments, it is essential that organizations and projects are adaptive to the environment.

In a software or IS project, adaptation happens in response to environmental constraints or the possibility of new opportunities [6]. Changing regulatory requirements for meeting standards or compliance can lead to changes in the tasks or the projects in order to meet the new requirements; hence, the projects have to adapt. If projects do not adapt, they could fail as the original requirements are outdated and project resources are wasted in unnecessary tasks. Factors assessing or contributing to the degree of adaptation can be: identifying whether the project tasks can be changed or modified to suit new conditions or needs, whether the governance is tight or not, and whether change is welcome or not.

## 39.3 Research Method

This research is been conducted by applying a design science (DS) approach [13]. DS attempts to establish a link between research and practice, and to have a stronger impact on practice [14]. DS focuses on the development and evaluation of a novel artefact. Since the scope of our study was to develop an initial proof of concept for the CAPEA tool, which is deemed an artefact; the DS approach was found to be appropriate for this research.

In the *development phase*, first, we identify the problem and explain the motivation for the development of the CAPEA tool. Second, we identify the main objective for developing the proposed solution, i.e., the CAPEA tool. Third, drawing from existing CAS literature, and using the "dimension and fact modelling technique" [15], we developed the CAPEA tool. In the evaluation phase, the CAPEA tool is used to assess the capability of two project methods to support the management of complex projects. This initial evaluation provides an indicative proof of concept and directions for further research. The main objective of this initial evaluation is to determine to what extent this tool is applicable and useful to achieve its design objectives.

## 39.4 The CAPEA Tool

In this section we present the CAPEA tool. We describe how we designed the tool using the CAS characteristics described in Sect. 39.2 and the "dimension and fact modelling technique [12]."

Project environments in organizations include project management methods, processes, people, practices, tools, and techniques. Depending on the characteristics of a project environment, adaptive, traditional, or hybrid project methods can be used. We therefore argue that a tool that can assess the capability of the *project environment* is needed to support the management of complex software projects. By undertaking this assessment, organizations could be in a position to better plan and manage their software projects, and consequently increase their rate of success In this paper we are assessing only specific project management methods, and we do not assess a specific organization and its entire environment. Once we finalise the low-level assessment factors, we will design and test the assessment tool for the entire environment with a number of organizations.

Following the design technique, the CAPEA tool consists of two key components: dimension and fact. The *dimension component* refers to the logical grouping of data objects, each dimension being represented as a separate table or logical schema containing the lower level factors. A total of eight high level assessment factors mirroring the CAS dimensions are captured in CAPEA tool (see Fig. 39.1). The dimension component provides an index of the eight high level factors along with their assessment and description. The dimension index is shown in Fig. 39.1.

The *fact component* refers to the actual assessment results (see Figs. 39.2 and 39.3), only, of a project method in this current state, against the factors indicated in the dimension index (see Fig. 39.1). A particular project method is characterized by key practices that are assessed against each factor listed in the CAPEA tool dimension index (see Fig. 39.1). The assessment results are aggregated and captured in the fact component (see Figs. 39.2, 39.3, 39.4, and 39.5 in Sect. 39.5).

**Assessment Factor** New Assessment Factor Assessm

| Action | Assessment Factor Name | Description |
|---|---|---|
| Edit \| Del | Emergence | Does it support the emergent unexpected behaviour of a system (e.g. System as a Whole) that may arise due to non-linear or unplanned interactions within a system? |
| Edit \| Del | Autonomy | Does it allow the individual systems and components to operate independently? |

**Fig. 39.1** CAPEA Tool Dimension Index based on assessment factors

New Assessment Details

| Factor Compliance Value | Assessment Factor | Importance | Weight | Possible Max. Score | Actual Assessment Score | Remarks |
|---|---|---|---|---|---|---|
| 1 | Autonomy | Very High | 100 | 5.00 | 5.00 | Whole Team |
| 1 | Interdependence | Very High | 100 | 5.00 | 5.00 | Whole Team |
| 0 | Eco-system | Very High | 100 | 5.00 | 0.00 | Does not mention |
| 0 | Context | Very High | 100 | 5.00 | 0.00 | Does not mention |
| 1 | Non-linearity | Very High | 100 | 5.00 | 5.00 | Sit Together, Pair Programming |
| 1 | Emergence | Very High | 100 | 5.00 | 5.00 | Weekly Cycle, Quarterly Cycle |
| 1 | Self-organization | Very High | 100 | 5.00 | 5.00 | Energized Work |
| 1 | Adaptation | Very High | 100 | 5.00 | 5.00 | Incremental Design |

**Fig. 39.2** XP evaluation screen example

| | |
|---|---|
| Assessment Number | TA-00005 |
| Technology | Complex Adaptive Project Environment |
| Assessor | CA |
| Assessment Date | |
| Description | |
| Assessment Result | |
| Recomendations | |
| Overall Assessment | |
| Max. Score | 40.00 |
| Actual Score | 30.00 |
| Assessment Score Percentage | 75.00 |

**Fig. 39.3** XP assessment summary screen example

New Assessment Details

| Factor Compliance Value | Assessment Factor | Importance | Weight | Possible Max. Score | Actual Assessment Score | Remarks |
|---|---|---|---|---|---|---|
| 1 | Autonomy | Very High | 100 | 5.00 | 5.00 | Scrum Master, Scrum Team |
| 1 | Interdependence | Very High | 100 | 5.00 | 5.00 | Scrum Master, Scrum Team, Product Backlog |
| 1 | Eco-system | Very High | 100 | 5.00 | 5.00 | Scrum of Scrums |
| 0 | Context | Very High | 100 | 5.00 | 0.00 | Does not mention |
| 1 | Non-linearity | Very High | 100 | 5.00 | 5.00 | Sprint |
| 1 | Emergence | Very High | 100 | 5.00 | 5.00 | Sprint |
| 1 | Self-organization | Very High | 100 | 5.00 | 5.00 | Daily Scrum Meeting |
| 1 | Adaptation | Very High | 100 | 5.00 | 5.00 | Sprint Review, Product Backlog, Sprint Planning |

**Fig. 39.4** Scrum evaluation screen example

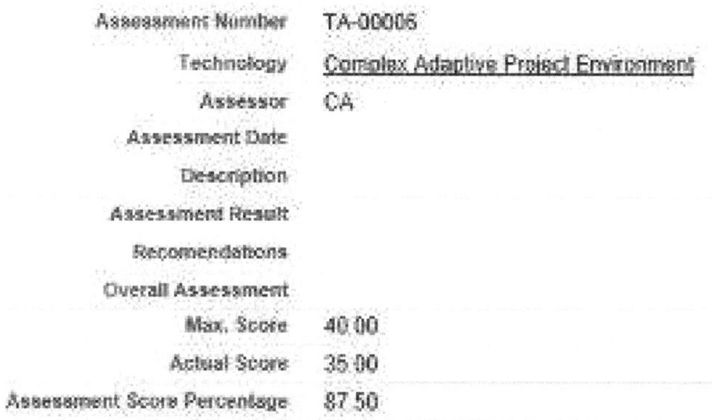


**Fig. 39.5** Scrum assessment summary screen example

If a project environment complies with the specific dimension, then the factor compliance value (FCV) would be 1 otherwise 0 (e.g., default value: 1). An assessor can capture the assessment score on the assessment scale value (ASV) between 1 and 5 (e.g., default value: 5); and their relevant importance (e.g., 5—Very High, High, Medium, Low, 1—Very Low; default value: Very High). An assessor can then assign a different weight (based on their relevant importance) to each assessment factor (e.g., default value: 100) for a particular project or situation-in-hand. The possible maximum score for each assessment factor can be calculated by multiplying the maximum assessment scale value by the assigned weight (e.g., 5 * weight). The actual score for each assessment factor can be calculated by multiplying FCV, ASV and the weight. The possible maximum and actual scores can be separately added for all the assessment factors in order to determine the total possible and total actual scores. The actual total score can be divided by the total maximum score in order to determine the final assessment score percentage. The final assessment score percentage would show how strongly (higher percentage value) or weakly (low percentage value) a project environment complies with the overall assessment criteria and therefore how well supports the management of a specific complex project.

Note that the CAPEA tool is in fact expandable and is not limited to the eight identified dimensions—organisations can add or remove dimensions, if found necessary in the future.

## 39.5 Evaluation of CAPEA Tool

In this section we evaluate the CAPEA tool by using it to assess two well-known agile methods, specifically, Extreme Programming (XP) and SCRUM. In doing so, we offer an initial proof of concept and highlight how organisations can use the tool. Note that the description of the two methods is beyond the scope of this paper as full descriptions are available and have been published in the public domain. We refer to the key practices employed in both XP and SCRUM in order to assess their suitability against the CAS characteristics by referring to the factors contributing to complexity. By understanding how well (i.e., interpreting the CAPEA score/s) a project method scores against the CAS characteristics, organization will be able to select the methods that are suited for their specific project contexts.

If a method practice (s) complies/y with the specific assessment factor, then the factor compliance value (FCV) would be 1, otherwise 0. For example, the Eco-system factor is not supported by the XP practices, therefore, zero (0) is assigned in the FCV column. By contrast, the Emergence factor is supported by the XP practices such as Weekly and Quarterly Cycles (see Remarks column on Fig. 39.2), therefore, one (1) is assigned in the FCV column. For the purpose of simplicity and brevity and in order to demonstrate the indicative proof of concept of the CAPEA tool usage, we only used the factor compliance values (FCV) 1 and 0.

Here, we used the default values for ASV as 5, Importance as Very High, and Weight as 100.

## 39.6 Limitations and Related Work

The CAPEA tool construct needs to be considered with a view of its limitations. In its current form, the tool mainly supports single assessor. It solely focuses on the CAS characteristics of the software project environment and assesses only project methods, not the entire project environment. The literature and practice are continuously progressing, the CAPEA tool should be considered as an evolving construct to be revised and extended by future research. Despite its current limitations, we believe the CAPEA tool is useful for providing the necessary criteria and test cases when assessing particular project methods from a CAS perspective.

The main purpose of this paper was to present the current CAPEA's construct and usability instead of making any recommendations about the use of some specific agile or non-agile method for a specific project. XP and SCRUM have been used as test examples for the evaluation of the CAPEA tool. It can be observed from our analysis (see Figs. 39.2, 39.3, 39.4, and 39.5) that the CAPEA tool: (1) is of acceptable quality and fit for purpose (suited for CAS-type projects) and (2) lays out a foundation how one might rationally approach the process when assessing the complexity support of a project method.

We compared the CAPEA tool with the IS success model [16], and found that the IS success model provides the support for mainly assessing the quality (e.g., system quality, information quality, service quality). We also compared the CAPEA tool with the 4-Dimensional Analytical Tool (4-DAT) [17] and found that the 4-DAT mainly provides support for assessing the agility of a project method and does not offer any support for assessing the complexity aspect as discussed here in the CAPEA tool. We would like to further investigate and extend Qumer and Henderson-Sellers' 4-DAT tool [17], retest it in, and present it to community for expert analysis as an ongoing contribution to both theory and practice.

## 39.7 Conclusions and Future Research

In this study, we developed an assessment tool that can allow organizations to assess the capability of their project environments to be adaptive and then tailor project management approaches to manage complex projects. We used a design science approach and CAS theory to develop the CAPEA assessment tool. We tested and provided an initial proof of concept by assessing the XP and SCRUM practices against the eight assessment factors. The overall result for XP method (75 %) indicates that the XP method for software development is well suited for managing CAS type projects. We found even stronger support for Scrum method, with 87.5 %. In conclusion, we believe that CAPEA tool can be a useful resource for organizations in their attempt to improve the success rate of their software and IS development projects. We believe that such a tool can assist organizations to improve the management of complex software development projects, and reduce costly consequences associated with failed projects.

This study represents a preliminary step towards the development of a more holistic and comprehensive assessment tool. Our future directions are focusing on: (1) identifying the low level factors affecting the complexity in software development projects to be used in the overall assessment of the project environment; (2) designing a tool to assess the degree of software project complexity; and (3) integrating the CAPEA and 4-DAT tools into the new overall tool. This will allow organizations to holistically and comprehensively assess the project environments and determine the best suited mix of project methods.

## References

1. Feldmann J (2012) Project management offices: a waste of money? http://www.informationweek.com/global-cio/interviews/project-managementoffices-a-waste-of-mo/240062641. Accessed 29 April 2013
2. Dr Dobb's Journal (2010) IT project success rates report. http://www.drdobbs.com/architecture-and-design/2010-it-project-successrates/226500046. Accessed 1 April 2013
3. The Standish Group (2010) Chaos summary for 2010. http://insyght.com.au/special/2010CHAOSSummary.pdf. Accessed 7 April 2013
4. Raduescu C, Gill AQ (2012) Handling the complexity of ISD projects with agile methods: a conceptual foundation. In: ISD 2013, Prato
5. Hobbs G, Scheepers R (2010) Agility in information systems: enabling capabilities for the IT function. Pac Asia J Assoc Inf Syst 2(4):1–27
6. Remington K, Pollack J (2007) Tools for complex projects. Gower Publishing Ltd., Aldershot
7. Axelrod R (1997) The complexity of cooperation: agent-based models of competition and collaboration. Princeton University Press, Princeton, NJ
8. Benbya H, McKelvey B (2006) Using coevolutionary and complexity theories to improve IS alignment: a multi-level approach. J Inf Technol 21(4):284–298
9. Holland JH (2006) Studying complex adaptive systems. J Syst Sci Complex 19(1):1–8
10. Lilliesköld J (2003) Coordinating dependencies in complex system development projects. In: Proceedings of the IEEE engineering management conference, IEMC'03, pp 400–404

11. Morrison K (2010) Complexity theory, school leadership and management: questions for theory and practice. Educ Manage Adm Leadersh 38(3):43–93
12. Harmsen F, Lubbers I, Wijers G (1995) Success-driven selection of fragments for situational methods: the S3 model. In: Pohl K, Peters P (eds) Proceedings of the second international workshop on requirements engineering: foundations of software quality (REFSQ'95), Aachener Beiträge zur Informatik, Band 13, pp 104–115
13. Henver AR, March ST, Park J, Ram S (2004) Design science in information systems research. MIS Q 28(1):75–105
14. Peffers K, Tuunanen T, Gengler CE, Rossi M, Hiu W, Virtanen V, Bragge J (2006) The design science research process: a model for producing and presenting information systems research. In: DESRIST 2006, Claremont, CA
15. Järvinen P (2001) On research methods. Juvenes-Print, Tampere
16. DeLone WH, McLean ER (2003) The DeLone and McLean model of information systems success: a ten-year update. J Manage Inf Syst 19(4):9–30
17. Qumer A, Henderson-Sellers B (2008) An evaluation of the degree of agility in six agile methods and its applicability for method engineering. Inf Softw Technol 50(4):280–295

# Chapter 40
# $ISO_2$: A New Breath for the Joint Development of IS and ISO 9001 Management Systems

**João Barata and Paulo Rupino da Cunha**

**Abstract** We present $ISO_2$, an approach for the joint development of information systems (IS) and ISO 9001 quality management systems (QMS). $ISO_2$ was outlined from 14 retrospective case studies, occurred between 2008 and 2012. We then validated and refined the approach through action research. We support the idea that IS and QMS synergies are more important than the perspective of one system merely supporting the other. The $ISO_2$ combines iterative development steps with a layered and incremental design framework, the $O_2$. The $O_2$ metaphor can provide a common abstraction level for the joint design. Over one million companies struggle with IS and QMS disintegration. Our findings offer new insights for the joint development of organizational systems.

**Keywords** $ISO_2$ • Information systems • Quality management systems • Joint development • Synergies

## 40.1 Introduction

"If you want to change the world, you change the metaphor" is an inspiring quote from Joseph Campbell [1]. There is a need to change the development of information systems (IS) and ISO 9001 quality management systems (QMS). The two

J. Barata (✉)
CTCV, Rua Coronel Veiga Simão, 3025-307 Coimbra, Portugal

CISUC, Department of Informatics Engineering, University of Coimbra, Pólo II, 3030-290 Coimbra, Portugal
e-mail: barata@dei.uc.pt

P.R. da Cunha
CISUC, Department of Informatics Engineering, University of Coimbra, Pólo II, 3030-290 Coimbra, Portugal
e-mail: rupino@dei.uc.pt

M. José Escalona et al. (eds.), *Information System Development: Improving Enterprise Communication*, DOI 10.1007/978-3-319-07215-9_40,

endeavors are conducted as separate projects, which are handled by distinct teams, using disconnected methodologies [2]. However, the development of both systems has synergies and often depends on each other [2]. We gathered indications during our research that an approach common to both can individually improve them, as well as the organizational outcome of their integration. How can we change the metaphor?

ISO 9001 is a standard for quality management, adopted by more than one million companies worldwide [3]. ISO 9001 requires the internal development of management procedures, work instructions, improvement plans and a demanding measurement system [4]. The external information flows are just as important. In addition, a company certified by ISO 9001 should assign a high priority to the customer relationship activities, create suppliers partnerships, and be prepared for external audits. Therefore, the QMS becomes a tool to manage the relations between the organization and its environment [5].

The information system development (ISD) must consider the influence of the business environment and internal characteristics of the company, such as its politics and procedures [6, 7]. Moreover, the IS has a significant impact in quality management and performance [8–10]. The IS becomes vital for "collecting, storing, analyzing and reporting information on quality to assist decision makers at all levels" [11]. The lack of involvement between the IS and the QMS is well known [12] and the IS and quality departments do not usually leverage the synergistic potential in combining their efforts [2]. Grounded on narrow perspectives, quality experts view IT as mere support, while the IS experts view the QMS as mere compliance. A joint development approach could reduce the pitfalls of the ISO 9001 QMS and the possibility of decreasing its benefits over time [13]. It also may provide simple collaboration tools that the IS and the QMS teams need.

Section 40.2 presents the background of our research, concerning ISO 9001, ISD and the potential synergies of the IS and the QMS. We then present the dual methodology used for the research. Section 40.4 presents the $ISO_2$ approach. Section 40.5 reports the results of $ISO_2$ adoption and the $O_2$ design framework. We particularly stress the IS and QMS design steps. The last section presents the conclusions and directions for future research.

## 40.2 Background

ISO 9001:2008 is a world-recognized standard for developing quality management systems. ISO 9001 [4] was published in 1987, and later revised in 1994, 2000 and 2008. ISO 9001 guides companies to improve business quality and adopt continuous improvement as a strategy [14]. The ISO 9001 comprise a model by which organizations of any type and sector of activity can establish, document, implement and optionally certify their QMS. As noted by [15], a "document" means information in any form or type of medium. The standard establishes requirements that each company must fulfill with the systems and approaches that it choose. ISO 9001 recommend a process approach to management and a continuous improvement using the "Plan-Do-Check-Act" cycle [4]. According to [16], the development of a documented ISO 9001 QMS has the following steps : (1) gaining management commitment; (2) employing external consultants; (3) conducting an awareness campaign; (4) creating a QMS manual; (5) developing a documentation system; (6) training employees on the system; (7) creating work processes and procedures; (8) conducting system wide reviews; and (9) pre-assessment audit.

The methodologies used to develop an IS are still a key research area. They include more technical perspectives such as the SDLC—Systems Development Life Cycle or the "waterfall model". Other methodologies consider both technical and managerial perspectives, such as RUP—Rational's Unified Process [17] or the ISO/IEC 12207 [18]. In contrast with the sequential waterfall model, the agile approaches advise a more iterative and incremental perspective in software development [19, 20]. The ISD research has also followed sociological perspectives, with its foundations in Checkland's SSM [21] and socio-technical approaches. An example of such approaches is the Multiview [22]. There are several ISD methodologies, but problems still exist. For instance, some methodologies may be too complex and inflexible, unfitting to all the possible situations [23]. Although ad-hoc and informal developments are observed in a number of cases, methodologies are essential for ISD and can be adapted or combined into specific situations [24]. The analysis, design and implementation of the IS consider the technology and the nature of the strategic and operational activities involved [14]. The ISD must deal with the problems of diversity, knowledge, and structure at distinct behavior levels such as the business, company, project, team, and the individual [6, 7].

A number of authors has suggested synergies between the IS and the QMS [2]. For example, [25] suggest that quality and IT plans should be simultaneously developed at the strategic level. Others like [26] claim that the IS and the QMS are capable of being combined into an integrated approach. [27] propose that the integration can occur at early stages of the design, while [28] identify the gaps between quality and the IS after the design. The lack of IS and QMS integration leads to inefficiency, week correspondence between procedures and practice [2]. The IS and the QMS teams must be involved in the improvement initiatives. [29] and [30] suggest that IS techniques and skills can improve process improvement actions and, conversely, the QMS can benefit the ISD. The development of the IS and the QMS considers organizational, social, and technological aspects that interact and support each other [31]. The IS and the QMS also require similar organizational cultures [32] and may be combined for a cultural change [33]. An example of the IS and QMS mutual benefits is presented by [34], in the company purchasing process. Other authors have found the mutual benefits of QMS and ERP implementations [9, 35]. Despite the several advantages, a joint development approach is absent from the literature. In fact, several barriers may be identified: the QMS does not provide a complete set of requirements for the IS; the level of detail and the distinct vocabulary between quality and IS practitioners are examples of the potential obstacles; continuous change and the internal politics developed in a QMS requires IS support, but may create difficulties for the IS implementation and management [29, 36].

## 40.3 Methods

This research adopts a dual methodology. In the first stage, we have used case studies, that are best suited when the frontiers between the phenomena and the context are not evident [37]. The retrospective case studies allows the identification of patterns indicative of dynamic processes [38, 39]. The data gathering techniques were the document collection and 28 semi structured interviews [40], carried out with the IS and the QMS manager of each company. The document analysis and observations have focused the documental structure of ISO 9001 and the IS that supports quality directly (e.g. document management systems) or indirectly, as a source of information for quality (e.g. complaints provided by a CRM or quality costs from an ERP system). Two distinct teams have developed the QMS and the IS. We also acted as consultants in 13 cases. The first version of the approach was designed from the retrospective case studies, as presented in Table 40.1.

In the second stage, we have selected action research to test and refine the approach. Adopting a cyclic process of theory building and refinement, this approach is suitable for increasing the understanding of an immediate social situation, with emphasis on its complex and multivariate nature [41–43]. Action research simultaneously aims to improve scientific knowledge and assist a practical problem, by joint collaboration [44]. We have followed the canonical action research, characterized by the five phases of *Diagnosing, Action planning, Action taking, Evaluating and Specifying learning* [45]. To evaluate our research, we have relied on the principles proposed by [46].

**Table 40.1** Retrospective case studies between 2008 and 2012

| Sector | Company size | IT scope (average duration: 1 year) |
|---|---|---|
| Ceramics #1 | Large (>250 employees) | Development of QMS software[a]; CRM acquisition[b] |
| Ceramics | Large (>250) | Dev. of QMS software |
| Ceramics | Medium (50–250) | Dev. of QMS software |
| Batteries | Large (>250) | Dev. of QMS software; acquisition of a process modeling software and statistical software |
| Agro food | Medium (50–250) | Dev. of QMS, production control software and computerized maintenance management system (CMMS) |
| Metal | Large (50–250) | Dev. of QMS software; CRM acquisition |
| Metal | Small (<50) | Dev. of QMS software |
| Paper | Medium (50–250) | Dev. of QMS software; dev. of B2B platform |
| Institute | Medium (50–250) | Dev. of QMS software |
| Institute | Large (>250) | Dev. of QMS software; dev. of B2B platform |
| Environment | Large (>250) | Dev. of QMS software; CMMS acquisition |
| Printer | Large (>250) | Dev. of QMS software; CMMS acquisition |
| Automotive | Large (>250) | Dev. of QMS software; dev. of CMMS |
| Plastics #14 | Large (>250) | Dev. of QMS and production software; ERP acquisition |

[a]The development of software applications for ISO 9001 requirements, such as document management, complaints and non conformity, action plans and others. The cases 1 to 4, 9 and 10 to 12 have also included the acquisition of at least one module of a QMS software package

[b]The acquisition only reports to the part of implementing an IT solution already on the market

## 40.4 Retrospective Case Studies and the $ISO_2$ Approach

An overview of the findings is provided in Table 40.2, (I) before, (II) during and (III) after the separated IS and QMS development.

The lack of integration in these cases occurs from the beginning, continues during the development and propagates the problems afterwards. The interaction of IS and QMS teams should be replaced by a partnership. The disconnected approach may compromise the ISD results and QMS benefits. Even worse, when the integration fails, each system may become superfluous to the other. All the interviewees

**Table 40.2** Findings from the retrospective case studies

| | |
|---|---|
| (I) before | The ISO 9001 certification was a top management decision, motivated by a combination of factors such as the internal improvement or the external company image. However, the development or acquisition of IT was in the majority of the cases (11), a quality manager's decision. In 12 of the cases, the development of the IS was planned after the QMS project started, therefore, only at this stage the IS team was involved |
| (II) during | In the prevalent scenario, the IS team supports the quality requirements by developing or buying software—a supplier role. The IS team defines the technologies and the preferred ISD approach. Curiously, when asked about the selected ISD method, nine of the teams could not identify a specific one. The QMS team establishes priorities, IS requirements and workflows. The QMS team has adopted a customer role. Independently, the QMS team creates documents and the IS team creates IT solutions, for the same processes and users. Top management involvement is not significant in this stage. In most cases, it is merely needed to approve the IT investments. We found that the IS team was not completely aware of ISO 9001 (13 cases), the standard was not used as an input for the ISD requirements—only the users and quality experts point of view. In the same cases, the IS team reported that wasn't well informed about the QMS processes or documents development. They also pointed out the lack of communication as a cause for delays in the IS implementation, late changes and misfit between quality procedures and the developed IS. The process model of the QMS was mostly reported (12 cases) as useless by the IS team |
| (III) after | In four of the cases, the IS manager also participates in the improvement teams. These cases present a closer relation between IS departments and top managers. In the ten remaining, the QMS managers monitor the information effectiveness, user satisfaction and improvement suggestions. The IS seems to have a more reactive role. Even after 3 years of certification (four of the cases), the IS interest in ISO 9001 seems to be on the part that directly concerns with IT (for the ISO 9001 audit). Ten of the IS development cases were still ongoing by the time of the final audit. Due to this delay, some users have started to develop their own tools. In 13 cases, surprisingly, the persons responsible for managing software validation (mostly the calculations) are the QMS managers. Both the IS and the QMS managers have complaints. The most common from the latter is that the IS does not correspond to their information needs (9). The majority (7) said that they prefer to build their own tools (e.g. spreadsheets, parallel records) than waiting for IS changes. The IS managers complains that QMS is a burocratic system (14) that does not correspond to practice (8). Additionally, part of the problem was precisely the parallel documents that QMS team develops (3) |

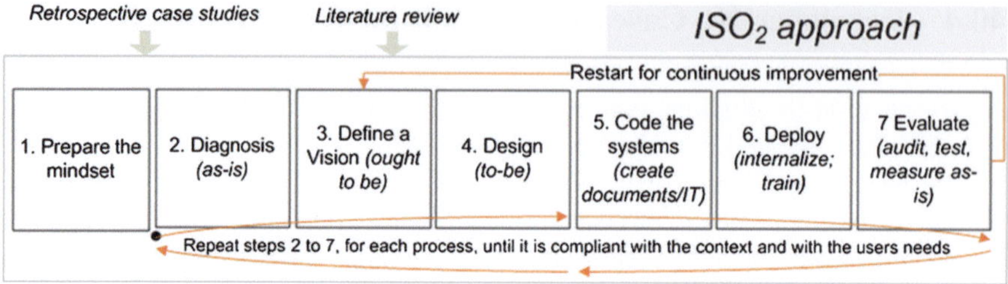


**Fig. 40.1** The steps of $ISO_2$ approach

**Table 40.3** $ISO_2$ steps

| Step | Description |
|---|---|
| 1 | ***Prepare the mindset:*** A common approach must be presented to all the stakeholders. We have learned from the retrospective cases that both systems must be aligned from the start and the decisions shared by IS and QMS developers. Three training actions of two hours each are proposed for (1) presenting the approach; (2) the QMS team presents the main cultural aspects of the standard, principles and requirements; and (3) the IS team presents the IS methods, the IT options and guidance for requirements analysis. This step may contribute for the team coordination, management commitment and an awareness campaign [4, 16] |
| 2 | ***Diagnosis (as-is):*** Identify current quality and IS practices, ISO 9001, and other contextual requirements [17, 18]. Define and assess the current processes by the users perspective [47] |
| 3 | ***Define a Vision (ought-to-be):*** Define quality and IS politics, create the quality manual [16]. Create a desired process map [4] |
| 4 | ***Design (to-be):*** Detail each process and indicators [4]. Establish the plan and objectives for each development [17, 18] |
| 5 | ***Code the systems:*** Develop the IT artifacts [48] and the QMS documents [16] |
| 6 | ***Deploy:*** Implement the systems, train, internalize, becoming daily practice [16–18] |
| 7 | **Evaluate**: Audit, test, validation and user acceptance [16–18, 47]. Restart to improve [4] |

agreed that a joint development approach could bring significant advantages. Regarding the benefits of that new approach, we highlight four statements that support "improving the communication process [tactical level by the QMS and ISD]", "encouraging the involvement of the top managers", "accomplishment of the project calendar" and "avoiding duplicated tasks that damage our [IS] internal image, creates systems that are more permissive to errors and harder to manage". As a result, we shaped a clear-cut version of $ISO_2$, represented in the Fig. 40.1.

$ISO_2$ consider the iterative nature of the development [21, 45], as proposed by the PDCA. Our frame of reference for action research is outlined in Table 40.3.

## 40.5 $ISO_2$ Action Research and the $O_2$ Framework

We have conducted action research in a private technological institute. The company wanted to certify the QMS and to develop quality modules integrated with their ERP. The modules included complaints management, non conformance and actions,

**Table 40.4** Findings from the action research

| Step | Description |
|---|---|
| 1 | ***Preparing the mindset—focusing on the awareness of synergies*** |
| | Due to the use of a common approach, the ISD and the QMS development could start simultaneously. The presence of the top manager and the mere existence of an approach successfully transmitted an idea of the relevance of the development to the participants. It was decided that both the IS and QMS teams would develop the same processes and "documents" at the same time, in the type of medium they prefer. The joint design should make the end users' satisfaction a main concern. Additionally, the design outcome should provide a predictable, continuous, reliable and complete information flow within the company and with their environment |
| 2 | ***Diagnosis (as-is)—focusing on the team designers and process users*** |
| | We have started by designing a global process map and then, for each process, carried out the diagnosis by observing the current practice and measuring the process acceptance by the users with a questionnaire [47]. We expected that the QMS team raised problems in sharing their "power" in information management. Surprisingly, they liked the idea because they could now focus on the principles of the standard: improvement and customer satisfaction |
| 3 | ***Define a Vision (ought-to-be)—focusing on the organization*** |
| | This step was faster than we expected. We involved the top manager in a brainstorming, with IS/QMS teams and the process owners. Due to step 1, the participants were focused on getting synergies from both the IS and the QMS. The questionnaire inputs were used for the new vision and the new process map was then communicated to all the organization |
| 4 | **Design (to-be)—focusing on the possibilities and restrictions of the design teams** |
| | We then quickly realized that the QMS design, although primarily represented as a sequence of steps in the QMS literature, are iterative and incremental. Developing documented procedures and forms was the main task of the QMS team. Developing or acquiring IT was the main purpose of the IS team. Since we were going to develop "documents", the challenge was to define an $ISO_2$ "shared document". We also found that the "process approach", by itself, was not sufficient, as we already suspected from the cases and the literature review [49, 50]. The QMS processes were too general to be used by the IS. A common abstraction level was necessary or the joint design would simply not work. Considering the ISO definition of "document" and the inclusion of IT in our approach, we have conceptualized the $ISO_2$ document as an IT artifact [48]: an application of IT that enables some processes in a human structure that itself is embedded within a context. We have named it $O_2$ artifact and its framework is presented in Fig. 40.2 |

audit, product design and development. Due to the complexity of the project, we have decided to focus our intervention on the first four steps of $ISO_2$, leading to the systems design. The lessons learned are summarized in Table 40.4.

An example is provided for the product design and development (D&D). The process and the $O_2$ artifacts were jointly designed, as exemplified in Fig. 40.3.

The figure illustrates the three main views of the $O_2$ framework. For the D&D process, two IT applications are identified: Innovation management (orange) and a Cloud project management platform (blue). The $O_2$ design is executed by:

1. For each process, identify the requirements according with the components of process tasks, people, IT and context needs (matrix lines). Consider the current and the planned. Take into account the outside-in, within and inside-out perspective (matrix columns) of the process;

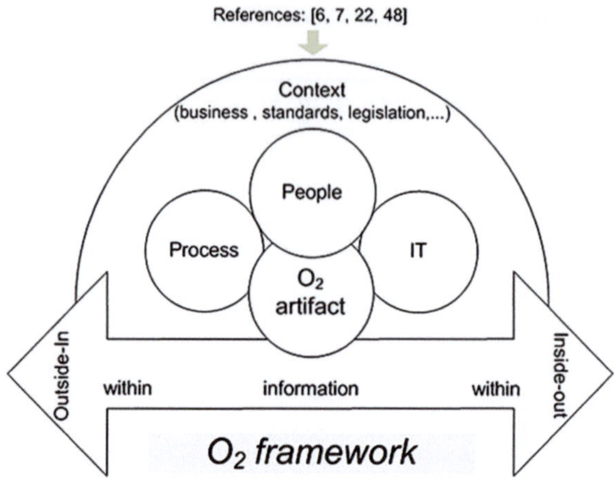


**Fig. 40.2** The $O_2$ framework. Represents the proposed level of abstraction for the IS and QMS teams. Each $O_2$ artifact is the practical result of the design cycles. Each $O_2$ may be linked with a structure of other N $O_2$ and N processes

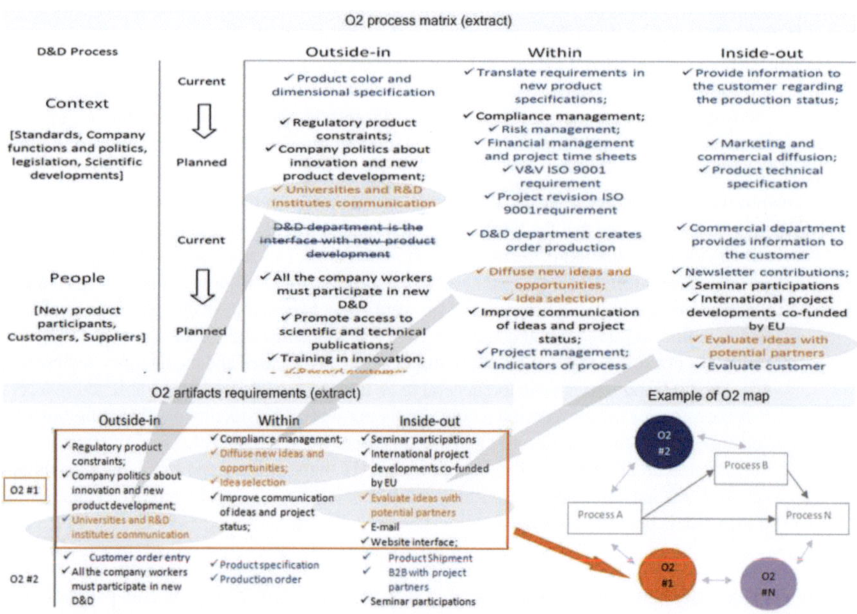


**Fig. 40.3** The $O_2$ matrix (on *top*), the $O_2$ artifacts (*bottom-left*) and an $O_2$ map (*bottom-right*)

2. Group the requirements by colors (color black represent a shared requirement), each one representing an $O_2$ artifact. Each one is a development project. It may be a new IT platform, a paper document, a part of an already existing system such as an ERP, or any other mean to allow the information (*oxygen*) flow, providing to each end user (*system cell*) the vital process information (*breathe*);
3. Repeat 2 to each process until an ecosystem of $O_2$ artifacts are designed;
4. Connect all the $O_2$ artifacts with the processes (*breathing system*).

**Table 40.5** Findings from the action research using a metaphor for collaboration

| Step | Description |
|---|---|
| 4 | ***Design (to-be)—focusing on the organization*** |
| | The teams acted as partners, understood that they could help each other and simplified the IS and the QMS. Interestingly, the QMS team found that when designing the $O_2$ artifacts, the process activities were easier to identify. Even more interesting, the process map has changed after the $O_2$ design. The joint IS and QMS may influence how the company wishes to operate. The $O_2$ framework had a major impact in our research and has become the focus of the following steps |
| 5 | ***Code the systems—focusing on each O2 artifact*** |
| | The coding and implementation was carried on by refining the $O_2$ concept and understanding its impact in the $ISO_2$ approach. Both the IS and the QMS developers have stated that the $O_2$ artifacts were simple to use and provided a proper guide for the development. The language was familiar to both teams and the metaphor had the desired effect, which is to be adopted simultaneously by the teams and to improve communication among the teams and with the end users |
| 6 | ***Deploy—focusing on the development results and the people usage of the O2 artifact*** |
| | A number of documented procedures and IT platforms were implemented at this point |
| | Contrarily to what we thought, the $O_2$ artifact was not helpful for the training to end users |
| | The $O_2$ was best fit for the step 5. Nevertheless, the platforms that were developed also incorporated the QMS procedures and rules, contributing to internalize the QMS practices |
| 7 | ***Evaluate—focusing on people satisfaction with the O2 artifact*** |
| | We have launched the same questionnaire of step 2 for each developed process. The process pain points were eliminated [47]. The auditors have recorded the integration as strong point of the QMS. One auditor said that "It's common that IT supports quality, what is uncommon is that we do not need to surf blindly in a jungle of disconnected software to find evidences of each requirement [...] for each process what we look for are those $O_2$ elements [...] QMS process maps usually represent what people do, scarcely how they do it". We add that why they do it is also essential. The O 2 artifact shows the organizational interfaces and the evolution from a plan to the real. The company achieved the ISO 9001 certification and the IS and QMS development was completed on schedule. After five months, 85 % of the quality preventive and improvement actions aim the IS or are achieved through IS joint developments. We did not yet started a new cycle to understand how both systems can now evolve combined |

The introduction of the $O_2$ artifact has completely changed our intervention. The action research progression is now presented in Table 40.5.

## 40.6 Conclusions and Future Work

From our knowledge, $ISO_2$ is the first approach meant for the joint development of IS and ISO 9001 QMS. According to the auditors and the developers, $ISO_2$ improves the results when compared with the practice of developing both systems independently. We combined IS and QMS methodologies in a new approach, coping with the ISD problems of diversity, knowledge, and structure [7]. A common abstraction level is determinant for the teams' communication and, eventually for the success of a joint development. The $O_2$ artifact is that construct. A process approach was followed by both teams. However, it was not sufficient for a joint development. $ISO_2$ was designed from practice, with a common and simple message. The developers found the $ISO_2$ suitable when developing the IS and the QMS from scratch or after a certification. A benefit of this approach is to focus the participants in the steps and the development outcomes, providing detail to the process layer. The $O_2$ matrixes are also a tool for the ISO 9001 auditors to connect requirements, processes, and IT. The study of a joint IS/QMS may contribute for the ISO 9001 revision, to be published in 2016.

In spite of the obtained insights, several limitations can be identified in this study. The $ISO_2$ approach is still under development and it requires a higher detail for the coding and implementation parts; the $O_2$ framework creates a structure of several $O_2$ artifacts, which are not yet reflected at this stage of the research; we have considered cases with the existence of internal IS and QMS departments and the majority were medium or large companies but the positive effect that we found may not be replicable in distinct client settings.

Several issues remain open. For instance, how both IS and QMS teams can deal with a stronger dependence of both systems and manage two integrated systems. The number of companies that adopt multiple standards, creating a system of systems with ISO 9001 in its core, has been increasing [3]. The auditors have pointed that $ISO_2$ could be adopted for managing organizational legislation awareness (outside-in), the internal application of the law (within) and how to comply with the report obligations (inside-out). The layers of the $O_2$ framework may be adapted or extended to include requirements and politics related with the environment management, health and safety, social responsibility, or other standards integration [51]. These are the challenges for the next action research cycle, in an aeronautical supplier with four certified management systems.

## References


1. Moyers B (2008) Fresh air [Radio broadcast]. http://www.npr.org/templates/story/story.php?storyId=96648963. Accessed 8 May 2013
2. Cunha PR, Figueiredo AD (2005) Quality management systems and information systems: getting more than the sum of the parts. In: Proceedings of the AMCIS 2005
3. ISO (2012) The ISO Survey of Certifications 2011. International Organization for Standardization, Geneva
4. ISO (2008) ISO:900 1:2008 quality management system—Requirement. International Organization for Standardization, Geneva
5. Singh PJ, Power D, Chuong SC (2011) A resource dependence theory perspective of ISO 9000 in managing organizational environment. J Operations Manage 29:49–64
6. Curtis B, Krasner H, Iscoe N (1988) A field study of the software design process for large systems. Commun ACM 31:1268–1287
7. Kautz K, Madsen S, Nørbjerg J (2007) Persistent problems and practices in information systems development. Inf Syst J 17:217–239

8. Perez-Arostegui MN, Benitez-Amado J, Tamayo-Torres J (2012) Information technology-enabled quality performance: an exploratory study. Ind Manage Data Syst 112:502–518
9. Sánchez-Rodríguez C, Martínez-Lorente AR (2011) Effect of IT and quality management on performance. Ind Manage Data Syst 111:830–848
10. Forza C (1995) The impact of information systems on quality performance: an empirical study. Int J Operations Prod Manage 15:69–83
11. Juran JM, Gryna FM (1993) Quality planning and analysis, 3rd edn. McGraw-Hill, New York, NY
12. Garimella KK (2006) The power of process: unleashing the source of competitive advantage. Meghan Kiffer Pr, New York, NY
13. Karapetrovic S, Fa MC, Saizarbitoria IH (2010) What happened to the ISO 9000 lustre? An eight-year study. Total Qual Manage Bus Excell 21:245–267
14. Ward J, Peppard J (2002) Strategic planning for information systems, 3rd edn. Wiley, Chichester
15. ISO (2008) Guidance on the documentation requirements of ISO 9001:2008. ISO/TC 176/SC 2/N525R2. International Organization for Standardization, Geneva
16. Bell M, Omachonu V (2011) Quality system implementation process for business success. Int J Qual Reliab Manage 28:723–734
17. Jacobson I, Booch G, Rumbaugh J (1999) The unified software development process. Addison-Wesley, Boston, MA
18. ISO (2008) ISO/IEC 12207:2008 Systems and software engineering—software life cycle processes. International Organization for Standardization, Geneva
19. Larman C, Basili VR (2003) Iterative and incremental developments. a brief history. Computer 3(6):47–56
20. Dybå T, Dingsøyr T (2008) Empirical studies of agile software development: a systematic review. Inf Softw Technol 50:833–859
21. Checkland P (1981) Systems thinking, systems practice. Wiley, Chichester
22. Avison D, Wood-Harper AT, Vidgen RT, Wood JRG (1998) A further exploration into information systems development: the evolution of Multiview2. Inf Technol People 11:124–139
23. Wood-harper T (1991) Information systems development research: an exploration of ideas in practice. Comput J 34:98–112
24. Avison D, Fitzgerald G (2003) Where now for development methodologies? Commun ACM 46:78–82
25. Jabnoun N, Sahraoui S (2004) Enabling a TQM structure through information technology. Competitiveness Rev 14:72–81
26. Wu X, Gu Y (2009) Influence mechanism of information technology on quality management and organizational performance. In: Proceedings of the IIS 2009
27. Bonazzi R, Hussami L, Pigneur Y (2010) Compliance management is becoming a major issue in IS design. In: D'Atri A, Saccà D (eds) Information systems: people, organizations, institutions, and technologies. Springer Physica, Heidelberg, pp 391–398
28. Gunasekaran N, Arunachalam VP, Devadasan SR (2006) TISIT: a model for integrating TQM with software and information technologies. TQM Mag 18:118–130
29. Spencer MS, Duclos LK (1998) TQM stress MIS. Am J Bus 13:59–64
30. Keith R (1994) MIS + TQM = QIS. Qual Prog 27:29–31
31. Jensen TJ (1991) A framework for quality system design in "one of a kind" manufacturing companies. Int J Prod Econ 22:121–130
32. Fok LY, Fok WM, Hartman SJ (2001) Exploring the relationship between total quality management and information systems development. Inf Manage 38:355–371
33. Philip G, McKeown I (2004) Business transformation and organizational culture: the role of competency, IS and TQM. Eur Manage J 22:624–636
34. Hemsworth D, Sánchez-Rodríguez C, Bidgood B (2008) A structural model of the impact of Quality Management Practices and purchasing-related Information Systems on purchasing performance: a TQM perspective. Total Qual Manage Bus Excell 19:151–164
35. Li L, Markowski C, Xu L, Markowski E (2008) TQM—a predecessor of ERP implementation. Int J Prod Econ 115:569–580

36. Attaran M (2004) Exploring the relationship between information technology and business process reengineering. Inf Manage 41:585–596
37. Yin R (1981) The case study crisis: some answers. Adm Sci Q 26:58–65
38. Leonard-Barton D (1990) A dual methodology for case studies: synergistic use of a longitudinal single site with replicated multiple sites. Organization Sci 1:248–266
39. Miller CC, Cardinal LB, Glick WH (1997) Retrospective reports in organizational research: a reexamination of recent evidence. Acad Manage J 40:189–204
40. Myers MD, Newman M (2007) The qualitative interview in IS research: examining the craft. Inf Organization 17:2–26
41. Baskerville R (1999) Investigating information systems with action research. Commun AIS 2:1–32
42. Baskerville R, Wood-Harper AT (1996) A critical perspective on action research as a method for information systems research. J Inf Technol 11:235–246
43. Avison D, Lau F, Myers MD, Nielsen PA (1999) Action research. Commun ACM 42:94–97
44. Hult M, Lennung S-Å (1980) Towards a definition of action research: a note and bibliography. J Manage Stud 17:241–250
45. Susman GI, Evered RD (1978) An assessment of the scientific merits of action research. Adm Sci Q 23:582–603
46. Davison R, Martinsons MG, Kock N (2004) Principles of canonical action research. Inf Syst J 14:65–86
47. Antunes A, Cunha PR (2013) Business processes the way they should be: tuning for low friction and sustainability. In: Proceedings of the ECIS 2013
48. Zhang P, Scialdone M, Ku M-C (2011) IT artifacts and the state of IS research. In: Proceedings of the ICIS 2011
49. Iden J (2012) Investigating process management in firms with quality systems: a multi-case study. Bus Process Manage J 18:104–121
50. Cardwell G (2008) The influence of Enterprise Architecture and process hierarchies on company success. Total Qual Manage Bus Excell 19:47–55
51. Jørgensen TH, Remmen A, Mellado MD (2006) Integrated management systems—three different levels of integration. J Clean Prod 14:713–722

# Chapter 41
# Incremental and Adaptive Software Systems Development of Natural Language Applications

**Elena Lloret, Santiago Escobar, Manuel Palomar, and Isidro Ramos**

**Abstract** Natural Language (NL) processing tools, such as tokenizers, part-of speech taggers or syntactic processors obtain knowledge from a set of documents (e.g., tokens, syntactic patterns, etc.) and produce the different elements that will take part on the discourse universe in a NL text (e.g., noun phrases, verbs, sentences, etc.). In this paper, we present how NL software systems development can be performed incrementally by using a high-performance specification language like Maude. A generic algebraic specification for NL is defined, including sorts and sub-sorts apart from equational properties, such as associativity and commutativity for built-in lists and sets. Then, the full discourse universe, available for NL processing, is described in terms of the algebraic specification by providing a non-deterministic but terminating set of transformation rules. Finally, and as a proof of concept, a set of documents for NL processing is given to Maude as an input term and successfully transformed into a proper document, exploring all the non-deterministic possibilities, as well as resolving the ambiguity in language. The main advantages of implementing NL in this manner are: generality, transparency, extensibility, reusability, and maintainability. To the best of our knowledge, this is the first attempt to represent and develop complex NL software systems with this formal notation, and based on the analysis conducted, this implementation constitute the basis for the design and development of more specific NL processing applications, such as text summarization.

**Keywords** Natural language applications • Software engineering • Incremental analysis • Adaptive systems • Software systems

---

E. Lloret (✉) • M. Palomar
Departamento de Lenguajes y Sistemas Informáticos, Universidad de Alicante, Apdo. de correos, 99, 03080 Alicante, Spain
e-mail: elloret@dlsi.ua.es; mpalomar@dlsi.ua.es

S. Escobar • I. Ramos
Departamentos de Sistemas Informáticos y Computación, Universitat Politêcnica de Valência, Valencia, Spain
e-mail: sescobar@dsic.upv.es; iramos@dsic.upv.es

M. José Escalona et al. (eds.), *Information System Development: Improving Enterprise Communication*, DOI 10.1007/978-3-319-07215-9_41,

## 41.1 Introduction

Software Engineering (SE) is a systematic and disciplined approach to developing software [1]. It applies both computer science and engineering principles and practices to the creation, operation, and maintenance of software systems. Important principles in SE are abstraction, modularity, incremental development, or functional independence, among others [2]. A popular approach is Formal Methods (FM), where a specific notation with a formal semantics, along with automatic and effective tools for reasoning, is used to specify, design, analyze, verify, certify, and even implement software [3].

On the other hand, Natural Language Processing (NLP) is a research area within computer science that concerns with the design and implementation of artificially intelligent systems that are capable of using language as fluently and flexibly as humans do [4]. Intuitively, NLP, as a computer science discipline, should create computer software following a SE methodology, thus, taking into account SE principles. However, when designing and developing NLP systems, SE aspects are generally neglected [5], resulting in highly specific applications that work only for restricted domains that are really difficult to reuse, scale, or adapt. Moreover, such applications heavily depend on hardware and software local specifications, making hard to separate the logic of the system from the computing requirements.

NLP applications could significantly benefit when taking into account SE fundamentals in the design and development of the components. Currently, natural language is modelled differently, depending on several issues, such as the purpose of its application, or the available resources. No attention is paid, however, to the structures selected for its modelling, and the manner in which NLP systems are developed. For this, it should be necessary to study in a more detailed manner, methods and models that could be employed in the software development process that allow the NLP applications to work independently from the environment characteristics, and produce good quality programs in a systematic, cost-effective and efficient way.

An example of these SE techniques could be the Model-Driven Development philosophy [6], which considers models as the main assets in the software development process. Models collect the information that describes the information system at a high abstraction level, allowing the development of the application in an automated way following generative programming techniques [7]. For the Model-Driven Architecture, we choose the high-performance specification language Maude [8]. The FM approach for software specification and implementation is perfect here by providing high-level specification languages for rapid software development and easy-to-check semantics that are also easily extended by repeatedly adding more algebraic properties. This would allow us to process natural language in an incremental manner with: (i) the possibility to extent and adapt the developing systems gradually, (ii) guaranteeing SE principles, such as transparency, maintainability, and reusability of the developed components, and (iii) a high-performance prototype, (iv) providing a formal and easy-to-check semantics.

The main goal of this paper is to study and analyze how natural language can be formally defined and modelled. Our approach follows an algebraic specification approach, via the Maude language [8]. As a result we obtain an implementation consisting of a set of transformation rules that will allow us to process natural language for each context and discourse. Achieving this objective, we will have an independent implementation for natural language that will be highly scalable and reusable, easy to adapt and extend and with the possibility to integrate more knowledge in an easy way through an incremental development. Moreover, this prototype may constitute the basis for developing wider NLP applications, such as text summarization. Specifically, this paper makes three contributions. First, it constitutes one of the first attempts to represent natural language in terms of a formal specification with a formal semantics, bringing NLP and SE/FM together. Second, it introduces the concept of incremental modelling for NLP, and shows how algebraic specification implemented through Maude can offer great advantages to the incremental development of NLP software applications. Third, it discusses how well-known problems of natural language, such as ambiguity, are resolved with this approach.

In the remainder of the paper, we first provide an analysis of previous work that bears some relation to ours (Sect. 41.2). Then, we focus on describing our formalization approach for natural language (Sect. 41.3) and its implementation (Sect. 41.4). Finally we summarize the main conclusions and outline future directions obtained from our research work (Sect. 41.5).

## 41.2 Related Work

Representation and modelling of natural language has always attracted the attention of NLP researchers. However, there is no consensus in the NLP research community about the form or the existence of a data structure that is able to describe natural language texts [9]. From a NLP perspective, natural language is modelled differently depending on: (i) the level of language analysis performed, and (ii) the specific task and/or application to be tackled.

Over the years, several mechanisms have been employed to represent natural language. Grammars were the ones which experimented a huge progress in the late 1990s, since they were a key element in the development of syntactic parsers. In those years, different types of grammars were developed, including probabilistic context-free grammars (lexicalized or unlexicalized) [10], definite clause grammars [11], or combinatory categorial grammars [12], among others. Most of them were derived from large corpora, such as Penn treebank. However, grammars had well-known limitations [13]. On the one hand, they were too big to be managed, whereas their coverage was somewhat limited. On the other hand, the degree of ambiguity was too great that exhaustive search was not feasible. Other existing approaches that also try to model natural language employ statistical techniques and probability, such as n-grams or bag-of-words [14], but their objective is to obtain knowledge, considering that a text is characterized by such elements, rather than proposing a reusable approach for defining language. Recent efforts have proposed to represent and model the semantics natural language using ontologies [15], as well to establish formal mechanisms (e.g., NLP Interchange Format) to facilitate the interoperability of NLP tools [16]. However, these are still in their early stages.

Platforms such as GATE[1] or UIMA[2] allow the integration of different NLP resources and tools, that were initially developed in a different way, to build complete NLP systems. Our aim greatly differs from the development of these tools. It focuses more on the definition of NLP tasks in an incremental and reusable way, so that complex NLP processes can be built on top of the basic ones with the addition of minimum knowledge.

To the best of knowledge there is not previous research work with the goal of modelling natural language from a SE perspective, and more concretely using FM, and in particular, algebraic specification. By describing natural language in an incremental manner and through a high-performance specification language like Maude, as we proposed in our approach, we have a structure flexible and adaptative enough to be used for developing NLP applications following a standard methodology.

## 41.3 Formalization of Natural Language

To set the basis for modelling natural language by means of a formal specification, we can follow one of these approaches: (i) a manual approach, where a linguistic expert defines all necessary knowledge and takes into account all its possible variations; (ii) a fully automatic approach, where existing NLP tools are employed for extracting and obtaining knowledge from a set of documents; and (iii) a hybrid approach, where (i) and (ii) are combined, in the sense that the output of NLP tools is revised by an expert human in order to correct the possible errors made by the automatic processing. Our proposed approach follows this latter approach, relying on the knowledge obtained from automatic NLP tools, and correcting the output of these tools, when necessary, in order to avoid the time-consuming process of (i), and the errors derived from automatic tools in (ii). One important difference is that we explore all possibilities, so returning all possible meanings to a human user for further inspection.

The process of knowledge extraction follows a top-down methodology. Using existing NLP tools that take a document as input and obtain information about their components, we employ such knowledge for describing natural language. Two levels of language analysis are involved in this process: (i) lexical-morphological analysis through a part-of-speech tagger, and (ii) syntactic analysis through a syntactic parser. The former is useful for identifying the types of tokens (i.e. words) included in a document, whereas the second determines the types of syntactic structures.

Then, our framework for describing natural language is as follows. First of all, a generic algebraic specification for NLP is defined, including sorts and subsorts apart of equational properties, such as associativity and commutativity for built-in lists

[1] http://gate.ac.uk/

[2] uima.apache.org/

and sets. This establishes a cornerstone for the assets on generality and extensibility of our incremental natural representation. Then, the full discourse universe, obtained from existing NLP tools and which must be available for processing natural language, is described in terms of the algebraic specification. For each lexical element, a transformation rule is defined. If a lexical element has some ambiguity, several transformation rules would be defined, providing a non-deterministic set of transformation rules. Also, for each syntactic association, a transformation rule is defined but in terms of the lexical elements rather the original text elements. This part is clearly non-deterministic, since multiple combinations of associations may be possible. The addition of extra transformation rules for lexical elements or syntactic associations provides the main assets on transparency, reusability, and maintainability of our incremental natural language processing. Finally, a set of documents for natural language processing is given to Maude as an input term and successively transformed into a proper document, exploring all the non-deterministic possibilities. Since, all the transformation rules are terminating, we know the search space is finite and, thus, manageable by existing formal methods tools, such as a reachability command within the specification language Maude.

## 41.4 The Rewriting Logic Semantics of Natural Language

This section contains the technical implementation details for representing natural language in an incremental and adaptive manner, using the Maude language. Maude [17] is a high-performance reflective language and system supporting both equational and rewriting logic specification and programming. Functional modules describe data types and operations on them by means of equational theories. Mathematically, such a theory can be described as a pair (Ó, $E \cup A$), where: Ó is the signature that specifies the type structure (sorts, subsorts, kinds, and overloaded operators); $E$ is the collection of equations declared in the functional module, and $A$ is the collection of equational attributes (associativity, commutativity, and so on) that are declared for the different operators. System modules allow the description of the dynamic behaviour of a system beyond data types and operations on them by allowing system transition rules that constitute a search space graph. Given an initial state and a search pattern, the `search` command performs the search through the rules and equations defined. These equations are, in fact, rules associated to deterministic actions and their application do not generate new states in the search space.

### 41.4.1 Incremental Implementation

The cornerstone of our framework is the built-in Maude sort for *Quoted Identifiers* (QID). A QID is denoted by symbol ' followed by letters and numbers, e.g. `'account 123`, and is a wrapper for strings in order to provide a

Maude representation for tokens of Maude syntax, which is essential for the meta-representation and meta-reasoning capabilities of Maude. We could have used the built-in Maude sort *String* for representing words but its treatment is much more inefficient than that of QID. The module QID-LIST defines a built-in data type of lists of quoted identifiers, and it is our most important tool because it provides an associative operator __ (this represents the symbol with empty syntax in Maude) with an identity operator `nil` for an empty list, allowing yuxtaposition of QID's. This is crucial for allowing us to define our types based on sequences of QIDs without having to look for the individual elements in a different data structure. That is, if we need to search for the word `'a` followed by the word `'tree` with zero or more words between them, for example to be replaced by `'a 'dog`, we just have to write the transition rule in Maude where `X:QidList`, `Y:QidList`, and `Z :QidList` can be substituted by the empty list or any sequence of QID's. This avoids writing code for searching for the word `'a` and then searching for the word `'tree` or restarting the search after finding `'a` but not finding `'tree`, thus easing the way for the specification of NLP.

```
 rl X:QidList 'a Y:QidList 'tree Z:QidList
=> X:QidList 'a Y:QidList 'dog Z:QidList.
```

From the knowledge obtained through the lexical-morphological and syntactic analysis, we begin with the definition of the most basic types, so complex types could be then generated from the basic ones. For the implementation of our prototype, a corpus of 567 English newswire documents provided by DUC[3] was employed. This corpus will constitute the discourse universe for our implementation.

From the part-of-speech tagger output (lexical-morphological analysis), the different categories were grouped into nine general ones (*determiner*; *pronoun*; *noun*; *preposition*; *adjective*; *adverb*; *verb*; *stopword*; *punctuation mark*). The set of tags from the part-of-speech tagger[4] was automatically mapped to the previously mentioned categories. For instance, the *verb* category would consider tags, such as *VVN*, *VBD*, *VBG*, *VBZ*, or *VBP*. Each of these categories will be a different module in Maude.

For each module, we first define a sort for the category and we specify whether it is also a subsort of another one. All these basic categories are subsort from the QID sort. The next step is to define the signature and semantics for each of the modules. The signature consists of the operations for obtaining the categories. The semantics of each module will contain the equations and rules by means of which this category is built. For this, we use a symbol representing each category, i.e., `pronoun`, `noun`, `adverb`, etc., and the original word is kept as an argument.

[3] http://www-nlpir.nist.gov/projects/duc/

[4] http://www.sketchengine.co.uk/tagsets/penn.html

We could have defined a symbol for each word in each category, e.g. `pronoun-he` or `noun-attack`, but it is easier to implement by keeping the original word as an argument. However, this forces us to make all the category symbols to be frozen for further rewriting (denoted by the labels `[frozen strat (0)]` in Maude), i.e., to avoid nested applications of the transformation rules in the form `'you -+ pronoun ('you) -+ pronoun (pronoun ('you))`, leading to an infinite loop.

We next provide an example of the signature and fragments of the semantics for the category *noun*.

```
mod NOUN is
 protecting QID .

 sort Noun .
 subsort Qid < Noun .

 op noun : Qid -> Noun [frozen strat (0)] .

 eq 'advantages = noun('advantages) .
 eq 'adventurers = noun('adventurers) .
 eq 'adventures = noun('adventures) .
 eq 'advice = noun('advice) .
 [...]
 rl 'account => noun('account) .
 rl 'act => noun('act) .
 rl 'aid => noun('aid) .
 rl 'bear => noun('bear) .
 [...]
 endm
```

As it can be seen, for this module, as well as for the remaining types (determiner, verb, adverb, adjective, stopword, and punctuation mark), we define that each type can take the form of a QID. The attribute *frozen* is also present in all of them and it is crucial in the definition, in order to ensure that the program terminates; otherwise, we could end up with an infinite loop. In the semantics of each type, the elements of our discourse universe that fall into the corresponding categories are included. Non-ambiguous terms (e.g., `eq.'advice = noun ('advice) .`) are represented through equations (eq.), whereas rules (rl) are used for the ambiguous ones (e.g., `rl'account => noun ('account) .`).

Having implemented the basic categories, we exploit the knowledge obtained from the syntactic parser (syntactic analysis) for defining types of language structures that are needed for creating sentences, and documents. In particular, the ones that we initially use for our research are: *noun phrase*, *verb phrase*, which will lead to the definition of the modules *SN*, *SV* in Maude. Below, the implementation of *SN* module is shown.

```
mod SN is
 protecting DETERMINER .
 protecting PRONOUN .
 protecting STOPWORD .
 protecting PREPOSITION .
 protecting ADJECTIVE .
 protecting NOUN .
 protecting QID-LIST .

 sort Sn .
 subsort QidList < Sn .
 op sn : QidList -> Sn [frozen strat (0)] .

 rl noun(A:Qid) => sn(A:Qid) .
 rl pronoun(A:Qid) => sn(A:Qid) .
 rl determiner(A:Qid) sn(B:QidList) => sn(A:Qid B:QidList) .
 rl preposition(A:Qid) sn(B:QidList) => sn(A:Qid B:QidList) .
 rl stopword(A:Qid) sn(B:QidList) => sn(A:Qid B:QidList) .
 rl adjective(A:Qid) sn(B:QidList) => sn(A:Qid B:QidList) .
 rl sn(A:QidList) sn(B:QidList) => sn(A:QidList B:QidList) .
endm
```

The *SN* module inherits the properties of some of the basic categories, as well as the built-in QID-LIST Maude module. The rules reflected in the definition of this type shows how a noun phrase could be produced. These are recursive rules, that will allow the construction of noun phrases following a recursive and transformational process starting from the basic types of words included in the text. Following an analogously process, but taking into account the elements that would constitute a verb phrase, we define the *SV* type.

With the *SN* and *SV* data types, we can define the signature and semantics for a sentence and a document. We define a sentence as a structure that contains a noun phrase, followed by a verb phrase, and ended with a punctuation mark. In addition, the syntactic analysis carried out also detected as sentences, grammatical structures involving either a noun or a verb phrase followed by a punctuation mark. Finally, we create a module for representing natural language documents. We assume that a document can be formed by a single sentence or a concatenation of them. It should be immediate to understand for the reader that all the modules are easily extensible and reusable, since all the lexical and syntactical information is defined by transformation rules/equations and all such rules/equations are local to the specific word or token being used, so there is no necessary global data or context information. Furthermore, these modules provide a lot of generality and transparency, since the set of new defined sorts is minimal.

### 41.4.2 Analysis and Discussion

Once all the necessary data types via Maude modules were implemented, we conduct an analysis of its performance. In this analysis, we want to check if natural language can be represented through our proposed types. Moreover, we want to check whether our approach would be computationally efficient, even though optimization is out of the scope of this paper.

We conduct a battery of tests in order to evaluate our implementation in a qualitative manner. Next, we provide several representative examples for the different types and discuss their results.

*Example 1* Basic categories. The following examples show cases where individual words are tested in order to check if they belong to the correct category.

```
search 'happy =>* adjective(Q:Qid) .
Solution 1 (state 0)
Q:Qid --> 'happy
No more solutions.

search 'happy =>* verb(Q:Qid) .
No solution.

search 'account  =>* verb(Q:QidList) .
Solution 1 (state 2)
Q:QidList --> 'account
No more solutions.

search 'account  =>* noun(Q:QidList) .
Solution 1 (state 1)
Q:QidList --> 'account
No more solutions.
```

As it can be seen, the token "happy" is correctly recognized as an adjective, and not as a verb. It is worth mentioning how ambiguity is tackled at the execution stage. Let's consider the term *"account"*. This word is ambiguous since it can act as a noun as well as a verb. At the execution time, we perform a search within a particular category, so for all the possible states in the specific category, the program will check whether the word *account* exists. In our examples, when we search this word either in the space of a verb or a noun, the word will be recognized.

Another issue that is also important to note is the number of states visited and the time spent for processing the terms. Maude is really fast, consuming less than 1 s of CPU time. The predominant use of rules and equations makes the program to be faster than if we had used only rules for implementing the semantics for all the types.

*Example 2* Language structures for creating sentences. These examples show how natural language structures can be also identified.

```
search 'the 'account =>* sn(Q:QidList) .
Solution 1 (state 5)
Q:QidList --> 'the 'account
No more solutions.


search
'the 'dog 'and 'the 'bear =>* sn(Q:QidList) .
Solution 1 (state 22)
Q:QidList --> 'the 'dog 'and 'the 'bear
No more solutions.
```

Concerning verb phrases (*SV*), our implementation deals with verb phrase that include prepositional phrases form with a preposition, a determiner and a noun.

```
search
'play 'with 'the 'cat =>* sv(Q:QidList) .
Solution 1 (state 20)
Q:QidList --> 'play 'with 'the 'cat
No more solutions.
```

Finally, we provide an example in order to ensure that noun phrases are not recognized as verb phrases.

```
search
'the 'dog 'and 'the 'bear =>* sv(Q:QidList) .
No solution.
```

From all these illustrative examples, we can highlight the fact that although the number of states and rewriting steps have increased with respect to the search performed for the basic categories, the time spent for obtaining the solution is marginal.

*Example 3* Sentence and document recognition. Provided that we have the vocabulary employed in our discourse universe, our model for representing natural language will be able to recognize sentences and documents. In these cases, the number of rewriting steps is higher than in the previous examples. The longer and more complex the sentence or the document is, the longer it takes for the system to analyze it. Moreover, for their definition, we only have used rules and not equations, leading to the fact that the search space is higher.

```
search 'the 'dog 'plays 'with 'the 'cat ';
   =>* sentence(Q:QidList) .
Solution 1 (state 81)
states: 82
rewrites: 158 in 0ms cpu
Q:QidList --> 'the 'dog 'plays
              'with 'the 'cat ';
No more solutions.
states: 95
rewrites: 191 in 0ms cpu

search 'nero 'specializes 'in 'sniffing
   'out 'bombs 'and 'narcotics ';
    =>* sentence(Q:QidList) .
Solution 1 (state 602)
states: 603
rewrites: 1805 in 12ms cpu
Q:QidList --> 'nero 'specializes 'in 'sniffing
              'out 'bombs 'and 'narcotics ';
No more solutions.
states: 616
rewrites: 1848 in 12ms cpu

search 'the 'dog 'plays 'with 'the 'cat ';
     'nero 'specializes 'in 'sniffing
     'out 'bombs 'and 'narcotics ';
     =>* document(Q:QidList) .
Solution 1 (state 58515)
states: 58516
rewrites: 288541 in 1928ms cpu
Q:QidList --> 'the 'dog 'plays 'with 'the
              'cat '; 'nero 'specializes 'in
              'sniffing 'out 'bombs 'and
              'narcotics ';
No more solutions.
states: 58549
rewrites: 288737 in 1932ms cpu
```

In this paper, we have focused on the presentation of the framework and further optimizations on the search space have to be considered, which is a typical topic of research in formal methods applied to verification. The key idea is that the transformation rules are terminating and, thus, the search space would always be finite. Although our approach has not been evaluated in a quantitative manner, we have verify the adequacy of our method through a set of representative examples, focusing also on the time spent for processing the different modules. More evaluation examples and their performance analysis can be found in [18].

## 41.5 Conclusion and Future Directions

In this paper we modelled natural language following an incremental and adaptive algebraic specification approach via a high-performance language (Maude). As far as we know, this is the first attempt to represent natural language in this manner. NLP tools were employed for obtaining and extracting knowledge from documents, and such knowledge was used to define sorts, subsorts and equational properties. Not only was a novel method to represent natural language proposed, but also our implementation was designed in a way that is highly scalable and reusable, thus taking into consideration SE principles, which are not often paid attention by the NLP research community.

An important advantage of this representation is that it would be very easy to add further knowledge, as well as to extend it to other languages. Furthermore, the development of more complex NLP applications, such as information retrieval or text summarization, could be quite straightforward, given that the necessary language analysis levels are performed.

Taking this preliminary study as a starting point, there are several interesting issues that will guide our research work in the future. In the short-term we plan to define more complex data types, as well as optimizing the performance of our initial prototype. The addition of new knowledge as well as new data types will constitute the basis for the development of complex NLP applications, e.g., text summarization, that we plan to tackle in the long-term. Moreover, we also plan to adopt the ideas for natural language deconstruction proposed in [19] that could benefit our approach by providing a flexible approach and determining which language structures should be modelled depending on the aim of the NLP application.

**Acknowledgements** E. Lloret and M. Palomar have been partially funded by the Spanish Government through the project TEXT-MESS 2.0 (TIN2009-13391-C04) and Técnicas de Deconstrucción en la tecnologías del Lenguaje Humano (TIN2012-3 1224) and by the Generalitat Valenciana through project PROMETEO (PROMETEO/2009/199). Moreover, S. Escobar has been partially supported by the EU (FEDER) and the Spanish MEC/MICINN under grant TIN 2010-21062-C02-02, and by Generalitat Valenciana PROMETEO201 1/052.

## References

1. Ghezzi C, Jazayeri M, Mandrioli D (2002) Fundamentals of software engineering, 2nd edn. Prentice Hall PTR, Upper Saddle River, NJ
2. Pressman RS (2001) Software engineering: a practitioner's approach, 5th edn. McGraw-Hill Higher Education, Columbus, OH
3. Hinchey M, Jackson M, Cousot P, Cook B, Bowen JP, Margaria T (2008) Software engineering and formal methods. Commun ACM 51(9):54–59
4. Dale R, Somers HL, Moisl H (eds) (2000) Handbook of natural language processing. Marcel Dekker, Inc., New York, NY
5. Leidner J (2003) Current issues in software engineering for natural language processing. In: Proceedings of the workshop on software engineering and architecture of language technology systems, pp 45–50

6. Frankel D (2002) Model driven architecture: applying MDA to enterprise computing. Wiley, New York, NY
7. Czarnecki K, Eisenecker UW (2000) Generative programming: methods, tools, and applications. ACM Press/Addison-Wesley Publishing Co., New York, NY
8. Clavel M, Durán F, Eker S, Lincoln P, Martí-Oliet N, Meseguer J, Talcott CL (eds) (2007) All about Maude—a high-performance logical framework, vol 4350, How to specify, program and verify systems in rewriting logic, Lecture Notes in Computer Science. Springer, Heidelberg
9. Collobert R, Weston J, Bottou L, Karlen M, Kavukcuoglu K, Kuksa P (2011) Natural language processing (almost) from scratch. J Mach Learn Res 12:2493–2537
10. Klein D, Manning CD (2003) Accurate unlexicalized parsing. In: Proceedings of the 41st annual meeting on association for computational linguistics—vol 1. Association for Computational Linguistics, Stroudsburg, PA, pp 423–430
11. Pereira F, Warren D (1980) Definite clause grammars for language analysis—a survey of the formalism and a comparison with augmented transition networks. Artif Intell 13:231–278
12. Steedman M, Baldridge J (2011) Combinatory categorial grammar. Wiley-Blackwell, Oxford, pp 181–224
13. Steedman M (2010) Some important problems in natural language processing. Technical report, University of Edinburgh
14. Huang F, Yates A, Ahuja A, Downey D (2011) Language models as representations for weakly-supervised nlp tasks. In: Proceedings of the fifteenth conference on computational natural language learning, pp 125–134
15. Bateman JA, Hois J, Ross R, Tenbrink T (2010) A linguistic ontology of space for natural language processing. Artif Intell 174(14):1027–1071
16. Chiarcos C (2012) A generic formalism to represent linguistic corpora in rdf and owl/dl. In: Proceedings of the eight international conference on language resources and evaluation (LREC'12)
17. Clavel M, Durán F, Eker S, Lincoln P, Martí-Oliet N, Meseguer J, Talcott C (2003) The Maude 2.0 system. In: Rewriting techniques and applications (RTA 2003), 2706, pp 76–87
18. Lloret E, Escobar S, Palomar M, Ramos I (2013) Natural language modelling using maude. Technical report, University of Alicante
19. Martínez-Barco P, Ferrández-Rodríguez A, Tomás D, Lloret E, Saquete E, Llopis F, Peral J, Palomar M, Gmez-Soriano JM, Romá MT (2013) Legolang: Técnicas de deconstrucción en la tecnolog´ıas del lenguaje humano. Procesamiento de Lenguaje natural (51)

# Chapter 42
# UCL: Universal Constraint Language

**Peter Piják, Jakub Malý, Martin Nečaský, and Irena Holubová (Mlýnková)**

**Abstract** Current software applications typically consist of a system of multiple components, each being modeled using a different type of meta-model. In addition, when developing the application, various integrity-constraint languages are used for particular meta-models (e.g. OCL for UML, Schematron for XML schemas, or SQL triggers for relational databases). Hence, the constraint expressions need to be converted to expressions over the different meta-models (i.e. modeled) too, which is a non-trivial task. In this paper, a new universal language called *Universal Constraint Language* (*UCL*) for expressing integrity constraints over various data models is introduced. It is formally defined and also its parser is implemented within the *DaemonX* project. In particular, expressing constraints in UCL for UML class diagrams and for XML schemas is supported. Thanks to preserving mutual relations between models in *DaemonX* we can also we can also translate the respective constraints between the models.

**Keywords** Constraint language • Model-driven architecture • Universal formalism

## 42.1 Introduction

With the progress of advanced software architectures, such as *Service-Oriented Architecture* (*SOA*) [1], the current software applications are no longer composed of consistent monolithic software. Information systems rather consist of several simpler

P. Piják (✉) • J. Malý • M. Nečaský • I.H. (Mlýnková)
Department of Software Engineering, Charles University in Prague, Malostranské nám. 25, 118 00 Praha 1, Czech Republic
e-mail: peter.pijak@gmail.com; maly@ksi.mff.cuni.cz; necasky@ksi.mff.cuni.cz; holubova@ksi.mff.cuni.cz

M. José Escalona et al. (eds.), *Information System Development: Improving Enterprise Communication*, DOI 10.1007/978-3-319-07215-9_42,

components, each being responsible for specific functionality part. This decomposition of software architecture brings many advantages but also problems.

An example of a typical information system is a runtime process that is implemented in an object-oriented language, reads/saves data from/to a relational database, and sends or receives SOAP messages (described in WSDL using XML). During the design of each component, different models are used—for classes of an object-oriented programming language we use *UML class diagrams* [2], for relational databases we use the *relational model*, for XML schemas we can use the XML conceptual meta-model *XSEM* [3]. In addition, the individual models are usually *semantically* related and interconnected. For example in UML class models a runtime object, which is in a relational database stored in columns of a table, or in an XML document represented as an element.

*Integrity constraints* (ICs) are expressions which are used to express the consistency and the accuracy of modeled data. When modeling a software system, various IC languages are used for particular parts of the models. And also the ICs can be semantically connected and, hence, the ICs must be translated into other languages. In complex systems consisting of numerous models and languages, it is impossible for a human expert to handle this task manually.

*Aims of the Paper.* In this paper, we introduce a new common language for expressing ICs called *Universal Constraint Language* (**UCL**) [4]. It is based on OCL [5] and satisfies the following criteria:

- UCL is able to express ICs over *different data models*.
- UCL is based on a general data model which is defined *formally* and *precisely*.
- UCL constraints can be *transformed to different* specific *constraint languages*, such as SQL triggers.
- If elements in separate models as parts of a complex software system are related and interconnected, then it is possible to *transform* UCL constraints over *one model to* UCL constraints over *another model automatically*.

We both define UCL formally and provide its implementation. The language was designed generally; however, its primary aim was usage within a general modeling and evolution management framework called *DaemonX* [11]. The framework currently serves as a proof of the concept and demonstrates the ability to use UCL with other meta-models as well as mutual transformations of constraints among these models. For instance, we may express a UCL expression over, e.g., UML class diagrams and XML. These two expressions may represent semantically the same condition of allowed data, but these expressions do not have to be equal. Names and relationship between elements can be different.

*Paper Outline.* The paper is structured as follows: In Sect. 42.2 we discuss the related work. In Sect. 42.3 we provide description of the DaemonX framework and extension of its data model for UCL purposes. In Sect. 42.4 we define UCL formally, we provide several examples, and we outline its relation to other data models. Finally, in Sect. 42.5 we conclude and outline possible future work.

## 42.2 Related Work

Probably the most popular language for expressing ICs is the *Object Constraint Language* (*OCL*) [5]. It was developed in 1995 at IBM and from version 1.1 (from 1997) it is a part of the official standard for UML. OCL is a based on the first-order predicate logic and it is a strong typed language. Each expression has a type which is defined statically before the interpretations. The language has a predefined set of primitive types (`integer`, `boolean`, `string` and `real`). Each class in the UML model represents a special type in OCL.

*Dresden OCL* [6] is a software platform for OCL developed at the Technische Universität in Dresden, Germany. It is probably the most complex tool with the support for OCL. *Dresden OCL2 Toolkit* is based on *NetBeans Metadata Repository* (*MDR*), is a standalone implementation of *MOF* [7]. *Dresden OCL for Eclipse* is based on Eclipse SDK. Its architecture is based on meta-model called *Pivot model.* Pivot model provides an abstraction for the evaluation of OCL expressions. It is based on *Eclipse Modeling Framework* (*EMF*). EMF includes a meta-model *ECore* [8] for describing models.

*Kent Object Constraint Language Library* [9] (*Kent OCL*, *KOCL*) is an OCL implementation from the University of Kent at Canterbury, England. It consists of an OCL parser, analyzer and code generator. It defines the *Bridge* meta-model over various meta-models, a more reduced meta-model than *ECore*. It provides its libraries for *Kent Modeling Framework project* (*KMF*), EMF and for Java.

There are also some studies which analyze derivation of OCL constraints over UML to other constraint languages for other meta-models, e.g. the relational model. In paper [10] mapping of OCL expressions to SQL code is presented. The result of the study is that almost all constructions of OCL (except for iterate construct) can be transformed to SQL. The authors propose a language mapping according to a family of patterns. Each pattern represents an idea of translating an OCL construct to an SQL query, in particular the *CHECK* constraint.

There are some frameworks that work with constraint in OCL. They support usage of OCL also for other data models (e.g. usage of OCL for XML schemas). But constraints in OCL over different models (e.g. constraint over UML models and constraints over XML schemas) are different. Another disadvantage of OCL is its complexity. We want to in a different way. We want to define a new meta-meta model that is not be so general, but using this model it is possible to define interconnection between different meta-models. And it is possible to mutually translate constraints over different models using this interconnection.

## 42.3 Architecture and Model

The UCL is a general-purpose language. However, it was primarily implemented for the purpose of usage in the *DaemonX* framework [11]. The *DaemonX* project is a plug-in-able framework for data and/or process modeling. It was developed by the

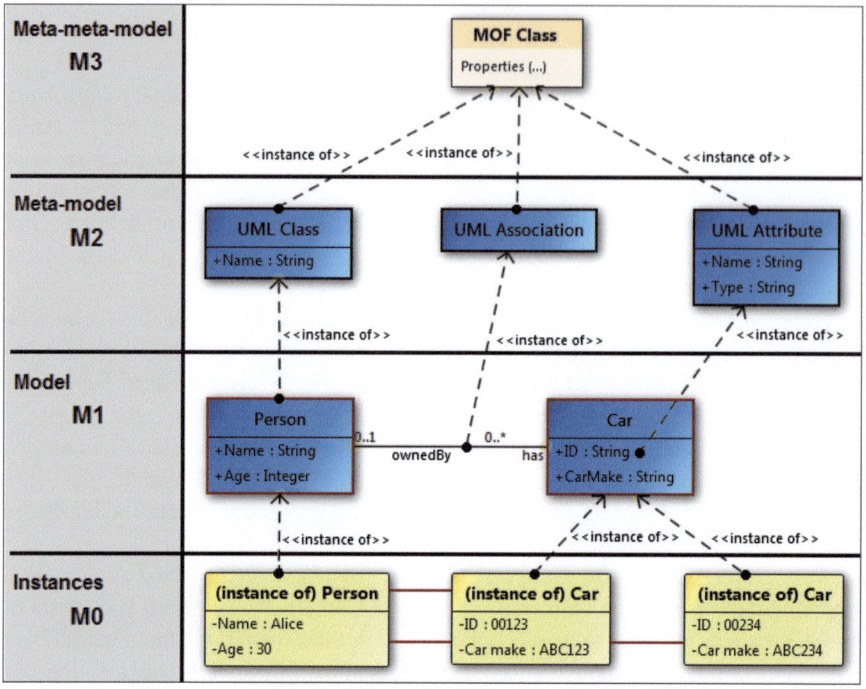


**Fig. 42.1** MOF four-layer meta-modelling architecture of *DaemonX*

*DaemonX* team at the Faculty of Mathematics and Physics of the Charles University in Prague as a student software project. *DaemonX* is currently available with the support for modeling of UML diagrams for general data structures, ER models for relational data, BPMN models for business processes, and XSEM model for XML data.

### *42.3.1 Architecture*

Expressions and types of UCL must be based on a specific meta-model called *UCL Data meta-model*. The aim is to use constraints in UCL for various meta-models. And, for two different interconnected models of different meta-models, it must be able to interconnect also the two mapped models of UCL meta-model. Our aim is to mutually transform constraints over different meta-models. UCL Data meta-model should consist of the widest possible types of data constructs and relationships between these constructs.

The architecture of *DaemonX* is based on the *Meta-Object Facility* (*MOF*). As depicted in Fig. 42.1, the considered layers are *instances* (layer M0), *user model of the system* (layer M1), *meta-model* (layer M2) and *meta-meta-model* (layer M3). As depicted in Fig. 42.2, at layer M3 we have defined the UCL Data meta-model (***A***),

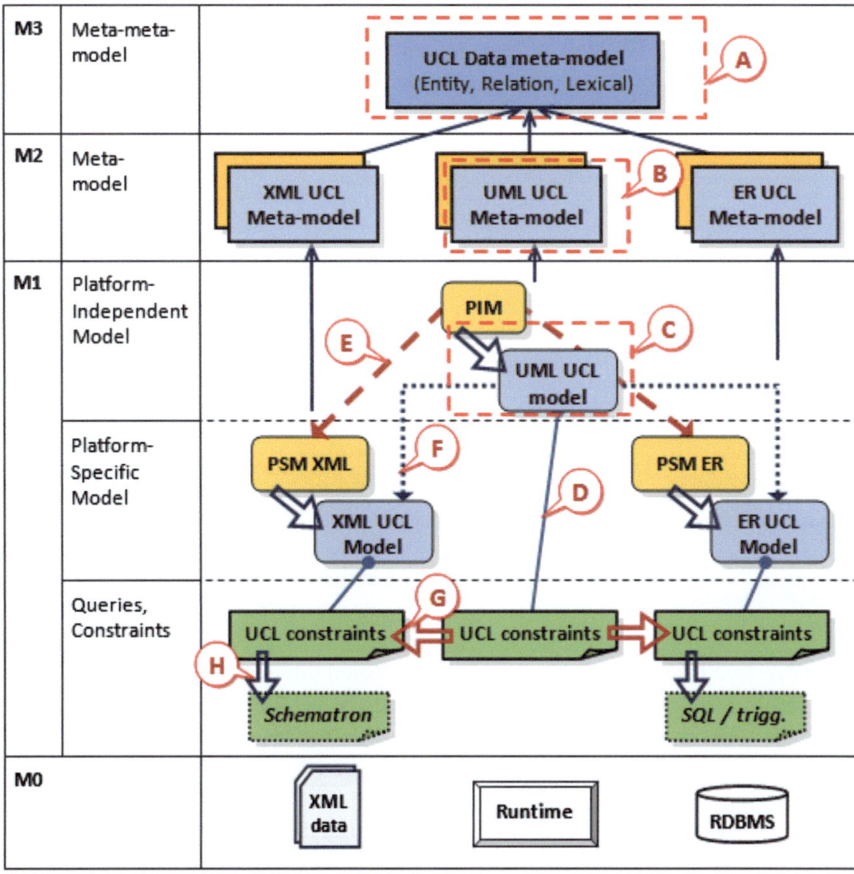


**Fig. 42.2** Integration of UCL within the *DaemonX* MOF architecture

the common data meta-model for purposes of UCL. At layer M2, there are also UCL meta-models. For each meta-model, we define a particular UCL Data meta-model (***B***). E.g. we have the UML meta-model; it defines what a class or an attribute is. Then at layer M2, we define UML-UCL-meta-model; it defines what is a class or an attribute in a view of UCL Data meta-model.

At layer M1, we create for each model its UCL model. E.g. for a UML model we create a UML UCL model (***C***). UCL constraints are based on individual UCL models. Let us have a UML model and we want to express UCL constraints over it. We do not express these constraints over the UML model but over the created UML UCL model of this UML model, i.e. UCL constraints are based on the UML UCL model (***D***). Individual models (at layer M1) represent parts of the complex software system. These models are semantically related. A software architect must create the mapping (interconnections) between elements in the individual models—in *DaemonX* the mapping is expressed using the platform-independent model (PIM) of MDA (***E***).

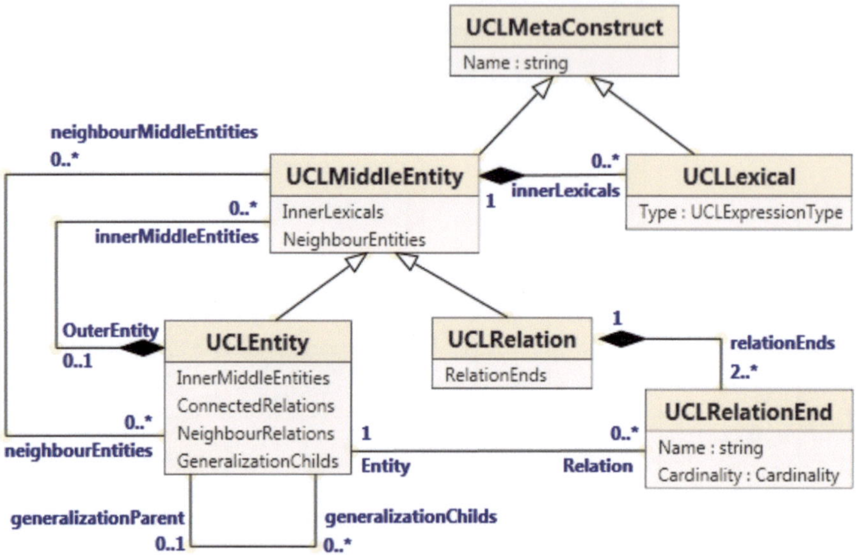


**Fig. 42.3** Constructs of UCL Data meta-model

The individual UCL models are then interconnected too in a similar way (***F***). Then, if we have UCL constraints over one model (e.g. over UML), we can automatically derive the UCL constraints over other models (***G***). It is also possible to derive UCL constraints for specific constraint languages; e.g. derive UCL constraints over XML to Schematron rules (***H***).

## *42.3.2 UCL Data Meta-Model*

UCL Data meta-model must be sufficiently general to be able to model many different data models. There are two ways how to proceed with the analysis of what and how many elements will form the UCL Data meta-model. The first option is that it will contain only a small number of elements that will be universal, but it will define different kinds of relationships and connections between these elements. The second option is that UCL Data meta-model will include a wider number of elements which will be more concrete and they will correspond directly to specific elements of selected modeling data models. The overall proposal of the UCL Data meta-model is depicted in Fig. 42.3.

Data meta-models typically contain simple elements that directly model the data values; e.g. an attribute of a class in UML, a column in a relational model, an attribute or a simple element in XML. These kinds of elements from layer M2 will be represented at layer M3 by the element *UCL lexical*. A UCL lexical will contain properties *Name* (string) and *Type* (enumeration of the basic data types *Integer*, *Real*, *Boolean* and *String*).

Data meta-models typically contain also complex elements that contain simple elements (represented by UCL lexicals); e.g. a UML class or a relational table. These kinds of elements from layer M2 will be represented at layer M3 by the element *UCL entity*. A UCL entity has property *Name*. It can contain inner UCL lexicals and inner UCL entities. This models the situation when some data meta-models can contain elements to group other elements (e.g. a package in UML) or that complex elements can contain other complex elements.

Next, the UCL Data meta-model defines relationships between UCL entities. It defines a *neighborhood* at an equivalent level and a *generalization*. The neighborhood between UCL entities can represent, e.g., the relationship between UML classes in the same UML package. Many data meta-models define both binary and *n*-ary links between their elements; e.g. a UML association or a foreign key in databases. These links are in UCL Data meta-model represented by element *UCL relation*. A UCL relation is an *n*-ary link between UCL entities. A UCL relation has property *Name* and it has two or more *UCL relation ends*. A relation end is connected to an UCL entity and it is assigned with *Name* and *Cardinality*. Consequently, a UCL entity can contain also inner UCL relations; e.g. when an UML association is inserted in a UML package. In UCL Data meta-model, there is also neighborhood between a UCL entity and a UCL relation; e.g. when a UML association class is connected to a UML association.

All elements of data meta-models and relationships between them can (and must) be mapped to the presented elements in UCL Data meta-model and their relationships.

## 42.4 UCL Formal Definition and Examples

Let $T_e$ be a finite set of basic types, $E$ be a finite set of all entities (*UCLEntities*), $R$ be a finite set of all relations (*UCLRelations*), $L$ be a finite set of all lexicals (*UCLLexicals*), and $R_e$ be a finite set of all relation ends (*UCLRelationEnds*) of all relations.

*Definition 1* A structure of *entities* is a tuple *Entities* = (*E, E.name, E.lex, E.inEn, E.inRel, E.conRel, E.accEnds, E.neEn, E.neRel,* $<_{E.gen)}$, where

- *E.name:* $E \rightarrow £$ assigns a name to each entity,
- *E.lex:* $E \rightarrow P(L)$ assigns a set of lexicals to each entity,
- *E.inEn:* $E \rightarrow P(E)$ assigns a set of inner entities to each entity,
- *E.inRel:* $E \rightarrow P(R)$ assigns a set of inner relations to each entity,
- *E.conRel:* $E \rightarrow P(R)$ assigns to each entity a set of relations which are connected to the entity,

- *E.accEnds:* $E \rightarrow P(R_e)$ assigns to each entity a set of relations ends which are accessible (via navigation) from the entity,
- *E.neEn:* $E \rightarrow P(E)$ assigns a set of neighboring entities to each entity,
- *E.neRel:* $E \rightarrow P(R)$ assigns a set of neighboring relations to each entity,
- the partial order $< E.gen \subseteq E \times E$ is a generalization hierarchy on entities.

Relations describe structural relationships between entities, whereas a relation can connect two or more entities. Relations connect entities through their relations ends. The number *n* of all relation ends of the relation is called the *degree* of the relation. The relation is called the *n*-ary relation respectively. An entity can be connected to any number of relations. Binary associations where both ends are connected to the same entity are called *recursive* or *self-relations*.

*Definition 2* A structure of *relations* is the tuple *Relations* = (*R, R.name, R.lex, R.neEn, R.ends*) and a structure of *relation ends* is the tuple *RelEnds* = ($R_e$, $R_e$*.relation*, $R_e$*.name*, $R_e$*.card*, $R_e$*.entity*), where

- *R.name:* $R \rightarrow (£\ U\ \lambda)$ assigns a name (possibly empty) to each relation,
- *R.lex:* $R \rightarrow P(L)$ assigns a set of lexicals to each relation,
- *R.neEn:* $R \rightarrow P(E)$ assigns a set of neighboring entities to each relation,
- *R.ends:* $R \rightarrow P(R_e)$ assigns a set of relation ends to each relation (each relation has at least two ends),
- $R_e$*.name:* $R_e \rightarrow £$ assigns a role name to each relation end,
- $R_e$*.relation:* $R_e \rightarrow R$ assigns to each relation end its (outer) relation,
- $R_e$*.entity:* $R_e \rightarrow E$ assigns a connected entity to each relation end,
- $R_e$*.card:* $R_e \rightarrow C$ assigns to each relation end the cardinality of the connection to the connected entity.

Lexicals describe properties of entities and relations. A lexical has a name and one of the basic types.

*Definition 3* A structure of *lexicals* is the tuple *Lexicals* = (*L, L.name, L.type*), where

- *L.name:* $L \rightarrow £$ assigns a name to each lexical,
- *L.type:* $L \rightarrow T_E$ assigns a basic type to each lexical.

We also need to define the *generalization* hierarchy over entities. Generalization is a partial order over the set of entities $E$. We denote it $<E.gen \subseteq E \times E$. The generalization is a taxonomic relation between two entities. If entities $e_1$, $e_2 \in E$ are in the generalization relationship $e_1 <_{E.gen} e_2$, then $e_1$ is the child entity of the parent entity $e_2$. A child entity inherits all lexicals, inner entities, neighbor entities and relations and all connected relations of the parent entity.

The model must satisfy some restrictions like that the names of two lexicals in one entity cannot be the same. These restrictions are necessary, because the UCL expressions over the model cannot be ambiguous. These restrictions are formally defined in [12]; we omit them for space limitations.

### 42.4.1 UCL Expressions

Similarly to OCL, UCL is based on terms and predicates of the first-order predicate logic. UCL is a typed language; each expression has a type defined statically before the interpretation. As mentioned before, the language has the predefined set of primitive types. Each UCL entity and each UCL relation in UCL model also represents a special type in UCL. Also collection types from other types can be created. All constraints in UCL are *invariant expressions*. An invariant is a constraint expression which expresses a condition which must be satisfied all the time for all instances of the defined context.

Consider the sample UML model in Fig. 42.4. Each UCL constraint is expressed in the specified *context* of a UCL entity or a UCL relation of the source model. This model is an instance of UCL Data meta-model. The context must be defined at the beginning of each block of UCL constraints:

```
Context Person
inv:
...
```

To specify context for non-root elements, we must specify the *full path* to the element from the root entity through all inner entities separated by a double colon:

```
Context Employees::Job …
Context RootEntity1: :Entity2: :Entity3: :TargetEntity …
```

To define ICs, UCL uses *invariant expressions*. An invariant is an expression of the Boolean type. It must return true for all instances of the context element at any time. It is defined in a context of an entity or relation. The keyword `self` in refers to instances of the context. The sample invariant `constraint1` does not refer to instances of the context entity because it does not contain the self expression. Invariant `constraint2` refers to instances of the entity `Person`:

```
Context Person
inv constraint1: 0 < 1
inv constraint2: self.age >= 0
```

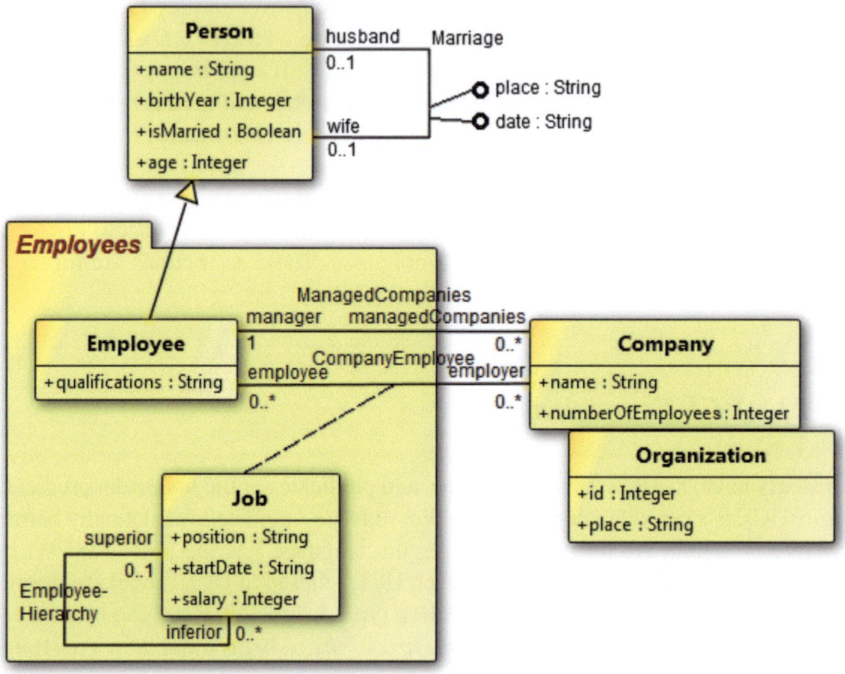


**Fig. 42.4** Sample UML model

## 42.4.2 Navigation Expressions

The main difference between UCL and OCL are the navigation expressions. In UCL the *simple step navigation* from an entity is a technique how to access other model element that is connected with the entity. It is possible to connect these elements via: (a) lexicals of the source entity, (b) inner entities of the source entity, (c) inner relations of the source entity, (d) neighboring entities of the source entity, or (e) neighboring relations of the source entity. Names of all these connected elements must be distinct. We can access these elements by the expression of the syntax: a dot operator (.) followed by the name of the connected element. For example:

```
-- in the context of Company:
self.name                  /* (a) lexical of Company */
-- in the context of Employees:
self.Employee              /* (b) inner entity of Employees */
self.EmployeeHierarchy     /* (c) inner relation of Employees */
-- in the context of Company:
  self.Organization        /* (d) neighbouring entity of Company */
-- in the context of Job:
  self.CompanyEmployee     /* (e) neighbouring relation of Job */
```

The value of this expression is the value of the target lexical (in case (a)) or an expression referring to the target model element (in the other cases). The type of the expression is the type of the target lexical (in case (a)) or the type of the target model element (in the other cases). Using these navigation expressions we can express calculations over the model. For example domain constraint of lexicals (e.g. "the age of persons is always greater than zero") or we can define complex constraints based on the relationship between elements which are connected by these kinds of navigation.

In the similar way as by the entities, we can also access model elements which are connected with a relation. It is possible to connect (a) lexicals of the source relation, (b) neighboring entities of the source relation, or (c) relation ends of the source relation which is connected to an entity. Names of these connected elements and names of these relation ends must be distinct according the formal definition of UCL Data meta-model. Examples of expressions are as follows:

```
-- in the context of Marriage:
self.place /* (a) lexical of Marriage */
-- in the context of CompanyEmployee:
self.Job   /* (b) neighboring entity of CompanyEmployee */
self.employer
           /* (c) relation end of the relation CompanyEmployee;
                  it is connected to the entity Company;
                  the type of expression is the entity Company */
```

The value of these expressions is the value of the target lexical (in case (a)) or an expression referring to the neighboring entity (in case (a)) or an expression referring to the connected entity (in case (c)). The type of the expression is the type of the target lexical (in case (a)) or the type of the target entity (in cases (b) and (c)). Concrete self-descriptive examples of expressions are as follows:

```
Context Employees
inv: (self.Job.position = "baker")
       => (self.Job.CompanyEmployee.employer.name <> "IBM")
Context Employees::EmployeeHierarchy
inv: self.inferior.salary <= self.superior.salary
Context Company
inv: (self.name = "ABC") => (self.Organization.place <> "USA")
Context Marriage
inv: (self.place = "Prague") => (self.husband.age >= 18)
```

Las but not least, starting from an entity, we can create a navigation expression to another entity using a connected relation. This navigation can be created using the name of an opposite relation end of a connected relation. For example, in the context of entity *Employee* we can navigate to entity *Company* using expression `self->employer`. Or from the entity *Company*, we navigate to the entity *Employee* using `self->manager`. The value of such navigation expressions is the target entity or a sequence of the target entities. If the cardinality is *(0, 1)* or

(*1, 1*), then the value is an instance of the target entity; else it is a sequence of instances of the target entity. The type of the first exemplar expression is *Sequence* (*Company)*, the type of the second expression is *Employee*.

A navigation expression from an entity to other entity through a connected relation expressed with the operator -> is called a *connected relation step expression*. The second type of a navigation expression through connected relation is *connected relation stop expression*. It is expressed with the operator :>. The value of this expression is not the opposite entity but the relation which connects the entities. Concrete examples of expressions are as follows:

```
Context Company
inv: self->manager /* expression refers to entity Employee */
inv: self:>manager /* refers to relation CompanyEmployee */
Context Employee
inv: self->employer /* refers to Sequence (Company) */
inv: self:>employer /* refers to Sequence (CompanyEmployee) */
Context Company
inv: self->manager.qualifications = "manager"
Context Person
inv: (self.isMarried and self:>husband.place = "StateX")
       => (self.age >= 21)
```

### 42.4.3 Relation of UCL to Other Data Models

In [4], we formally define how we can map meta-model for UML class diagrams and XML model XSEM to UCL Data meta-model. This mapping is necessary when we want to use UCL expressions over these models. The mapping is a function that assigns all possible types of element in the UML/XSEM models (e.g. a package, a class, an attribute, an association, an element) to possible elements in UCL Data meta-model (i.e. entity, relation, lexical). Then the mapping assigns all possible relations between elements (e.g. a UML attribute of a UML class in an inner lexical of its parent entity).

We also define a mapping between two different data models. As we have mentioned before, when we model a complex information system, the individual elements in the different models are semantically related. This interconnection must be created manually by software architects. Using this mapping we can automatically derive UCL constraints over one model to the other related model if the created mapping satisfies specific conditions. For the space limitations we refer the interested reader to [4].

## 42.5 Conclusion

The aim of this paper was to introduce a new constraint language UCL which is much simpler than OCL, but has universal usage for any kind of data model. Having this feature and the respective relations among the models, the language enables simple transformation of constraints between the models.

As a future work we plan to implement transformation of UCL constraints to other languages and full support of evolution management among ICs. In the latter case it is a non-trivial task, since a change in one model can influence numerous other models which may influence also ICs. And such situations must be identified precisely, fully, and preferably automatically.

**Acknowledgement** This work was supported by the GACR grant no. P202/1 1/P455 and the GAUK grant no. 541013.

## References

1. Service Oriented Architecture—SOA. IBM. http://www-01.ibm.com/software/solutions/soa/
2. Object Management Group (2010) Unified modelling language specification, version 2.3. http://www.omg.org
3. Nečaský M (2009) Conceptual modelling for XML, vol 99 of dissertations in database and information systems series. IOS Press/AKA Verlag, Amsterdam
4. Piják P (2011) Universal constraint language. Master's thesis, Charles University in Prague. http://www.ksi.mff.cuni.cz/~holubova/dp/Pijak.pdf
5. Object Management Group (2010) Object constraint language OMG available, specification version 2.2. http://www.omg.org
6. Dresden OCL. http://www.dresden-ocl.org/index.php/DresdenOCL
7. Object Management Group. MetaObject facility. 2.4—Beta 2. ISO/IEC 19470:2005
8. ECore meta-model. Package org.eclipse.emf.ecore
9. Akehurst D, Linington P, Patrascoiu O (2003) OCL 2.0: implementing the standard. Computer Laboratory, University of Kent, Tonbridge
10. Demuth B, Hussmann H (1999) Using OCL constraints for relational database design. In: UML'99—the unified modelling language, Proceedings of the 2nd international conference, LNCS 1723, Springer, pp 598-613
11. DaemonX-Team (2001) DaemonX. http://DaemonX.codeplex.com/

# Index




M. José Escalona et al. (eds.), *Information System Development: Improving Enterprise Communication*, DOI 10.1007/978-3-319-07215-9,

MIX
Papier aus verantwortungsvollen Quellen
Paper from responsible sources
FSC® C105338

If you have any concerns about our products,
you can contact us on
**ProductSafety@springernature.com**

In case Publisher is established outside the EU,
the EU authorized representative is:
**Springer Nature Customer Service Center GmbH**
**Europaplatz 3, 69115 Heidelberg, Germany**

Printed by Libri Plureos GmbH
in Hamburg, Germany